CRC Handbook of Organic Analytical Reagents

Editors

K. L. Cheng, Ph.D.
Professor of Chemistry
University of Missouri - Kansas City
Kansas City, Missouri

Keihei Ueno, Dr. Eng.
Professor of Chemistry
Kyushu University
Fukuoka, Japan

Toshiaki Imamura
Senior Research Chemist
Dojindo Laboratories
Kumamoto, Japan

CRC Press, Inc.
Boca Raton, Florida

Library of Congress Cataloging in Publication Data
Main entry under title:

Handbook of organic analytical reagents.

Bibliography
Includes index.
1. Chemical tests and reagents. 2. Organic
compounds. I. Cheng, K. L. II. Ueno, Keihei,
1920-   . III. Imamura, Toshiaki.
QD77.H35   547.3'01   81-38552
ISBN 0-8493-0771-6   AACR2

Direct all inquiries to CRC Press, Inc., 2000 Corporate Blvd., N.W., Boca Raton, Florida, 33431.

© 1982 by CRC Press, Inc.

International Standard Book Number 0-8493-0771-6

Library of Congress Card Number 81-38552
Printed in the United States

PREFACE

Since the well-known four volumes of *Organic Analytical Reagents* by Professor Welcher* appeared in 1947, a similar up-to-date monograph is lacking. In this handbook, it is not the intention merely to report new organic analytical reagents nor to include all organic reagents, but rather to provide selectively both classical and recent ones which we believe to be important. We also limit ourselves to those for metals and some anions; with a few exceptions those for organic analysis are not included at this time.

The editors have both the research and applied analytical chemists in mind. We briefly introduce a chapter about the theoretical background for using the organic analytical reagents. The remaining chapters will present information of direct interest to the analyst on the preparation, properties, and analytical applications of organic reagents.

The organic analytical reagents have been widely used in gravimetric, volumetric, and spectrometric analyses. They are also used in column chromatography and TLC for elution as well as for detection. Because of their complex formation with metal ions and anions, they are often used in combination with many instrumental techniques, such as spectrophotometry, fluorimetry, electrometry, atomic absorption, ion exchange, etc.

In the early days of analytical chemistry, the common analytical reagents were simply the available inorganic compounds lacking sensitivity and selectivity. At the end of the 19th century, the value of organic reagents was immediately recognized. There are four criteria in the evaluation of an analytical method:
1. Selectivity
2. Sensitivity
3. Simplicity
4. Actual state

The so-called "actual state" refers to the information about the state of its existence in the sample, for instance, the oxidation state, molecular structure, ionic state, etc. The organic reagents have been regarded as playing an important role in the selectivity, sensitivity, and simplicity, but a limited role in the "actual state". However, this limitation may be improved through future research.

Since the introduction of EDTA and many metallochromic indicators to chemical analysis, a great number of volumetric and photometric analyses have been done with organic reagents. In the past three decades, EDTA and other complexing agents have been widely used as masking agents to increase the selectivity of analytical methods. This trend will continue. The reports of unusually high sensitivity of Thio-Michler's ketone (TMK) for many platinum metals and of the highly selective functional groups have stimulated exciting research with organic reagents. We have suggested some new directions with organic reagents in future analytical research. It is hoped that an analytical chemist will find this handbook a convenient source of information for many useful organic analytical reagents and will stimulate further research with organic reagents, paying special attention to develop simpler analytical methods.

The authors (K.U. and T. I.) wish to thank to Miss Izumi Uchiyama for preparing the major part of the typescript and to Dr. Hiroshi Nakamura, Mr. Takashi Hayashita and Hajime Ootsuka for their laborious works of preparing the camera-ready artwork.

K. L. Cheng,
Keihei Ueno,
and
Toshiaki Imamura

* D. Van Nostrand Co., Inc., N.Y.

EDITORS

Dr. Kuang Lu Cheng is Professor of Chemistry, University of Missouri, Kansas City. He received his B.S. degree from the Northwestern College, China in 1941, his M.S. in 1949, and his Ph.D. in 1951 from the University of Illinois.

Dr. Cheng has been a postdoctoral fellow at the University of Illinois and an instructor at the University of Connecticut, Department of Chemistry. Dr. Cheng is a member of the American Chemical Society, Electrochemical Society, American Microchemical Society, Society for Applied Spectroscopy, International Water Resources Association, and American Institute of Physics. He is a fellow of the American Association for Advancement of Science and of the Chemical Society of London. He is also a member of Sigma Xi, Sigma Pi Sigma, and Phi Lambda Upsilon.

Dr. Cheng received the RCA Achievement Award in 1963. He was elected titular member of the Division of Analytical Chemistry, International Union of Pure and Applied Chemistry (1969 to 1979). He received the N.T. Veatch Award for Distinguished Research and Creative Activity and the Certificate of Recognition from the U.S. Office of Naval Research in 1979, and that from the College of Engineering, Texas A & M University, 1981.

His areas of research interest include photoelectron spectroscopy, surface chemistry, ion selective electrodes, chemical separations, and organic reagents.

Dr. Keihei Ueno is one of the most active analytical chemists in the development and applications of organic analytical reagents. He graduated from Kyushu University in 1944, majoring in applied chemistry at Faculty of Engineering. After serving in the Navy during World War II, he joined Dojindo Laboratories in 1946 as a Research Fellow and was promoted to Director of Research in 1949. He received his Dr. of Engineering degree from Kyushu University in 1953, and has been a postdoctoral Fellow at Clark University, Worcester, Mass. During this period, he engaged in the development and production of EDTA, metal indicators, and various photometric reagents. In 1959, he was called to Kyushu University as a Professor of analytical chemistry in the Department of Organic Synthesis, Faculty of Engineering. Since then, he has made a great contribution to the development of analytical chemistry in the field of organic analytical reagents and separation chemistry. He received a Society Award in 1967 from Japan Society for Analytical Chemistry for his distinguished achievement in the progress of organic analytical reagents and another Society Award in 1977 from Japan Chemical Society for his great contribution on the separation chemistry.

He was a Vice President of Japan Society for Analytical Chemistry in 1971 and is an associate member for IUPAC (Analytical Chemistry Division, Commission on Analytical Reactions and Reagents).

He has published more than 130 research papers and 5 monographs including *Compleximetric Titration* (1956) and *Introduction to Chelate Chemistry* (1969). He also edited several series of books, including *Chelate Chemistry* in six volumes (1975 to 1977).

Mr. Toshiaki Imamura has broad experience in organic analytical reagents. He graduated from Kyushu University in 1953, majoring in chemistry, at the Faculty of Science. He joined Dojindo Laboratories in 1957 as a research chemist and has been engaged in the research, development, and manufacture of various organic analytical reagents in Japan's most famous reagent chemical firm. He is now acting as an advisor to the Technical Information Department of the company.

Since 1976, he has been assigned to Ariake Technical College as lecturer to teach analytical chemistry. In addition to the activities in his profession, he has been an active member of Japan Society for Analytical Chemistry and was appointed as Secretary (1966 to 1974) and Vice Chairman (1973) of Kyushu Regional Division of the Society.

He is an author or a joint author of many papers including review articles on various fields of organic analytical reagents. He also published a textbook on volumetric analysis and a monograph on chelating agents.

ADVISORY BOARD

TABLE OF CONTENTS

A THEORETICAL SURVEY OF
ORGANIC ANALYTICAL REAGENTS

CHEMICAL BONDS

When any two things are fixed to each other, we say that a bond exists. In chemical compounds, positively charged nuclei from two or more elements become fixed relative to each other. They do not touch, but are simply firmly fixed at some very small distance from each other and it is said that a chemical bond is present. There are actually five classes into which bonds are generally divided, although the boundaries between them are not well defined.

Ionic bonding — An electrostatic attraction is developed when each of two atoms or groups of atoms attain an electronic structure such that resulting electrostatic forces are of sufficient strength to form a chemical bond. Oppositely charged ions attract each other, and a bond, called an ionic bond, exists between them. The expression of "ionic bond" does not refer to a thing, but the existence of a force. It is a strong force and important in many molecules.

Covalent bonding — Electron transfer between like atoms does not happen. Even if one of the atoms were able to accept an electron or two and acquire an inert gas structure, the other, by giving up an electron, would move further away from such a condition of stability. Outer electron shells of two atoms may penetrate each other. Some of the outer electrons may, in a sense, be shared between the two nuclei. Instead of electron transfer between two like atoms, a sharing of electrons occurs. For instance, one electron from the outer shell of one chlorine is said to pair with one electron from the outer shell of the other. The pair of electrons is considered to be shared equally by both nuclei, forming an octet. The bond that exists because of this cooperative action is called a covalent bond.

Let us look at the shape of the electron clouds in a carbon atom. The four electrons in the L shell interact rather strongly with one another, changing the shapes of the s and p orbital until they are practically identical. Each electron cloud concentrates itself away from those of the other three because of the strong electrostatic repulsion. This means that each cloud is sausage-like and points away from the nucleus; the four arrange themselves with the largest possible angle, 109.5°, between each pair.

Coordinate covalent bonding — Each electron in a shared pair need not come from two separate atoms. Both electrons may be provided by one of the involved atoms. The resulting bond, indistinguishable from covalent bond, is sometimes called a coordinate covalent bond. The lone pair electrons from nitrogen or oxygen are often donated to metal ions. The coordinate covalent bonding is extremely important in the coordination chemistry.

Multiple covalent bonding — Often, for two nuclei to be bonded together covalently, two pairs or three pairs of electrons need to be shared between them. Since one shared pair between two nuclei is called a single bond, two shared pairs of electrons yield a double bond (such as carbon dioxide and ethylene).

Metallic bonding — As its name implies, metallic bonding is confined to metals and many near metals found in Groups I, II, and III of the periodic table. Since the metallic bond is of little significance in chemistry of organic analytical reagents, we shall not get into its details. Generally it is explained that in metallic bonding the positively charged ions are held together by their attraction to the cloud of negative electrons in which they are embedded rather than like ball bearings in a liquid glue. One might say that a metallic bond is a sort of ionic bond in which the free electron is donated to all the other atoms in the solid. Thus, the bonds are not directional, and their most important characteristic is the freedom of the valence electrons to move.

van der Waals' bonding — When nonpolar aromatic compounds are in contact with a nonpolar polystyrene resin, the former are strongly adsorbed on the latter by a force attributed to the van der Waals' force, a secondary bond.

The strong interactions between atoms lead to energies in the range of 50 to 300 kcal/mol. Such compounds possess considerable stability. The van der Waals' forces are weaker intermolecular attractions which give rise to bond energies of 2 to 20 kcal/mol. These intermolecular attractive forces were initially called simply the van der Waals' forces, but as molecular structure became better understood, it was recognized that the forces had three components, each arising from a different structural feature of the constituent molecules. Known as Keesom, Debye, and London forces, they arise from dipole-dipole, dipole-induced dipole, and instantaneous dipole-induced dipole (or dispersion force) interactions, respectively. van der Waals' forces play an important role in the charge transfer complex formation and the chromatographic separations.[1,2]

Hydrogen bonding — There is another secondary bond which can be made between atomic groups that have no electrons to spare. A covalent compound composed of polar molecules such as water exhibits this kind of intermolecular bond, dipole-dipole attraction. Water molecules in a drop of water are attracted to each other in such a way that the electron-dense oxygen side of one molecule is next to the hydrogen-dense side of another molecule. As a result, a hydrogen atom of one molecule is attracted to the oxygen atom of another molecule. This particular attraction of dipoles so that a hydrogen atom is placed between two polar atoms or molecules (forming a bridge) is known as a hydrogen bond. This is a special case of dipole-dipole interaction.

Mixed bonds — Both the secondary bonds (the hydrogen bond and the van der Waals' bond) differ distinctively from the three primary bonds (ionic, covalent, and metallic). It is interesting to examine the isoelectronic compounds with the same total number of electrons: methane, ammonia, water, and hydrogen fluoride. In methane the carbon atom can form four covalent bonds, one with each hydrogen atom. In hydrogen fluoride at the extreme, the hydrogen donates its electron to the fluorine to complete the $2p_z$ orbital of fluorine. The atoms are bonded ionically. In the intervening compounds, water and ammonia, neither of these pictures is quite correct in itself, but a mixture of the two applies. In such cases, it is probable that a portion of an electronic charge may be transferred while the rest of it is shared equally to form a partial covalent bond. This means merely that the electron clouds are distorted roughly, so that there is a predominance of electronic charge density away from the hydrogen atom. The shift of the "charge center" increases as we go through the series from the C−H bond to the F−H bond.

Bond strength — Since there is a smooth transition between covalent and ionic and metallic bonds, there is also a smooth variation in bonding strength. Some distinguished characteristics are obvious when the pure covalent, hydrogen, and van der Waals' bonds are concerned. The strength of a bond is best measured by the energy required to break it. The heat of vaporization has been used to measure the bond strength. Figures in Table 1 indicate that the van der Waals' bond and the hydrogen bond are the weakest. Next comes the metallic bond, followed by the ionic and covalent bonds whose strength are nearly comparable.

LIGAND FIELD THEORY

During the past 35 years, our understanding of the chemistry of transition metals has made rapid progress largely because of the development of ligand field theory.

Transition metals have an incomplete d orbital (one to nine d electrons) in the second outmost electron level. The outer electron level usually contains two s electrons. The electronic configuration of the transition metals can therefore be described as $(n - 1)d^{1-9}ns^2$, where n is the principal quantum number of the outer electron level and $(n - 1)$ is the principal quantum number of the second outmost level. For example, iron has a $3d^6 4s^2$ configuration which means that it has six d electrons in the $n = 3$ level and two s electrons in the $n = 4$ level.

Table 1

Bond	Molecule	Heat of vaporization ΔH (kcal/mol)
van der Waals'	He	0.02
	N_2	1.86
	Methane	2.4
Hydrogen		
O--H--O	Phenol	4.4
F--H--F	HF	6.8
Metallic	Zn	28
	Fe	94
Ionic	NaCl	153
	NaI	121
	MgO	242
Covalent	Si	85
	SiC	283
	SiO_2	405

A general chemical property of transition metals is that they show various oxidation states, depending upon the number of d electrons used in complex formation. This is attributed to the fact that both d and s electrons with similar energies can act as valence electrons. The two s electrons always act as valence electrons; then the number of d electrons acting as valence electrons decides the oxidation state. In the case of manganese, it has a $3d^5 4s^2$ configuration and several oxidation states; $2+$ in MnO, corresponding to use of the two s electrons only; $3+$ in Mn_2O_3, using two s electrons and one d electron; $4+$ in MnO_2, using two s electrons and two d electrons; $6+$ in MnO_4^{2-}, using two s electrons and four d electrons; and $7+$ in MnO_4^-, using two s electrons and all five d electrons.

Another important property of transition metals, resulting from their incomplete d level, is that they tend to form complexes, such as the ferricyanide ion and the cuprammonium ion. In any complex, the metal acts as a central atom to which are attached a definite number of negative ions or neutral molecules which are called coordinating groups or ligands. For example, the ferrous ion contains a central atom of Fe(II) with ligands of three neutral o-phenanthroline molecules. In addition, the transition metals do not closely resemble one another chemically; they vary from reactive metal such as scandium to chemically inert metals such as gold.

In a particular type of energy level, there may be many different orbitals. Two important characteristics of any orbital are energy and the direction in which the electron cloud is concentrated. In a d electron level, there are five separate d orbitals.

In an isolated ion, all five d orbitals are of the same energy. Orbitals with the same energy are said to be degenerate. Regarding the directional characteristics of the orbitals, the normal three coordinates naturally at right angles are used, the x and z axes are in the plane of the paper, and the y axis is perpendicular to the plane of the paper. The five d orbitals which are concentrated in different directions may be distinguished by subscripts as d_{z^2}, $d_{x^2-y^2}$, d_{xz}, d_{xy}, and d_{yz}. They fall into two groups. The $d_r(e_g)$ orbitals are those directed along the axes, the d_{z^2} and $d_{x^2-y^2}$ orbitals. The $d_\varepsilon d_{2g}$ orbitals are those directed between the axes, the d_{xy}, d_{xz}, and d_{yz} orbitals.

The transition metal ion in its complex is not isolated, but is surrounded by ligands which are either neutral polar molecules such as ethylenediamine in the cupric ethylenediamine complex or negative ions such as CN^- in the ferricyanide complex. When a polar molecule acts as a ligand, the negatively charged end of the molecule is directed toward the positively charged transition metal ion. Then the transition metal ion in its

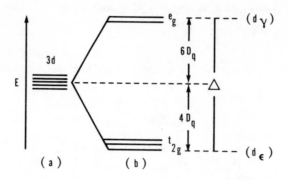

FIGURE 1.   The 3 d orbital splitting in an octahedral field.
(A) A field-free atom or ion (fivefold degenerate); (B) an ion
perturbed by an octahedral ligand field.

complex is surrounded by negative charges which produce an electrostatic field acting
on the central metal ion. The ligand field theory is concerned with the effect of this
electric field on the d orbitals of the central ion.

The ligand field has no effect on the directional properties of the d orbitals of the
transition metal ion, but it increases the energies of some of the d orbitals. The number
of ligands and their arrangements around the transition metal ion are the factors to
determine the energies of which d orbitals are altered.

Ligand field theory states that in a tetrahedral complex, the energy of the three t_{2g}
orbitals is increased, whereas the energy of the two e_g orbitals is unaffected. In an
actahedral complex, the energy of the two e_g orbitals is increased, whereas the energy
of the three t_{2g} is unaffected. The five d orbitals in a complex are not of the same
energy, but are split into two sets with different energies. Since metal ion electrons
located in orbitals along the x,y,z bonding axes (d_{x^2} and $d_{x^2-y^2}$) will suffer an electrostatic repulsion from electron pairs of the ligand molecules, the d orbitals located between the bonding axes (d_{xy}, d_{yz}, and d_{xz}) are energetically more stable. The five d
orbitals in an octahedral field may be described on an energy level diagram as shown
in Figure 1.

The two sets of orbitals from the degenerate 3d orbitals are referred to as e_g (high
energy) and t_{2g} (low energy). The energy difference between them is called the orbital
separation shown by Δ. Two factors, the nature of the ligands and the charge on the
transition metal ion, influence the magnitude of the orbital separation which is of the
order of 20 to 50 kcal/mol. A case in which the orbital separation is large is called the
strong field. One in which the orbital separation is small is called the weak field. The
strength of the field determines how the d electrons of the transition metal ion are
distributed among the e_g and t_{2g} orbitals.

The distribution of electrons between the d orbitals of differing energy can be de-
duced from two simple principles:

1.   Electrons occupy separate orbitals in preference to both electrons occupying the
      same orbital because two electrons in different orbitals are farther apart than if
      they were in the same orbital, resulting in less mutual electrostatic repulsion.
2.   Electrons occupy orbitals of lower energy in preference to those of higher energy.

When the transition metal ion has one, two, three, eight, or nine d electrons, the
electron distribution is the same whether the field is strong or weak. However, when
it has four, five, six, or seven d electrons, the electron distribution is different in the
strong-field and weak-field cases. The different electron configurations show different

numbers of unpaired electrons. Magnetic susceptibility measurements are used to determine which of two possible electron configurations is present in a particular complex.

LIGAND FIELD STRENGTH AND SPECTRA OF COMPLEXES

The energy that is able to promote an electron in the transition metal complexes from a t_{2g} to an e_g orbital is often in the range of energy of the visible radiation. Absorption of radiation can induce these electronic transitions, and the absorption spectra of many complexes are the result of this energy promotion. As the magnitude of orbital separation increases, the energy required to promote the transition also increases and the color of the absorbed light changes. Study of the UV and visible spectra yields an orbital separation information. The sequence of ligand field strengths was derived from spectroscopic measurement of orbital separation and is often known as the spectrochemical series: $I^- <$ $Br^- <$ $-SCN^- <$ $Ce^- <$ $NO_3^- <$ $F^- <$ urea $\sim$ $ONO^- \sim$ $OH^- <$ $HCOO^- <$ $C_2O_4 <$ $H_2O <$ $-NCS^- <$ $CH_2(NH_2)COOH <$ $EDTA^{4-} <$ pyridine $\sim$ $NH_3 <$ ethylenediamine $\sim$ diethylenetriamine $<$ triethylenetetramine $<$ dipyridyl $<$ o-phenanthroline $<$ $-NO_2^- \ll$ CN^-.

Other transition metal complexes may have square planar, tetragonal, or tetrahedral symmetries resulting in different splitting patterns for the five d orbitals of the transition metal ion. However, the electron transitions from low-energy d orbitals (t_{2g}) to unfilled higher-energy d orbitals (e_g) are still involved in the absorption of radiation. A molecular orbital treatment of transition metal complexes presents basically the same conclusion as the ligand field theory for the interpretation of electronic spectra.

ELECTRONIC SPECTRA AND MOLECULAR STRUCTURE

When atoms or molecules absorb radiation, thereby becoming excited, they usually remain in the excited state for a very short time. The atom or molecule then emits a photon of energy and returns to the ground or unexcited state with different absorption maximum and intensity. There are two major approaches to spectrophotometric analysis. One is to measure the radiant energy absorbed by the ion or molecule itself. The other approach is used with species that do not absorb significant amounts of radiation. A suitable reagent is then added to convert them to new molecules that absorb some type of radiation energy intensely or emit certain radiation energy of longer wavelength (fluorescence or phosphorescence).

The UV, visible, and IR spectroscopic methods are commonly used in analytical chemistry. The spectral regions covered are 185 to 400 nm (UV), 400 to 800 nm (visible), and 800 nm to 30 μm (IR). Serious difficulties preclude routine measurements at wavelengths shorter than 185 nm and longer than 30 μm. The absorption in the visible spectrum is responsible for the color of certain organic compounds and metal complexes. Molecular electronic transitions require the amount of energy possessed by the photons in the UV and visible regions and become the chief area of interest in chemical analysis.

Electrons in an organic molecule can be classified into three orbitals, namely, those involved in head-on atomic orbital overlap, to form σ bonds, known as σ electrons, those involved in parallel atomic orbital overlap, to form π bonds, known as π electrons, and those not involved in bonding, known as nonbonding or ν electrons. The π electrons of organic molecules have two possible molecular levels, π and π^*. In the ground state, both π electrons are in the bonding π orbital with opposite spins, while the antibonding π^* orbital is empty. When photons of right energy interact with the molecule (absorption), one of these π electrons is promoted to the antibonding π^* orbital, then the molecule is said to be in an excited state. Similar transitions can occur for σ and ν electrons which are associated with UV and visible spectroscopy.

Much less energy is needed to cause transitions in molecular vibrational and rotational states. Thus, the difference in energy between molecular quantitized levels (ΔE) has a general order as $E_{electronic} > E_{vibrational} > E_{rotational}$. Consequently, rotational transitions are from much weaker photons (longer wavelength radiations) than those for vibrational transitions (IR radiations), and electronic transitions require very much stronger radiations (shorter wavelengths, UV and visible). When radiation interacts with molecules, it can increase the quantized rotational, vibrational, and electronic energy levels of the molecules. This increase takes place only at the absorption of correct energy, and the light absorption also only takes place with the molecules of certain structure or certain functional groups. Therefore, a great many of spectrophotometric and fluorometric methods are based on the molecular absorption or emission of radiation by the complexes.

MOLECULAR STRUCTURE AND SOLUBILITY

One of the factors should be noted in the selection of organic reagents for photometric determination, titration, and precipitation for their solubility. The solubility of a molecule in an aqueous medium is mainly controlled by its polarity; sometimes it is referred to as the hydration center. A molecule containing $-OH$, $-SO_3H$, $-OOH$, etc. is very polar, tending to be soluble in water. A molecule containing a long aliphatic chain or multiple rings tends to be less soluble in water, but in nonpolar solvents.

In the gravimetric method, the precipitant should be relatively soluble in water, but its precipitate should be very sparingly soluble in water. In the solvent extraction, the extractant should be less soluble in water, but very soluble in certain organic solvents. The solubility of the organic reagents used is often very important as successful analytical reagents. The characteristic analytical reactions of organic reagents depend largely on the presence of certain functional groups, and the remainder of the molecular structure to which they are joined is generally of secondary importance. Hence, there are in many instances, a series of organic compounds containing the same functional group, but in which the remainder of the molecule is different. Because of the remarkable effect of structure and composition of organic molecules on solubility, it is possible to prepare compounds with modified structure to improve the character of organic reagents, either through increasing their solubility in water or in organic solvents or decreasing the solubility of the reaction products.

When a solute dissolved in water is shaken with an immiscible organic solvent, there is a competition between the two solvents for the solute. There are no completely satisfying explanations for the preference of solute in water than in a particular organic solvent, and vice versa. A qualitative rule that often works is that "like dissolves like". This also agrees with the fact that a polar solute tends to be soluble in polar solvent and the nonpolar solute tends to be soluble in nonpolar solvents.

An interesting example of modifying the molecular structure may be found in the application of salicylaldhyde as an organic reagent for hydrazine. Fichter and Goldach[3] used benzaldehyde as an organic reagent for gravimetric determination of hydrazine, and Feigl and Schwarz[4] proposed salicylaldehyde instead of benzaldehyde. Because of its low solubility in water, benzaldehyde must be used in an alcoholic solution, but owing to the phenolic hydroxyl group, salicylaldehyde is rather soluble in water and is thereby a better precipitating reagent.

Other situations affecting the solubility are summarized as follows.

The relative position of the $-OH$ group with respect to a group, such as $=CO$, $-NO_2$, or $-N=N-$, which is capable of coordinating with the hydrogen atom of the $-OH$

group, affects the solubility. Compounds in which the −OH group is ortho to the coordinating group generally show rather low solubility in water as compared to the para and meta isomers.[5] The ortho position may cause the formation of chelation through hydrogen bonding.

The solubility of a compound in water is always decreased by substituting the O atom of molecule with a S atom which has a lower ability for binding water than O atom. For instance, phenol is rather soluble in water, but thiophenol is practically insoluble in water.

The N atom of a primary amine group is an active center for binding water molecules to organic compounds. Through protonation, substituted ammonium compounds are formed from amines; they are strongly polar and very soluble in water. Protonation can also decrease the water solubility of organic compounds. Many organic compounds containing −COOH are less soluble in strongly acidic medium than in less acidic solutions. Ethylenediaminetetraacetic acid (EDTA) is not very soluble in water; its solubility is increased by addition of base. Pure EDTA acid may be obtained by acidification of its neutral solution.[6]

The hydration of water-soluble compounds may be decreased by addition of less polar solvent. To a rather concentrated metal EDTA complex solution, addition of acetone or ethanol causes the precipitation of the complex.[7]

In neutral or acidic aqueous solution, phenol is largely in the molecular form, and the distribution strongly favors organic phase when the latter is shaken with the aqueous solution. However, a strongly basic solution converts the phenol to the negatively charged phonolate anion and results in extraction from the organic phase into the aqueous phase. A more-charged ion (more polar) tends to stay in an aqueous phase, and a less-charged ion or neutral molecule tends to stay in an organic phase. The selenium diaminobenzene in 0.1 N HCl is not much extracted by benzene or toluene, but is easily extracted into toluene at pH above 5 due to removal of positively charged proton on the amino groups.[8] The chloranil amino acid charge transfer complexes may be extracted by hexanol at 1 N HCl because of removal of negative charge on carboxylic group.[2] Most EDTA and NTA (nitrilotriacetic acid) complexes are very soluble in water with a few exceptions, for instance, the Mg-EDTA complex and NaCaNTA complex are slightly soluble at pH 5 to 6.[9-10]

The water solubility of an organic compound depends in part on its ability to coordinate with water molecules and thereby increase its chemical similarity to the water solvent molecules. The hydration of compounds of high molecular weight generally does not change its solubility much as compared to compounds of low molecular weight, but in extreme cases the compounds of high molecular weight can decrease to a considerable extent the solubility of these reactions.[5] Replacement of the methyl groups of dimethylglyoxime with two furil groups increases the sensitivity of the nickel reaction threefold. This may not be purely the weighting factor; other structural changes may contribute to the sensitivity enhancement.

KINETIC REACTIONS INVOLVING ORGANIC ANALYTICAL REAGENTS

In chemical analysis of a mixture which contains an interfering substance, often it is separated or masked with some reagent that offers a favorable difference in equilibrium constant between the two substances. Analytical chemists can accomplish the same purpose by utilizing favorable differences in reaction rates. Most analytical reactions are fast, but many slow reactions used in chemical analysis are common, particularly the use of organic analytical reagents for anions and for organic molecules. EDTA reacts rapidly with most metal ions at room temperature with a few excep-

.tions. It reacts very slowly with Cr (III) at room temperature, but it forms a purple CR (III) EDTA complex rapidly upon heating.[11,12] It can also be catalyzed by addition of bicarbonate.[11] Chromium may be selectively determined based on the rate reactions with EDTA. The coordination kinetics of EDTA complexes has been reviewed by Margerum,[13] in which he discussed the sexadentate and pentadentate functions of EDTA as revealed by UV spectroscopy during the formation of metal chelates. Some of these could be the geometric isomers of pentadentate species.

When cobalt (II) EDTA or cobalt (II) 1-(2-pyridylazo)-2-naphthol (PAN) formed at pH above 5 is oxidized to their cobaltic complexes with H_2O_2, the cobaltic complexes are very stable in strong acid medium. However, they are formed simply by mixing Co (II) salt and EDTA or PAN in an acidic medium or adding H_2O_2. These cobaltic complexes are kinetically stable in a very strong acid medium due to formation of inert complexes.[14]

Chloranil forms charge transfer complexes with amino acids at pH 9 with help of heating at 60°C for 30 to 60 min. Once the complexes are formed, they are stable in acid medium (1 *M*).[2]

Diaminobenzidine forms a yellow-colored compound piazselenol at pH 3, and Bismuthiol II forms a yellow compound with Te (IV). These two reactions are slow, requiring at least 30 min for color development. Both Se (IV) and Te (IV) are stabilized by hydration which prevents other ligands to come in easily to contact Se (IV) or Te (IV).[12,15]

Catalysts have been used to make end points detectable; these have been reviewed by Mottola.[25] For instance, the oxidation of Malachite Green by peroxidate is catalyzed by Mn (II) in dilute dilutions. Trace amounts of EDTA can be determined by their ability to complex with some of the Mn present and hence to slow the reaction.[27] The kinetics of the formation of complexes has been reviewed by Alimarin.[26]

STABILITY OF COORDINATION COMPOUNDS

The solution stability of a coordination compound (different from its thermal stability or its stability towards redox) deals with an equilibrium of the type:

$$M + nL \rightleftharpoons ML_n$$

By applying the mass action law, an equilibrium constant expression may be written:

$$\beta^{\circ}_n = \frac{A_{ML_n}}{A_M \cdot A^n_L}$$

where A_M is the activity of the species M, and β°_n is the overall thermodynamic stability constant for the reaction.

As an example, EDTA acid equilibrium is affected by pH. Increasing acidity weakens the MY complex by protonation of Y and thus decreasing the proportion of unprotonated EDTA (Y^{4-}).

$$M + Y \rightleftharpoons MY$$

$$H^+ \Updownarrow$$

$$HY, H_2Y, H_3Y, \text{ etc.}$$

The effect of hydrogen ions on the equilibrium may be calculated using α_Y, the fraction of all forms of uncomplexed EDTA present as Y^{4-}.

$$\alpha_Y = \frac{[Y^{4-}]}{[H_6Y^{2+}] + [H_5Y^+] + [H_4Y] + [H_3Y^-] + [H_2Y^{2-}] + [HY^{3-}] + [Y^{4-}]} = \frac{[Y^{4-}]}{[Y']}$$

The method of calculating α_Y may be found in Reference 16. For any given ligand, such as EDTA, α_Y may be calculated and plotted as a function of pH.

The effect of other weak competing ligands on the metal EDTA equilibrium is to shift the equilibrium to the left,

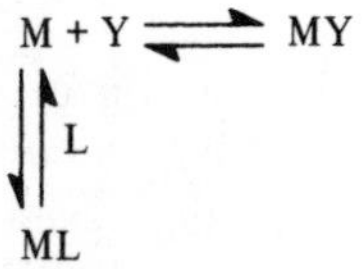

$$M + Y \rightleftharpoons MY$$
$$\updownarrow L$$
$$ML$$

The effect of a competing ligand on the equilibrium may be calculated by using β_M which is the fraction of all forms if M and ML in solution present as the free metal ion:

$$\beta_M = \frac{[M]}{[M']}$$

where [M] is the concentration of free metal ion M^{n+}, and [M'] is the analytical total concentration of metal ion in solution. For a given concentration of free, unbound L, the value of β_M is calculated from the stepwise formation constants of ML_n as follows:

$$\frac{1}{\beta_M} = 1 + k_1[L] + k_2[L]^2 + \ldots k_n[L]^n$$

where k_1, k_2, and k_n are the products of stepwise formation constants for the metal ligand complexes. Once the β_M and α_Y are known, the conditional stability constants ($K_{M'Y'}$) may be calculated with the following equation:

$$K_{MY} = \frac{[MY]}{[M][Y]}$$

$$[Y] = \alpha_Y[Y'] \text{ and } [M] = \beta_M[M']$$

$$K_{MY} = \frac{[MY]}{[M'][Y']\beta_m\alpha_y}$$

$$K_{MY} \cdot \alpha_Y \cdot \beta_M = K_{M'Y'} = \frac{[MY]}{[M'][Y']}$$

The conditional stability constants have more practical significance in analytical chemistry. The methods for determining stability constants can be found in many books.[17,18] The most frequently used is the pH titration method.[17] The use of spectrophotometric methods are also common.[19] Recently the ion-selective electrode method[20,50] and the charge-transfer-complex method have been proposed.[2]

In pH methods, the concentration of ligand is obtained by measuring the hydrogen ion through a pH meter. Equilibrium mixtures allow the competing reactions to take place between metal ion and protons in the ligand. It is then possible to calculate the concentration of the ligand. From the pH, concentration of all its protonated forms of ligands, and acid dissociation constants of the protonated ligands, an equation allows one to calculate the equilibrium constants.

The pM methods depend upon a different principle. The metal ion concentration is measured directly by a redox reaction through an electrode which may be a mercury cathode, a platinum, or an ion-selective electrode.[20,21]

In ion exchange methods and solubility method, the metal ion concentration and the total metal ion concentration of all species are obtained from changes affected by the complexing agent in the uptake of catex or in the solubility of a moderately soluble salt of a metal ion.[22,23]

In the distribution method, the distribution of the metal between two immiscible solvents is determined. In the organic phase, the uncharged complex is measured, and in the aqueous phase, the amount of uncomplexed metal ions is determined or measured by difference.[24]

One of the most frequently used methods is the optical method which is more suitable when only one or two complexes are involved in the equilibrium and when these are optically very different from the metal free ligand. The principle of these methods is to measure the small degree of dissociation spectrophotometrically. The most popular method is the Job's method of continuous variation; the mole ratio method and the third method requiring plots of $e/[M]_t$ vs. $[L]_t$ are occasionally used.[17]

Conductivity methods are useful only in special cases when the complexes are uncharged.

CHARGE TRANSFER COMPLEX

There are many complexes which exhibit spectra owing to charge transfer transitions.[28] The absorption is characterized by high intensity and evidently fully allowed transitions. The common feature is that the upper state may be considered as one in which an electron has been transferred from one atom to another rather than elevated to high orbitals of the same sort at the ground state. Mulliken[29] reported the early examples of recognition of electron transfer transition in the spectra of alkali halides and termed them electron donor-acceptor complexes or adducts.

Commonly there are two classes of donors (D): v and π electrons, where v stands for nonbonding lone pair v orbital and π stands for bonding π orbital. There are also three classes of acceptors (A): π, antibonding orbital, v, vacant orbital, and σ, antibonding σ orbital.[28,29]

There are three major treatments for the description of charge transfer transitions.

1.   Valence-bond treatment — Mulliken's valence bond description concerns intermolecular charge-transfer complexes expressed in terms of quantum mechanics.[29] He provided an explanation for many previous observations, and his work has major stimulus to the extensive development which have taken place in this field since 1952.

Relatively weak interaction between an electron donor and an electron acceptor can be described in terms of wave functions in the following form:

$$\psi_N(AD) = a\psi_0(A,D) + b\psi_1(A^- - D^+)$$

$$\psi_E(AD) = a^*\psi_1(A^- - D^+) + b^*\psi_0(A,D)$$

The acceptor and the donor may be atoms or atom-ions, molecules, or molecule-ions. ψ_O has been called by Mulliken the "no bond" function. It corresponds to the structure of a complex in which binding results from dipole-dipole and dipole-induced-dipole interactions, hydrogen bonding, van der Waals' forces, and London forces. The ψ_1 wave function has been called for "dative" function, this corresponding to the structure of the complex where one electron has been completely transferred from the donor to the acceptor. The ψ_N and ψ_E stand for ground states and excited states, respectively.

2. Simple molecular-orbital treatment — Dewar and co-workers,[30] using simple molecular-orbital theory to describe the transition energy in the charge-transfer, proposed that an excited state may be considered as an electron transferred from a filled orbital in the donor to an empty orbital in the acceptor. For a single charge-transfer band, it is generally assumed to involve the highest occupied molecular orbital (HOMO) of the donor and lowest empty molecular orbital (LEMO) of the acceptor. A more detailed treatment was given by Murrell,[31] who has attempted to demonstrate a relationship between the stability of a charge-transfer complex and the intensity of the charge-transfer band. A semiempirical linear combination of molecular orbital (LCMO) calculation has been used by Flurry[32] to predict the ground state stabilization energies of charge transfer complexes. Other molecular orbital calculations have also been attempted.

3. Free-electron model treatment — The free-electron model for conjugated molecules that was developed by many workers, including Kuhn[33] and Bayliss,[34] treats π electrons in a conjugated molecule as a free-electron gas which moves in the potential field of the molecule. This concept has been applied to charge transfer transition by Shuler[35] and Boeyens.[36] Both have used the simplified one-dimensional case.

The analytical applications of chloranil charge transfer complexes for determining amino acids and related compounds have been reported by Al-Sulimany and Townshend[37] and Lin and Cheng.[2]

CHELATION AND CHELATE EFFECT

Ligands which possess two or more donating groups may share more than one pair of electrons with a single metal ion by coordinating to two or more sites around the central metal ion. These ligands are generally known as multidentate ligands, commonly called bidentate, tridentate, etc. Multidentate ligands complex metal ions to form complexes that are specifically called chelates. Instead of a linear structure, they have a chelate ring structure. For example, Copper (II) ion and ethylenediamine form a chelate, $Cu(en)_2^{2+}$ (en = ethylenediamine). Both of the chelate rings of $Cu(en)_2^{2+}$ are five-membered rings. A stable chelate usually contains a five- or six-membered ring.

A common case is where the chelating ligand has at least one acidic group ($-OH$) or donor atom, as well as one or more basic donor atoms such as nitrogen. During the chelation, the acidic group loses a proton and becomes anionic donor, thus resulting in charge neutralization. The basic nitrogen donates a pair of electrons to the metal ion. An example of such a chelating agent is 8-hydroxyquinoline (Oxine).

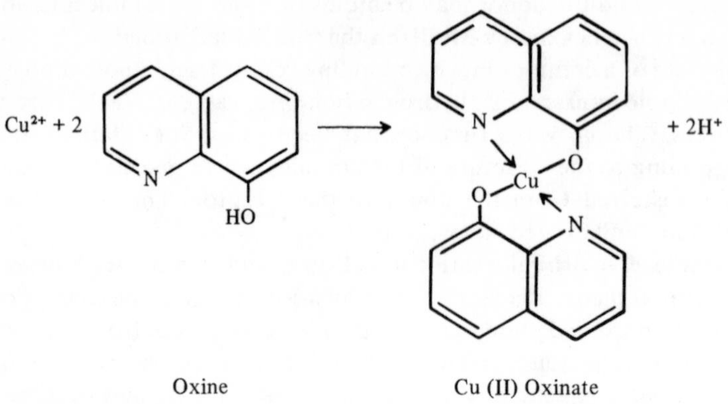

Oxine   Cu (II) Oxinate

Chelate effect — One of the most striking properties of chelates is their unusual stability. They resemble the aromatic rings of organic chemistry. Not all organic compounds can form chelates with metal ions. It is important to recognize the chelating groups (see Table 2). A chelating ligand must possess two acidic or two coordinating groups or one acidic and one coordinating group. Almost all acidic organic compounds contain $-OH$, $-SH$, or $-NH-$ group in some form. Molecules containing N, O, and S atoms frequently form coordinate bonds with metal ions in forming chelating rings. These groups must be located in the molecule in such positions that the metal ion will be involved in the ring formation of five or six atoms. An organic compound containing more than two donor groups is capable of forming multiple rings with the metal ion resulting in more stable chelate structure. Table 3 shows the chelate effect on stability. The EDTA forms much more stable complexes than NTA or IDA. However, DTPA and TTHA do not form much more stable complexes with copper because copper cannot accept many electrons from donor groups due to coordination number limitation. Therefore, EDTA is the most suitable chelating structure for metal ions. The number of rings formed with a particular metal ion depends largely on the pH of solution and whether the ligand or the metal ion is in large excess. Our recent experimental results suggest that when EDTA is in excess, 1 to 1 complex is formed, however, when the metal ion is in excess, complexes of 1 to 1, 1 to 2, or 1 to 3. . . are probably formed.[38]

MIXED LIGAND COMPLEXES

A central ion reacts with ligand forming complexes,

$$M + L_1 \rightleftharpoons ML_1$$

$$ML_1 + L_2 \rightleftharpoons ML_1 L_2$$

$$ML_1 L_2 + L_3 \rightleftharpoons ML_1 L_2 L_3 \ldots$$

where M is the central metal ion, and L is the ligand. L_1, L_2, L_3...may be the same ligand and may be different. If different, the ML_1L_2 and $ML_1L_2L_3$ are mixed ligand complexes. Strictly speaking, many complexes may be considered as mixed complexes because they contain water molecule ligand besides the other principal ligand. Usually water molecule is not considered as a ligand.

Mixed ligand complexes are characterized by their extreme stability and show different properties from the simple complexes. Formation of mixed ligand complexes often

Table 2
COMMON CHELATING GROUPS

Name	Chelating Groups
o-Diphenols	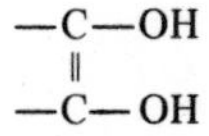 $-C-OH$ $\parallel$ $-C-OH$
o-Dithiophenols	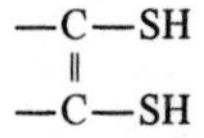 $-C-SH$ $\parallel$ $-C-SH$
o-Hydroxyaldoximes	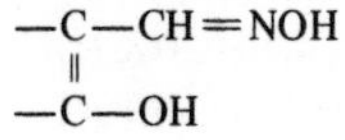 $-C-CH=NOH$ $\parallel$ $-C-OH$
Glycine	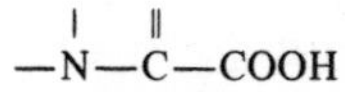 $\overset{\mid}{-N}-\overset{\parallel}{C}-COOH$
β-Diketones	$-CO-CH_2-CO-(+CS_2)$
Isonitrosoketones	$-CO-C(=NOH)-$
Mandelic Acid	$Ar-CH(OH)-COOH$
α-Acyloinoxime	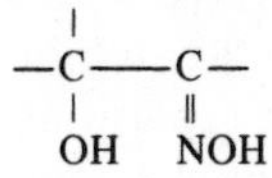 $\overset{\mid}{-C}\underset{OH}{\quad}\overset{\quad}{-}\overset{\parallel}{C}\underset{NOH}{\quad}-$
Oxine	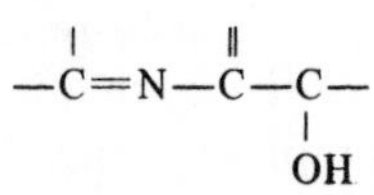 $\overset{\mid}{-C}=N-\overset{\parallel}{C}-\overset{\mid}{\underset{OH}{C}}-$
α-Dioximes	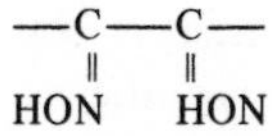 $-\overset{\parallel}{\underset{HON}{C}}\quad-\overset{\parallel}{\underset{HON}{C}}-$
Ferroine	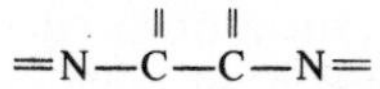 $=N-\overset{\parallel}{C}-\overset{\parallel}{C}-N=$
Cuproine	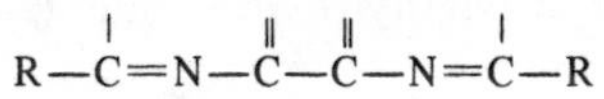 $R-\overset{\mid}{C}=N-\overset{\parallel}{C}-\overset{\parallel}{C}-N=\overset{\mid}{C}-R$
PAN	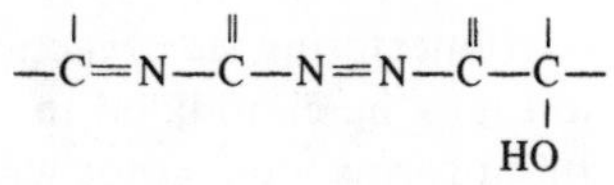 $-\overset{\mid}{C}=N-\overset{\parallel}{C}-N=N-\overset{\parallel}{C}-\overset{\mid}{\underset{HO}{C}}-$

provides special sensitivity and selectivity of analytical importance. The process of complex formation and the properties of mixed ligand complexes containing inorganic ligands are common with the platinum group metals. In the mixed ligand complexes, ML_1L_2, both L_1 and L_2 may be organic ligands, or L_1 is inorganic ligand and L_2 is organic ligand. Investigations have been emphasized in the improved extractibility (synergy) and coextraction of elements as mixed complexes and in the changes in the physicochemical properties of systems such as optical and solubility properties. Many new analytical procedures have been proposed based on the mixed ligand complexes.

Alimarin and Shlenskaya[39] gave an excellent review on mixed ligand complexes. The principal properties of mixed ligand complexes may be summarized in the following:

Table 3

STABILITY CONSTANTS OF COPPER CHELATES AS A FUNCTION OF
NUMBER OF DONATING GROUPS

Chelating ligand		Number of donating groups	Log K			
			K_1	K_2	K_3	K_4
	NH_3	1	4.2	3.5	2.9	2.1
en	$NH_2CH_2CH_2NH_2$	2	10.8		9.3	
den	$N(CH_2CH_2NH_2)_2$	3	16.0			5.4
tren	$N(CH_2CH_2NH_2)_3$	4	19.8			
trien	$(N_2HCH_2CH_2NHCH-)_2$	4	20.4			
PENTEN	$(H_2N)(CH_2CH_2)_2(NHCH_2CH_2)_3$	5	22.4			
GLY	NH_2CH_2COOH	2	8.6	7.0		
IDA	$NH(CH_2COOH)_2$	3	10.6	5.7		
NTA	$N(CH_2COOH)_3$	4	12.7			
EDTA	$[CH_2N(CH_2COOH)_2]_2$	6	18.8			
DTPA	$N(CH_2CH_2N)_2(CH_2COOH)_5$	8	21.1	6.8		
TTHA	$N(CH_2CH_2N)_3(CH_2COOH)_6$	10	27.6			
	$(2Cu^{2+} + L^{6-} \rightarrow Cu_2L^{2-})$					

1. Enhancement of properties of similar elements — The coordination number of central ion or atom plays the most important role in the formation of mixed ligand complexes. Other factors involve the number of electron donating groups in the primary ligand and the stability of metal secondary ligand complexes. New and selective methods for separating elements of similar properties, such as lanthanides and transuranium elements, metals of the platinum groups, zirconium, hafnium, and others, have been reported based on the mixed ligand complexes.

   Of a large number of the lanthanides in the mixed ligand M, *o*-phenanthroline-dibenzoylmethane complexes, only Eu and Sm exhibit fluorescence when their mixed ligand complexes are excited by UV radiation at 365 nm. Their fluorescent spectra are rather different. It is believed that the tendency of the lanthanides to show higher coordination numbers and the nature of the splitting of the 4f electrons in the field of symmetry created by different ligands favor the enhancement of the individual chemical properties of lanthanides in their mixed ligand complexes. Cheng reported a selective color reaction for hafnium with Xylenol Orange (XO) or Methylthymol Blue (MTB) as a primary ligand and fluoride as a secondary ligand.[40] Hf may be determined in the presence of Zr which is masked by fluoride.

2. Elimination of interfering side reactions — Hydrolysis or polymerization of central metal ion may be eliminated in the presence of a secondary ligand. For example, niobium forms a red color with xylenol orange in the presence of tartrate at pH 5; tartrate is a secondary ligand and helps to keep niobium in solution.[41] An interaction of Nb and Ta with 1-(2-pyridylazo)-2-resorcinol (PAR)[39,42] and that of In with 1-(2-pyridylazo-2-naphthol (PAN) in the presence of acetate[43] facilitated their solvent extraction.

3. Mixed ligand complexes as intermediates in analytical reactions — Many intermediate forms involving mixed ligand reactions are necessary steps in the catalytical and redox reactions. It is also not surprising that mixed ligand complexes play an important role in ligand exchange and dissociation processes.

4. Optical property changes — Copper (II) reacts with triethylenetetramine (trien) forming a blue-colored complex; addition of halide salt (except fluoride) to the solution enhances the color intensity due to the formation of a more intense copper-trien-X mixed ligand complex (X = halide). The enhancing strength is found

in the order $I^- > Br^- > Cl^- > F^-$.[44] Pribil[45] reported the Th mixed ligand complex. Belcher and West[46] found that on the interaction of F^- with solutions of the chelate of Ce(III), La (III), or PR (III) with alizarin-complexan (alizarin fluorine blue), a mixed compound containing F, with individual optical characteristic, is formed. This reaction is highly sensitive and selective for fluoride.

STERIC HINDRANCE AND SELECTIVITY

Because of steric hindrance of molecular structure, many organic reagents have exhibited their high selectivity. The mutual repulsion of parts of the same or different molecules can affect reaction rates and mechanisms; it can affect the intrinsic stability of a particular compound; it can favor a particular conformation; it can influence special physical properties, such as optical absorption, solubility, crystallization, etc. Irving and Pettit[47] have written an excellent review.

The most stable complexes are likely to be those formed when vacant orbitals of the metal ion are arranged in such directions that overlap fills ligand orbitals (or vice versa) without serious distortion; such structures may be difficultly formed because of steric hindrance. For instance, if unidentate ligands are large, their packing around a central metal ion will be handicapped if many ligand molecules are to be fitted in a limited space. The situation becomes much more serious if the organic ligand can fit in more than one donating group to the central metal ion. The well-known effect of substituents ortho to the nitrogen atoms in 2,2′-dipyridyl or 1,10-phenanthroline in preventing the color reaction with Fe (II) (the ferroin reaction) provides a specific color reaction for Cu (I). The Fe (II)-(1,10-phenanthroline)₃ complex (not the mono or bis complex) is low spin and has a high stability constant. It is intensely colored. The Cu(I)-(1,10-phenanthroline)₂ lack ligand-field stabilization and are less stable. In 2,9-dimethyl-1,10-phenanthroline, the methyl groups hinder the formation of the tris Fe(II) complex, and the bis Fe(II) complex is high spin and colorless. On the other hand, the methyl groups do not affect the formation of the tetrahedral Cu(I) bis complex, but actually increase the stability of the complex because they raise the basicity of the ligand. The difference in size of the central metal ion may also be important. Cu(I) ion has the greater radius, so that the ligands approach each other less closely so that the steric effect is not pronounced. A similar effect has been observed, the smaller radius of Al^{3+} ion may expel Al from formation of complex with 2-methyl-8-hydroxyquinoline and with similarly substituted acridines which react with larger ions like Fe^{3+}, Cu^{2+}, Cr^{3+}, Zn^{2+}, and Ga^{3+}.[48] Thus, methods of using 2-methyl-8-hydroxyquinoline were developed to determine Zn, Fe, Cu, etc. in the presence of Al.

The "packing effect" may be illustrated in the case of porphyrins and phthalocyanines, the central cavity has an important effect on the stability of the complex. Large ions such as Pb(II) and Hg(II) may be unable to fit into the central cavity of the porphrin nucleus[49] whereas with smaller ions, the stability of the complexes would be expected to decrease with decreasing metal ion radius as the rigidity of the porphyrin nucleus would progressively reduce the bond strength between the central ion and the ligand. Another example of a geometric factor is in complex formation of potassium with 18 crown compounds[50] with high selectivity.

SOME PROPERTIES OF ORGANIC REAGENTS

1. Purity — Except for a few, most organic compounds obtained commercially are not very pure. Purification may be required depending upon each individual case. For example, chloranil as a charge-transfer reagent for amino acids should be purified before using.[2]

2. Solubility — EDTA acid is not soluble in water; it is necessary to neutralize with a base to prepare its solution. 8-Hydroxyquinoline is slightly soluble in water; it is usually dissolved in glacial acetic acid and diluted with water. If the ligand or its complexes are not soluble in water, solvent extraction will be suitable for separation.

3. Vapor pressure — One complex may have higher vapor pressure than others. Certain methoxy or ethoxy derivatives are more volatile than their parent compounds. Based on difference in vapor pressure of the ligands or their complexes, many substances may be separated by gas chromatography.

4. Stability — As discussed previously, many chelates are very stable in terms of complex formation. However, many complexes are thermally stable and can be distilled without decomposition; for example, phthalocyanine complexes and acetylacetonates. Many organic reagents are light sensitive or air sensitive such as 3,3'-diaminobenzidine; proper care must be taken. The bismuth dithiocarbamate and tellurium dithiocarbamate are light sensitive, but silver salt of 1,2,3-benzotriazole is light stable.

5. Polarity — The polarity of a molecule determines its solubility in a solvent. As an indicator, it is usual to prepare it with a sulfonic acid group. On the other hand, a nonpolar molecule will be favorable in solvent extraction. Besides, separations based on the polar or nonpolar attraction of molecules have been widely used.

APPLICATIONS OF ANALYTICAL ORGANIC REAGENTS

Organic reagents have many applications in chemical analysis. In the past, they have been used for gravimetric, volumetric, and photometric determinations. They are also used in instrumental analysis as well. We have constantly found new organic reagents and developed better uses with them.

Gravimetric analysis — The organic reagents were found very beneficial in the gravimetric analysis. 8-Hydroxyquinoline and dimethylglyoxime are good examples. 2,3-Diaminonaphthalene has been found to be a very desirable reagent for selenium.

Volumetric analysis — The most important organic reagents used in volumetric analysis are EDTA and its analogs. Its use has drastically changed the modern analytical laboratory in the past 30 years.

Photometric analysis — Many organic reagents form colored products with metal ions. Their uses have advanced the trace analysis in the parts per million ranges. Dithizone was the most popular photometric reagent for determining trace amounts of metals. The discovery of Thio-Michler's ketone (TMK) introduced probably the most sensitive photometric agent known, $\lambda_{max} = 2.1 \times 10^5$ for Pd and $\lambda_{max} = 1.2 \times 10^5$ for Hg.[51]

Instrumental analysis — Though the organic reagents are not widely used in the instrumental analysis, they are uniquely used in special functions. In NMR, tetramethylsilane (TMS) is universally used as a standard. All chemical shifts are measured relative to TMS. Many lanthanide chelates have been used as NMR shift reagents. In the mass spectrometry, organic substances, such as methane and ethane ions, have been used as reagent ions in the mass spectrometric chemical ionization. EDTA and the complexing organic reagents have been used in the atomic absorption spectrometry for the purposes of masking and increasing sensitivity. Dialkyl phosphonic acids have been used in the preparation of liquid membranes in the ion selective electrodes for calcium and bivalent ions. Other ion selective membranes are Valiminomycine for K and Fe-1,10-phenanthroline for perchlorate. Many iminodiacetic acid and its derivatives are used in the preparation of chelating ion exchange resins. Sephadex and other organic reagents are commonly used in chromatographic separations. The above are some examples that demonstrate the important role played by the organic reagents in instrumental analysis.

FUTURE RESEARCH INVOLVING ORGANIC ANALYTICAL REAGENTS

1. Synthesis of new organic reagents — Unquestionably we are constantly looking for new and better organic reagents. In particular, we need reagents for anions and less common elements. Development and research of methods for analyzing iron and copper with new reagents which are inferior to the existing ones should be discouraged.

2. Simple, sensitive, and selective reagents — Efforts toward developing simpler, more sensitive, and more selective methods employing new or current organic reagents will be continuing.

3. Functional groups study — Many new functional groups may be found. The functional groups for Fe(II), Cu(I), Ag, Se, Ni, etc. are well known. The research along this area may prove to be fruitful.

4. Structural study — Study of ligands and their complexes by modern techniques have been useful; the research in this area will be continued in order to understand the relationship between their structures and chemical reactions.

5. Kinetic study — In the past two decades, the kinetic study involving organic reagents has been very successful. It is expected that more interesting results will come.

6. Organic chelating resins — Several new organic chelating resins have been reported. The development of this area will help achieve better separations and preconcentration of trace substances.

7. Group extractants — EDTA can complex practically all polyvalent metal ions in an aqueous solution. It would be desirable to have ligands capable of complexing a number of metal ions or anions in organic solvents just like EDTA in water.

8. Tailor-made molecules — Many derivatives of dimethylglyoxime (DMG) and EDTA have been synthesized and proposed. EDTA forms generally 1:1 complexes, however, TTHA (triethylenetetramine-hexa acetic acid) forms 2:1 (metal to ligand) for some metal ions, depending on the coordination number of the metal. It is possible to titrate a mixture of metal ions without separation using EDTA and TTHA in two titrations. We may synthesize many ligands with specially designed molecules. Another example may be cited that Thio-Michler's ketone (TMK) gives an unusually high sensitivity for Hg and Pd. The mechanism and reactions for such a high sensitivity should be studied, and the synthesis of derivatives of TMK should be attempted.

9. Computers and Fourier Transform — Computers and Fourier Transform technique have been successfully applied to the spectroscopy, for instance, the Fourier Transform multinuclear NMR spectrometers and Fourier Transform Infrared spectrometers are commercially available. It might be possible to apply the Fourier Transform to the UV and visible spectrometry in order to gain high resolution and to analyze mixtures spectrometrically using organic reagents.

10. Preconcentration — Preconcentration often increases the sensitivity of the reaction and lowers the detection limit of organic reagents. The preconcentration of trace amounts of metals on the surface of glass and nonpolar resin which have been coated with a layer of ligand such as dithiocarbamate or oxine have been reported. The glass beads containing metal and ligand are then examined by a physical means such as photoelectron spectroscopy.

11. Charge transfer complex study — This may prove to be an exciting field in studying organic reagents. Efforts for improving the existing methods using charge transfer reactions and applying new charge transfer reactions to the chemical analysis should be encouraged.

12. Simplicity study — There are huge, expensive, and sophisticated instruments which can do certain analyses that organic analytical reagents cannot do. However, organic reagents offer simple and inexpensive methods with reasonable accuracy. This is a big advantage. The majority of analyses performed everyday are those using organic reagents. For example, most people use pocket or watch calculators for everyday computations, not the IBM computers requiring terminals. Efforts to develop simple tests with organic reagents and to have portable test kits to meet the needs would be very rewarding.

13. Reagents for fluorescence and chemiluminescence — Research in this area has been always welcome.

14. Surface — The use of organic reagents to test the surface of materials may offer an attractive method for characterization of materials on the spot with a test kit. Many sophisticated instruments cannot be handy for their performance.

15. Oxidation states study — At the present, photoelectron spectroscopy is the only technique available for determining the oxidation states of the elements. Organic reagents may be able to show reactions for elements with different oxidation states in a limited scope. This is not impossible.

The above are some of our thoughts suggesting possible future research directions involving organic analytical reagents. To keep the idea in mind that we always need improved, simple, sensitive, and selective methods for which the organic analytical reagents will continually play an important role.

REFERENCES

1. Cheng, K. L., and Guh, H. Y., *Mikrochim. Acta*, 1, 55, 1978.
2. Lin, B. Y. and Cheng, K. L., *Anal. Chim. Acta*, 120, 335, 1980.
3. Fichter, Fr. and Goldach, A., *Helv. Chim. Acta*, 15, 1513, 1932.
4. Feigl and, F. and Schwarz, R., *Rec. Trav. Chim.*, 58, 474, 1939.
5. Welcher, F. J., The development of organic reagents in inorganic analysis, in *Analytical Chemistry*, West, P. W., MacDonald, A. M. G., and West, T. S., Eds., Elsevier, Amsterdam, 1963, 106.
6. Blaedel, W. J. and Meloche, V. W., *Elementary Quantitative Analysis*, Harper & Row, New York, 1963, 577.
7. Cheng, K. L., Carver, J. C., and Thomas, T. A., *Inorg. Chem.*, 12, 1720, 1973.
8. Cheng, K. L., *Anal. Chem.*, 28, 1738, 1976.
9. Bricker, C. E. and Parker, G. H., *Anal. Chem.*, 29, 1470, 1957; Brunisholz, G. B., Chimia (Switz.), 11, 363, 1957.
10. Lin, E. and Cheng, K. L., *Mikrochim. Acta*, 1, 337, 1961.
11. Irving, H. and Thomlinson, W. R., *Chemist-Analyst*, 55, 14, 1966.
12. Cheng, K. L., *Talanta*, 14, 875, 1967.
13. Margerum, D. W., *J. Phys. Chem.*, 63, 336, 1959; Margerum, D. W., Pausch, J. B., Nyssen, G. A., and Smith, G. F., *Anal. Chem.*, 41, 233, 1969.
14. Cheng, K. L. and Bray, R. H., *Anal. Chem.*, 27, 782, 1955; Cheng, K. L., *Anal. Chem.*, 30, 1035, 1958; Chang, F. C. and Cheng, K. L., *Mikrochim. Acta*, II, 219, 1979.
15. Cheng, K. L., *Talanta*, 8, 201, 1961; Wang, J. C. and Cheng, K. L., *Microchem. J.*, 15, 607, 1970.
16. Schwarzenbach, G. and Flaschka, H., (Translated by H. Irving), *Complexometric Titrations*, Methuen, London, 1969.
17. Bjerrum, J., *K. Dan. Vidensk. Selsk. Mat.-Fys. Medd.*, 21(4), 1944.

18. Rossotti, F. J. C. and Rossotti, H., *The Determination of Stability Constants*, McGraw-Hill, New York, 1961; Christian, S. D., *J. Chem. Educ.*, 45, 713, 1968.
19. Job, P., *Anal. Chim.*, 9, 113, 1928; Harvey, A. E. and Manning, D. L., *J. Am. Chem. Soc.*, 72, 4488, 1950.
20. Chao, E. and Cheng, K. L., *Anal. Chem.*, 48, 267, 1976; Talanta, 24, 247, 1-250, 1977.
21. Schmid, R. W. and Reilley, C. N., *J. Am. Chem. Soc.*, 78, 5513, 1956; Reilley, C. N. and Schmid, R. W., *Anal. Chem.*, 30, 947, 1958.
22. Druger, P. and Schubert, J., *J. Chem. Educ.*, 30, 196, 1953.
23. Keefer, R. M., Reiber, H. G., and Bisson, C. S., *J. Am. Chem. Soc.*, 62, 2951, 1940; Colmann-Porter, C. A. and Monk, C. B., *J. Chem. Soc.*, 4363, 1952.
24. Rydberg, J., *Acta Chem. Scand.*, 4, 1503, 1950; Irving, H., Rossotti, F. J. C., and Williams, R. J. P., *J. Chem. Soc.*, 1906, 1955.
25. Mottola, H. A., *Talanta*, 16, 1267, 1969.
26. Alimarin, I. P., *Pure Appl. Chem.*, 34, 1, 1973.
27. Mottola, H. A. and Freiser, H., *Anal. Chem.*, 39, 1294, 1967.
28. Slifkin, M. A., *Charge Transfer Interactions of Biomolecules*, Academic Press, London, 1871.
29. Mulliken, R. S., *J. Am. Chem. Soc.*, 70, 600, 1950; *J. Chem. Phys.*, 23, 397, 1955; *Rec. Trav. Chim.*, 75, 845, 1956.
30. Dewar, M. J. S. and Rogers, H., *J. Am. Chem. Soc.*, 84, 395, 1962.
31. Murrell, J. N., *Q. Rev. (London)*, 15, 191, 1961; *J. Am. Chem. Soc.*, 81, 5037, 1959.
32. Flurry, R. L., Jr., *J. Phys. Chem.*, 69, 1927, 1966.
33. Kuhn, H., *J. Chem. Phys.*, 17, 1198, 1948.
34. Bayliss, N. S., *Q. Rev., (London)*, 6, 319, 1952.
35. Shuler, K. E., *J. Chem. Phys.*, 20, 1865, 1952.
36. Boeyens, J. S. A., *J. Phys. Chem.*, 71, 2969, 1967.
37. Al-Sulimany, F. and Townshend, A., *Anal. Chim. Acta*, 66, 195, 1973.
38. Cheng, K. L., unpublished result.
39. Alimarin, I. P. and Shlenskaya, V. I., The analytical chemistry of mixed ligand complexes, in *International Symposium on Analytical Chemistry, Birmingham, U.K., 1969,* Butterworths, London, 1970, 461.
40. Cheng, K. L., *Anal. Chim. Acta*, 28, 41, 1963.
41. Cheng, K. L., *Talanta*, 9, 987, 1962.
42. Alimarin, I. P. and Si-i, Khan, *Zh. Anal. Khim.*, 22, 79, 1967.
43. Zolotov, Y. A., Seryakova, I. V., and Vorobieva, G. A., *Talanta*, 14, 737, 1967.
44. Cheng, K. L., *Anal. Chem.*, 34, 1392, 1962.
45. Pribil, R. and Vesely, V., *Talanta*, 12, 191, 1965.
46. Belcher, R., Leonard, M. A., and West, T. S., *Talanta*, 2, 92, 1959.
47. Irving, H. and Pettit, D. L., Steric hindrance in analytical chemistry with special reference to the reactions of analogs of 8-hydroxy quinoline, in *International Symposium on Analytical Chemistry, Birmingham, U.K., 1962,* West, P. W., MacDonald, A. M. G., and West, T. S., Eds., Elsevier, Amsterdam, 1963, 122.
48. Irving, H., Butler, E. J., and Ring, M. F., *J. Chem. Soc.*, 1489, 1949.
49. Barnes, J. W. and Dorough, G. D., *J. Am. Chem. Soc.*, 69, 1860, 1947.
50. Freiser, H., Ed., *Ion-Selective Electrodes in Analytical Chemistry,* Plenum Press, New York, 1978.
51. Cheng, K. L., and Goydish, B. L., *Microchem. J.*, 10, 158, 1966.

ABBREVIATIONS

AA (AcAc)	acetylacetone
AABN	AzoazoxyBN, 2-[(2″-hydroxy-1-naphthyl-1″-azo)-2′-phenylazo]-4-methylphenol
ALC (AFB)	Alizarin fluorine blue, Alizarin Complexone, 3-[di(carboxymethyl)amino-methyl]-1,2-dihydroxyanthraquinone, 1,2-dihydroxyanthraquinon-3-yl-methylamino-*N,N*-diacetic acid
ALC-SS (AFBS)	sulfonated Alizarin Complexone
iso-AmOH	iso-amylalcohol
AMBB	dialkylmonomethylbenzylammonium bromide
ABMC	dialkylmonomethylbenzylammonium chloride
APANS (APNS)	Thoron, Thorin, 2-(2-hydroxy-3,6-disulfo-1-naphthylazo)-benzenearsonic acid
APDC	ammonium pyrrolidinedithiocarbamate
BAL	2,3-dimercapto-1-propanol
BG	Bindschedler's Green
Bipy	2,2′-bipyridine
BPA (NBPHA,BPHA)	*N*-benzoyl-*N*-phenylhydroxylamine, *N*-phenylbenzohydroxamic acid
BPR	Bromopyrogallol Red, 5,5′-dibromopyrogallolsulfonephthalein
5-Br-DMPAP	2-(5-bromo-2-pyridylazo)-5-dimethylaminophenol
BT (EBT)	Erio T, F-241, Eriochrome® Black T
BTA (BFA)	benzoyltrifluoroacetone, 4,4,4-trifluoro-1-phenyl-1,3-butanedione
BTA	*N*-benzoyl-*N-O*-tolylhydroxylamine, *N-O*-tolylbenzoyhydroxamic acid
iso-BuOH	iso-butanol
BzA (BA)	benzoylacetone
5-Cl-PADAB	4-(5-chloro-2-pyridylazo)-1,2-diaminobenzene
CPA	*N*-cinnamoyl-*N*-phenylhydroxylamine, *N*-phenylcinnamohydroxamic acid
CPB	cetylpyridinium bromide
CPC	cetylpyridinium chloride
CTAB	cetyltrimethylammonium bromide
CTMC	cetyltrimethylammonium chloride
CTMB	cetyltrimethylammonium bromide
CYDTA (CDTA, DCTA, CDyTA)	(1,2-cyclohexylenedinitrilo)-tetraacetic acid
DAB	3,3′-diaminobenzidine
DAM (DAPM, MDAP)	1,1′-diantipyrinylmethane, 4,4′-methylenediantipyrine
DAN	2,3-diaminonaphthalene
DBM (DBzM)	dibenzoylmethane

DDC (DDTC)	diethyldithiocarbamate
dien (den)	diethylenetriamine
DMF	dimethylformamide
DMG	dimethylglycoxime
DMSO	dimethylsulfoxide
DOMBC	dodecyloctylmethylbenzylammonium chloride
DPM (Hthd, TMTD)	dipivaroylmethane
DSNADNS R-8 (Beryllon II)	2-(3,6-disulfo-8-hydroxynaphthylazo)-1,8-dihydroxynaphthalene- 3,6-disulfonic acid
DTPA	diethylenetriaminepentaacetic acid
EDTA	ethylenediaminetetraacetic acid
en	ethylenediamine
Eu-DPM (Eu-THD)	tris(dipivaloylmethanato)- europium(III)
Eu-FOD	tris(heptafluorobutanoylpivaroylmethanato)-europium(III)
Eu-TFMC	tris[3-(trifluoromethylhydroxymethylene)-d-camphorato]-europium(III)
FOD (Hfod, HFM, HPM, Hhdod)	heptafluorobutanoylpivaroylmethane, 1,1,1,2,2,3,3-heptafluoro-7,7-dimethyl-4,6-octanedione
FTA	furoyltrifluoroacetone, 4,4,4-trifluoro-1-(2-furyl)-1,2-butanedione
GCR	Glycinecresol Red
GEDTA (EGTA)	ethyleneglycol bis(2-aminoethylether)- N,N,N',N'-tetraacetic acid, glycoletherdiaminetetraacetic acid
GHA	glyoxalbis(2-hydroxyanil), 2,2'-(ethanediylidenedinitrilo)-diphenol, N,N'-bis(o-hydroxyphenyl)ethylenediimine, di-(o-hydroxyphenylimino)ethane
gly	glycine
GTB	Glycinethymol Blue
HDEHP (DEHP, D₂EPHA)	di(2-ethylhexyl)-phosphoric acid
HDTB	hexyldecyltrimethylammonium bromide
HDTMB	hydroxydodecyltrimethylammonium bromide
HFA (Hhpd)	hexafluoroacetylacetone, 1,1,1,5,5,5-hexafluoro-2,4-pentanedione
HIDA	N-(2-hydroxyethyl)-iminodiacetic acid
HNB	Hydroxynaphthol Blue
HOMO	highest occupied molecular orbital
IDA	iminodiacetic acid
INT	3-(p-iodophenyl)-2-(p-nitrophenyl)-5-phenyl-2H-tetrazolium chloride
IPT	β-isopropyltropolone
LCMO	linear combination of molecular orbital
LEMO	lowest empty molecular orbital
MIBK (IBMK)	methyl iso-butyl ketone
MTB	Methylthymol Blue
MTT	3-(4,5-dimethyl-2-thiazolyl)-2,5-diphenyl-2H-tetrazolium bromide

MX	murexide
MXB	Methylxylenol Blue, 3,3′-bis[N,N-di(carboxymethyl)-aminomethyl]-p-xylenosulfonphthalein
NaTPB	sodium tetraphenylborate
NN	Calcon carboxylic acid, Patton and Reeder's dye
Neo-TB	3,3′-(4,4′-biphenylylene)-bis[(2,5-diphenyl)-2H-tetrazolium chloride]
Nitro-TB	3,3′-(3,3′-dimethoxy-4,4′-biphenylene)-bis[2-p-nitrophenyl)-5-phenyl-2H-tetrazolium chloride]
NTA	nitrilotriacetic acid
3-OH-PAA	3-hydroxypicolinaldehyde azine
PAN (β-PAN, o-β-PAN)	1-(2-pyridylazo)-2-naphthol, α-pyridylazo-β-naphthol
α-PAN	2-(2-pyridylazo)-1-naphthol
ϱ-PAN	4-(2-pyridylazo)-1-naphthol
PAPH	(pyridine-2-aldehyde)-2′-pyridylhydrazone
PAQH	(pyridine-2-aldehyde)-2′-quinolylhydrazone
PAR	4-(2-pyridylazo)resorcinol
PC	Phthalein Complexone, 3,3′-bis[N,N-di(carboxymethyl)-aminomethyl]-o-cresolphthalein, o-cresolphthalein-3,3′-bis[methyliminodiacetic acid], Metal Phthalein
PDT	pyridyldiphenyltriazine
penten	pentaethylenehexamine
PDTS	pyridyldi(4-sulfophenyl)-triazine
phen	1,10-phenanthroline
PMBP	1-phenyl-3-methyl-4-benzoyl-5-pyrazolone
PPDT (PAT)	3-(4-phenyl-2-pyridyl)-5,6-diphenyl-1,2,4-triazine
PPKO	phenyl-2-pyridylketoxime, 2-benzoylpyridine oxime
PR	Pyrogallol Red, pyrogallol sulfonaphthalein
Pr-DPM (Pr-THD)	tris(dipivaroylmethanato)-praseodymium(III)
Pr-FOD	tris(heptafluorobutanoylpivaroylmethanato)-praseodymium(III)
Pr-TFMC	tris[3-(trifluoromethylhydroxy-methylene)-d-camphorato)]-praseodymium(III)
PTA (TPM, TAPM, Htdhd, Hfhd)	pivaroyltrifluoroacetone, 1,1,1-trifluoro-5,5-dimethyl-2,4-hexanedione
PV	Pyrogallol Red, Catechol Violet
PVC	poly(vinyl chloride)
QAQH	(quinoline-2-aldehyde)-2′-quinolylhydrazone
SABF	N,N-bis(salicylidene)-2,3-diaminobenzofurane
SAPH	Manganon, o-salicylidene aminophenol
SATP	salicylideneamino-2-thiophenol
Semi-MTB	3-[N,N-di(carboxymethyl)-aminomethyl]-thymolsulfonphthalein
Semi-XO	3-[N,N-di(carboxymethyl)-aminomethyl]-o-cresolsulfonaphthalein
SPADNS (SPANS)	3-(4-sulfophenylazo)-4,5-dihydroxy-2,7-naphthalenedisulfonic acid

STTA	thiothenoyltrifluoroacetone, 1,1,1-trifluoro-4-mercapto-4-(2-thienyl)-but-3-en-2-one
TAA (TFA,ATA)	trifluoroacetylacetone, 1,1,1-trifluoro-2,4-pentanedione
TAC	2-(2-thiazolylazo)-*p*-cresol, 2-(2-thiazolylazo)-4-methylphenol
TAM	2-(2-thiazolylazo)-5-dimethylaminophenol
TAN (*o*-TAN, β-TAN)	1-(2-thiazolylazo)-2-naphthol
TAR	4-(2-thiazolylazo)-resorcinol
TB	3,3′-(3,3′-dimethoxy-4,4′-biphenylylene)-bis[(2,5-diphenyl)-2H-tetrazolium chloride]
TBP	tri-n-butylphosphate
TBPO	tri-n-butylphosphine oxide
Nitro-TB	3,3′-(3,3′-dimethoxy-4,4′-biphenylylene)-bis[2-(ϱ-nitrophenyl)-5-phenyl-2H-tetrazolium chloride]
TDEB	tridodecylethylammonium bromide
TEHPO	tris(2-ethylhexyl)phosphine oxide
TFMC	3-(trifluoromethylhydroxymethylene)-d-camphor
THF	tetrahydrofuran
TIOA	tri-isooctylamine
TLC	thin layer chromatography
TMK	Thio-Michler's ketone, 4,4′-bis(dimethylamino)-thiobenzophenone
T(3-MPy)P	α,β,γ,δ-tetra(3-*N*-methylpyridyl)-porphin
T(4-MPy)P	α,β,γ,δ-tetra(4-*N*-methylpyridyl)-porphin
TMS	tetramethylsilane
TNTB	3,3′-(3,3′-dimethoxy-4,4′-biphenylylene)-bis[2,5-bis(ϱ-nitrophenyl)-2H-tetrazolium chloride]
TOA (TNOA)	tri-*n*-octylamine
TOPO	tri-*n*-octylphosphine oxide
TPAC	tetraphenylarsonium chloride
TPB	tetraphenylborate
TPC	3,3′-bis[*N*,*N*-di(carboxymethyl)-aminomethyl]-thymolphthalein, Thymolphthalexone
TPPC	tetraphenylphosphonium chloride
TPPS₃	α,β,γ,δ-tetraphenylporphin trisulfonic acid
TPPS₄	α,β,γ,δ-tetraphenylporphin tetrasulfonic acid
TPP	α,β,γ,δ-tetraphenylporphin
TPTZ	tripyridyltriazine
tren	nitrilotriethylamine, triaminotriethylamine
trien	triethylenetetramine
TTA	2-thenoyltrifluoroacetone, 4,4,4-trifluoro-1-(2-thienyl)-1,3-butanedione
TTHA	triethylenetetraminehexaacetic acid
VBB	Variamine Blue B
XO	Xylenol Orange

O,O-DONATING CHELATING REAGENTS

PHENYLFLUORONE

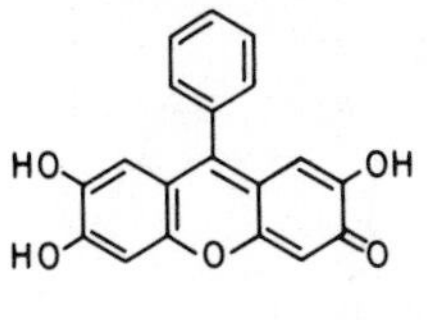

$C_{19}H_{12}O_5$
mol wt = 320.30

H₃L

Synonyms

2,3,7-Trihydroxy-9-phenyl-6-fluorone, 2,6,7-trihydroxy-9-phenylisoxanthene-3-one

Source and Method of Synthesis

Commercially available. It is synthesized by the reaction of 1,2,4-benzenetriol with benzotrichloride.[1]

Analytical Uses

Sensitive and selective photometric reagent for Ge and Sn; also used as a photometric reagent for Co, Fe, In, Mo(VI), Nb, Ni, Ti, and Zr.

Properties of Reagent

It is an orange crystalline powder, mp >300°. A greenish fluorescence appears in aqueous alcoholic solution above pH 8. In strongly alkaline solution, the reagent decomposes gradually.

Very slightly soluble in water (3 × 10⁻⁷ M in 20% aqueous ethanol, 25°), slightly soluble in cold ethanol, but easily soluble in acidified alcohol (HCl) and sulfuric acid. It can be recrystallized from warm acidified alcohol by addition of ammonia. Commercial samples often have unsatisfactory purity, but can be purified as described in Purification and Purity of Reagent.

Phenylfluorone behaves as an ampholyte, and acid dissociation constants in 25% ethanol have been determined as below.[2]

$$H_4L^+ \xrightleftharpoons{pKa_1 = 2.3} H_3L \xrightleftharpoons{pKa_2 = 5.8} H_2L^- \xrightleftharpoons{pKa_3 = 11.3} HL^{2-} \xrightleftharpoons{pKa_4 = 12.3} L^{3-}$$

Yellow   Yellow Orange   Orange-Pink

The absorption spectrum in aqueous solution is illustrated in Figure 1.

Complexation Reactions and Properties of Complexes

Phenylfluorone reacts with a number of metal ions to form colored insoluble chelates. In 0.1 M acidic solution, Ge(IV) forms a red 1:2 chelate according to the following equation:

$$Ge(OH)_4 + 2H_3L \; K \rightleftharpoons [Ge(OH)_2(H_2L)_2] + 2H_2O.$$

The equilibrium constant in 20% ethanol solution at 25° and $\mu = 5$ was determined approximately as 8 (±4) × 10¹².[4] The coordinating structure of Ge(IV) in the chelate may be represented as below.

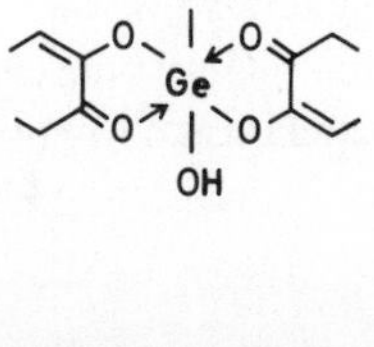

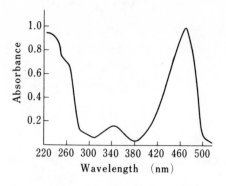

FIGURE 1. Absorption spectrum of phenyl-fluorone in 1 *N* HCl (water-ethanol); dye concentration, $5 \times 10^{-4}\%$. (From Petrova, G. S., Ryabokabilko, Yu. S., Lukin, A. M., Visokova, N. N., and Poponova, R. V., *Anal. Lett.*, 5, 695, 1972. With permission.)

Table 1
REACTION OF PHENYLFLUORONE WITH 100 μg OF METALS

Metal ion	HCl (1:99)[a]	HCl (1:99) + H_2O_2[a]	HCl (1:99) + oxalate[a]	HCl (1:99) + H_2O_2 + oxalate[a]
Fe(III)	S	S	—	—
Ge(III)	S	S	S	S
Hf(IV)	S	S	—	—
Mo(VI)	S	—	L	—
Nb(V)	S	—	—	—
Sb(III)	S	S	S	S
Sn(IV)	S	S	L	S
Ta(V)	S	—	L	—
Ti(IV)	S	—	L	—
V(V)	S	—	—	—
W(VI)	S	—	—	—
Zr(IV)	S	L	—	—

[a] S, strong color; L, light color; — , little or no color.

From Luke, C. L., *Anal. Chim. Acta,* 37, 97, 1967. With permission.

The result of investigation on the color reaction of phenylfluorone with common elements in (1:99) hydrochloric acid showed that only 12 metals were found to react with the reagent, as summarized in Table 1. In the presence of hydrogen peroxide and oxalate, the complexation is highly selective for Ge(IV), Sb(III), and Sn(IV).

As the chelate is insoluble in water, the colored chelate has to be kept in solution for the photometric purpose, either by stabilizing the dispersed particles with polyvinyl alcohol, gum arabic, or gelatin or by extracting the chelate with immiscible solvent, such as carbon tetrachloride, benzyl alcohol, or methyl iso-butyl ketone (MIBK). In the photometric determination, the sensitivity and selectivity are appreciably increased by the solvent extraction from strong acid solution.

The spectral characteristics of phenylfluorone chelates are summarized in Table 2, and the absorption spectrum of Ge chelate is shown in Figure 2.

Table 2
SPECTRAL CHARACTERISTICS OF PHENYLFLUORONE CHELATES

Metal ion	Condition	Chelate Ratio	λ_{max}(nm)	$\varepsilon(\times 10^4)$	Range of determination (ppm)	Ref.
Co(II)	pH 4.5 ∿ 5.0, NO₂⁻, Zephiramine®	ML₂	645	4.8	∿2	6
Fe(III)	0.001 NNaOH	ML₂	580	8.6	∿1	7
Ge(IV)	1.2 NHCl, gelatin, gum arabic or polyvinyl alcohol	ML₂	510	8.7	—	8
Ge(IV)	0.5 NHCl, extraction with benzyl alcohol	ML₂	505	14.5	0.05 ∿ 0.5	9
Ge(IV)	1 ∿ 1.5 NHCl, cetyltrimethyl ammonium chloride	ML₂	505	17.1	0.02 ∿ 0.16	10
In	pH 5.5, gelatin	ML₂	540	4.88	0.2 ∿ 1.4	11
Mo(VI)	pH 2, gum arabic	—	550	—	0.33 ∿ 1.67	12
Mo(VI)	pH 1.5, cetylpyridinium chloride	ML₂	540	9.6	0.4 ∿ 6	13
Nb	0.8% H₃PO₃	ML₂	520	3.7	—	14
Ni	pH 8.5 ∿ 10.0, cetyltrimethyl ammonium bromide, pyridine	ML₂	620	10.4	0 ∿ 0.33	15
Sb(III)	H₂SO₄, gelatin, thiourea	—	540	3.42	—	16
Sn(IV)	pH 1.2 ∿ 2.0 (HCl), tartarate, gum arabic or polyvinylalcohol	ML₂	510	5.93	∿2	17,18
Sn(IV)	pH 1.8 (H₂SO₄), oxalate, cetylpyridinium chloride	ML₂	530	3.96	0 ∿ 0.5	19
Ta	pH 4.5	—	530	6.39	0 ∿ 0.5	14
Ti	pH 0.7 ∿ 1.0 (HCl), BPA, Cl⁻, extraction with CHCl₃ + AmOH (1:1)	MLX₂Cl	540	9.0	0.05 ∿ 0.1	20
Ti	CPA, Cl⁻, extracted with CHCl₃	MLX₂Cl	550	—	—	21
Zr(IV)	0.1 NHCl, stabilized with cyclohexanol	—	540	13	∿1	22
Zr(IV)	pH 4.6, F⁻, cetylpyridinium chloride	ML₂	560	4.6	0 ∿ 7	13

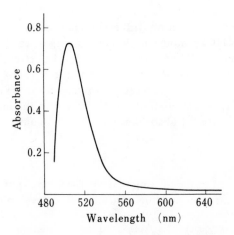

FIGURE 2. Absorption spectrum of Ge-phenylfluorone in water. 1.15 N HCl solution containing 5 ml of 0.5% gum arabic, 15 ml of 0.04% phenylfluorone, and 22.5 µg Ge in 50 ml. Observed 90 min after mixing. (From Oka, Y., Kanno, T., and Shiba, K., *Bunseki Kagaku*, 3, 389, 1954. With permission.)

Purification and Purity of Reagent

The reagent may be purified by extraction of 1 g crude sample with 50 mℓ ethanol in Soxlet apparatus for 10 hr to remove impurities.[3,24]

The purity can be checked by observing the absorption spectra of $5 \times 10^{-4}\%$ water-ethanol (1 N HCl) solution, (λ_{max},462 nm,$\varepsilon = 4.06 \times 10^4$).[3,24]

The presence of impurities can also be checked by the paper electrophoresis of 0.1% solution in a mixture of 6 N HCl (3 mℓ)-ethanol (97 mℓ) using an aqueous solution containing 0.05% oxalic acid and 30% acetic acid as an electrolyte. Pure sample should give one yellow spot.[3]

Analytical Application

Use as a Photometric Reagent

Phenylfluorone has been most widely used as a photometric reagent for Ge. Some other elements, such as Co, Fe, In, Mo(VI), Nb, Ni, Sn(IV), Ta, and Zr, can also be determined with this reagent. In the determination of Ge, the sensitivity can be appreciably increased either by extracting the germanium chelate plus excess reagent into an immiscible organic solvent such as benzyl alcohol or by extracting germanium chloride from HCl solution into carbon tetrachloride or MIBK, then forming the phenylfluorone chelate directly in the organic phase by the addition of an alcoholic solution of the reagent. In contrast to these extraction methods, the sensitivity of photometry in aqueous phase is about one half of the extraction method, but the sensitivity and reaction rate can be improved by the combined use of cationic surfactants such as cetyltrimethylammonium chloride.[10] The representative methods for Ge are summarized in Table 3.

In the determination of Sn, on the contrary, it is not possible to form colored tin chelate by adding methanol solution of phenylfluorone to MIBK extract of tin chloride. However, tin phenylfluorone chelate, formed in dilute acid solution, can be quantitatively extracted with MIBK. After extraction, the color of the chelate tends to fade slowly just as it does in aqueous solution. The rate of fading increases with an increase in acid concentration of the solution being extracted.[5]

The conditions for the photometric determination of metals using phenylfluorone may be referred to in Table 2.

Determination of Ge after extraction with carbon tetrachloride[29,30]

Place 20 mℓ 9 N HCl sample solution containing 0.5 to 10 μg Ge in a separatory funnel and add 10 mℓ carbon tetrachloride. After shaking for 2 min, separate the organic layer. Place 5 mℓ aliquot in a 10-mℓ volumetric flask, and add 1 mℓ of reagent solution (0.05 g phenylfluorone plus 0.43 mℓ concentrated HCl in 100 mℓ ethanol), and dilute it to the mark with ethanol. After 5 min, measure the absorbance at 508 nm against the reagent blank.

Other Analytical Uses

Phenylfluorone is used as a detecting reagent for Ge by spot test reaction (detection limit 1:5 $\times$ 10^5)[M1] or on paper chromatogram, after developing with 10% HBr butanol.[31,32]

Other Reagents with Related Structure

p-Phenylene-bis(fluorone) was recently introduced as a highly selective reagent for Te(IV) in the presence of Se[33] and for Al.[34]

Beside this reagent, many derivatives of phenylfluorone have been prepared and tested as chromogenic reagents for metals. A short survey on this subject may be found in Reference M3.

Table 3
PHOTOMETRIC DETERMINATION OF GERMANIUM WITH PHENYLFLUORONE

Method	λ_{max}	$\varepsilon(\times 10^5)$	Interference	Refs.
Complex formation in aqueous solution $(0.2 \sim 1.5\ N\mathrm{HCl}$ or H_2SO_4); stabilized with polyvinylalcohol	505	0.87	Element except Mo, Nb, Sb(III) can be masked with EDTA	25-27
Complex formation in 1.15 NHCl, in the presence of polyvinylalcohol and cetyltri-methylammonium chloride	505	1.7	Not investigated	10
Complex formation in 0.5 NHCl, followed by extraction with benzylalcohol	505	1.4	Bi, Fe(III), Mo(VI), Sb(III), Sn(II), 100-fold As(III), Ti	9
Extraction of $GeCl_4$ into CCl_4 from $8 \sim 9$ NHCl, followed by complex formation	508	1.6	V(V)	28,29
Extraction of $GeCl_4$ into MIBK from 7.5 NHCl, followed by complex formation in the presence of ethanol	504	$\simeq 1.6$	Fe(III), Mo(VI), Sb(III), (V), Sn(II), and Zr can be separated by Cupferron extraction; Nb and W can be separated by thiocyanate extraction	30

MONOGRAPHS

M1. Yardley, J. T., Phenylfluorone, in *Organic Reagents for Metals,* Vol. 1, Johnson, W. C., Ed., Hopkin & Williams, Chadwell Heath, England 1955, 134.

M2. Budesinsky, B. W., Aromatic ortho-dihydroxy compounds as reagents for inorganic analysis, in *Chelates in Analytical Chemistry,* Vol. 5, Flaschka, H. A. and Barnard, A. J., Jr., Eds., Marcel Dekker, New York, 1976, 244.

M3. Sandell, E. B. and Onishi, H., *Photometric Determination of Traces of Metals,* 4th ed., Part 1, John Wiley & Sons, New York, 1978, 316.

REFERENCES

1. Kehrman, F. and Gunther, M., *Ber.,* 45, 2884, 1912.

2. Sivanova, O. V., Ivankovich, G. S., and Lagutrina, G. S., *Zh. Anal. Khim.,* 26, 1874, 1971; *J. Anal. Chem. USSR,* 26, 1675, 1971.

3. Petrova, G. S., Ryabokobilko, Yu. S., Lukin, A. M., Visokova, N. N., and Poponova, R. V., *Anal. Lett.,* 5, 695, 1972.

4. Schneider, W. A., M.S. thesis, Minnesota University, Minneapolis, 1953.

5. Luke, C. L., *Anal. Chim. Acta,* 37, 97, 1967.

6. Sakuraba, S. and Kojima, M., *Nippon Kagaku Kaishi,* 345, 1976.

7. Minczewski, J. and Stolarczyk, U., *Chem. Anal. (Warsaw),* 9, 1135, 1964; *Anal. Abstr.,* 13, 1764, 1966.

8. Schneider, W. A., Jr. and Sandell, E. B., *Mikrochim. Acta,* 263, 1954.

9. Hillebrandt, A. and Hoste, J., *Anal. Chim. Acta,* 18, 569, 1958.

10. Shijo, Y. and Takeuchi, T., *Bunseki Kagaku,* 16, 51, 1967.

11. Stolarczyk, U. and Minczewski, J., *Chem. Anal. (Warsaw),* 9, 151, 1964; *Chem. Abstr.,* 61, 23c, 1964.

12. Black, A. H. and Bonfiglio, J. D., *Anal. Chem.,* 33, 431, 1961.

13. Mori, I., Yamamoto, S., and Enoki, T., *Bunseki Kagaku,* 23, 1061, 1973.
14. Pilipenko, A. T. and Eremenko, O. M., *Ukr. Khim. Zh.,* 29, 538, 1963; *Anal. Abstr.,* 11, 2575, 1964.
15. Sakuraba, S. and Kojima, I., *Nippon Kagaku Kaishi,* 208, 1978.
16. Gurkina, T. V. and Konovalova, K. M., *Tr. Kaz. Nauchno-Issled. Inst. Miner. Syr'ya,* 249, 1961; *Chem. Abstr.,,* 58, 9616c, 1963.
17. Ishibashi, M., Shigematsu, T., Yamamoto, Y., and Inoue, Y., *Bunseki Kagaku,* 7, 473, 1958.
18. Nakamura, H., Miura, T., and Hashimoto, M., *Bunseki Kagaku,* 13, 264, 1964.
19. Yamasaki, M., Mori, I., and Enoki, T., *Bunseki Kagaku,* 22, 112, 1973.
20. Pilipenko, A. T., Shpak, E. A., and Zulfigarov, O. S., *Geokhim. Anal. Metody Izuch. Veschestv. Sostava. Osad, Porod Rud,* 2, 92, 1974; *Chem. Abstr.,* 87, 94903t, 1977.
21. Pilipenko, A. T., Shpak, E. A., and Bakardzhieva, D. I., *Ukr. Khim. Zh.,* 42, 1073, 1976; *Chem. Abstr.,,* 86, 100126q, 1977.
22. Kimura, K. and Sano, H., *Bull. Chem. Soc. Jpn.,* 30, 80, 1957.
23. Oka, Y., Kanno, T., and Shiba, K., *Bunseki Kagaku,* 3, 389, 1954.
24. Funasaka, W., Ando, T., Fujimura, K., and Hanai, T., *Bunseki Kagaku,* 17, 86, 1968.
25. Oka, Y. and Kanno, T., *Nippon Kagaku Zasshi,* 76, 874, 1955.
26. Luke, C. L. and Compbell, M. E., *Anal. Chem.,* 28, 1273, 1956.
27. Burton, J. D. and Riley, J. P., *Mikrochim. Acta,* 587, 1959.
28. Yamauchi, F. and Murata, A., *Bunseki Kagaku,* 9, 959, 1960.
29. Bansho, K. and Umezaki, Y., *Bunseki Kagaku,* 16, 715, 1969.
30. Senise, P. and Sant' Agostino, L., *Mikrochim. Acta,* 572, 1959.
31. Ladenbauer, I. M., Bradacs, L. K., and Hecht, F., *Mikrochim. Acta,* 388, 1954.
32. Ladenbauer, I. M. and Hecht, F., *Mikrochim. Acta,* 397, 1954.
33. Nazarenko, V. A., Shitareva, G. G., Poluektova, E. N., and Yakovleva, T. P., *Zh. Anal. Khim.,* 29, 1850, 1974.
34. Fedin, A. V. and Vakar, G. P., *Zh. Anal. Khim.,* 30, 2125, 1975.
35. Fedin, A. V. and Kravchuk, S. I., *Zh. Anal. Khim.,* 29, 1734, 1974.
36. Fedin, A. V. and Vakar, G. P., *Izv. Vyssh. Uchebn. Zaved. Khim. Khim. Tekhnol.,* 20, 645, 1977; *Anal. Abstr.,* 33, 5B57, 1977.

PYROCATECHOL VIOLET

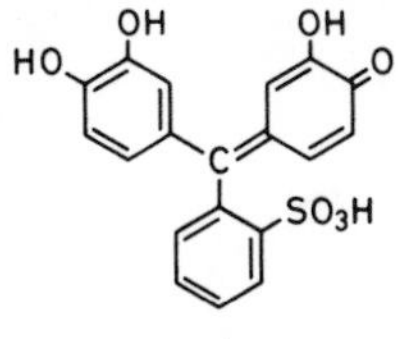

$$C_{19}H_{14}O_7S$$
mol wt = 386.38

H₄L

Synonyms

3,3',4-Trihydroxyfuchsone-2''-sulfonic acid, catechol sulfonephthalein, Catechol Violet, PV

Source and Method of Synthesis

Commercially available. Synthesized by the condensation of *o*-sulfobenzoic anhydride with pyrocatechol.[1]

Analytical Uses

As a metal indicator in the chelatometric titration of heavy metals. As a photometric reagent for various metal ions including Al, Bi, Ge, Sn, Y, and Zr. The sensitivity can be improved by carrying out the reaction in the presence of cationic surfactant and also by extracting the colored anionic chelate as an ion-pair with a large cation into immiscible solvent.

Properties of Reagent[1,2]

It is a dark reddish-brown crystalline powder with blue-green metallic luster. It is very hygroscopic and should be stored in well-closed containers. Commercial products occasionally have acetic acid odor which is due to the incomplete removal of the recrystallization solvent. It is freely soluble in water and aqueous alcohol, less soluble in cold absolute alcohol or glacial acetic acid, and insoluble in nonpolar solvents, such as benzene, xylene, and ether.

An aqueous solution of Pyrocatechol Violet is yellow, and the color of solution changes with pH as a result of proton dissociation of the reagent.[2]

$$H_5L^+ \xrightleftharpoons[]{pKa_1} H_4L \xrightleftharpoons[]{pKa_2\ <1} H_3L^- \xrightleftharpoons[]{pKa_3\ =\ 7.82} H_2L^{2-} \xrightleftharpoons[]{pKa_4\ =\ 9.76} HL^{3-} \xrightleftharpoons[]{pKa_5\ =\ 11.7} L^{4-}$$

Red   Yellow   Violet   Red-Violet

The scheme of proton dissociation may be written as shown in Figure 1, and the absorption spectra of the reagent at different pH are illustrated in Figure 2. An aqueous solution of the reagent is very stable in acidic range, but is gradually decolorized above pH 7.5.

Complexation Reactions and Properties of Complexes

Pyrocatechol Violet forms colored chelates with various metals, mostly in weak acidic and basic solution, as summarized in Table 1. Blue chelates are formed above the indicated pH. The main species of the colored chelate may be [ML²⁻], [M₂L], and [MHL⁻], where M represents a divalent metal, as illustrated below:[4]

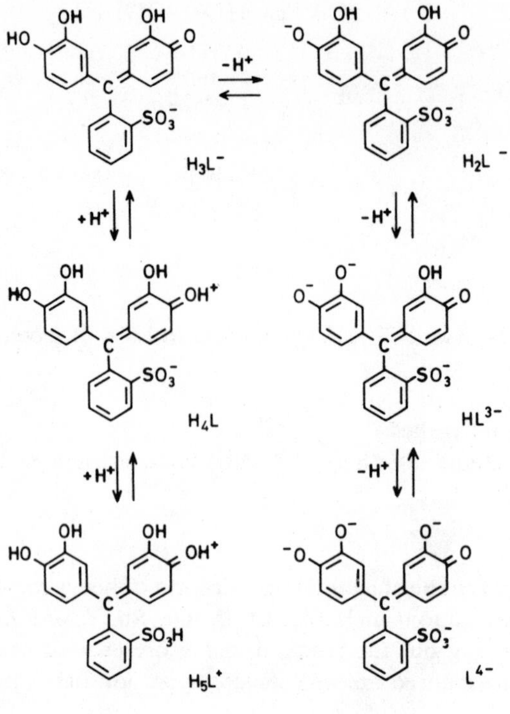

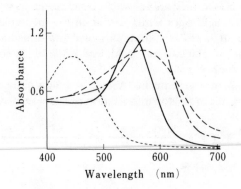

FIGURE 1.   Proton dissociation scheme of Pyrocatechol Violet.

FIGURE 2.   Absorption spectra of Pyrocatechol Violet in 4×10^{-5} M aqueous solution:

$[H_4L]$ ————————	pH <0
$[H_3L^-]$ ··················	pH 2.8
$[H_2L^{2-}]$ —·—·—·—·—	pH 8.01
$[HL^{3-}]$ — — — — — — —	pH 10.69

(From Ryba, O., Cifka, J., Malat, M., and Suk, V., *Collect. Czech. Chem. Commun.*, 21, 349, 1956. With permission.)

Table 1
COLOR REACTION OF PYROCATECHOL VIOLET WITH METALS[R1,4]

Metal ion[a]	pH	Color[b]	Metal ion[a]	pH	Color[b]
Al	2.5 ∿ 4.0	Reddish purple	Nd(III)	6.5	
Be	8.0		Ni	7.5	
Bi	<1.5	Bluish purple	Pb	5.5	
BO_3^{2-}	6.0	Pink	Pr	6.5	
Cd(II)	8.0		Sb(III)	<1.5	Reddish purple
Ce(III)	6.5		Sm	6.5	
Co(II)	7.5		Sn(II)	3.5	
Cu(II)	5.0		Sn(IV)	0.5	
Er(III)	6.0		Th	<3.0	Red
Fe(II)	7.5		Ti(IV)	<1.5	Purple
Fe(III)	<2.0	Purple (oxidizes	UO_2^{2+}	5.0	
Ga	2.5	the indicator)	WO_4^{2-}	<3.0	Red
In	3.5		WO_4^{2-}	>3.0	Purple
La	7.0		Y	6.3	
Mg	9.5		Zn	7.0	
Mn(II)	8.5		Zr	<1.0	Purple
MoO_4^{2-}	<3.0	Reddish purple			

[a]   Concentration of metal ions, $\simeq 5 \times 10^{-3}$ M.
[b]   Color of metal chelates is blue unless otherwise indicated.

From Cifka, J., Ryba, O., Suk, V., and Malat, M., *Collect. Czech. Chem. Commun.*, 21, 1418, 1956. With permission.

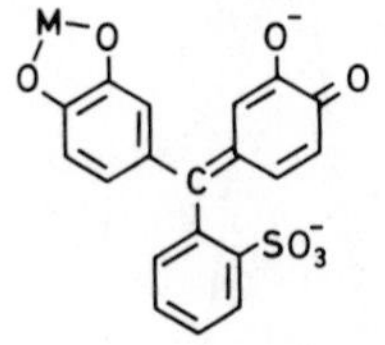

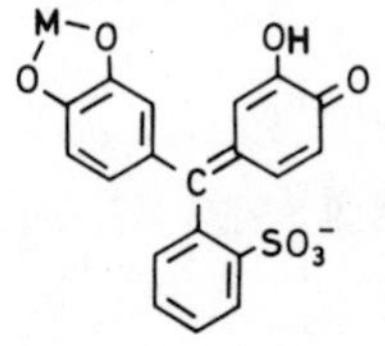

MHL⁻ (blue-violet)

ML²⁻ (blue-violet)

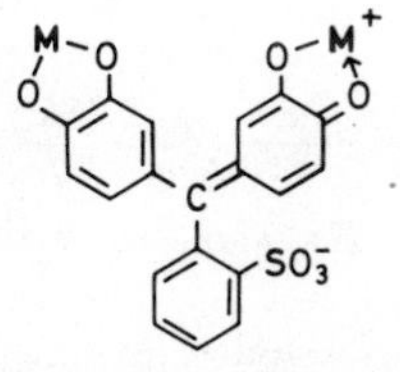

M₂L (blue)

The chelate stability constants for these species are summarized in Table 2.

A typical spectral change upon chelate formation is shown in Figure 3. Although the wavelength of absorption maxima and absorption intensity vary with metal ions, the similar absorption spectra are observed on the other metals.

In the presence of cationic surfactants, however, a pronounced bathochromic shift and an intensification of the absorption band are observed, as exemplified by Sn(IV)-Pyrocatechol Violet-CTAB (cetyltrimethylammonium bromide) system (see Figure 4). Such effects are considered to be due to the formation of the chelate of higher order (higher ligand to metal ratio) on the interface of cationic micelle[8] and are utilized for the highly sensitive photometry of trace of metals.

Table 2
CHELATE STABILITY CONSTANTS OF PYROCATECHOL VIOLET[4,5]

Metal ion	log K_{ML}	log K_{M2L}	log K_{MHL}[a]
Al(III)	19.13	4.95	—
Bi(III)	27.07	5.25	—
Cd(II)	8.13	—	5.86
Co(II)	9.01	—	6.53
Cu(II)	16.47	—	11.18
Ga(III)	22.18	4.65	—
In(III)	18.10	4.81	—
Mg(II)	4.42	—	3.67
Mn(II)	7.13	—	5.37
Ni(II)	9.35	—	6.85
Pb(II)	13.25	—	10.19
Th(IV)	23.36	4.42	—
Zn(II)	10.41	—	7.21
Zr(IV)	27.40	4.18	—

Note: $\mu = 0.2$ (NaClO$_4$)

[a] $K_{MHL} = [MHL^{n-3}] / [ML^{n-4}][H^+]$.

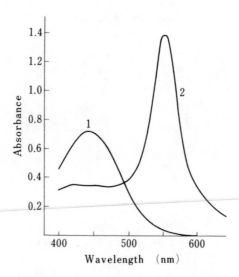

FIGURE 3.   Absorption spectra of Sn(IV)-Pyrocatechol Violet system. (1) Aqueous solution containing 4×10^{-5} M of reagent at pH 4, 1-cm cell; (2), (1) plus 50 μg Sn per 25 ml. (From Tanaka, K. and Yamayoshi, K., *Bunseki Kagaku*, 13, 540, 1964. With permission.)

The negatively charged colored chelates can also be extracted into immiscible solvent as an ion-pair with large cations, such as long chain alkylammonium ions, alkylpyridinium ion, or diphenylguanidinium ion. Examples will be given in Table 5.

Purification and Purity of Reagent

Commercially available samples vary in quality. Impure material may be purified

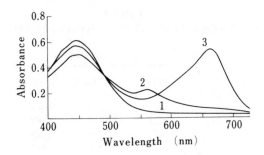

FIGURE 4.   Absorption spectra of Sn(IV)-Pyrocatechol Violet-Cetyltrimethylammonium bromide system. (1) 5 mℓ of 10^{-3} M Catechol Violet solution diluted to 100 mℓ at pH 2.2, against distilled water, 1-cm cell; (2), (1) plus 5 mℓ of 10^{-4} M Sn(IV) solution before dilution to 100 mℓ; (3), (2) plus 2 mℓ of 0.1% cetyltrimethylammonium bromide solution. (From Dagnall, R. M., West, T. S., and Young, P., *Analyst*, 92, 27, 1967. With permission.)

Table 3

CHELATOMETRIC TITRATION USING PYROCATECHOL VIOLET INDICATOR[R1,11]

Metal ion	pH	Buffer	Color change at end point	Remarks
Bi	2 ∿ 3	HNO₃	Blue→yellow	Fe(III), Hg(II), Th, Zr are titrated together
Cd	10	NH₃-NH₄Cl	Greenish blue →reddish purple	—
Co(II)	9.3	NH₃-NH₄Cl	Greenish blue→ reddish purple	—
Cu(II)	5 ∿ 6	HNO₃,NH₃	Blue→yellow	—
Cu(II)	6 ∿ 7	Pyridine	—	Traces of heavy metal do not interfere
Fe(III)	5 ∿ 6	Pyridine	Blue→yellow	—
Ga	3.8	Acetate	Blue→yellow	Titrate slowly
In	5	Pyridine	Blue→yellow	Titrate hot solution
Mg	10	NH₃-NH₄Cl	Greenish blue →reddish purple	—
Mn(II)	9.3	NH₃-NH₄Cl	Greenish blue→ reddish purple	Add NH₂OH
Ni	9.3	NH₃-NH₄Cl	Greenish blue → reddish purple	—
Pb	5	Hexamine	Blue→yellow	—
Th	2.5 ∿ 3.5	HNO₃	Blue→yellow	Same interference in Bi
Zn	10	NH₃-NH₄Cl	Greenish blue →reddish purple	—

by recrystallization from glacial acetic acid. Pure material in acetate buffer (pH 5.2 to 5.4) should be lemon yellow with no green tinge[9] (λ_{max} 445 nm, $\varepsilon = 1.4 \times 10^4$).[10]

Analytical Applications
Use as Metal Indicator in Chelatometric Titration
Metal ions that can be titrated with EDTA using Pyrocatechol Violet as an indicator are listed in Table 3.[R1,11] A 0.1% aqueous solution is used and can be stored for many

Table 4

APPLICATION OF PYROCATECHOL VIOLET AS PHOTOMETRIC REAGENT

Metal ion	Condition[a]	Metal chelate			Range of determination (ppm)	Ref.
		Ratio[b]	λ_{max} (nm)	$\varepsilon(\times 10^4)$		
Al(III)	pH6.0	ML$_2$	580	6.8	0 $\sim$ 0.4	6
Al(III)	pH 9.7 $\sim$ 10.2, CTAC	ML$_2$X$_5$	670	0.53	0.27 $\sim$ 54	13
Al(III)	pH 10.0, CPC	MLX$_5$	700	3.8	—	52
B	pH 8.55 $\sim$ 8.65	ML	494	0.165	0.2 $\sim$ 2.0	14,15
Be	pH 8.0, CPC	MLX$_5$	680	1.5	—	52
Bi(III)	pH 3.8, acetate	—	610	—	4 $\sim$ 24	16
Bi(III)	pH 5, CPC	MLX$_5$	680	1.0	—	52
Cd(II)	pH 10, CPC	MLX$_5$	710	1.2	—	52
Co(II)	pH 10, CPC	MLX$_5$	720	1.1	—	52
Cu(II)	pH $\simeq$7	—	630	—	4 $\sim$ 24	17
Cu(II)	pH 7, CPC	MLX$_5$	720	1.8	—	52
Cr(III)	pH 5 $\sim$ 6, acetate(90°)	—	605	—	—	18
Fe(III)	pH 6.2 $\sim$ 7.0	ML$_2$	610	6.2	0 $\sim$ 0.6	19
Ga(III)	pH 6.05 $\sim$ 6.4	ML$_2$	580	$\sim$7.3	0.56 $\sim$ 3.1	20
Ge(IV)	pH 3.4 $\sim$ 3.6, gelatine	ML$_2$	650	5.0	$\sim$0.3	21
Ge(IV)	pH 3 $\sim$ 3.5, 1.10-phenanthroline	ML$_2$X	640	3.7	0.16 $\sim$ 1.6	22
Ge(IV)	1 NHCl, CPC	MLX$_5$	680	7.0	—	52
In(III)	pH 5.3, gelatine	ML$_2$	600	6.4	0.1 $\sim$ 0.7	23
Mo(IV)	pH 2 $\sim$ 6	ML	530	—	1 $\sim$ 10	24
Mo(VI)	pH 3.0 $\sim$ 5.0, CTAB	MLX$_2$	675	4.6	9.6 $\sim$ 96	25
Mo(VI)	pH 2, CPC	MLX$_5$	670	8.2	—	52
Nb(III)	pH 2.2 $\sim$ 2.3	ML	575	2.5	0.2 $\sim$ 6	26
Nb(III)	pH 0.8 $\sim$ 1.6, CPB	ML$_4$	565	5.3	0.19 $\sim$ 1.9	27
Pd(II)	pH 8, CPC	MLX$_5$	700	1.7	—	52
Sb(IV)	pH $\sim$ 6, CPC or CTAB	L/X <1	530	3.0	6.1 $\sim$ 61	25
Sn(IV)	pH 2.5 $\sim$ 3.5	ML$_2$	555	6.5	0.24 $\sim$ 1.6	28
Sn(IV)	pH 2.2, CTAB	ML$_2$X$_4$	662	9.56	0.2 $\sim$ 2	7
Sn(IV)	pH 4, CPC	MLX$_5$	670	7.8	—	52
Ta(III)	pH 1.5, EDTA	—	580	—	$\sim$110	29
Ta(III)	$\sim$ 1.5 NHCl, CTAB	—	605 $\sim$ 610	$\sim$1	0.18 $\sim$ 1.8	30
Th	pH 3.8, acetate	—	480	—	4 $\sim$ 24	31
Ti(IV)	pH 3.3 $\sim$ 3.5, gelatine	ML$_2$	690	—	0.06 $\sim$ 0.71	32
Ti(IV)	pH 2.0, CPC	MLX$_5$	745	7.5	—	52
V(V)	pH 2 $\sim$ 6	ML	420	—	1 $\sim$ 6	24
W(VI)	pH 2 $\sim$ 6	ML	540	—	1 $\sim$ 6	24
W(VI)	pH 1.0, CPC	MLX$_5$	670	6.4	—	52
Y(III)	pH 8.7, H$_3$BO$_3$	MLX	607	1.9	0 $\sim$ 7.8	33
Y(III)	pH 8.4 $\sim$ 9.0, gelatine	—	665	2.59	0 $\sim$ 1.8	34
Zn(II)	pH 9.0, CPC	MLX$_5$	690	1.3	—	54
Zr(IV)	pH 3.5 $\sim$ 5.0	ML$_2$	650	3.26	0 $\sim$ 2.0	35
Zr(IV)	pH 6.0, CPC	MLX$_5$	700	1.3	—	52
Rare earths	pH 8.5 $\sim$ 9.0	ML	665	0.9 $\sim$ 2.5	0 $\sim$ 5	36

[a] Key for cationic surfactants, CTAC, cetyltrimethylammonium chloride; CTAB, cetyltrimethylammonium bromide; CPC, cetylpyridinium chloride; CPB, cetylpyridinium bromide.

[b] X indicates the quarternary ammonium ion used.

months.[12] As the aqueous solution is stable and the color change is not blocked by traces of heavy metals, it is recommended for use in place of Eriochrome® Black T or murexide. The color change with Bi and Th is so sharp that these metals can be used as a back titrant in the titration of other metals.[R1,11]

Table 5

ION PAIR EXTRACTION OF PYROCATECHOL VIOLET CHELATES

Metal ion	Condition	Metal chelate			Extraction		Range of determination (ppm)	Ref.
		Ratio[a]	λ_{max} (nm)	$\varepsilon(\times 10^4)$	Cation (X)	Solvent		
Al(III)	pH 6.0	ML_2X_3	597	8.4	Ethyltridecylammonium bromide	Benzene or xylene	0 ~ 0.26	37
Al(III)	pH 7.4 ~ 10	ML_2X_5	594	6.5	Cetylpyridinium iodide	Butanol	0.03 ~ 0.6	50
Bi(III)	pH 6	MLX_2	630	2.6	Diphenylguanidine	n-Butanol	0.5 ~ 5	38
Cu(II)	pH 7.5 ~ 8.5	ML_2X_2	663	7.9	Tridodecylethylammonium bromide	Benzene	0 ~ 0.65	39
Fe(III)	pH 5.7 ~ 6.4	ML_2X_2	623	7.76	Tridodecylethylammonium bromide	MIBK	0.01 ~ 0.17	40
Ga(III)	pH 5.4 ~ 6.5	ML_2X_2	600	8.7	Cetylpyridinium bromide	n-Butanol	0.01 ~ 0.28	41
In(III)	pH 6.2 ~ 8.0	ML_2X_2	600	7.2	Cetylpyridinium chloride	n-Butanol	0.03 ~ 0.46	42
Mo(VI)	0.2 ~ 0.6 N HCl	ML_2X_2	560	6.25	Dialkylmethyl benzylammonium chloride	CHCl$_3$	0.01 ~ 1	43
Nb(III)	pH 4 ~ 5	ML_2X_3	553	4.4	Tridodecylethylammonium bromide	CCl$_4$	0.12 ~ 1.4	44
Sb(III)	pH ≃4[b]	—	555	6.5	Tri-n-octylamine	Toluene	0.2 ~ 1	45
Sn(IV)	pH 3.6	ML_2X_2	581	7.3	Tridodecylethylammonium bromide	Xylene	0 ~ 0.6	46
VO^{2+}	pH 2.4 ~ 3.4	ML_2X_2	450	1.72	1,10-Phenanthroline	Butanol	0.5 ~ 4	51
Zr(IV)	pH 5, acetate	MLX_3	586	3.9	Tridodecylethylammonium bromide	Butyl acetate	0.1 ~ 2	47
Rare earths	pH 7.5 ~ 9	MLXn	630 ~ 670	2.4 ~ 3.0	Cetylpyridinium chloride or ethylene-bis(decyl-) oxycarbonylmethyl) dimethylammonium chloride	CHCl$_3$	—	48

[a] Formula does not indicate the state of protonation, but only indicates the ratio of three components.

[b] After separation as bromide or iodide into toluene.

Use as Photometric Reagent

Metal ions that can be determined by the direct photometry with Pyrocatechol Violet in aqueous solution are summarized in Table 4. As described previously, the sensitivity is substantially improved by the combined use of cationic surfactant. Such examples are also included in Table 4. In general, Pyrocatechol Violet is not so selective a reagent and suffers from interferences with many elements. Therefore, these methods are applicable after appropriate separation or with the combined use of suitable masking agents.

Anionic chelates with Pyrocatechol Violet can be extracted into organic solvent as an ion-pair with an appropriate cation, thus attaining sensitivity and selectivity of the determinations. Examples are summarized in Table 5.

Photometric determination of Sn(IV) with the combined use of Pyrocatechol Violet and cetyltrimethylammonium bromide[7]

Transfer 10 mℓ aliquot of sample solution (0.01 to 0.1 mg Sn) and 5 mℓ of 10^{-3} M Pyrocatechol Violet solution in a 100-mℓ beaker. Add 2 mℓ of 0.1% cetyltrimethylammonium bromide solution and dilute to about 50 mℓ. Adjust the pH of solution to 2.2 with dilute ammonia solution by using a pH meter. Transfer the solution to 100-mℓ volumetric flask and dilute to the mark. After 10 min, measure the absorbance in 1-cm cell against a reagent blank at 662 nm.

This method was successfully applied to the determination of Sn in steel after separating by solvent extraction as tin iodide.[49]

REVIEW

R1. Suk, V. and Malat, M., Pyrocatechol Violet: indicator for chelatometric titrations, *Chemist-Analyst*, 45, 30, 1956.

REFERENCES

1. Vodak, Z. and Leminger, O., *Chem. Listy*, 48, 552, 1954; *Collect. Czech. Chem. Commun.*, 19, 925, 1954.
2. Ryba, O., Cifka, J., Malat, M., and Suk, V., *Chem. Listy*, 49, 1786, 1955; *Collect. Czech. Chem. Commun.*, 21, 349, 1956.
3. Saginashvili, R. M. and Petrashen, V. I., *Zh. Anal. Khim.*, 22, 984, 1967.
4. Cifka, J., Ryba, O., Suk, V., and Malat, M., *Chem. Listy*, 50, 888, 1956; *Collect. Czech. Chem. Commun.*, 21, 1418, 1956.
5. Ryba, O., Cifka, J., Jezbova, D., Malat, M., and Suk, V., *Chem. Listy*, 51, 1462, 1957; *Collect. Czech. Chem. Commun.*, 23, 71, 1958.
6. Tanaka, K. and Yamayoshi, K., *Bunseki Kagaku*, 13, 540, 1964.
7. Dagnall, R. M., West, T. S., and Young, P., *Analyst*, 92, 27, 1967.
8. Ueno, K., *Bunseki Kagaku*, 20, 736, 1971.
9. Lukyanov, V. A. and Knyazeva, E. M., *Zh. Anal. Khim.*, 23, 536, 1968; *J. Anal. Chem. USSR*, 23, 455, 1968.
10. Mustafin, I. S., Molot, L. A., and Arkhnglskaya, A. S., *Zh. Anal. Khim.*, 22, 1808, 1967.
11. Ueno, K., *Chelatometric Titration*, 3rd ed., Nankodo, Tokyo, 1972.

12. Dougan, W. K. and Wilson, A. L., *Analyst,* 99, 413, 1974.
13. Chester, J. E., Dagnall, R. M., and West, T. S., *Talanta,* 17, 13, 1970.
14. Hiiro, K., *Bull. Chem. Soc. Jpn.,* 34, 1743, 1961.
15. Hiiro, K., *Nippon Kagaku Zasshi,* 83, 81, 1962.
16. Svach, M., *Fresenius Z. Anal. Chem.,* 149, 325, 1956.
17. Svach, M., *Fresenius Z. Anal. Chem.,* 149, 417, 1956.
18. Golubtsova, R. B. and Yaroshenko, A. D., *Zavod. Lab.,* 36, 147, 1970; *Chem. Abstr.,* 73, 10370s, 1970.
19. Ishito, T. and Ichinohe, S., *Bunseki Kagaku,* 21, 1207, 1972.
20. Akhmedli, M. K., Bashirov, E. A., Gluschenko, E. L., and Zykova, L. I., *Zh. Anal. Khim.,* 21, 1022, 1966.
21. Nazarenko, V. A. and Vinarova, L. I., *Zh. Anal. Khim.,* 18, 1217, 1963.
22. Ganago, L. I. and Semenovich, I. A., *Zh. Anal. Khim.,* 29, 1964, 1974.
23. Malat, M. and Hrachovcova, M., *Collect. Czech. Chem. Commun.,* 29, 1503, 1964.
24. Majumdar, A. K. and Savariar, C. P., *Naturwissenschaften,* 45, 84, 1958.
25. Bailey, B. W., Chester, J. E., Dagnell, R. M., and West, T. S., *Talanta,* 15, 1359, 1968.
26. Maltsev, V. F., Pashchenko, E. N., and Volkova, N. P., *Zh. Anal. Khim.,* 21, 1205, 1966.
27. Nakashima, R., Sasaki, S., and Shibata, S., *Bunseki Kagaku,* 22, 723, 1973.
28. Ross, W. J. and White, F. C., *Anal. Chem.,* 33, 421, 1961.
29. Babko, A. K. and Shtokolo, M. I., *Ukr. Khim. Zh.,* 30, 220, 1964; *Anal. Abstr.,* 12, 2767, 1965.
30. Nakashima, R., Sasaki, S., and Shibata, S., *Bunseki Kagaku,* 22, 729, 1973.
31. Svach, M., *Fresenius Z. Anal. Chem.,* 149, 414, 1956.
32. Malat, M., *Fresenius Z. Anal. Chem.,* 201, 262, 1964.
33. Serdyuk, L. S. and Silich, U. F., *Zh. Anal. Khim.,* 18, 166, 1963.
34. Young, J. P., White, J. C., and Bell, R. G., *Anal. Chem.,* 32, 928, 1960.
35. Young, J. P., French, J. R., and White, J. C., *Anal. Chem.,* 30, 422, 1958.
36. Takano, T., *Bunseki Kagaku,* 15, 1087 (1966).
37. Shijo, Y., *Nippon Kagaku Kaishi,* 1912, 1974.
38. Shestidesyatnaya, N. L., Milyaeva, N. M., and Katelyansbaya, L. I., *Zh. Anal. Khim.,* 30, 522, 1975; *J. Anal. Chem. USSR,* 30, 442, 1975.
39. Shijo, Y., *Bull. Chem. Soc. Jpn.,* 47, 1642, 1974.
40. Shijo, Y., *Bull. Chem. Soc. Jpn.,* 50, 1013, 1977.
41. Ishito, T., *Bunseki Kagaku,* 21, 752, 1972.
42. Ishito, T., and Tonosaki, K., *Bunseki Kagaku,* 20, 689, 1971.
43. Kohara, H., Ishibashi, N., and Abe, K., *Bunseki Kagaku,* 19, 48, 1970.
44. Shijo, Y., *Bull. Chem. Soc. Jpn.,* 50, 1011, 1977.
45. Tsukahara, I., Sakakibara, M., and Tanaka, M., *Anal. Chim. Acta,* 92, 379, 1977.
46. Shijo, Y., *Nippon Kagaku Kaishi,* 1658, 1974.
47. Shijo, Y., *Bull. Chem. Soc. Jpn.,* 49, 3029, 1976.
48. Serdyuk, L. S., Karaseva, L. B., Albota, L. A., and Denisenko, V. P., *Zh. Anal. Khim.,* 32, 2361, 1977; *Org. Reagenty Anal. Khim., Tezisy Dokl. Vses. Konf.,* 4th, 2, 22, 1976; *Chem. Abstr.,* 87, 161134h, 1976.
49. Ashton, A., Fogg, A. G., and Burns, D. T., *Analyst,* 98, 202, 1973.
50. Tananaiko, M. M. and Vdovenko, O. P., *Zh. Anal. Khim.,* 32, 1121, 1977.
51. Ganago, L. I. and Bukhteeva, L. N., *Zh. Anal. Khim.,* 32, 1537, 1977.
52. Chernova, R. K., Kharlamova, L. N., Belousova, V. V., Kulapin, E. G., and Sumina, E. G., *Zh. Anal. Khim.,* 33, 858, 1978; *J. Anal. Chem. USSR,* 33, 667, 1978.

PYROGALLOL RED AND BROMOPYROGALLOL RED

(1)   X = H
(2)   X = Br

H_4L

Synonyms

Reagents covered in this section are listed in Table 1, together with their synonyms.

Source and Method of Synthesis

Commercially available. PR (1) is prepared by the condensation of *o*-sulfobenzoic anhydride with pyrogallol.[1] BPR (2) is a bromination product of (1).

Analytical Uses

Metal indicator for the chelatometric titration of Bi, Co(II), Ni, and Pb. Also used as photometric reagents for various heavy metals. BPR (2) is more widely used as analytical reagent than PR (1). The deeply colored ternary complex formation of (2) with 1,10-phenanthroline and Ag is utilized for the photometric determination of Ag and for the indirect determination of anions, such as halides and cyanide.

Properties of Reagents

(1) and (2): Dark red crystalline powder with metallic luster. Only slightly soluble in water and alcohol and insoluble in nonpolar organic solvents. Aqueous solution is orange-red in strongly acidic, red in neutral, and violet in alkaline range. Acid dissociation scheme of the reagents is similar to that of Pyrocatechol Violet (p.35) and can be written as below. Their dissociation constants are summarized in Table 2.

$$H_5L^+ \underset{Ka_1}{\rightleftharpoons} H_4L \underset{Ka_2}{\rightleftharpoons} H_3L^- \underset{Ka_3}{\rightleftharpoons} H_2L^{2-} \underset{Ka_4}{\rightleftharpoons} HL^{3-} \underset{Ka_5}{\rightleftharpoons} L^{4-}$$

Orange-Red   Red   Violet

Absorption spectra of the reagents in visible region changes with pH as illustrated in Figures 1 and 2.[2]

Complexation Reaction and Properties of Complexes

(1) and (2) are structurally related with phenylfluorone (p.29) and react with various metals to form water-soluble colored chelates. Aqueous solution of (1) or (2) (red in neutral range) turn to blue or violet upon complex formation with metal ions, such as Al, Cu(II), Ge, In, Mo(VI), Sb(V), V(V), and W, for (1) and Bi, Ge, In, Sb(V), Th, Ti, UO_2^{2+}, V(V), and Zr for (2). As these chelates are also soluble in water, (1) and (2) are used as metal indicators in the chelatometry and as photometric reagents for metals.

Interestingly, Ag(I) forms a deep blue ternary complex with (1) or (2) in the presence of excess 1,10-phenanthroline at pH $\simeq 7$. The reaction is faster and more sensitive with (2) than with (1).[3] The shift in wavelength of maximum absorption from 560 nm of free (2) to 635 nm of Ag-phen-BPR complex may be explained on the reaction scheme shown in Figure 3.

Table 1
PYROGALLOL RED (1) AND BROMOPYROGALLOL RED (2)

Number/Reagent	Synonyms	Molecular formula	Mol wt
(1) Pyrogallol Red	Pyrogallol sulfonephthalein, PR	$C_{19}H_{12}O_8S$	400.36
(2) Bromopyro-gallol Red	5,5′-Dibromopyrogallol sulfonepth-thalein, BPR	$C_{19}H_{10}O_8Br_2S$	558.15

Table 2
ACID DISSOCIATION CONSTANTS[2]

Compounds	pKa_2	pKa_3	pKa_4	pKa_5
PR (1)	2.56	6.28	9.75	11.94
BPR (2)	0.16	4.39	9.13	11.27

Note: $\mu = 0.2$ (KCl).

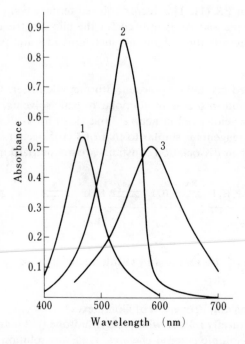

FIGURE 1.   Absorption spectra of Pyrogallol Red (1) in aqueous solution (2×10^{-5} *M*). (1) pH 0.80; (2) pH 7.90 to 8.65; (3) pH 14.0. (From Suk, V., *Collect. Czech. Chem. Commun.*, 31, 3127, 1966. With permission.)

Since no proton release was observed during the reaction, silver does not seem to react directly with the hydroxyl groups of BPR. However, the λ_{max} of the ternary complex (635 nm) is very close to that of the fully ionized reagent molecule (630 nm). Therefore, it is likely that the association of two complex ions [Ag(phen)$_2$$^+$], with BPR produces a change in the BPR molecule which is closely allied to release of the available

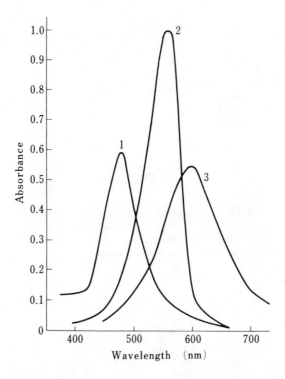

FIGURE 2. Absorption spectra of Bromopyrogallol Red
(2) in aqueous solution (2×10^{-5} M). (1) pH -0.51; (2) pH
5.65 to 7.45; (3) pH 12.80. (From Suk, V., *Collect. Czech.
Chem. Commun.*, 31, 3127, 1966. With permission.)

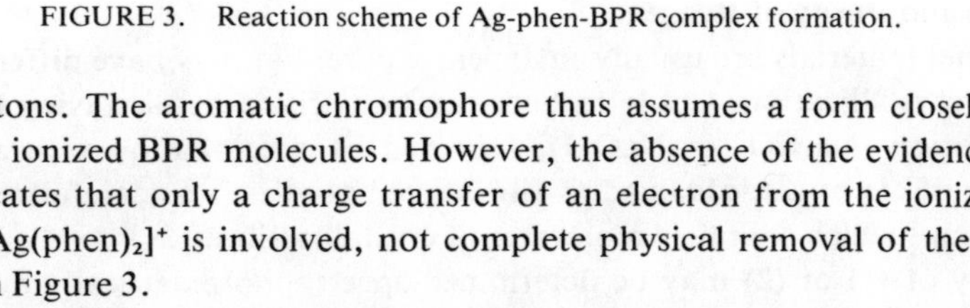

FIGURE 3. Reaction scheme of Ag-phen-BPR complex formation.

bonded protons. The aromatic chromophore thus assumes a form closely similar to
that of fully ionized BPR molecules. However, the absence of the evidence of proton
release indicates that only a charge transfer of an electron from the ionized phenolic
groups to $[Ag(phen)_2]^+$ is involved, not complete physical removal of the protons, as
illustrated in Figure 3.

The ternary complex is only slightly soluble in water and precipitates at once if concentrated solutions are mixed or on standing if dilute solutions are used. The complex
can also be extracted into nitrobenzene.

Examples of absorption spectra of binary chelate and ternary chelate with (2) are
shown in Figures 4 and 5, respectively. The spectral characteristics of (1) seem to be
similar to those of (2).

These anionic chelates of (1) or (2) can be extracted into higher alcohol or solvents
of higher dielectric constant as an ion-pair with large cations as summarized in Table
7.

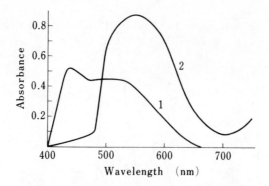

FIGURE 4.   Absorption spectra of Ge(BPR)₂ chelate.
(1) BPR, 3×10^{-5} *M*; (2) Ge (BPR)₂ (pH 2 to 3 stabilized
with 0.5% gelatin). (From Popa, Gr. and Paralescu, I.,
Talanta, 15, 272, 1968. With permission.)

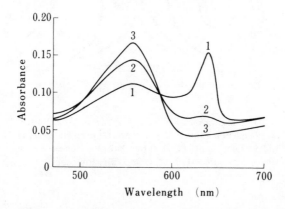

FIGURE 5.   Absorption spectra of Ag(phen)₂BPR. (1)
10^{-5} *M* BPR (2), 5 m*l*; 10^{-3} *M* 1,10-phenanthroline, 1 m*l*;
20% ammonium acetate, 1 m*l*; and water to make 50 m*l*;
(2), (1) plus 10^{-5} *M* AgNO₃,5m*l*; (3), (1) plus 10^{-5} *M*
AgNO₃, 10 m*l*. 4-cm cell. (From Dagnall, R. M. and West,
T. S., *Talanta*, 11, 1533, 1964. With permission.)

Purification and Purity of Reagent

Commercial materials are usually sufficiently pure, but may have different amounts
of water of crystallization. Crude reagent can be purified by dissolving it in aqueous
alkaline solution (Na₂CO₃ or NaOH), followed by precipitation with acidification.
Polyhydrate of (1) or (2), obtained as fibrous red needles, was carefully dried to give
dark red clumps which contain 4 mol water of crystallization.

The purity of (1) or (2) may be determined spectrophotometrically by its aqueous
solution.

PR (1) pH 7.9 to 8.6(H₂L²⁻) λ_{max}, 542 nm; ε, 4.3×10^4
BPR (2) pH 5.6 to 7.5(H₂L²⁻) λ_{max}, 558 nm; ε, 5.45×10^4

Analytical Applications

Use as Metal Indicator in Chelatometric Titration

Metal ions that can be titrated with EDTA using (1) or (2) as indicator are listed in
Tables 3 and 4, respectively. A 0.05% aqueous ethanol (50 v/v %) solution is used
and can be kept for many months without deterioration.

Table 3
CHELATOMETRIC TITRATION USING PYROGALLOL
RED (1) AS INDICATOR

Metal ion	pH	Buffer	Color change at end point	Ref.
Bi	2 ∿ 3	HNO₃	Red → orange-red	6
Co(II)	10	NH₃-NH₄Cl	Blue or violet → red	6
Ni	10	NH₃-NH₄Cl	Blue or violet → red	6
Pb	∿5	AcOH-AcONa	Violet → red	7

Table 4
CHELATOMETRIC TITRATION USING BROMOPYROGALLOL RED (2) AS
INDICATOR

Metal ion	pH	Buffer	Color change at end point	Remark and Ref.
Bi	2 ∿ 3	HNO₃	Wine-red → orange-red	8
Cd	10	NH₃-NH₄Cl	Blue → red	8
Co(II)	9.3	NH₃-NH₄Cl	Blue → red	8
Mg	10	NH₃-NH₄Cl	Blue → red	8
Mn(II)	9.3	NH₃-NH₄Cl	Blue → red	8 (NH₂OH · HCl added)
Ni	9.3	NH₃-NH₄Cl	Blue → red	8
Pb	∿5	AcOH-AcONa	Violet → red	7
Rare earths	∿7	AcONa	Blue → red	8 (ascorbic acid added)

Use as Photometric Reagent

Metal ions that can be determined by the direct photometry with (1) or (2) in aqueous solution are summarized in Tables 5 and 6, respectively.

Several metal chelates with (2) can be extracted with appropriate cation (or conjugate acid of amine) into organic solvent as an ion-pair. Examples are summarized in Table 7.

The color reaction of Ag-phen-BPR system is supressed by the presence of complexing anions, such as cyanide, thiocyanide, sulfide, and halides. Based on this fact, an indirect photometric method was developed for the determination of such anions.[34,35]

Photometric determination of Ag using BPR (2) and 1,10-phenanthroline[3]

Place not more than 40 mℓ of sample solution (1 to 10 μg Ag) in a 50-mℓ volumetric flask containing 1 mℓ of 0.1 M EDTA solution, 1 mℓ of 10^{-3} M 1,10-phenanthroline aqueous solution, 1 mℓ of 20% ammonium acetate solution, and 2 mℓ of 10^{-4} M BPR solution. Dilute the solution to 50 mℓ with water and measure the absorbance immediately, or within 30 min, in a 4-cm cell at 635 nm against a reagent blank. If iron(II) is present, add sufficient phenanthroline to complex it completely and to react with Ag. When U(VI), Th, or Nb are present, add fluoride (for the first two) or hydrogen peroxide (for the last) before the color reaction.

Table 5

APPLICATION OF PYROGALLOL RED (1) AS A PHOTOMETRIC REAGENT

Metal ion	Condition (pH)	Ratio	Metal chelate λ_{max} (nm)	$\varepsilon(\times 10^4)$	Range of determination (ppm)	Ref.
Ag	7	—	390	1.0	—	3
Cu(II)	6	ML	582	0.95	0.25 ~ 2.3	9
Ga	4.5	ML	530	1.4	0.28 ~ 4.8	10
Ge	5	ML$_2$	500	—	0.28 ~ 2	11
In	4.7	ML$_2$	580	—	—	12
Mo(VI)	5.0 ~ 5.3[a]	ML$_3$	587	8.15	0.04 ~ 0.3	13
Sb(III)	4.5	ML	520	0.25	0.3 ~ 3.0	14
Sb(V)	2 ~ 3 (60°)	ML	500	0.98	—	15
Ti(III)	2.7, ethanol	—	610	7.6	0.6 ~ 3.12	16
W(VI)	[b]	ML	576	6	~1	17

[a]   Dodecanoltrimethylammonium bromide added.
[b]   Hexadecyltrimethylammonium chloride added.

Table 6

APPLICATION OF BROMOPYROGALLOL RED (2) AS PHOTOMETRIC REAGENT

Metal ion	Condition (pH)	Ratio	Metal chelate λ_{max} (nm)	$\varepsilon(\times 10^4)$	Range of determination (ppm)	Ref.
Ag(I)	4 ~ 12	ML	410,550	—	—	18
Ag(I)	7, 1,10-phenan-throline	M(phen)$_2$L$_2$	635	5.1	0.02 ~ 0.2	3
Bi(III)	2.0, gelatin		635	5.0	0.1 ~ 5	19
Ce(III)	9, EDTA,NTA or DTPA	MLX	640 ~ 650	1.6	—	20
Cu(II)	5	ML	619	1.25	—	21
Ge(IV)	2 ~ 3, gelatin	ML$_2$	550	2.05	0.2 ~ 3	5
In(III)	3.4 ~ 3.6	ML	610 ~ 620	—	0.5 ~ 16	22
Mo(VI)	0.16 ~ 0.24 N HCl, Zephira-mine®	ML	629	5.56	0.08 ~ 1	23
Nb(V)	5.8, gelatin, EDTA	ML$_3$	610	6	—	24
Sb(V)	2 ~ 3 (60°)	ML$_2$	500	1.29	—	15
Sb(III)	5.6 ~ 6.8	ML	560	3.9	0.1 ~ 1	25
Sc	6.1	ML	610	2.4	0.12 ~ 2.4	26
Th	2	ML	590	2.5	0.4 ~ 2	27
Th	5	ML$_2$	630	6.3	0.08 ~ 2.4	27
Ti(IV)	2.2, gelatin	ML$_2$ or ML$_3$	530	9	0.02 ~ 0.5	28
UO$_2^{2+}$	6.5 ~ 7	ML	620	0.88	0.2 ~ 5	29
UO$_2^{2+}$	6 ~ 7[a]	ML	650	2.9	~15	30
VO^{2+}	4.4	ML$_2$	540	0.17	3 ~ 7	31
W(VI)	1.0 ~ 1.3 N HCl, Zephira-mine®	ML	621	6.5	0.08 ~ 1.4	32
Zr	5.0 ~ 5.4	ML$_2$	670	4.5	0.16 ~ 1.4	33

[a]   In presence of butyltriphenyl phosphonium bromide, polyvinyl alcohol, pyridine, and 10% ethanol.

Table 7

ION PAIR EXTRACTION OF BROMOPYROGALLOL RED (2) CHELATES

Metal ion	Condition (pH)	Metal chelate			Extraction		Range of determination (ppm)	Ref.
		Ratio	λ_{max}(nm)	ε ($\times 10^4$)	Amine (X)	Solvent		
Ge(IV)	$0.5 \sim 4.5$	ML_3	550	3.8	Diphenylguanidine	Hexylalcohol	—	39
Nb	5	ML_3	610	2.5	Di-n-octylmethylamine	Isopentylacetate	~ 5.4	36
Ti(IV)	2.5	MLX_4	630	3.75	4-Aminoantipyrine		$0.2 \sim 2$	37
Ti(IV)	—	MLX_2	—	—	Gelatin or diphenylguanidine		—	37
V(V)	4(60°)	MLX_3	610	0.85	Diphenylguanidine	Isoamylalcohol + chloroform(1:1)	—	38

REFERENCES

1. Vodak, Z. and Leminger, O., *Chem. Listy,* 50, 943, 1956; *Collect. Czech. Chem. Commun.,* 21, 1522, 1956.
2. Suk, V., *Collect. Czech. Chem. Commun.,* 31, 3127, 1966.
3. Dagnall, R. M. and West, T. S., *Talanta,* 11, 1533 1964.
4. Bailey, B. W., Chester, J. E., Dagnall, R. M., and West, T. S., *Talanta,* 15, 1359, 1968.
5. Popa, Gr. and Paralescu, I., *Talanta,* 15, 272, 1968.
6. Suk, V., Malat, M., and Jenichova, A., *Chem. Listy,* 49, 1798, 1955; *Collect. Czech. Chem. Commun.,* 21, 418, 1956.
7. Jenickova, A., Malat, M., and Suk, V., *Chem. Listy,* 50, 1113, 1956; *Collect. Czech. Chem. Commun.,* 21, 1599, 1956.
8. Jenickova, A., Suk, V., and Malat, M., *Chem. Listy,* 50, 760, 1956; *Collect. Czech. Chem. Commun.,* 21, 1257, 1956.
9. Bashirov, E. A., Akhmedli, M. K., and Abdullaeva, T. E., *Azerb. Khim. Zh.,* 122, 1966; *Chem. Abstr.,* 65, 14417g, 1966.
10. Srivastava, K. C., *Chim. Anal. (Paris),* 53, 525, 1971.
11. Akhmedli, M. K., Gasanov, D. G., and Alieva, R. A., *Uch. Zap. Azerb. Gos. Univ. Ser. Khim. Nauk.,* 3, 3, 1966; *Chem. Abstr.,* 68, 26604d, 1968.
12. Bashirov, E. A. and Ayubova, A. M., *Azerb. Khim. Zh.,* 128, 1971; *Chem. Abstr.,* 78, 66514n, 1973.
13. Takeuchi, T. and Shijo, Y., *Bunseki Kagaku,* 15, 473, 1966.
14. Naruskevadus, L., Kazlauskas, R., Skadauskas, J., and Karitonaite, N., *Nauch. Tr. Vyssh. Uchebn. Zaved. Lit. SSR Khim. Tekhnol.,* 95, 1972; *Anal. Abstr.,* 26, 2616, 1974.
15. Skadauskas, J., Naruskevicius, L., and Kazlauskas, R., *Nauchn. Konf. Khim. Anal. Pribalt. Resp. B, SSR [Tezisy Dokl.],* 1, 29, 1974; *Chem. Abstr.,* 86, 25545v, 1977.
16. Ramonaite, S., Abromaityte, D., and Gecaite, G., *Liet. TSR Aukst. Mokyklu Mokslo Darb. Chem. Chem. Technol.,* 12, 73, 1970; *Chem. Abstr.,* 76, 67706x, 1972.
17. Shijo, Y., and Takeuchi, T., *Bunseki Kagaku,* 22, 1341, 1973.
18. West, T. S., *Analyst,* 87, 630, 1962.
19. Suk, V. and Smetanova, M., *Collect. Czech. Chem. Commun.,* 30, 2532, 1965.
20. Turkina, L. A., Kirilov, A. I., and Vlasov, N. A., *Org. Reagenty Anal. Khim. Tezisy Dokl. Vses. Konf.,* 4th, 2, 21, 1976; *Chem. Abstr.,* 87, 161133g, 1977.
21. Bashirov, E. A., Akhmedli, M. K., and Abdullaeva, T. E., *Azerb. Khim. Zh.,* 114, 1967; *Chem. Abstr.* 67, 121996w, 1967.
22. Talipan, Sh. T., Abdulaeva, Kh. S., and Gorkovaya, G. P., *Uzb. Khim. Zh.,* 6, 16, 1962; *Anal. Abstr.,* 11, 71, 1964.
23. Deguchi, S., Iizuka, M., and Yashiki, M., *Bunseki Kagaku,* 23, 760, 1974.
24. Belcher, R., Ramakrishna, T. V., and West, T. S., *Chem. Ind.,* 531, 1963.
25. Christopher, D. H. and West, T. S., *Talanta,* 13, 507, 1966.
26. Shimizu, T., *Talanta,* 14, 473, 1967.
27. Vasilenko, V. D., Shanga, M. V., and Balbas, V. I., *Zh. Anal. Khim.,* 22, 1818, 1967.
28. Suk, V., Nemcova, I., and Malat, M., *Collect. Czech. Chem. Commun.,* 30, 2538, 1965.
29. Lukyanov, V. F. and Duderova, E. P., *Zh. Anal. Khim.,* 16, 60, 1961.
30. Sucmanova-Vondrova, M., Havel, J., and Sommer, L., *Collect. Czech. Chem. Commun.,* 42, 1812, 1977.
31. Mushran, S. P., Prakash, O., and Awasthi, J. N., *Microchem. J.,* 14, 29, 1969.
32. Deguchi, S. and Mamiya, T., *Bunseki Kagaku,* 25, 60, 1976.
33. Sakai, T. and Funaki, Y., *Bull. Chem. Soc. Jpn.,* 42, 2272, 1969.
34. Dagnall, R. M., El-Ghamry, M. T., and West, T. S., *Talanta,* 15, 107, 1968.
35. Deguchi, S., Abe, R., and Okumura, I., *Bunseki Kagaku,* 18, 1248, 1969.
36. Ramakrishina, T. V., Rahim, S. A., and West, T. S., *Talanta,* 16, 847, 1969.
37. Gambarov, D. G., *Azerb. Khim. Zh.,* 99, 1975; *Chem. Abstr.,* 85, 201548p, 1976.
38. Gordeeva, M. N. and Ryndina, A. M., *Vestn. Leningr. Univ. Fiz. Khim.,* 153, 1974; *Chem. Abstr.,* 82, 118586s, 1975.
39. Nazarenko, V. A. and Makrinich, N. I., *Zh. Anal. Khim.,* 24, 1694, 1969.

CHROMAZUROL S®

$C_{23}H_{16}O_9Cl_2S$
Mol wt = 539.34

H₄L

Synonyms

3″-Sulfo-2″,6″-dichloro-3,3′-dimethyl-4-hydroxyfuchson-5,5′-dicarboxylic acid, So-lochrome Brilliant Blue, Polytrope Blue B, C.I. 43825, C.I. Mordant Blue 29

Source and Method of Synthesis

Chromazurol S® is a trade name of dyestuff manufactured by Geigy Co.; dyestuff grade as well as reagent grade materials are commercially available. They are mostly trisodium salt. It is synthesized by the condensation of sulfo-*o*-dichlorobenzaldehyde with 2,3-cresotic acid, followed by oxidation.[1]

Analytical Uses

It is used as a metal indicator for chelatometric titration of Al, Ba, Ca, Fe, Mg, Ni, and Th and also as a photometric reagent for various metal ions, such as Al, Be, Ce, Co, Cr, Cu, Fe, Ga, Hf, In, La, Mn, Mo, Ni, Pd, Rh, Sc, Th, Ti, V, Y, Zn, Zr, and F⁻. Recent investigations include the highly sensitive photometry of the above cations in the presence of cationic surfactant.

Properties of Reagent

The commercial trisodium dihydrate salt is red-brown powder. It is hygroscopic and easily dissolves in water. Aqueous solution is red-orange at pH 3 to 4 and yellow at pH 4. In the acidity of pH <0 (1.2 to 2.0 *N*HCl), the free dye (H₄L) precipitates which again dissolves in the more acidic solution (8 ∿ 9 *N*HCl).

Proton dissociation constants were determined as indicated (μ = 0.1 (NaClO₄), 25°),[2] and absorption spectra in aqueous solution at various pH are illustrated in Figure 1.

Complexation Reaction and Properties of Complexes

Chromazurol S® reacts with various metal ions, such as Al, Be, Ce, Co, Cr, Cu, Fe, Ga, Hf, In, La, Mn, Mo, Ni, Pd, Rh, Sc, Th, Ti, V, Y, Zn, and Zr, to form colored soluble chelates. In contrast to red-orange or yellow color of the reagent, the color of chelate ranges from blue to violet. The lowest pH values at which color reaction is observed are summarized in Table 1. The metal to ligand ratio varies from metal to metal and is also dependent upon the solution conditions, such as pH and metal to reagent ratio. As an example, the proposed structures for three types of Iron (III) chelate are illustrated in Figure 2.

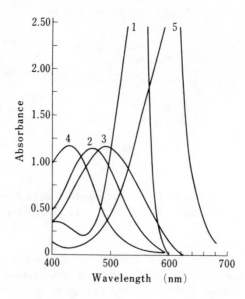

FIGURE 1. Absorption spectra of Chroma-
zurol S® in water. (1) 1.0 M H$_2$SO$_4$; (2) pH 1.5;
(3) pH 3.5; (4) pH 8.0; (5) 9 N NaOH. Dye,
0.01%. (From Martynov, A. P., Novak, V. P.,
and Reznik, B. E., *Zh. Anal. Khim.*, 32, 519,
1976; *J. Anal. Chem. USSR,* 32, 416, 1976. With
permission.)

Table 1
COLOR REACTION OF CHROMAZUROL S®
WITH VARIOUS METAL IONS[4]

Metal ion	pH	Metal ion	pH	Metal ion	pH
Be	2.5[a]	Nd	6.5	U (IV)	2.5
Ce (III)	5.0	Ni	7.5	UO$_2^{2+}$	2.0[a]
Co (II)	8.5	Pb	6.5	VO^{2+}	4.0
Ga	1.5	Pr	6.5	Y	5.5
In	2.5	Sr	11.0	Zn	7.5
La	4.5	Th	1.0[a]	Zr	1.0[a]
Mn (II)	7.5	Ti (IV)	1.0[a]		

[a] The chelate is wine red at these pH, but changes to violet
at higher pH.

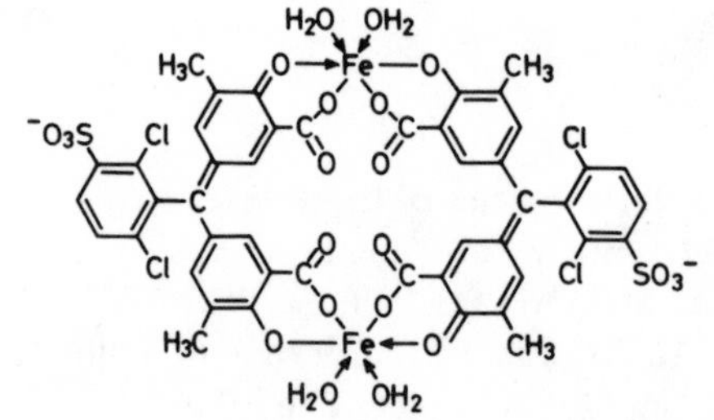

FIGURE 2. Proposed structures for iron (III) Chromazurol S® chelate. (A) Dimer, [Fe(H$_2$O)$_2$] L$_2^{2-}$; (B)
monomer, Fe(H$_2$O)$_4$HL; and (C) binuclear, [Fe(H$_2$O)$_4$]$_2$L^{2-}. (From Langmyer, F. J. and Klausen, K. S.,
Anal. Chim. Acta, 29, 149, 1963. With permission.)

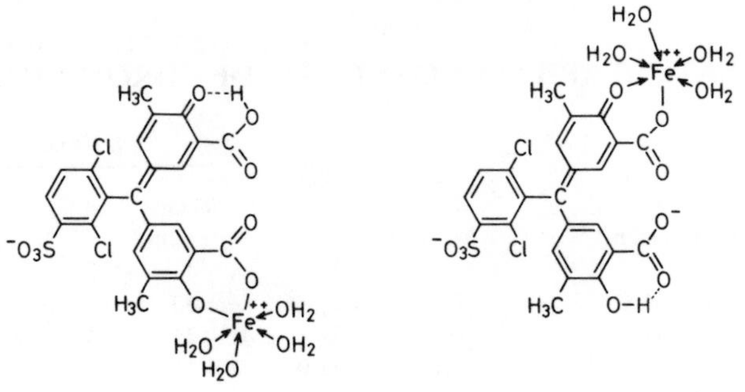

FIGURE 2B

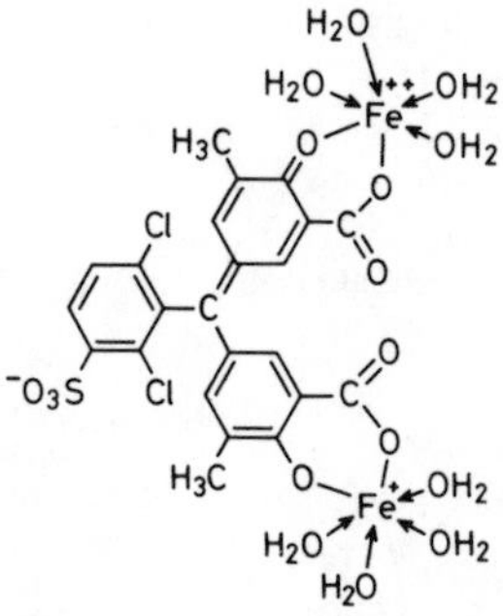

FIGURE 2C

In most cases of the photometry using Chromazurol S®, it is possible that the observing solution is an equilibrium mixture of two or more chelate species.

Due to the uncertainty in defining the chelate species in the equilibrated solution and also due to the uncertainty of the reagent purity, very few values on the chelate stability constant are reliable. However, reported values may be useful to evaluate the chelating behavior of Chromazurol S® qualitatively. Those values are summarized in Table 2.

As to the absorption spectra of Chromazurol S® chelate, that of Al chelate is illustrated in Figure 3 as an example.[6] Although the values for λ_{max} may shift from metal to metal, the shape of the curve does not differ greatly.

In the presence of the cationic surfactant, such as Zephiramine® or cetyltrimethylammonium chloride, the spectra of free dye and the metal chelate changes markedly. Figure 4 illustrates the absorption spectra of Chromazurol S® and its aluminum chelate in the presence of cationic micelle of Zephiramine®.[13]

A similar spectral change of the chelates is also observed in the presence of water miscible solvent. Figure 5 illustrates the effect of organic solvents on the absorption spectrum of copper chelate.[14]

Purification and Purity of Reagent

Most of the samples which are available as reagent grade are not pure enough to be used for the physicochemical studies, although they can be used as a photometric reagent for metals or a metal indicator for chelatometric titrations. Accordingly, many of the previous reports on the values of metal-ligand ratio should be reinvestigated. A recommended procedure of purification is as follows.[3]

Table 2
CHELATE STABILITY CONSTANTS OF CHROMAZUROL S®

Metal ion	Equilibrium[a]	log K	pH	Condition Temp (°C)	μ	Ref.
Al (III)	M + 2CAS⇌M(CAS)₂	2.04	4.2 ∿ 5.5	30	0.1(NaClO₄)	6
Be (II)	M + HL ⇌MHL	4.66	—	25	0.1(NaClO₄)	2
Cd (II)	M + 2CAS⇌M(CAS)₂	9.1	11.0	25	—	24
Co (II)	M + 2CAS⇌M(CAS)₂	6.18	9.38	20	0.1(NaClO₄)	7
Cu (II)	M + HL ⇌MHL	4.02	—	25	0.1(KCl)	8
Fe (III)	M + CAS ⇌M(CAS)	4.3	2.7	20	0.1(KCl)	5
Ga (III)	M + L ⇌ML	13.1	—	25	0.1(NaCl + NH₄Cl)	9
Ga (III)	ML + L ⇌ML₂	12.9	—	25	0.1(NaCl + NH₄Cl)	9
Ga (III)	ML₂ + L ⇌ML₃	11.8	—	25	0.1(NaCl + NH₄Cl)	9
Th (IV)	M + CAS ⇌M(CAS)	4.2 ∿ 4.8	4.5	30	0.15(NH₄NO₃)	10
VO²⁺	M + CAS ⇌M(CAS)	4.0 ∿ 4.6	4.0	25	—	11,12

[a] CAS represents the protonated dye species at given pH value.

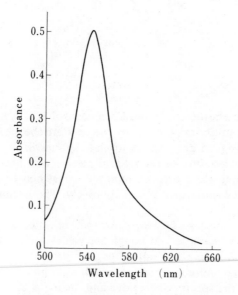

FIGURE 3.   Absorption spectra of aluminium Chromazurol S® chelate in aqueous solution; Al, 7.41×10^{-6} M; dye, 1.48×10^{-5} M; and pH 5.6. (From Nishida, H., *Bunseki Kagaku*, 19, 972, 1970. With permission.)

Dissolve 40 g of the commercial sample into 240 mℓ of water and remove the insoluble matter by filtration. Add 50 mℓ concentrated HCl to the filtrate with stirring. Filter the precipitate, wash it with 2 N HCl, and dry. Dissolve the dried cake again into 250 mℓ of water and repeat the precipitation twice more on heating on a water bath to 70°. After the third precipitation, dry the final product under vacuum over solid KOH and then over P₂O₅ to a constant weight (yield 50% as H₄L·2H₂O).

Purity of Chromazurol S® can be determined by observing the absorbance of a

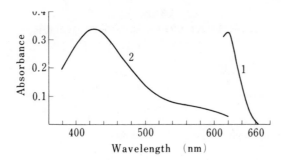

FIGURE 4. Absorption spectra of aluminium Chromazurol S® chelate in the presence of cationic micelle. (1) Al, 2.96×10^{-4} M; dye, 3.4×10^{-4} M; Zephiramine®, 2.72×10^{-3} M; pH $4.0 \sim 5.1$; (2) Dye, 3.42×10^{-5} M, Zephiramine®, 5.42×10^{-3} M; pH 4.68. (From H. Nishida, and T. Nishida, *Bunseki Kagaku,* 21, 997, (1972), and H. Nishida, *Bunseki Kagaku,* 20, 410, 1971. With permission.)

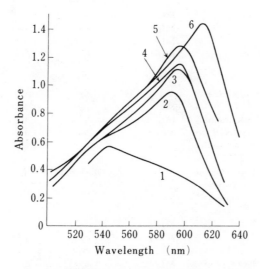

FIGURE 5. Absorption spectra of copper Chromazurol S® chelate in the mixture of 80 v/v solvent and water. Cu, 2.56×10^{-5} M; dye, (1) 2.2×10^{-3} M and (2) $\sim$(6) 2.6×10^{-4} M; Solvent, (1) water, (2) methanol, (3) ethanol, (4) 1-propanol, (5) 2-propanol, and (6) dioxane; pH, (1) 6.4 and (2) $\sim$(6) 0.01 M hexamine; reference, reagent blank; and temperature, 25°. (From Nishida, H. and Nishida, T., *Bunseki Kagaku,* 25, 55, 1976. With permission.)

solution of known concentration at the wavelength appropriate to the acidity[3] (see Table 3).

More accurate assay can be done by the potentiometric titration of free acid ($H_4L \cdot 2H_2O$) with NaOH solution.

Analytical Applications

Use as a Photometric Reagent for Metals

More than 30 metal ions have been determined with the use of Chromazurol S®.

Table 3
SPECTRAL CHARACTERISTICS OF CHROMAZUROL S®

Condition	λ_{max}(nm)	$\varepsilon(\times 10^4)$
10 NHCl	540	7.80
pH 7 ∿ 9 (acetate, borax)	430	2.24
pH >13 (≃1 NNaOH)	600	7.56

Table 4
APPLICATION OF CHROMAZUROL S® AS A PHOTOMETRIC REAGENT

	Condition		Metal chelate				
Metal ion	(pH, buffer, auxilary reagents)		Ratio (M:L)	λ_{max} (nm)	$\varepsilon(\times 10^4)$	Range of determination (ppm)	Ref.
Al	<3	—	ML				
Al	5.6 ∿ 6.8	—	ML₂	550	—	0.05∿0.3	15
Al	5.6	—	ML₃	585	5.1	0.02∿0.4	6
Al	5.9	Cetyltrimethyl-ammonium chloride	ML₃∿₄	620	10.8	0.01∿0.08	16
Al	4.5 ∿ 5.1	Zephiramine®	—	620	10.8	0.01∿0.3	13
Be	4.6	—	ML	569	0.40	0.04∿0.32	17
Be	5.1	Zephiramine®	ML₂	610	9.9	0.001∿0.04	18,19
Be	4.4 ∿ 5.0	Polyoxyethylene-dodecylamine + EDTA	ML₂	605	10	∿0.08	20,21
Be	5	Cetyltrimethyl-ammonium bromide + EDTA	—	605 ∿ 615	9.5	∿0.08	22,23
Cd	9.5 ∿ 11.5	—	ML₂	540	—	—	24
Ce(III)	6.4	—	ML₁∿₂	510	1.83	0.2∿4.8	25
Co(II)	10	—	ML₂	567	3.3	0.08∿2	8
Co(II)	10.9	Hydroxydodecyl-trimethylam-monium bromide + pyridine	ML₅	654	10.9	0.02∿0.32	26
Cr(III)	3.6	—	ML₂∿₃	570	—	0.04∿0.4	27
Cu(II)	6.7	—	ML	600	2.1	0.9∿2.2	28
Cu(II)	0.1 M hexamine	80% Dioxane	ML₂	610	6.3	3∿1.6	14
Cu(II)	8.6	Hydroxydodecyl-trimethylam-monium bromide + pyridine	ML₄X₄	592	11.9	0.02∿0.02	29
Cu(II)	7	Zephiramine®	ML₂	620	14.0	0.025∿6	13
Cu(II)	10 ∿ 10.6	2.2′-Bipyridine	ML₂X	510	2.06	0.6∿2.2	57
Fe(III)	<3	—	ML	—	—	—	30
Fe(III)	5.3	Acetate	ML₂∿M₂L	575	4.15	0.07∿1.4	31
Fe(III)	3.0	Zephiramine®	ML₂	634	15	0.04∿0.5	30
Fe(III)	3.1 ∿ 3.8	Cetyltrimethyl-ammonium chloride	ML₆X₆	630	14.7	0.02∿0.24	32

Table 4 (continued)
APPLICATION OF CHROMAZUROL S® AS A PHOTOMETRIC REAGENT

Metal ion	Condition (pH, buffer, auxiliary reagents)		Metal chelate Ratio (M:L)	λ_{max} (nm)	$\varepsilon (\times 10^4)$	Range of determination (ppm)	Ref.
Ga	4.3	Acetate	ML_2	547	4.95	0.1~0.6	33
Hf	1.4	Acetate	M_2L_5	550	4.5	0.16~3.4	34
In	5.8	Acetate	$ML_{1\sim2}$	555	0.71	0.4~10	33
La	6.5	—	ML	510	1.3	0.2~5.6	35
Lu	—	Pyridine	$ML_{1\sim2}$	582	2.55	—	36
Mn	10	Borate	$ML_{1\sim2}$	555	0.31	0.8~12	37
Mo(V)	3.7	—	$ML_{1\sim2}$	600	1	0~8	38
Ni	9.7	—	$ML_{1\sim2}$	567	3.93	0.06~1	39
Ni	10.5 ~ 11.3	Hydroxydodecyl-trimethylam-monium bromide + pyridine	$ML_3X_{3\sim5}$	639	17.5	0.01~0.2	40
Pd(II)	6.1	—	$ML_{1\sim2}$	610	4.34	0.1~2.8	41
Pd(II)	6.1	Zephiramine®	$ML_{1\sim2}X_3$	635	11.7	0.02~0.5	41
Rh(III)	4.0	—	ML_2	570	1.3	0.7~7.0	42
Ru(III)	4.0	—	ML_2	540	0.5	0.45~7.0	42
Sc	5.6	—	ML_2	550	2.7	0.4~2.7	43
Sc	5.5	Zephiramine®	$ML_{3\sim4}$	625	13.7	0.01~0.3	44
Th	5.6	—	ML_2	554	3.2	0.8~7.0	45
Th	5.6 ~ 6.1	Cetyltrimethyl-ammonium chloride + pyridine	ML_4	631	17.4	0.15~0.8	46
Ti(IV)	4.5 ~ 5.0	H_2O_2	ML_2	560	4.4		
UO_2^{2+}	4 ~ 5.5	Cetyltrimethyl-ammonium chloride + pyridine	ML_2	625	10	0.14~1.3	49
VO^{2+}	4.4	NH_2OH,Zephir-amine®	$ML_{3\sim5}$	610	6.83	0.04~0.48	50
Zn	8.3	Zephiramine®	ML_2	510	4.4	0.16~1.4	51
Zr	1.4	Acetate	M_2L_5	550	4.76	0.16~3.0	34
Rare earth	5.4	Cetylpyridinium bromide	M_2LX_2	540 ~ 620	(3.6 for Nd)	0.7~2	52
Rare earth	6.5	—	$ML_{1\sim2}$	520 or 600	—	0.2~5.6	35

In recent years, the molar absorptivity of the metal chelate was found to increase to as high as 10^5 in the presence of cationic micelle; thus many approaches to the highly sensitive photometric methods to various metal ions have been reported. The only drawback with the use of Chromazurol S® as a photometric reagent is a lack of selectivity for metal ions. Selected examples on the photometry using Chromazurol S® are summarized in Table 4.

Photometric Determination of Beryllium[17]

Transfer an aliquot of the slightly acidic sample solution (1 ~ 80 μg of Be) to a 25-mℓ volumetric flask. On a separate aliquot, determine the amount of 10% sodium hydroxide required to neutralize the free acid to methyl orange indicator. Add 1.0 mℓ of 1% ascorbic acid, 2.00 mℓ of 10% EDTA, and 5.00 mℓ of acetate buffer (pH 4.60, 238 g of sodium acetate trihydrate

Table 5
CHELATOMETRIC TITRATION OF METAL IONS USING CHROMAZUROL S® AS AN INDICATOR

Metal ion	pH	Buffer	Color change at end point		Remarks	Ref.
Al (III)	4	Acetate	Violet ➤	yellow	Titrate at 80°	53
Cu (II)	6 ∿ 6.5	Acetate	Blue ➤	green		54
Fe (III)	2 ∿ 3	Monochloroacetic acid-acetic acid	Blue green ➤	orange yellow	Titrate at 60°	55
Ni (II)		Pyridine + NH₃	Blue violet ➤	yellow		4
Th (IV)	1 ∿ 2	HNO₃-NH₃	Red violet ➤	yellow	Photometric titration	4, 56
Rare earths		Pyridine + NH₃	Blue violet ➤	yellow		4

and 102 mℓ of glacial acetic acid in 1ℓ). To this mixture, add the predetermined amount of 10% sodium hydroxide, and dilute it to about 20 mℓ with water. Add 2.00 mℓ of Chromazurol S® solution (0.165 g in 100 mℓ water) and dilute to the volume. Measure the absorbance against a reagent blank at 569 nm in 1-cm cells. The sensitivity is 0.00022 μg Be cm² on the Sandell scale.

If Al, Cr, or Zr is present, special precaution must be taken. To determine Be in bronze or Al alloys, proceed as above, except for the addition of an amount of Cu or Al to the calibration standards, approximately equal to that in the sample aliquot.

More Sensitive Photometric Determination of Be Using Surfactant[20]
Transfer an aliquot containing less than 2 μg of Be to a beaker and add 1.00 mℓ of 1% EDTA and 5.00 mℓ of 5 *M* sodium perchlorate solution. Adjust pH of the solution to 4.5 with acetate buffer and evaporate the solution to 5 to 10 mℓ.* After cooling to room temperature, add 1.00 mℓ of 0.25% Chromazurol S® solution and 2.00 mℓ of 0.5% polyoxyethylene-dodecylamine (dissolve 0.5 g in water containing 0.5 mℓ of 2 *N* HCl, then adjust pH to 4.5 with acetate buffer and dilute to 100 mℓ.). Transfer the solution to a 25-mℓ volumetric flask and dilute it to the volume with water. After standing 15 min, measure the absorbance at 605 nm against a reagent blank in 1-cm cells. The sensitivity is 0.0001 μg Be cm² on the Sandell scale.

Anions such as perchlorate, nitrate, and chloride do not interfere and diverse cations on 1 to 0.5 mg level can be tolerated.

Use as a Metal Indicator in Chelatometry
Chromazurol S® has been recommended as a metal indicator in chelatometry as summarized in Table 5. Although the color change is not so sharp as in the case of Xylenol Orange, it is useful in the titration of Al, Cu, and Fe where Xylenol Orange or Methyl Thymol Blue does not function properly. For the indicator use, a 0.1 ∿ 0.4% aqueous solution. It can be kept for many months.

Other Reagent of Related Structure
Aluminon
Ammonium salt of aurine tricarboxylic acid.

---

* By this procedure, the interference from Al can be masked.

Red-brown amorphous powder. Commercial samples are of various grades of purity.[58] It was used for the detection and the photometric determination of Al,[59] Be,[60] and rare earths.[61]

Eriochrome Cyanine R

2″-Sulfo-3,3′-dimethyl-4-hydroxyfuchson-5,5′-dicarboxylic acid, trisodium salt ($C_{23}H_{16}O_9S$, mol wt = 468.43). A brick-red powder; easily soluble in water and alcohol, to give an orange solution which turns to violet at pH >9.[62] In acidic or neutral

range, it forms red-violet or blue-violet soluble chelates with Al, Be, Ca, Cu, Fe, Mg, Nb, Ta, U, and Zr, and it has been known as a detecting and photometric reagent for Al (pH 6.0,λ_{max},535 nm).[63] A highly sensitive method for Be in the presence of cationic micelle (Zephiramine®) has been proposed (pH 6.7 to 7.2,λ_{max},595 nm,$\varepsilon = 10^5$, 18 to 55 ppb).[64]

Fluoride can be determined indirectly by observing the color fading resulting from the reaction of Al-dye or Zr-dye with F⁻.[65,66]

REFERENCES

1. Geigy Co., British Patent 15204, 1907; U.S. Patent 877054; German Patent 199943.
2. Baldwin, W. G. and Stranks, D. R., *Austral. J. Chem.*, 21, 603, 1968.
3. Martynov, A. P., Novak, V. P., and Beznik, B. E., *Zh. Anal. Khim.*, 32, 519, 1977; *J. Anal. Chem. U.S.S.R*, 32, 416, 1977.
4. Malat, M. and Tenorova, M., *Chem. Listy*, 51, 2135, 1957; *Chem. Abstr.*, 52, 2662a, 1958.
5. Langmyhr, F. J.,and Klausen, K. S., *Anal. Chim. Acta*, 29, 149, 1963.
6. Nishida, H., *Bunseki Kagaku*, 19, 972, 1970.
7. Nishida, H., *Bunseki Kagaku*, 19, 34, 1970.
8. Semb, A. and Langmyhr, F. J., *Anal. Chim. Acta*, 35, 286, 1966.
9. Nishida, H., *Bunseki Kagaku*, 24, 261, 1975.
10. Srivastava, S. C., Sinha, S. N., and Dey, A. K., *Bull. Chem. Soc. Jpn.*, 36, 268, 1963.
11. Mukherji, A. K. and Dey, A. K., *J. Inorg. Nucl. Chem.*, 6, 314, 1958; *Anal. Chim. Acta*, 18, 324, 1958.
12. Sanyar, P., Sangal, S. P., and Mushran, S. P., *Anal. Chim. Acta*, 40, 217, 1967.
13. Nishida, H. and Nishida, T., *Bunseki Kagaku*, 21, 997, 1972.
14. Nishida, H. and Nishida, T., *Bunseki Kagaku*, 25, 55, 1976.
15. Kakida, Y., Goto, H., and Hosoya, M., *Nippon Kinzoku Gakkaishi (Tokyo)*, 24, 32, 1960.
16. Shijo, Y. and Takeuchi, T., *Bunseki Kagaku*, 17, 61, 1968.
17. Pakalns, P., *Anal. Chim. Acta*, 31, 576, 1964.
18. Horiuchi, Y., and Nishida, H., *Bunseki Kagaku*, 18, 180, 1969.
19. Horiuchi, Y. and Nishida, H., *Bunseki Kagaku*, 18, 1401, 1969.
20. Nishida, H., Nishida, T., and Ohtomo, H., *Bull. Chem. Soc. Jpn.*, 49, 571, 1976.

21. Nishida, H., *Bunseki Kagaku,* 27, 77, 1978.
22. Marczenko, Z. and Kalowska, H., *Chem. Anal. (Warsaw),* 23, 935, 1977; *Anal. Abstr.,* 34, 5B54, 1978.
23. Mulwani, H. R. and Sathe, R. M., *Analyst,* 102, 137, 1977.
24. Sangal, S. P. and Dey, A. K., *Bull Chem. Soc. Jpn.,* 36, 1347, 1963.
25. Horiuchi, Y. and Nishida, H., *Bunseki Kagaku,* 17, 824, 1968.
26. Shijo, Y., Takeuchi, T., and Yoshizawa, S., *Bunseki Kagaku,* 18, 204, 1969.
27. Malot, M. and Hrachovcova, M., *Collect. Czech. Chem. Commun.,* 29, 2484, 1964.
28. Ishida, R. and Sawada, T., *Bunseki Kagaku,* 16, 590, 1967.
29. Shijo, Y. and Takeuchi, T., *Bunseki Kagaku,* 15, 1063, 1966.
30. Nishida, H., *Bunseki Kagaku,* 20, 410, 1971.
31. Horiuchi, Y. and Nishida, H., *Bunseki Kagaku,* 16, 769, 1967.
32. Shijo, Y. and Takeuchi, T., *Bunseki Kagaku,* 17, 1519, 1968.
33. Horiuchi, Y. and Nishida, H., *Bunseki Kagaku,* 16, 1146, 1967.
34. Horiuchi, Y. and Nishida, H., *Bunseki Kagaku,* 16, 20, 1967.
35. Horiuchi, Y. and Nishida, H., *Bunseki Kagaku,* 17, 1233, 1968.
36. Akhmedli, M. K. and Granovskaya, P. B., *Org. Reagenty Anal. Khim. Tezisy Dokl. Vses. Konf.,* 4th, 2, 3, 1976; *Chem. Abstr.,* 87, 177044q, 1977.
37. Horiuchi, Y. and Nishida, H., *Iwate Daigaku Kogakubu Kenkyu Hokoku,* 20, 35, 1967; *Chem. Abstr.,* 69, 40966a, 1968.
38. Horiuchi, Y. and Nishida, H., *Bunseki Kagaku,* 18, 1092, 1969.
39. Horiuchi, Y. and Nishida, N., *Bunseki Kagaku,* 16, 576, 1967.
40. Shijo, Y. and Takeuchi, T., *Bunseki Kagaku,* 17, 1192, 1968.
41. Horiuchi, Y. and Nishida, H., *Bunseki Kagaku,* 16, 1018, 1967.
42. Saxena, K. K. and Dey, A. K., *Indian J. Chem.,* 7, 75, 1969; *Chem Abstr.,* 70, 84011k, 1969.
43. Ishida, R. and Hasegawa, H., *Bull. Chem. Soc. Jpn.,* 40, 1153, 1967.
44. Horiuchi, Y. and Nishida, H., *Bunseki Kagaku,* 17, 1486, 1968.
45. Ishida, R., *Nippon Kagaku Zasshi,* 86, 1169, 1965.
46. Shijo, Y. and Takeuchi, T., *Bunseki Kagaku,* 18, 469, 1969.
47. Matubara, C. and Takamura, K., *Microchem. J.,* 22, 505, 1977.
48. Nishida, H., *Bunseki Kagaku,* 19, 30, 1970.
49. Shijo, Y., and Takeuchi, T., *Bunseki Kagaku,* 20, 297, 1971.
50. Horiuchi, Y. and Nishida, H., *Bunseki Kagaku,* 18, 850, 1969.
51. Horiuchi, Y. and Nishida, H., *Bunseki Kagaku,* 17, 756, 1968.
52. Vekhande, C. and Munshi, K. N., *Indian J. Chem.,* A14, 189, 1976; *Anal. Abstr.,* 32, 6B54, 1976.
53. Theis, M., *Fresenius Z. Anal. Chem.,* 144, 106, 1955.
54. Theis, M., *Fresenius Z. Anal. Chem.,* 144, 275, 1955.
55. Sommer, L. and Kolarik, Z., *Collect. Czech. Chem. Commun.,* 22, 203, 1957.
56. Svach, M., *Fresenius Z. Anal. Chem.,* 149, 417, 1956.
57. Ishida, R. and Tonosaki, K., *Nippon Kagaku Kaishi,* 1496, 1975.
58. Smith, W. H., Sager, E. E., and Stewers, I. J., *Anal. Chem.,* 21, 1334, 1949.
59. Sandell, E. B., *Colorimetric Determination of Traces of Metals,* 3rd ed., Interscience, New York, 1959, 228.
60. McCloskey, P., *Microchem. J.,* 12, 40, 1967.
61. Sinha, S. N., Sangal, S. P., and Dey, A. K., *Chemist-Analyst,* 56, 59, 1967.
62. Suk, V. and Miketukova, V., *Collect. Czech. Chem. Commun.,* 24, 3629, 1959.
63. Hill, U. T., *Anal. Chem.,* 28, 1419, 1956, *Anal. Chem.,* 31, 429, 1959.
64. Kohara, H., Ishibashi, N., and Fukamachi, K., *Bunseki Kagaku,* 17, 1400, 1968.
65. Dixon, E., *Analyst,* 95, 272, 1970.
66. Analytical Methods Committee, *Analyst,* 99, 413, 1974.

TIRON

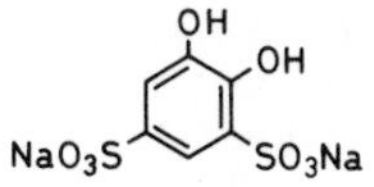

$$OH$$

$C_6H_4O_8S_2Na_2 \cdot H_2O$

mol wt = 332.21

Na_2H_2L

Synonyms

1,2-Dihydroxybenzene-3,5-disulfonic acid, disodium salt; Pyrocatechol-3,5-disulfonic acid, disodium salt

Source and Method of Synthesis

Commercially available. Synthesized by the sulfonation of pyrocatechol with chlorosulfonic acid, followed by hydrolysis.[1]

Analytical Uses

Widely used as a metal indicator in the chelatometric titration of Fe(III). The reagent was named as Tiron because it has been used as a photometric reagent for titanium and iron.[2] Later, although it was also suggested as a photometric reagent for various elements, better reagents are now available.

Properties of Reagent

It is usually supplied as disodium monohydrate colorless needles. Easily soluble in water, slightly soluble in alcohol, and insoluble in acetone and nonpolar organic solvents. Aqueous solution of disodium salt is colorless and stable for more than a year, but easily oxidized in alkaline solution.[3] Ionization constants of two hydroxyl groups are $pKa_1 = 7.66$, $pKa_2 = 12.6$ ($20°$, $\mu = 0.1$ (KCl)).[4]

Complexation Reaction and Properties of Complexes

The reaction of Tiron with metal ions are essentially the same as those of catechol, but Tiron forms more stable chelates. Although the complexation reactions with metal ions have been investigated rather extensively, only a limited number of elements, Cu(II), Fe(III) (blue to red), Os (red-violet), and Ti (orange), and rare earths (wine red) give colored chelates which are soluble in water. One of the drawbacks of Tiron as a reagent is that the metal-ligand ratio and, hence, absorption spectra of the chelate are dependent upon the solution pH, as exemplified in a iron-Tiron system.[4,5]

$$[Fe(HL)] \text{ and } [FeL]^- \rightleftarrows [FeL_2]^{5-} \rightleftarrows [FeL_3]^{9-}$$

pH 1~3   pH 3~5   pH >7

Blue-Green   Violet   Deep Red

The absorption spectra of this system are illustrated in Figure 1, and the coordination structure of $[FeL_3]^{9-}$ may be shown as below.

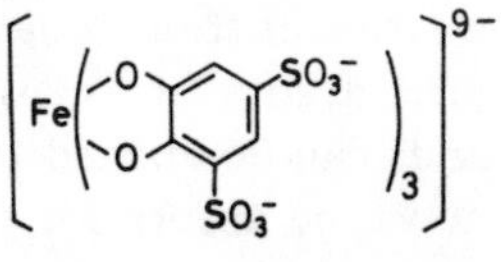

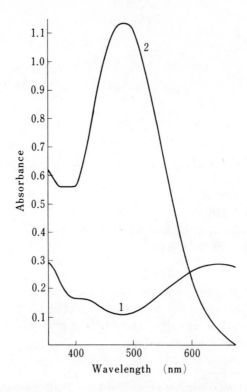

FIGURE 1.   Absorption spectra of iron(III)-Tiron system (Fe, 10 ppm). (1) pH 9.36 $[FeL_3]^{9-}$; (2) pH 4.0 $[FeL_2]^{5-}$. (From Yoe, J. H. and Jones, A. L., *Ind. Eng. Chem. Anal. Ed.,* 16, 111, 1944. With permission.)

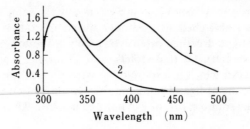

FIGURE 2.   Absorption spectra of Nb and Ta-Tiron chelates. (1) Nb-Tiron, Nb 60 ppm, neutral; (2) Ta-Tiron, Ta 250 ppm, acidic. (From Ackermann, G. and Koch, S., *Talanta,* 9, 1015, 1962. With permission.)

In some cases, the spectral changes were observed only in a near UV range, but they can be used for the photometric purpose. Examples are shown in Figure 2. The rare earth-Tiron chelate shows rather stronger (four to ten times) absorption bands due to f-f transitions or charge transfer, in comparison with the corresponding aquo- or chloro- complexes, and these bands can be utilized for the specific photometry of respective rare earths. Figure 3 shows the absorption spectra of some rare earth-Tiron chelates.[9]

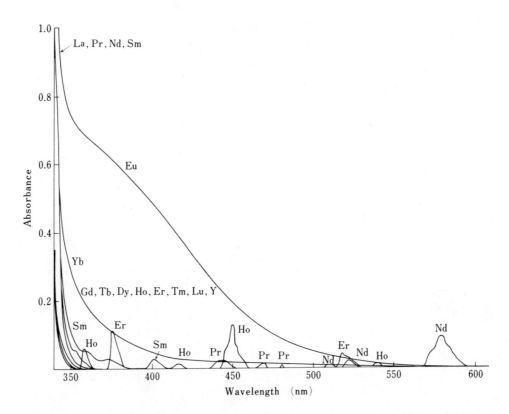

FIGURE 3.   Absorption spectra of rare earth-Tiron chelates. Rare earth, 3.3×10^{-3} M; Tiron, 3×10^{-2} M; and pH 6.4 ∿ 6.6. (From Taketatsu, T. and Yamauchi, T., *Talanta*, 18, 647, 1971. With permission.)

Terbium-Tiron chelate (M_2L_3 at pH 9) shows fluorescence at 530 to 560 nm when exited with mercury lamp. It is quenched completely by Er, Eu, Ho, Nd, Pr, Sm, and Yb, while its intensity is decreased to one half by the presence of Ce and Dy, but is not interfered with La, Lu, and Y.[10] In the presence of EDTA, Dy, and Tb form mixed ligand chelate with Tiron (MLY) at pH 12 to 13 which is also fluorescent and is not quenched by other members of the rare earths.[11]

The chelate stability constants were determined with various metal ions in aqueous solution, as summarized in Table 1. In the case of the rare earth ions, the rare earth-EDTA chelates (MY^-) react with Tiron (H_2L^{2-}) to form mixed ligand chelates (MYL^{5-}). The formation constants of these chelates:

$$MY^- + L^{4-} \underset{\longleftarrow}{\overset{K_f}{\rightleftharpoons}} MYL^{5-}$$

are summarized in Table 2. Coordination number of the rare earths must be greater than six.[13]

Purification and Purity of Reagent

Tiron (disodium salt) from commercial supply sources is usually substantially pure, but impure samples may be purified by recrystallization from hot water. The purity of the reagent can be checked by measuring the molar absorptivity in the aqueous solution ($\varepsilon = 6.99 \times 10^3$ at λ_{max} 260 nm, pH 10.8).[14]

Table 1
STABILITY CONSTANTS OF TIRON CHELATES[12]

Metal ion	$\log K_{ML}$	$\log K_{ML_2}$	$\log K_{ML_3}$	Temp (°C)	μ	Ref.
Al(III)	19.02	12.08	2.4	25	→0	
Ba(II)	4.10	—	—	20	0.1(KCl)	
Be(II)	12.88	9.37	—	20	0.1(KNO₃)	
Ca(II)	5.80	—	—	20	0.1(KCl)	
Cd(II)	7.69	5.60	—	25	1 (NaClO₄)	
Ce(III)	—	—	$\beta_3$3.75	—	1 (NaOAc)	
Co(II)	9.49	—	—	20	0.1(KCl)	
Cu(II)	14.53	—	—	20	0.1(KCl)	
Fe(III)	20.4	15.1	10.8	25	0.1(KNO₃)	
Ga(III)	∿18.4	15.50	—	30	0.2(NaNO₃)	32
Ge(II)	—	$\beta_2$12.63	—	—	—	20
Hf(IV)	24.66	—	$\beta_3$66.92		0.2(HClO₄)	
In(III)	17.00	13.85	—	30	0.2(NaNO₃)	32
Mg(II)	6.86	—	—	20	0.1(KCl)	
Mn(II)	8.6	—	—	25	0.1(KNO₃)	
Ni(II)	9.96	—	—	20	0.1(KCl)	
Pb(II)	11.95	6.33	—	25	1 (NaClO₄)	
Sr(II)	4.55	—	—	25	0.1(KCl)	
UO₂²⁺	15.90	—	—	25	0.1(KNO₃)	
VO²⁺	17.2	—	—	25	0.1(KNO₃)	
Zn(II)	10.19	8.33	—	30	1.0(HClO₄)	
Zr(IV)	24.15	—	—		1.0(HClO₄)	

Table 2
FORMATION CONSTANTS OF MIXED LIGAND
CHELATES, RARE EARTH - EDTA - TIRON[13]

Metal ion	$\log K_f$	Metal ion	$\log K_f$
Dy	7.31 ± 0.04	Nd	6.45 ± 0.05
Er	7.45 ± 0.04	Pr	6.14 ± 0.06
Eu	6.90 ± 0.08	Sm	6.69 ± 0.07
Gd	7.12 ± 0.03	Tb	7.18 ± 0.05
Ho	7.26 ± 0.03	Y	7.19 ± 0.05
La	5.72 ± 0.03	Yb	7.45 ± 0.05

Analytical Applications

Use as a Metal Indicator

Tiron is used as a metal indicator in the chelatometric titration of Fe(III) (at pH 2 to 3 buffered with *p*-chloroaniline 40°). Color change at the end point is from blue to pale yellow.[4] It is also used in the titration of Ge.[20] Tiron can be used as a titrant for the titration of the rare earth ions using Eriochrome® Cyanine R as an indicator.[31]

Use as a Photometric and Fluorimetric Reagent

Because of the relatively low molar absorptivity of Tiron chelates and less specificity to metal ions, Tiron is, nowadays, not so widely accepted as a photometric reagent. Examples are summarized in Tables 3 and 4.

Table 3
APPLICATION OF TIRON AS A PHOTOMETRIC REAGENT

		Metal chelate					
Metal ion	Condition (pH)	Ratio	λ_{max} (nm)	$\varepsilon(\times 10^3)$	Range of determination (ppm)	Interference	Ref.
Al(III)	5.5 ∿ 7	ML	310	13	0 ∿ 2	—	15
B	7.4 ∿ 7.5	ML	306	1.4	0.4 ∿ 3.2	$Cl^-,F^-,NO_3^-,SO_4^{2-}$	16
Co(II)	9.6	—	340	—	0.5 ∿ 10	Cr,Cu,Fe,Ni,Ti	17
Cu(II)	6.1 ∿ 6.9	ML₂	375	0.25	10 ∿ 160	Many elements[b]	18
Fe(III)	4.7	ML₃	560	3.9	0.2 ∿ 10	Many elements	2
Fe(III)	9.5	ML₃	480	5.3	0.2 ∿ 10	Ce,Fe,Mn,Ti, U,W	6
Ga(III)	2.5 ∿ 2.9	ML	310	4.9	7 ∿ 35	Al does not interfere	19
Mo(VI)	6.6 ∿ 7.5	—	390	—	0.2 ∿ 10	Ce,Fe,Mn,Ti, U,W	21,22
	2.2[a]	ML	320 ∿ 30	6.5	1 ∿ 16	Fe(III),Ti(IV), V(V)	23
Nb(V)	8.2	ML₁∿₂	405	5.1	∿ 60	Oxalate,Tartarate	7,24
Os	4.9 ∿ 5.5	MLCl₄	470	5.7	2 ∿ 32	Ce,Fe,M,Nb,Ta, Ti,VO₂²⁺,W	25
Sc	6.0	ML₁∿₃	310	8	0.06 ∿ 5.0	Al,Ce,Fe,Mo,Th, Ti,Zr	14
Ta(V)	6 NHCl	—	320	0.11	∿ 250	Oxalate,Tartarate	7
Ti(IV)	4.7	—	410 ∿ 5	14.5	0 ∿ 4	Many elements Fe can be masked with sodium di-thionite	2
UO₂(II)	3	ML	373	—	1.3 ∿ 130	—	26
V(IV)	7	ML₂	310	7.5	0 ∿ 5	Cu,Fe,Mo,W,Cl⁻, CO₃²⁻	27
V(IV)	2.9 ∿ 4.1[a]	VOL₂	585	10.3	0 ∿ 3.6	Ti(IV)	33
W	5.0	ML	313	7.6	1 ∿ 20	After ion exchange separation	28

[a] Ion-pair extraction with 1,3-diphenylguanidine into iso-pentanol + chloroform.
[b] A green color appears in the presence of Ca, Ba, Sr, and Pb due to the mixed complex formation.

Reagents with Related Structure
Catechol (Pyrocatechol)

Colorless crystals, mp 105°, $pKa_1 = 9.46$, $pKa_2 = 12.7$. It forms colored chelates with Fe(III) (violet), Mo(VI) (orange-red), Ti (orange), V (blue), Nb, and Ta (yellow), but most of the reactions are of little practical importance.

Pyrogallol

Colorless crystals, when pure, mp 133 to 134°, $pKa_1 = 9$. Complexing behavior is similar to that of catechol.

Table 4
APPLICATION OF TIRON AS A PHOTOMETRIC REAGENT FOR RARE EARTHS

Metal ion	Method[a]	Condition (pH)	Metal chelate Ratio	Metal chelate λ_{max} (nm)	Range of determination (ppm)	Remarks	Ref.
Ce	Ab	9.5 $\sim$ 11.5	—	510	2.4 $\sim$ 13.1	Al,Fe and Ti interfere	29
Er	Ab	4.8	M_2L_3	376	$\sim$5	Eu interferes	8
Eu	Ab	6.4 $\sim$ 6.6	ML	390,420	0 $\sim$ 700	—	9
Ho	Ab	4.8	M_2L_3	450	$\sim$4	Eu interferes	8,30
Nd	Ab	4.5	M_2L_3	578	$\sim$4	—	8
Nd	Ab	12.0	ML_2	571	$\sim$2.5	—	8
Nd	Ab	Alkaline	—	574	—	Ce can be masked with Na stannate	30
Pr	Ab	Alkaline	—	445,471	—	Ce can be masked with Na stannate	30
Sm	Ab	Alkaline	—	405.6	—	Ce can be masked with Na stannate	30
Dy	Fl	12 $\sim$ 13 (EDTA)	MLY	Exit.UV; flu. 572	0.01 $\sim$ 1(as Dy_2O_3)	—	11
Tb	Fl	$\sim$9	M_2L_3	Exit.UV; flu.530 $\sim$ 560	$4 \times 10^{-7} \sim$ (as Tb_4O_7)	—	10
Tb	Fl	12 $\sim$ 13 (EDTA)	MLY	Exit.UV; flu.546	$1 \times 10^{-5} \sim 1$ (as Tb_4O_7)	—	11

[a]   Ab, absorption photometry; Fl, fluorimetry.

MONOGRAPH

M1. **Budesinsky, B. W.,** Aromatic ortho-dihydroxy compounds as reagents for inorganic analysis, in *Chelates in Analytical Chemistry,* Vol. 5, Flaschka, H. A. and Barnard, A. J., Jr., Eds., Marcel Dekker, New York, 1976, 163.

REFERENCES

1. **Pollak, J. and Gebauer-Fulnegg, E.,** *Monatsh. Chem.,* 47, 112, 1926.
2. **Yoe, J. H. and Armstrong, A. R.,** *Ind. Eng. Chem. Anal. Ed.,* 19, 100, 1947.
3. **Atkinson, G. F. and McBryde, W. A. E.,** *Can. J. Chem.,* 35, 477, 1957.
4. **Schwarzenbach, G. and Willi, A.,** *Helv. Chim. Acta,* 34, 528, 1951.
5. **Harvey, A. E., and Manning, D. L.,** *J. Am. Chem. Soc.,* 72, 4488, 1950.
6. **Yoe, J. H. and Jones, A. L.,** *Ind. Eng. Chem. Anal. Ed.,* 16, 111, 1944.
7. **Ackermann, and Koch, S.,** *Talanta,* 9, 1015, 1962.
8. **Taketatsu, T. and Toriumi, N.,** *Talanta,* 17, 465, 1970.
9. **Taketatsu, T. and Yamauchi, T.,** *Talanta,* 18, 647, 1971.
10. **Poluektov, N. S., Alakaeva, L. A., and Tishchenko, M. A.,** *Zh. Anal. Khim.,* 25, 2351, 1970.
11. **Tishchenko, M. A., Alakaeva, L. A., and Poluektov, M. S.,** *Ukr. Khim. Zh.,* 39, 475, 1973; *Anal. Abstr.* 26, 2025, 1974.

12. Sillén, L. G. and Martell, A. E., *Stability Constants of Metal Ion Complexes*, Suppl. 1, The Chemical Society, London 1964, 1971.
13. Afghan, B. K. and Israeli, J., *Talanta*, 16, 1601, 1969.
14. Hamaguchi, H., Onuma, N., Kuroda, R., and Sugishita, R., *Anal. Chim. Acta*, 9, 563, 1962.
15. Yotsuyanagi, T., Goto, K., and Nagayama, M., *Nippon Kagaku Zasshi*, 88, 1282, 1967.
16. Hiiro, K., *Bull. Chem. Soc. Jpn.*, 35, 1097, 1962.
17. Bhaskare, C. K. and Deshmukh, S. K., *Fresenius Z. Anal. Chem.*, 277, 127, 1975.
18. Majumdar, A. K. and Savariar, C. P., *Anal. Chim. Acta*, 21, 53, 1959.
19. Reksc, W., *Chem. Anal. (Warsaw)*, 14, 795, 1969; *Anal. Abstr.*, 19, 2116, 1970.
20. Koryukova, V. P., Oleinik, L. K., and Andrianov, A. M., *Zavod. Lab.*, 39, 21, 1973; *Chem. Abstr.*, 78, 131721u, 1973.
21. Yoe, J. H. and Will, F., *Anal. Chim. Acta*, 6, 450, 1952.
22. Will, F. and Yoe, J. H., *Anal. Chim. Acta*, 8, 546, 1953.
23. Busev, A. I. and Rudzit, G. P., *Zh. Anal. Khim.*, 19, 569, 1964.
24. Land, J. E. and Morris, N. M., *J. Less-Common Metals*, 8, 266, 1965.
25. Majumdar, A. K. and Savariar, C. P., *Anal. Chim. Acta*, 21, 146, 1959.
26. Sarma, B. and Savariar, C. P., *J. Sci. Ind. Res.*, 16B, 80, 1957; *Chem. Abstr.*, 51, 11168d, 1957.
27. Nishikawa, S., Nakagawa, Y., Satake, M., and Matsumoto, T., *Bunseki Kagaku*, 15, 944, 1966.
28. Shimizu, T., Kato, K., Oyama, S., and Hosohara, K., *Bunseki Kagaku*, 15, 120, 1966.
29. Ruschel, R., *Mikrochim. Acta*, 344, 1960.
30. Tserkasevich, K. V. and Poluektov, N. S., *Zh. Anal. Khim.*, 19, 1309, 1964; *J. Anal. Chem. USSR*, 19, 1223, 1964.
31. Poluektov, N. S., Alakaeva, L. A., and Tishchenko, M. A., *Zh. Anal. Khim.*, 26, 181, 1971.
32. Athavale, V. T., Sathe, R. M., and Mahadevan, N., *J. Inorg. Nucl. Chem.*, 30, 3107, 1968.
33. Wakamatsu, Y. and Otomo, M., *Bull. Chem. Soc. Jpn.*, 47, 761, 1974.

CUPFERRON

$$C_6H_9N_3O_2$$
mol wt = 155.16

HL (NH₄L)

Synonym

N-Nitrosophenylhydroxylamine, ammonium salt

Source and Method of Synthesis

Commercially available. Obtained by the nitrosation of phenylhydroxylamine with $NaNO_2$ in hydrochloric acid or with amylnitrite in ammonia.[1]

Analytical Uses

As precipitating and solvent extraction reagents for Cu(II), Fe(III), Sn(IV), Ti(IV), U(IV), V(V), Zr, and some other elements.

Properties of Reagent

Cupferron is a trivial name for the ammonium salt of *N*-nitrosophenylhydroxylamine.[2] Pale yellow crystalline flakes or powder, mp 163 to 164°, sublimes above 30° (2 × 10⁻² Torr).[3] Soluble in water (12 g/100 mℓ at 25°)[R1] and alcohol. It can be recrystallized from ethanol. The reagent is unstable to visible and UV light and to air. To minimize this defect, the reagent is usually stored in a brown bottle with a few lumps of ammonium carbonate as a preservative.[R1]

The free acid (HL) is a white unstable solid (mp 51°) and decomposes spontaneously into nitrosobenzene, benzenediazonium nitrate, 4,4'-dinitrodiphenylamine, and other products. It is very slightly soluble in water, but easily soluble in various organic solvents. It is monobasic acid, pKa = 4.16 (μ = 0.1 $NaClO_4$, 25°),[4] K_D($CHCl_3$/H_2O) pH >3 with HCl or $HClO_4$ = 142 (room temperature)[5] K_D (ethyl acetate/water) = 285,[6] and K_D (CCl_4/H_2O) = 2300 (15°).[7]

Complexation Reactions and Properties of Complexes

Cupferron is a bidentate ligand with coordination sites of the oxygen of nitroso group and the negative-charged oxygen. Most of metal ions in the periodic table are precipitated with Cupferron from aqueous solution. As an analytical reagent, Cupferron is not so selective, but the selectivity is somewhat greater in strongly acidic solutions than in weakly acidic solutions.[36] Under the former condition, only Fe(III), Hf, Ga, Nb, Sn, Ta, Ti, V, and Zr precitate. As the acidity is reduced, other ions are precipitated. The conditions for the precipitation of metal Cupferrates and their solubilities are summarized in Table 1. As the precipitates of metal, Cupferrates are not so stable on drying, they are usually ignited to oxides for gravimetry.

The metal Cupferrates which are uncharged and coordination saturated, as illustrated below, can be extracted into inert solvents.

Thus, Cupferron has been widely used as a solvent extraction reagent for the separation of metal ions. Chloroform is the preferred solvent for most extractions. Extraction behaviors of metal ions with Cupferron have been investigated in detail, and pH for

Table 1

PRECIPITATION OF METAL CUPFERRATES

Metal ion	Condition	Solubility (18°)		
		H_2O^{35} (Mg/l)	1 NHCl^{35} (M/l)	0.01 N acetic acid^{33} (M/l)
Ag	—	150.0	—	—
Al	Weak acid	0.9	6×10^{-4}	—
Bi	HNO_3	8.4	6×10^{-4}	—
Cd	—	40.0	+	5.46×10^{-3}
Co(II)	—	77.0	+	1.07×10^{-2}
Cr(III)	—	+	+	—
Cu(II)	Acetic acid	0.7	5×10^{-4}	1.3×10^{-3}
Fe(II)	—	0.02	2×10^{-4}	—
Fe(III)	Strong acid	0.02	2×10^{-4}	—
Hg(I)	—	0.3	±	—
Hg(II)	—	+	+	—
Mn(II)	—	+	+	7.55×10^{-3}
Ni	—	52.0	+	1.06×10^{-2}
Pb	—	25.0	+	—
Sb(III)	—	5.5	10^{-4}	—
Sn(IV)	H_2SO_4	2.4	10^{-4}	—
Ti	H_2SO_4	—	—	—
Zn	—	32.0	+	6.3×10^{-3}

the quantitative extraction, the values of $pH_{1/2}$, and effective extraction constants on a chloroform-water system are summarized in Table 2.

Since most of the metal Cupferrates are not highly colored, Cupferron is not of practical importance as photometric reagent for the determination of trace of metals.

Purification and Purity of Reagent

Crude material may be purified by recrystallization from ethanol. Purity of the reagent is checked by observing the molar absorptivity of aqueous solution (pH 5.0 ± 0.2) at 281 nm, against the solvent blank. The value should be not less than 7990.[12]

Analytical Applications

Use as a Solvent Extraction Reagent[M3,M4]

Cupferron is extremely useful for the group separation of Fe(III), Sn(IV), Ti(IV), U(IV), V(V), Zr(IV), and few other metals from the rest of elements. In some cases, a major constituent is removed by extraction with Cupferron, leaving trace elements in aqueous phase for the subsequent analysis. In other cases, trace analytes are separated from matrix elements by extraction as metal Cupferrates prior to the determination. The extracted metals are recovered either by the wet digestion (HNO_3 + H_2O_2) of the residue after evaporating the solvent or by the back extraction with a strong acid solution.

Selectivity of extraction can be improved by proper selection of acidity in the aqueous phase and of masking agents, such as EDTA or tartarate. It is common practice to add aqueous Cupferron solution or solid Cupferron to the sample solution before extraction with organic solvent. Table 3 summarizes some typical examples of separation by Cupferron extraction. Other possibilities may be found by referring to Table 2.

Chloroform-Cupferron extraction of Al is unusual. According to Sandell, K'_{ex} de-

Table 2
SOLVENT EXTRACTION OF METAL CUPFERRATES

Metal ion[a]	Extraction ($CHCl_3$)		
	pH for quantitative extraction[R1]	$pH_{1/2}$[b]	log K'_{ex}[c]
Ag	35	—	—
Al	3.5	2.51	−3.5
Be	3.8	2.07	−1.54
Bi*	2	−0.4	5.08
Cd	4.5	—	—
Ce(IV)	0 ∿ 1	—	4.6[d]
Co(II)	4.5	3.18	−3.56
Cr(III)	3	—	—
Cu(II)*	2	0.03	2.66
Fe(III)*	0	<0	9.8[e]
Ga*	4.5	−0.3	4.92
Hf*	0 ∿ 1	0	>8
Hg(II)	5	0.85	0.91
In	3	0.5	2.42
La(III)	4	3.4	−6.22
Mn(II)	7[f]	—	—
Mo(VI)*	1.5	<0	—
Nb(V)	3.5	<0	—
Ni	9	—	—
Pa(V)	0 ∿ 1	—	—
Pb	3	2.06	−1.53
Pd(II)*	0	<0	—
Pm(III)	5	—	—
Pu(IV)	2	—	7
Sb(III)*	0 ∿ 1	<0	∿7
Sc	3	0.2	3.34
Sn(II)*	0 ∿ 1	—	∿ 6
Sn(IV)*	0 ∿ 1	—	—
Ta(V)	3	—	—
Th	2.5	0.2	4.44
Ti(IV)*	4	<0	—
Tl(III)	1.5[b]	—	∿ 3
U(IV)	0.1	—	∿ 8[g]
U(VI)	3[b]	—	—
V(IV)	0 ∿ 1	—	—
V(V)	2.5	<0	—
W(VI)	0.3[b]	—	—
Y	5	2.9	−4.74
Zn	9	7.4	—
Zr	3	<0	—

[a] Elements easily extracted quantitatively are marked with *.

[b] In the presence of 0.05 M Cupferron solution.[8]

[c] Effective extraction constant.

$$K'_{ex} = \frac{[ML_n]_{org}\,[H^+]^n}{\Sigma[M]\,[HL]^n_{org}}$$

depends on the ionic strength and on the concentration of complexing anions as well as on $[HL]_{org}$ if charged Cupferrates are formed. Cited values are obtained under following conditions: initial HL concentration in aqueous phase, 0.005 M for Bi, Ga, Mo, Sc, Th, Tl, and Y and 0.05 M for Be, Co, Cu, Hg, La, and Pb; μ = 0.1 ($NaClO_4$) at room temperature.[M3.8]

[d] In butyl acetate.[9]

[e] In 1 M $HClO_4$.[10]

[f] Partial extraction.

[g] In ethyl ether.

Table 3
SOME EXAMPLES OF CUPFERRON EXTRACTION

Metal ion				
Extracted	From	Condition	Solvent	Ref.
Al,Be,Ce(III),(IV), Hf,Nb,Sn(II),(IV), Ta,Ti,U(IV),(VI), V,Zr,rare earths	—	pH 5.5 ∿ 5.7, EDTA, citrate	4-Methylpentan-2-one	13
Ti	Bi,Cd,Co(II),Cu,Mn(II), Mo,Ni,V(V),W,Zn	Slightly alkaline solution, EDTA, tartarate,S^{2-}	CHCl₃	14
Ti	Steel	pH 5.5, EDTA, Ce, interfere	MIBK	34
Bi,ᵃFe(III),Ga, Mo(VI),Sn(IV),Ti, U(IV),V(V),Zr	Ag,As(III)(IV),Cd,Co, Hg,Ni,Pb,Se(VI), Te(IV), U(VI),Zn	0.1 NHNO₃	CHCl₃	15
Mo,W	Bio-material (after wet digestion and dithizone extraction)	6 NH₂SO₄	iso-AmOH	16
Fe(III),Mo(VI), Sn(IV),Ti(IV), V(V),Zr	U(VI)	Dil mineral acid	CHCl₃ or ethylacetate	17
Bi	Ag,Al,As,Cd,Co,Cr,Hg, Mn,Ni,Pb, Sb,Zn	1 NHCl	CHCl₃	34
Fe	Al,Cr,Mn,U	Ti,V,Zn interfere	CHCl₃	34
Ga	Al,Cr,In, rare earths	2 NH₂SO₄,Fe,Ge, Nb,Ta,Ti,V,Zr interfere	CHCl₃	34
Th	Trivalent metal	pH 0.3 ∿ 1, Hf, La U(VI) interfere	Benzene + iso-AmOH(1:1)	34
Actinoids,Am	Pu(III)	0.5 ∿ 1.5 NHCl	CHCl₃	34

ᵃ Bi may be back-extracted with 2 NH₂SO₄, rest of the elements remaining in the organic phase.

pends on the pH and the times of standing before extraction.[M3] K'_{ex} was found to decrease with increasing pH and Cupferrate ion concentration when the extraction was soon after the addition of Cupferron to the aqueous aluminum solution. Extraction of Fe(III) in the presence of Al may be possible if the aqueous solution is made to 1 *M* or stronger in hydrochloric acid and the extraction is carried out immediately after Cupferron has been added.

To overcome the instability of aqueous Cupferron solution, Fe(III)-,[18] Cu(II)-,[18] or Al-Cupferrate[19] have been used for the extraction of Mo, Nb, Sb, Sn, Ta, Ti, V, and Zr, based on the exchange extraction.

Use as a Precipitating Reagent[R1]

Beside gravimetric determination of metals, Cupferron has been used as a precipitating reagent for the group separation of metals.[46] Precipitation behaviors of metal ions are summarized in Table 1, and some examples of separation by precipitation are shown in Table 4.

Other Uses

Cupferron has been used as an extraction photometric reagent for Ti(IV) (extraction with MIBK from pH 5.5 to 5.7, ε = 2100 at 425 nm)[13], Fe(III) (extraction with isoamylacetate from strongly acid solution, at 460 nm), Nd (ion-pair extraction with diphenylguanidine into ChCl₃, 521.5 nm)[31], and V(V) (after exchange extraction with

Table 4
SOME EXAMPLES OF CUPFERRON PRECIPITATION

Metal ion	Condition	Remarks	Ref.
Ga(III)	$2 N H_2SO_4$	Separation from Al,Cr,Sc,U(VI), Zn,rare earths	20
Ga(III)	pH 4.5 ∿ 5.5, EDTA, magnesia mixture after coprecipitate with Sn or Ti	Ta and almost all the ions except Be,Ti,U,PO_4^{3-} (in Nb-Mo stell), Nb to Ta(30:1 ∿ 1:30)	32
Hf(IV)	6 ∿ 10 v/v % H_2SO_4	Separation from Al,Be,Cr,U; Th is ppt; similarly at pH 1 ∿ 8	21
Ti(IV)	H_2SO_4, tartaric acid, EDTA[a]	Si,U,V,W,PC_4^{3-} interfere	22
U(IV)	1 ∿ 2 M HCl, $HClO_4$ or H_2SO_4, at 0 ∿ 5°, after reduction with NH_2OH or S_2O_4	Bi,Fe,Ga,Hf,Mo,Nb,Pa,Po,Sn, Ta,Ti,V,Zr, and quadrivalent actinides interfere	17
U(VI)	pH ∿ 7, EDTA, tartarate	Be and a large amount of Ti, Zr,F^-,CO_3^{2-} interfere	23

[a] *In situ* preparation of Cupferron from phenylhydroxylamine and $NaNO_2$.

AlL_3 at pH 1.5 to 2 in $CHCl_3$-ethanol, $\varepsilon = 5100$ at 400 nm).[19] It was also used as a titrant in the potentiometric or amperometric titration of Zr.[25,26]

Other Reagents with Related Structure
Neocupferron

Ammonium salt of N-nitroso-1-naphthylhydroxylamine ($C_{10}H_7N(NO)NH_4$), mol wt = 205.21. Pale yellow crystalline powder, mp 125 to 126° (with decomposition). Soluble in water (6.0 g/100ml at 25°).[R1] Physical properties and chelating behavior are similar to those of Cupferron: p$Ka = 4.1$ and K_D ($CHCl_3/H_2O$) = 1.3 × 10³.[M3] It is claimed to be more stable than Cupferron in aqueous solution, and the weighing effect of naphthyl group may be expected for gravimetry.

N-Nitroso-N-cyclohexylhydroxylamine[27]

Cyclohexyl analogue of Cupferron. Ammonium salt is named as "Hexahydro Cupferron"; mp 140° and decomposition 250°. It is more stable than Cupferron in the solid state as well as in aqueous solution (a 10⁻⁴ M solution in 6N HCl decomposed ∿ 50% after 4 days, whereas Cupferron decomposed to the same extent in only 65 min); fairly soluble in water (11.49 g/100 ml), p$Ka = 5.58$. The chelating behavior is very similar to that of Cupferron.[28,29]

Other alkyl analogues, N-nitroso-N-cyclooctyl-, N-nitroso-N-cyclododecyl-, and N-nitroso-N-isopropyl-hydroxylamine are also reported.[30]

MONOGRAPHS

M1. **Smith, G. F.** *Cupferron and Neocupferron,* G. F. Smith Chemical Co., Columbus, Ohio, 1938.
M2. **Welcher, F. J.,** *Organic Analytical Reagents,* Vol. 3., Van Nostrand, Princeton, N.J., 1947.
M3. **Sandell, E. B. and Onishi, H.,** *Photometric Determination of Traces of Metals,* 4th ed., John Wiley & Sons, New York, 1978.
M4. **Morrison, G. H. and Freiser, H.,** *Solvent Extraction in Analytical Chemistry,* John Wiley & Sons, New York, 1957.

REVIEW

R1. **Shendrikar, A. D.,** Substituted hydroxylamines as analytical reagents, *Talanta,* 16, 51, 1969.

REFERENCES

1. **Marvel, C. S. and Kamm, C.,** *Org. Synth.,* 4, 19, 1925.
2. **Baudish, O.,** *Chem. Ztg.,* 33, 1298, 1909.
3. **Honjo, T., Imura, H., Shima, S., and Kiba, T.,** *Anal. Chem.,* 50, 1545, 1978.
4. **Dyrssen, D.,** *Svensk. Kem. Tidskr.,* 64, 217, 1952; *Chem. Abstr.,* 384h, 1953.
5. **Kemp, D. M.,** *Anal. Chim. Acta,* 27, 480, 1962.
6. **Stander, C. M.,** *Anal. Chem.,* 32, 1296, 1960.
7. **Hagiwara, E.,** *Tech. Rep. Tohoku Univ.,* 18, 16, 1953.
8. **Stary, J. and Smizanska, J.,** *Anal. Chim. Acta,* 29, 545, 1963.
9. **Hagiwara, Z.,** *Kogyo Kagaku Zasshi,* 57, 266, 1954.
10. **Sandell, E. B. and Cammings, P. F.,** *Anal. Chem.,* 21, 1356, 1949.
11. **Dyrssen, D. and Dahlberg, V.,** *Acta Chem. Scand.,* 7, 1186, 1953.
12. *Anala R,* Standards for Laboratory Chemicals, 7th ed., Analar Standards, England, 1977.
13. **Cheng, K. L.,** *Anal. Chem.,* 30, 1941, 1958.
14. **Donaldson, E. M.,** *Talanta,* 16, 1505, 1969.
15. **Bode, H. and Henrich, G.,** *Fresenius Z. Anal. Chem.,* 135, 98, 1952.
16. **Allen, S. H., and Hamilton, M. B.,** *Anal. Chim. Acta,* 7, 483, 1952.
17. **Grimaldi, F. S., et al.,** Collected papers on method of analysis for uranium and thorium, *U.S. Geol. Surv. Bull.,* 1006, 184, 1954.
18. **Zolotov, Yu. A., Spivakov, B. Ya., and Gavrilina, G. N.,** *Zh. Anal. Khim.,* 24, 1168, 1969.
19. **Sasaki, Y. and Kawae, Y.,** *Bunseki Kagaku,* 25, 108, 1976; *Bunseki Kagaku,* 27, 366, 1978.
20. **Gastinger, E.,** *Fresenius Z. Anal. Chem.,* 140, 244, 252, 1953.
21. **De, A. K. and Sahn, C.,** *Sep. Sci.,* 2, 11, 1967.
22. **Heyn, A. H. and Dave, N. G.,** *Talanta,* 13, 33, 1966.
23. **Korkisch, J.,** *Mikrochim. Acta,* 401, 1967.
24. **Beckwith, R. S.,** *Chem. Ind.,* 663, 1954.
25. **Ivanova, Z. I. and Rivina, V. Ya.,** *Metody Khim. Anal. Stokov Vod Predpr. Khim. Prom.,* 163, 1971; *Chem. Abstr.,* 78, 37597m, 1973.
26. **Khadeev, V. A.,** Deposited DOC., *Viniti 152,* 77, 6, 1977; *Chem. Abstr.,* 90, 132251a, 1979.
27. **Buscarons, F. and Canela, J.,** *Anal. Chim. Acta,* 67, 349, 1973.
28. **Buscarons, F. and Canela, J.,** *Anal. Chim. Acta,* 70, 113, 1974.
29. **Canela, J. and Garcia-Lodona, A. M.,** *Quim. Anal.,* 28, 144, 1974; *Anal. Abstr.,* 28, 2B30, 1975.
30. **Buscarons, F. and Canela, J.,** *Anal. Chim. Acta,* 71, 468, 1974.
31. **Poluektov, N. S., Laver, R. S., Gava, S. A., and Mishchenko, V. T.,** *Ukr. Khim. Zh.,* 35, 238, 1969.; *Chem. Abstr.,* 71, 7046u, 1969.
32. **Majumdar, A. K. and Chowdhury, J. B. R.,** *Anal. Chim. Acta,* 19, 18, 1958.
33. **Treadwell, W. D. and Ammann, A.,** *Helv. Chim. Acta,* 21, 1249, 1938.
34. **Holzbecher, Z., Divis, L., Kral, M., Sucha, L., and Ulacil, F.,** *Handbook of Organic Reagents in Inorganic Analysis,* Ellis Horwood, Chichester, England, 1976.
35. **Pinkus, A. and Martin, F.,** *J. Chem. Phys.,* 24, 83, 137, 1927.
36. **Cheng, K. L.,** *Chemist-Analyst,* 50, 126, 1961.

CHLORANILIC ACID AND ITS METAL DERIVATIVES

$C_6H_2O_4Cl$
Mol wt = 208.99

H₂L

Synonym

3,6-Dichloro-2,5-dihydroxy-p-benzoquinone

Source and Method of Synthesis

Free acid and its metal derivatives, such as Ba, Hg(II), La, Sr, Th, and Zr chloranilates, are commercially available. Chloranilic acid is prepared by the alkaline hydrolysis of chloranil.[1] Metal chloranilates are prepared by the reaction of chloranilic acid with the respective inorganic salt in hot water.

Analytical Uses

Free acid is used as a precipitating reagent for heavy and multivalent metal ions and as a photometric reagent for metal ions based on the precipitation reaction. Metal chloranilates are used as photometric reagents for anions, based on the metathesis of metal chloranilate to form an insoluble salt of the anion to be determined and the subsequent liberation of highly colored acid chloranilate ion.

Properties of Chloranilic Acid

Orange red crystalline powder, mp 283 to 284°. It is relatively strong dibasic acid, and the pH of saturated aqueous solution is about 2;[2] pKa_1 0.81 ± 0.01, pKa_2 2.72 ± 0.05 (25°, μ = 0.15).[3] It has a tendency to sublime.

It is slightly soluble in water (3.0 g/ℓ at room temperature),[2] giving a purple solution which is similar to the color of permanganate solution. Easily soluble in aqueous alkali or in hot water. Insoluble in most organic solvents except alcohol. It reacts with various multivalent and heavy metal ions to give insoluble brown to violet precipitates.

Of the three species (H₂L, HL⁻, L²⁻) in aqueous solution, acid chloranilate ion (HL⁻) shows the spectrum of highest absorbance which is observed in the range of pH 1 to 2. Figure 1 illustrates the absorption spectrum of HL⁻ in visible region. The peak intensity is highest at pH 1 to 2, decreases to one half of its original intensity at pH 5, and does not change further in the range of pH 5 to 12.[2] Another strong peak is observed in UV region (λ_{max}, 332 nm).[5]

The aqueous solution is not so stable against light and decolorizes gradually. The solution is recommended to be stored in a cool dark place and preferably at low pH.[6]

When the aqueous solution of chloranilic acid is mixed with metal ion solution, the color change is observed due to chelate formation, and in some cases, insoluble metal chloranilate precipitates from the solution. At the same time, the purple color of the solution fades out. The precipitation reaction is often nearly quantitative, and the color of precipitate varies from brown to violet. The multivalent and heavy metal chloranilates are much less soluble than chloranilic acid. This characteristic provides the bases for their use in metal analyses. The reactions with various metal ions are summarized in Table 1.

Some of the metal chloranilates are soluble in water, and the following equilibrium constants were determined in the aqueous solution.

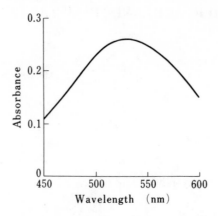

FIGURE 1.   Absorption spectrum of chloranilic acid in water; 0.01% HL⁻ (pH 2.2). (Reprinted with permission from Bertlacini, R. J. and Barney, J. E., Jr., *Anal. Chem.*, 29, 281, 1957. Copyright 1957 American Chemical Society.

Table 1
REACTION OF CHLORANILIC ACID WITH METAL IONS IN AQUEOUS SOLUTION[R1,6]

Metal ion	Visual change	Metal ion	Visual change
Ag	Ppt	Mg	Color fades
Al	No change; prevents the ppt of Ca complex	Mn(II)	Ppt
Be	Deep red	Mo(VI), W(VI)	Color fades
Bi	Green, ppt	Na,K	Color fades slightly
Ca,Ba,Sr	Ppt	Ni(II)	Violet
Cd,Zn	Ppt	Pb	Ppt
Co(III)	Ppt	Pd	Brown-Yellow
Cr(III)	Complex formation	Sn(IV)	Ppt. (blue in ethylether)[33]
Cr(VI)	Yellow	Th	Bright violet; pp at higher conc
Cu(II)	Color fades, ppt at higher conc	Ti	Color fades
Fe(II)	Color deepens	Tl(I)	Color fades slightly
Fe(III)	Color deepens, ppt at higher conc	U(IV)	Brown
Hg(I),(II)	Ppt	Zn	Ppt
		Zr,Hf	Violet-red; ppt at higher conc
		Rare earths	Ppt

$$M^{m+} + L^{2-} \xrightleftharpoons{K_{ML}} ML^{m-2}$$

$$2M^{m+} + L^{2-} \xrightleftharpoons{\beta_2} M_2L^{2m-2}$$

$\left. \begin{array}{l} \log K_{ML} \text{ for Fe (III) 5.81, Ni (II) 4.02} \\[2mm] \log \beta_2 \quad \text{ for Fe (III) 9.84} \end{array} \right\} (25°, \mu = 0.15)^3$

$$M^{4+} + H_2L \xrightleftharpoons{K} ML^{2+} + 2H^+$$

log K for Hf (IV) 3.73 (25°, $\mu = 3$ (HClO$_4$))[7]

Zr (IV) 5.76 (25°, $\mu = 2$ (HClO$_4$))[8]

$$ML^{2+} + H_2L \xrightleftharpoons{K} ML_2 + 2H^+$$

log K for Zr (IV) 3.78 (25°, $\mu = 2$ (HClO$_4$))[8]

Properties of Metal Chloranilates

Most of commercial products of Ba, Hg(II), La, Sr, and Th chloranilates are anhydrous crystals of a dull brown or black powder. Barium salt is also supplied as crystalline hydrate which shows a metallic sheen. Particle size differs from metal to metal and even from lot to lot on the same metal derivative, depending upon the condition of preparation.

These metal derivatives are almost insoluble in water as well as in most organic solvents, with the exception of certain solvents having a more polar nature, such as ethylenediamine, methyl cellosolve, acetic acid, pyridine, tetrahydrofuran, etc.

Solubility of Ba chloranilate in water is 2.2×10^{-4} mol/ℓ and in aqueous ethanol (1:1) is 5.2×10^{-6} mol/ℓ. Although the solubility of the other metal derivatives is not available, their solubilities may be approximately in a similar order of magnitude to that of Ba chloranilate.

The solubility of Ca chloranilate in water is pH dependent, showing the minimum solubility at pH 4.4.[9] The solubility product is reported to be 8.1×10^{-9} (pH 3) and 1.8×10^{-9} (pH 7), being comparable with that of calcium oxalate.[10]

When insoluble Ag chloranilate is shaken with the aqueous solution containing chloride ion, silver chloride precipitates and reddish purple acid chloranilate ion is released as shown below:

$$\underset{\text{solid}}{Ag_2L} + 2Cl^- + H^+ = \underset{\text{solid}}{2AgCl} + HL^-$$

The chloride ion concentration is proportional to the amount of chloranilate ion liberated which is determined photometrically at λ_{max} of 530 nm ($\varepsilon \approx 200$) or at 332 nm. This principle can be applied in the determination of other anions.[11]

In the reaction,

$$\underset{\text{solid}}{ML} + X^- = \underset{\text{solid}}{MX} + L^-$$

where X^- is the anion to be determined and ML is the insoluble metal chloranilate, ML must be much less soluble than MX, so that the reaction is quantitative. MX must be only sparingly soluble, so that blank will not be too high. These requirements can be met by proper selection of metal chloranilate, the solvent system, and pH. Some important metal chloranilates and their reactions with respective anions are listed in Table 2.

Purification and Purity of Reagent

Chloranilic acid is best purified by recrystallization from hot water. Commercial products are usually of high purity. Trace metal impurities may be determined on the sulfated ash. It can be assayed by an acidimetry in aqueous solution or by a gravimetry as metal chloranilates[16] (precipitating as calcium salt at pH 4.4, then weighing after drying at 105°).

Metal chloranilates for anion determination should be in a strict stoichiometry and be contaminated neither by excess cation as a soluble salt nor by excess chloranilic acid. The former can be checked by washing the sample, followed by the analysis of cations on the filtrate. The latter can be determined by the following procedure.[16]

To 0.30 g of sample, add 10 ml of 0.05 $M KH_2PO_4$ solution and 50 ml of 95% ethanol. Dilute the mixture to 100 ml with water. Shake it well by a mechanical shaker for 10 min and filter using a dry filter paper. Measure the absorbance of the filtrate at 530 nm against water.

It is also desirable to have the material of a uniform particle size of reasonable fineness, so that the reproducible result can be obtained in the metathesis reaction with the solid reagent.

Analytical Applications

Use as a Precipitating Reagent

Certain cations are precipitated by chloranilic acid and can be determined by gravimetry or some other means. Examples are summarized in Table 1.

Use as a Photometic Reagent for Cations

The method depends on the quantitative loss of absorbancy of a solution of excess chloranilic acid of known concentration by precipitation of an insoluble metal chloranilate. The absorbance is measured in the supernatant liquid after centrifugation or in the filtrate after filtration.

Alternatively, the isolated insoluble metal chloranilate is redissolved in strong acid or in alkaline EDTA solution, and the regenerated chloranilate ion is determined spectrophotometrically. This procedure is claimed to give a higher degree of accuracy.

Another method utilized the absorbance of highly colored soluble chelates of chloranilic acid with metal ions such as Zr(IV) or Mo(VI).

Table 3 summarizes some representative photometric procedures for cations. Aqueous solution of chloranilic acid or sodium chloranilate in appropriate concentration is used for general purposes, the former being more stable than the latter. The solution should be kept in a cool dark place. The standard solution for spectrophotometry should be prepared freshly every time.

Use as a Photometric Reagent for Anions

Certain anions can be determined by absorption photometry using insoluble metal chloranilates, based on the principle described above. In many cases, the concentration of freed acid chloranilate ion (HL⁻) is measured at 530 nm ($\varepsilon \approx 200$). Examples of the photometric determination of anions using metal chloranilates are summarized in Table 2.

Table 2

METAL CHLORANILATES AND THEIR REACTIONS WITH ANIONS

Metal chloranilate	Appearance (formula weight)	Reaction with anions			Range of determination (ppm)	Ref.
		Anion	Condition	Stoichiometry		
BaL	Violet-brown crystals (334.31 for BaL) (398.36 for BaL · H_2O)	SO_4^{2-}	pH 4 ~ 4.6, in 50% ethanol	$SO_4^{2-} + BaL + H^+ \rightarrow HL^- + BaSO_4\downarrow$	0.3 ~ 100	4, 5, 24—27
		BO_3^{3-}	pH 8, NH_4Cl-tartarate buffer	$H_3BO_3 + BaL + tart \rightarrow HL^- + Ba$-borotartarate complex	0.3 ~ 10	28
SrL	Black-brown powder (294.59)	F^-	pH 4, in 50% iso-PrOH	$2F^- + SrL + H^+ \rightarrow HL^- + SrF_2$	5 ~	5
Ag_2L	Grey-green powder (422.71)	Cl^-	—	$2Cl^- + Ag_2 + H^+ \rightarrow HL^- + 2AgCl\downarrow$	Not stoichiometric	11
HgL	Dark-green powder (407.56)	Cl^-	pH 2, in 50% methylcellosolve	$2Cl^- + HgL + H^+ \rightarrow HL^- + HgCl_2$	0.05 ~ 200	5, 29, 30
ThL_2	Black-brown powder (645.98)	F^-	pH 7, in 25% methylcellosolve	$6F^- + ThL_2 + 2H^+ \rightarrow 2HL^- + ThF_6^{2-}$ or $2F^- + ThL_2 + H^+ \rightarrow HL^- + ThLF_2\downarrow$	0.01 ~ 100	15, 31
						15
CuL	Red-brown powder (270.52)	S^{2-}	—	$S^{2-} + CuL_2 + 2H^+ \rightarrow 2HL^- + CuS\downarrow$	0 ~ 10, CN^-, oxalate, tartarate, citrate interfere	32
La_2L_3	Dark-grey powder (898.73)	F^-	pH 6.5 ~ 7, in 50% ethanol or in 50% methylcellosolve	$6F^- + La_2L_3 + 3H^+ \rightarrow 3HL^- + 2LaF_3\downarrow$	2 ~ 200	12, 14
		PO_4^{3-}	pH 7	$2PO_4^{3-} + La_2L_3 + 3H^+ \rightarrow 3HL^- + 2LaPO_4\downarrow$	3 ~ 300	13

Table 3
PHOTOMETRIC DETERMINATION OF METAL IONS WITH CHLORANILIC ACID

Metal ion	Condition	Species to be determined	Range of determination (ppm)	Ref.
Ba, Sr, Zn	pH 5 ∿ 7, after filtering ppt	Chloranilic acid at 530 nm	—	17
Ca	In weak acidic soln (HCl, acetic acid), after filtering ppt	Chloranilic acid at 550 nm	—	18
Ca	After centrifuging, ppt, wash with iso-PrOH, then dissolve into 5% alkaline EDTA	Chloranilic acid at 520 nm	40 ∿ 400	2
Mo(IV)	Extraction with MIBK from 6 MHCl + 0.4 MHF, then back extraction into water	Mo chloranilate at 350 nm	0.5 ∿ 9.6	19, 20
Nb	0.18 MH$_2$SO$_4$, Ta can be separated by extraction with MIBK from 6 MH$_2$SO$_4$ + 2 MHF	Nb chloranilate at 340 nm	0.2 ∿ 4.0	35
W(VI)	pH 3 ∿ 4 pH <2	M$_2$L at 335 nm ML	—	34
Zr	1 MHClO$_4$ + 1 MLiClO$_4$; accompanying metals should be separated by extraction from 4 MHCl with 7.5% tri-n-octylamine in benzene	Zr chloranilate at 340 nm	0.02 ∿ 2	21—23

REVIEWS

R1. **Broad, W. C., Ueno, K., and Barnard, A. J., Jr.,** Chloranilic acid and its metal salts as analytical reagents, *Bunseki Kagaku,* 9, 257, 1960.

R2. **Hart, W. G.,** Spectrophotometric uses of chloranilic acid and its metallic salts, *Organic Chem. Bull.,* 33, 1, 1961.

R3. **Bark, L. S.,** Use of chloranilates in colorimetric analysis, *Ind. Chem.,* 40, 153, 1964; *Anal. Abstr.,* 12, 3166, 1964.

REFERENCES

1. **Conant, J. B. and Fieser, L.,** *J. Am. Chem. Soc.,* 46, 1866, 1924.
2. **Ferro, P. V. and Ham, A. B.,** *Tech. Bull. Regist. Med. Technol.,* 27, 160, 1967, *Am. H. Chin. Pathol.,* 28, 208, 1957; *Chem. Abstr.,* 52, 2152c, 1958; *Am. H.Chin Pathol.,* 28, 689, 1957; *Chem. Abstr.,* 52, 8262c, 1958.
3. **Cabbiness, D. K. and Amis, E. S.,** *Bull. Chem. Soc. Jpn.,* 40, 435, 1967.
4. **Bertolacini, R. J. and Barney, J. E., Jr.,** *Anal. Chem.,* 29, 281, 1957.
5. **Bertolacini, R. J. and Barney, J. E., Jr.,** *Anal. Chem.,* 30, 202, 1958.
6. **Frost-Jones, R. E. U. and Yardley, J. T.,** *Analyst,* 77, 468, 1952.
7. **Uarga, L. P. and Veatch, E. C.,** *Anal. Chem.,* 39, 1101, 1967.
8. **Thamer, B. J. and Voigt, A. F.,** *J. Am. Chem. Soc.,* 73, 3197, 1951.
9. **Breyer, B. and McPhillips, J.,** *Nature (London),* 172, 257, 1953; *Analyst,* 78, 666, 1953.
10. **Korolef, F.,** *Fin. Kemist Samfundets Medd.,* 60, 56, 1951; *Chem. Abstr.,* 46, 7929e, 1952.

11. Coutinho, A. B. and Ameida, M. D., *An. Asoc. Quim. Bras.*, 10, 82, 1951.
12. Hayashi, K., Danzuka, T., and Ueno, K., *Talanta*, 4, 126, 1960.
13. Hayashi, K., Danzuka, T., and Ueno, K., *Talanta*, 4, 244, 1960.
14. Fine, L. and Wynne, E. A., *Microchem. J.*, 3, 515, 1959.
15. Hensley, A. L. and Barrney, J. E., Jr., *Anal. Chem.*, 32, 828, 1960.
16. Tech. Staff, Purity Specification of Organic Reagents, Dojindo Labs, Kumamoto, Japan.
17. Barreto, A., *Rev. Quim. Ind. (Rio de Janeiro)*, 15, 16, 1946.
18. Tyner, E. H., *Anal. Chem.*, 20, 76, 1948.
19. Waterburg, G. R. and Bricher, C. E., *Anal. Chem.*, 29, 129, 1957.
20. Holzbecher, Z., Divis, L., Kral, M., Sucha, L., and Ulacil, F., *Hankok of Organic Reagents in Inorganic Analysis*, Ellis Horwood, Chichester, England, 1976.
21. Menis, O., USAEC Rep., ORNL-1626, 27, 1954.
22. Thamer, B. J. and Voigt, A. F., *J. Am. Chem. Soc.*, 73, 3197, 1951.
23. Floh, B., Abrao, A., and Federgruen, L., *Publ. IEA*, 427, 10, 1976; *Chem. Abstr.*, 87, 12658le, 1977.
24. Klipp, R. W. and Barney, J. E., Jr., *Anal. Chem.*, 31, 596, 1959.
25. Stoffyn, P. and Keane, W., *Anal. Chem.*, 36, 398, 1964.
26. Schafer, H. N. S., *Anal. Chem.*, 39, 1719, 1967.
27. Gales, M. E., Jr., Kaylor, W. H., and Longbottom, J. E., *Analyst*, 93, 97, 1968.
28. Srivastava, R. D., Van Buren, P. R., and Gesser, H., *Anal. Chem.*, 34, 209, 1962.
29. Borney, J. E., Jr. and Bertlacini, R. J., *Anal. Chem.*, 29, 1187, 1957.
30. Hammer, C. F. and Graig, J. H., *Anal. Chem.*, 42, 1588, 1970.
31. West, C. D., Birke, R. L., and Hume, D. N., *Anal. Chem.*, 40, 556, 1968.
32. Meditsch, J. de O. and Coimbra, L. A., *Rev. Quim. Ind. (Rio de Janeiro)*, 39, 14, 1970; *Chem. Abstr.*, 74, 60626w, 1971.
33. Yoshimura, T., Noguchi, T., Inove, K., and Hara, H., *Bunseki Kagaku*, 15, 918, 1966.
34. Poirier, J. M. and Verchere, J. F., *Talanta*, 26, 349, 1979.
35. Krizonova, M., Reis, A., and Hainberger, L., *Fresenius Z. Anal. Chem.*, 261, 400, 1972.

β-DIKETONES

Synonyms

β-Diketones which are important as analytical reagents are listed in Table 1, together with their synonyms, physical properties, and pKa values of enolic proton.

Source and Method of Synthesis

All commercially available. They are prepared by the Claisen condensation of corresponding alkyl methyl ketone with methyl or ethyl carboxylate in the presence of base.[M6,M8,R1,1-5]

Analytical Uses

Reagents such as (1)-(4), (7), (8), (10), and (11) are used as solvent extraction reagent for metals and, in some cases, as chromogenic reagent for transition elements. Many of the metal chelates with (4)-(7), (9), (10), and (11) are so volatile that they can be analyzed by gas chromatography or fractional sublimation. The rare earth chelates with (4), (5), (6), and (11) are used as "shift reagent" in proton magnetic resonance spectroscopy.

Properties of Reagents

Physical data and ionization constants of the enolic proton of β-diketone reagents are summarized in Table 1. The solubility data and distribution coefficients of β-diketones which are important in solvent extraction are shown in Table 2.

These reagents do not show any absorption band in visible region, but show strong absorption in UV region, the intensity and shape of which are dependent upon the nature of solvent because the ratio of keto to enol form in the tautomeric mixture is greatly

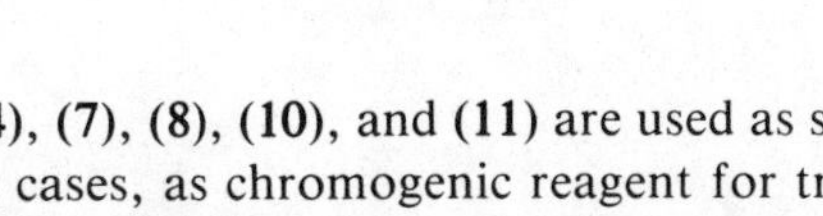

influenced by the solvent polarity. UV absorption spectra of AA (1), TFA (5), and HFA (6) in chloroform are shown in Figure 1.

Complexation Reactions and Properties of Complexes

The enol form of β-diketone, after deprotonation, behaves as a bidentate univalent anion, forming the metal chelate with the general structure as shown below.

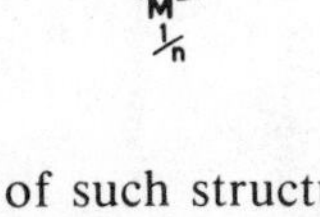

Most of the β-diketone chelates of such structure belong to the coordination-saturated uncharged chelates which are characterized by their relatively lower melting points, relatively higher vapor pressures, and good solubility in various organic solvents. The melting points, sublimation temperatures, and other physical properties of metal β-diketonates may be found in various references: for AA (1) chelates;[R2,2,15-24,33,39] for BzA (2) chelates;[R2,15,16,21,25] for DPM (4) chelates;[R2,19,21,22,25-28,30,38] for TAA (5) chelates;[R2,2,15,16,19,23,25,49] for HFA (6) chelates;[R2,2,19,21,22,31,34,40] for TTA (7) che-

Table 1
IMPORTANT β-DIKETONES AS ANALYTICAL REAGENTS

$$R_1 - \underset{\underset{O}{\|}}{C} - CH_2 - \underset{\underset{O}{\|}}{C} - R_2$$

Number	β-Diketone	R_1	R_2	Synonyms	Molecular formula	Mol wt	Physical properties	pKa of enol proton	Ref.
(1)	Acetylacetone	CH_3	CH_3	2,4-Pentanedione, AA, AcAc	$C_5H_8O_2$	100.12	Colorless liquid, bp 139° (760 Torr), mp −23.2° d = 0.976 (25°)	9.0, 25° $\mu \to 0$	R1, 6
(2)	Benzoylacetone	 ⬡	CH_3	1-Phenyl-1,3-butanedione, acetylbenzoylmethane, BzA, BA	$C_{10}H_{10}O_2$	162.19	Colorless crystal, mp 59∼60°	9.8, 20° $\mu \simeq 0.1$ 50% dioxane	M2, 7
(3)	Dibenzoylmethane	 ⬡	 ⬡	1,3-Diphenyl propane-1,3-dione DBM, DBzM	$C_{15}H_{12}O_2$	224.26	Colorless crystal, mp 76∼8°	13.75, 30°, 75% dioxane	8
(4)	Dipivaroylmethane	$C(CH_3)_3$	$C(CH_3)_3$	2,2,6,6-Tetramethyl-3,5-heptanedione, DPM, H Thd, TMTD	$C_{11}H_{20}O_2$	184.28	Colorless oil, bp 93∼94° (35 Torr)	11.75, $\mu = 0.1$	M12, 3
(5)	Trifluoroacetylacetone	CH_3	CF_3	1,1,1-Trifluoro-2,4-Pentanedione, TAA, TFA, ATA	$C_5H_5O_2F_3$	154.09	Colorless oil, bp 107°, d = 1.27, $n_D^{21} = 1.3893$	6.7, 20°, $\mu \simeq$ 0.1, 50% dioxane	7
(6)	Hexafluoroacetylacetone	CF_3	CF_3	1,1,1,5,5,5-Hexafluoro-2,4-pentanedione, HFA, Hhpd	$C_5H_2O_2F_6$	208.06	Colorless liquid, irritant odor, very easily solvated in donating solvent^a	4.42, 25°, $\mu = 0.1$, NaClO$_4$	9

	Name			Formula	MW	Properties		Ref.
(7)	Thenoyltrifluoro-acetone	(2-thienyl ring, S)	CF_3	 $C_8H_5O_2SF_3$	222.18	Colorless needles, bp 103 $\sim$ 104° (9 Torr), mp 42 $\sim$ 43.2°, exists as monohydrate in water	6.18, 25°, $\mu \to 0$	10
	4,4,4-Trifluoro-(2-thienyl)-1,3-butane-dione, TTA							
(8)	Furoyltrifluoro-acetone	(2-furyl ring, O)	CF_3	$C_8H_5O_3F_3$	206.12	Colorless oil, bp 89 $\sim$ 90° (14 Torr), mp 19 $\sim$ 21°	5.40, 20°, $\mu = 0.1$, 75% dioxane	11
	4,4,4-Trifluoro-(2-fu-ryl)-1,3-butanedione, FTA							
(9)	Pivaroyltrifluoro-acetone	$C(CH_3)_3$	CF_3	$C_8H_{11}O_2F_3$	196.17	Colorless oil, bp 68 $\sim$ 71° (6 Torr), d = 1.13, n_D^{23} = 1.4196	8.98	R2
	1,1,1-Trifluoro-5,5-di-methyl-2,4-hexane-dione, PTA, TPM, TAPM, Htdhd, Hfhd							
(10)	Benzoyltrifluoro-acetone	(phenyl ring)	CF_3	 $C_{10}H_7O_2F_3$	216.16	Colorless solid, bp 224°, mp 39 $\sim$ 40.5°	9.2, 30°, 75% dioxane	8
	4,4,4-Trifluoro-1-phenyl-1,3-butane-dione, BFA, BTA							
(11)	Heptafluoro-butanoylpivaroyl-methane	$C(CH_3)_3$	C_3F_7	$C_{10}H_{11}O_2F_7$	296.18	Colorless oil, but turns to dark-or-ange in a few days, bp 33° (2.7 Torr), irritant odor	6.7, 24°, $\mu = 0.1$ (tetra-methyl-ammonium perchlorate)	12
	1,1,1,2,2,3,3-Hepta-fluoro-7,7-dimethyl-4,6-octanedione, FOD, Hfod, HFM, HPM, Hhdod							

a In water, it forms dihydrate R-C(OH)₂CH₂C(OH)₂R (white solid); in methanol, it forms dimethanolate (white solid, mp 81 $\sim$ 81.5°); and in ethylenegly-col, it forms cyclic-monoadduct (white solid, mp 82 $\sim$ 83.5°).

Table 2

SOLUBILITY DATA AND DISTRIBUTION COEFFICIENTS OF β-DIKETONES[M9,M12,13]

β-Diketone	Solubility (g/100 ml)				log K_D		
	Water	Benzene	CHCl$_3$	CCl$_4$	Benzene/water	CHCl$_3$/water	CCl$_4$/water
(1) AA	17.3	Miscible	Miscible	Miscible	0.57, 0.74	1.21	0.50
(2) BzA	3.9×10^{-2}	63.3	40.5	30.8	3.14	3.44, 3.60	2.82
(3) DBM	1.3×10^{-4}	53.8	40.4	29.2	5.35	5.40	4.51
(4) DPM	1.7×10^{-2}	—	—	—	—	—	—
(5) TAA	—	—	—	—	0.28	0.53	0.32
(7) TTA	2×10^{-2}	—	—	—	1.62	1.73	1.30
(10) BFA	3×10^{-1}	—	—	—	—	—	—

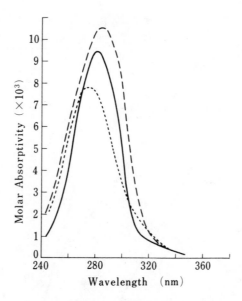

FIGURE 1. UV absorption spectra of AA (1), TAA (5), and HFA (6) in chloroform; AA, ———; TAA, ——-; HFA, · · · · · . (From Belford, R. L., Martell, A. E., and Calvin, M., *J. Inorg. Nucl. Chem.*, 2, 11, 1956. With permission.)

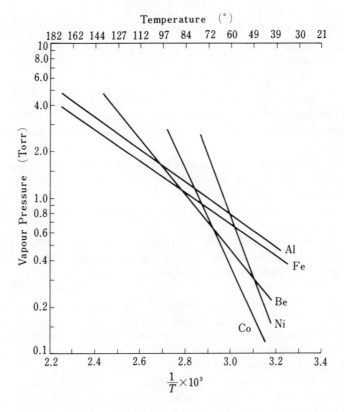

FIGURE 2. Vapor pressure - temperature curve for AA (1) metal chelates. (From Berg, E. W. and Truemper, J. T., *Anal. Chim. Acta*, 32, 245, 1965. With permission.)

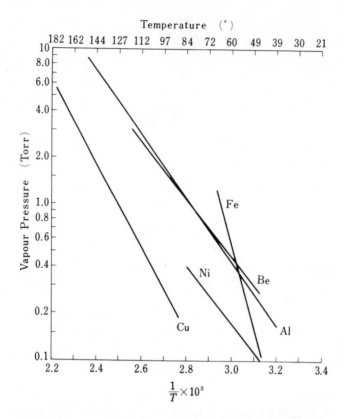

FIGURE 3.   Vapor pressure - temperature curve for TAA (5) metal chelates. (From Berg, E. W. and Truemper, J. T., *Anal. Chim. Acta*, 32, 245, 1965. With permission.)

lates;[R2,15,16,21,25] for FTA (8) chelates;[R2,15,16,21,25,39] for PTA (9) chelates;[22,25,29,33,35,36,38] for BFA (10) chelates;[R2,15,16,18,21] for FOD (11) chelates.[R2,5,22,28,33,37,38] The vapor pressure-temperature relationships of some of the volatile metal β-diketonates are illustrated in Figures 2 to 6.

The optimum pH range for the chelate formation is dependent upon the nature of metal ions and β-diketones. In principle, the ligand with lower pKa value complexes metal ion at lower pH region. The chelate stability constants of metal β-diketonate are observed in aqueous dioxane, and the representative values are summarized in Tables 3 and 4.

The metal β-diketonates are usually insoluble in water, but soluble in various organic solvents. Thus, β-diketones are widely accepted as solvent extraction reagents. The values of K_{ex} and $pH_{1/2}$ for selected β-diketones are tabulated in Table 5.

In the cases of uncharged, but coordination-unsaturated chelates, such as $NiL_2 \cdot 2H_2O$ or $CoL_2 \cdot 2H_2O$, the presence of coordinated water diminishes the distribution of the chelate into organic phase. However, the extractability and, in some cases, rate of extraction can be improved by the combined use of uncharged auxiliary ligand such as pyridine or 1,10-phenanthlorine or by the selection of coordinating solvents such as MIBK or butanol, instead of using inactive solvents.

As AA (1) is liquid, most metal ions can be extracted by themselves or by its solution in various organic solvents (benzene, carbon tetrachloride, or chloroform). However, the solubility of some uncharged metal acetylacetonates in the aqueous phase can not be overlooked, especially when undiluted acetylacetone is used as extractant, because the relatively high solubility of (1) in water increases the solubility of the chelates in water. So it is more advantageous to use its solution in organic solvents.

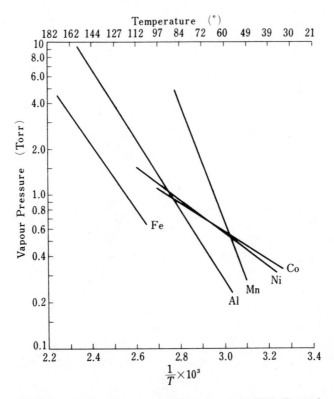

FIGURE 4.   Vapor pressure - temperature curve for TTA (7) metal chelates. (From Berg, E. W. and Truemper, J. T., *Anal. Chim. Acta,* 32, 245, 1965. With permission.)

Extraction equilibrium is reached more slowly with BzA (2), DBM (3), and TTA (7) than with AA (1). This tendency is more marked with TTA and especially so when the reagent anion concentration is low (in more acid solution or at low total reagent concentration). The rate is also dependent upon the kind of metals. However, (7) has been widely used in separations of the actinoid and lanthanoid elements. The values of log K_{ex} are Am, −7.46; Bk, −6.8; Cf, −7.1; Fm, −7.1; Pa(IV), 6.72; Pu(III), −4.44; Pu(IV), 6.85; Ce, −9.45; Dy, −7.03; Er, −7.2; Ho, −7.25; Lu, −6.77; Nd, −8.58; Pm, −8.05; Pr, −8.85; Sm, −7.68; Tb, −7.51; Tm, −6.96; and Yb, −6.72.[M3]

Solutions of metal β-diketonates give absorption bands in the UV or visible region, but their colors are usually not so strong that the higher photometric sensitivity can not be expected. The selectivity is also low unless proper masking agents are used. Absorption spectra of Cu(II) and uranyl chelates are illustrated in Figures 7 and 8, respectively, as examples.

Purification and Purity of Reagents

β-Diketones included in this section are all well-defined compounds and can generally be purified by fractional distillation at reduced pressure. The crude samples are purified through their metal chelate such as Cu(II) which can easily be prepared by mixing a solution of cupric acetate with an alcoholic solution of β-diketone. Copper chelate, after purification through recrystallization, is treated with aqueous mineral acid, and the liberating ligand is extracted with ether and then purified by distillation.

The purity of β-diketone can be checked most conveniently by observing their gas chromatograms or by the titration in nonaqueous solvent (in methanol with 0.1 *M* (Bu)₄NOH, using Crystal Violet indicator).[171]

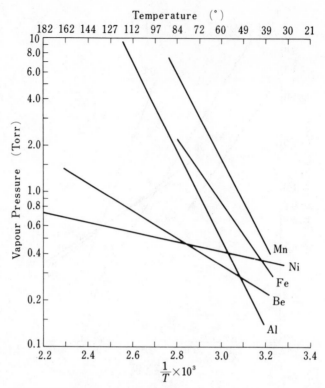

FIGURE 5.   Vapor pressure - temperature curve for BFA (10) metal chelates. (From Berg, E. W. and Truemper, J. T., *Anal. Chim. Acta,* 32, 245, 1965. With permission.)

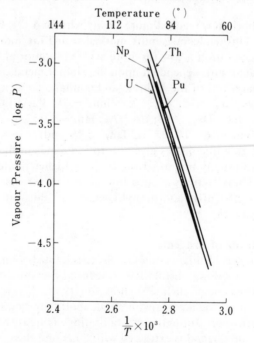

FIGURE 6.   Vapor pressure - temperature curve for FOD (11) metal chelates. (Reprinted with permission from Swain, H. A. and Karraker, D. G., *Inorg. Chem.*, 9, 1766, 1970. Copyright 1970 American Chemical Society.)

Table 3
CHELATE STABILITY CONSTANT OF METAL β-DIKETONATES[10]

Metal ion	AA (1)				BzA (2)				DBM (3)			
	log K_{ML}	log K_{ML_2}	log K_{ML_3}	Temp; dioxane, %	log K_{ML}	log K_{ML_2}	log K_{ML_3}	Temp; dioxane, %	log K_{ML}	log K_{ML_2}	log K_{ML_3}	Temp; dioxane, %
Ba	—	β_2 9.0	—	30; 75	—	β_2 9.4	—	30; 75	6.10	5.40	—	30; 75
Be	9.0	7.7	—	30; 50	12.02	11.36	—	30; 75	13.62	12.41	—	30; 75
Cd	7.79	6.75	—	30; 75	7.79	6.75	—	30; 75	8.67	7.96	—	30; 75
Ce(III)	7.86	8.53	6.91	30; 75	10.09	9.33	7.62	30; 75	10.99	10.54	8.85	30; 75
Co(II)	9.22	7.86	—	30; 75	9.74	8.28	—	30; 75	10.35	9.70	—	30; 75
Cu(II)	12.46	11.20	—	30; 75	9.42	8.41	—	30; 75	12.98	12.00	—	30; 75
Fe(II)	9.71	8.48	—	30; 75	—	—	—	—	11.15	10.35	—	30; 75
Mg	3.63	2.54	—	30; (→0)	7.84	6.20	—	30; 75	8.54	7.67	—	30; 75
Mn(II)	4.18	3.07	—	30; (→0)	8.66	7.12	—	30; 75	9.32	8.47	—	30; 75
Ni	5.92	4.46	2.11	30; (→0)	10.30	8.52	—	30; 75	10.83	9.89	—	30; 75
Pb	8.6	6.77	—	30; 75	8.84	.51	—	30; 75	9.75	9.04	—	30; 75
UO$_2^{2+}$	9.32	7.60	—	30; 50	12.15	11.12	—	—	10.83	9.89	—	30; 75
Y	7.73	6.00	4.77	—; 75	8.24	6.74	5.59	30; 75	—	—	—	30; 75
Zn	9.52	8.05	—	30; 75	9.62	8.28	—	30; 75	10.23	9.42	—	30; 75

Table 4

CHELATE STABILITY CONSTANT OF METAL β-DIKETONATES[10]

Metal ion	BFA (10) log K_{ML_2}	BFA (10) Temp; dioxane, %	Hfod (11) log K_{ML}	Hfod (11) log K_{ML_2}	Hfod (11) log K_{ML_3}	Hfod (11) Temp; dioxane, %	HTTA (7) log K_{ML}	HTTA (7) log K_{ML_2}	HTTA (7) Temp; dioxane, %	Ref.
Ba	β_2 15.4	30; 75	—	—	—	—	—	β_2 10.6	30, 75	
Cd	β_2 7.6	—; H₂O Corr	—	—	—	—	—	7.09	—	
Co(II)	β_2 10.50	—; H₂O Corr	—	—	—	—	7.81	—	30, 75	41
Cu(II)	β_2 18.8	30; 75	—	—	—	—	8.23	8.08	30, 75	41
Eu(III)	—	—	11.9	β_2 18.4	β_3 24.2	24, —[a]	6.9[b]	—	—	
Fe(II)	β_2 10.32	—; H₂O Corr	—	—	—	—	—	—	25; (0.1)	
Gd(III)	—	—	12.7	β_2 18.7	β_3 25.3	24; —[a]	—	—	—	
Mg	β_2 7.52	—; H₂O Corr	—	—	—	—	—	—	—	
Mn(II)	β_2 8.20	—; H₂O Corr	—	—	—	—	—	—	—	
Ni	β_2 15.4	30; 75	—	—	—	—	7.93	7.33	30, 75	41
Pr(III)	—	—	—	β_2 18.0	β_3 24.0	24, —[a]	—	—	—	
Pu(IV)	—	—	—	—	—	—	8.0	—	25, (→ 0)	
Sm(III)	—	—	—	β_2 18.4	β_3 24.2	24, —[a]	—	—	—	
Tb(III)	—	—	13.4	β_2 19.4	β_3 24.9	24, —[a]	8.1	—	25, (→ 0)	
Th	—	—	—	—	—	—	8.71	—	—	
U(IV)	—	—	—	—	—	—	7.75	7.92	30, 75	41
Zn	β_2 7.30	—; H₂O Corr	—	—	—	—	—	6.39	30, 75	41

[a] $\mu = 0.1$ ((C₂H₅)₄N Cl).

[b] Value for Fe(III).

[c] Chelate stability constants of the rest of β-diketone are[10] TAA(5), log β_2 for Ba(8.0 in 75% dioxane), Cu(II) (12.2 in 50% dioxane), and Ni (14.2); HFA (6), log β_2 for Ba(8.0 in 75% dioxane), Cu(6.4, $\mu = 0.1$), and log K_{ML} for Zn(1.6, $\mu = 0.1$); FTA (8), log β_2 for Co(II) (11.7 in 75% dioxane), Cu(II) (9.38, $\mu = 0.1$),[b] Ni(8.60 in 75% dioxane), and Zn(5.28, $\mu = 0.1$).[9]

Table 5
EXTRACTION EQUILIBRIUM DATA FOR SOME METAL
β-DIKETONATES IN BENZENE[M3]

Metal ion	AA (1)		BzA (2)		DBM (3)		TTA (7)	
	log K_{ex}	pH$_{1/2}$[a]	log K_{ex}	pH$_{1/2}$[a]	log K_{ex}	pH$_{1/2}$[a]	log K_{ex}	pH$_{1/2}$[a]
Ag	No	No	−7.81	8.9	−8.58	9.9	—	—
Al	−6.48	3.30	−7.60	3.60	−8.92	4.00	−5.23	2.48
Ba	No	No	No	No	—	12	−14.4	8.0
Be	−2.79	2.45	−3.88	2.94	−3.46	2.33	−3.2	2.33
Bi	No	No	—	9.2	—	10.5	−3.3[M12]	1.8
Ca	No	No	−18.3	10.1	−18.0	9.9	−12.0	6.7
Cd	No	No	−14.1	8.1	−14.0	—	−9.43	3.38
Co(II)	—	—	−11.1	6.60	−10.8	6.40	−6.7	4.1
Cu(II)	−3.93	2.90	−4.17	3.00	−3.80	2.90	−1.32	1.38
Eu	—	—	−18.9	7.3	—	—	−7.66	3.29
Fe(III)	−1.39	1.60	−0.50	1.20	−1.93	1.70	3.3	−0.24
Ga	−5.51	2.90	−6.34	3.10	5.76	2.90	−7.57	3.26
Hf	—	4	—	—	—	—	7.8	−1.5
Hg(II)	—	—	—	3.7	—	3.9	—	—
In	−7.20	3.95	−9.30	4.10	−7.61	3.60	−4.34	2.20
La	—	—	−20.5	7.96	−19.5	7.42	−10.5	4.24
Mg	—	9.4	−16.7	9.38	−14.7	8.5	—	—
Mn	—	—	−14.63	8.30	−13.7	7.80	−1	—
Ni	—	—	−1.21	6.9	−11.0	6.4	—	5
Pb	−10.15	6.2	−9.61	5.6	−9.45	5.6	−5.2	3.34
Pd(II)	<−2	<0	1.2	0.4	—	1.8	—	2.8[85]
Sc	−5.83	2.95	−5.99	3.10	−6.04	3.05	−0.77	0.99
Sr	—	—	−20	11.5	−20.9	11.1	−14.1	7.8
Th	−12.16	4.10	−7.68	2.9	−6.38	2.60	0.8	0.48
Ti(IV)	—	—	—	2.4	—	2.5	—	—
Tl(I)	—	—	—	—	—	—	−5.2	5.9
Tl(III)	—	1.3	—	4.0	—	3.8	−5.7[M12]	2.78
U(IV)	5.2	2.7	—	—	—	—	5.3	0.58
U(VI)[b]	—	—	−4.68	3.82	−4.12	3.56	−2.26	1.79
Y	—	—	−17.0	6.86	—	—	−7.39	3.2
Zn	−10.69[c,9]	—	−10.8	6.50	−10.7	6.4	—	—
Zr	—	3.4	—	3.4	—	3.3	9.15	−1.5

Note: No, not extracted.

[a] $[HL]_{org} = 0.10$ M.
[b] Extracted species is $UO_2L_2 \cdot HL$.
[c] Carbontetrachloride.
[d] Methyl propylketone.

Analytical Applications
Use as Extraction Reagent[M1,M4,M10,M14,R7]

The possibilities of separation of metals by solvent extraction with β-diketones can be roughly evaluated from Table 5. However, the extraction condition on the individual cases may further be influenced by many factors, such as the hydrolytic tendency of metal ion, rate of extraction, the concentration of reagent, etc.

Extractability of some representative metal ions with acetylacetone (1) as a function of pH is illustrated in Figures 9 and 10. Similar curves with thenoyltrifluoroacetone (7) are shown in Figure 11.

Use as Photometric Reagent
As discussed previously, β-diketones are generally not so sensitive and selective

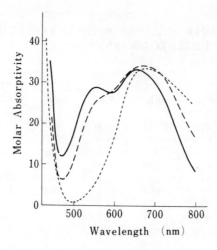

FIGURE 7. Absorption spectra of copper(II) β-diketonates; ——, *bis* (acetylacetonato) Cu(II); ---, *bis* (trifluoroacetylacetonato) Cu(II); ·····, *bis* (hexafluoroacetylacetonato) Cu(II); 0.01 M in chloroform. (From Belford, R. L., Martell, A. E., and Calvin, M., *J. Inorg. Nucl. Chem.*, 2, 11, 1956. With permission.)

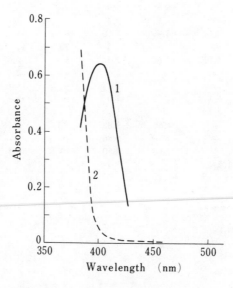

FIGURE 8. Absorption spectra of *bis* (dibenzoylmetanato) UO₂. (1) UO$_2^{2+}$, 4 × 10⁻⁵ M; ligand, 2 × 10⁻² M, pH 6.0; (2) reagent blank. (From Shigematsu, T., Tabushi, M., Matsui, M., and Munakata, M., *Bull. Chem. Soc. Jpn.*, 41, 1610, 1968. With permission.)

chromogenic reagent for metals. However, in some cases, they are used as extraction photometric reagents, and such examples are summarized in Table 6. Thenoyltrifluoroacetone (**7**) has been most widely used for such purpose.[173] Some rare earth elements are also determined by fluorimetry, such examples being summarized in Table 7.

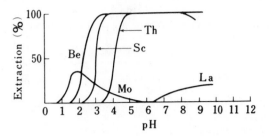

FIGURE 9. Extractability of Be, La, Mo(VI), Sc, and Th by 0.1 M acetylacetone in benzene as a function of pH. (From Stary, J. and Hladky, E., *Anal. Chim. Acta*, 28, 227, 1963. With permission.)

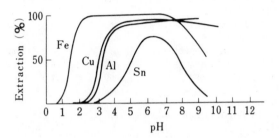

FIGURE 10. Extractability of Al, Cu(II), Fe(III), and Sn(II) by 0.1 M acetylacetone in benzene as a function of pH. (From Stary, J. and Eladky, E., *Anal. Chim. Acta*, 28, 227, 1963. With permission.)

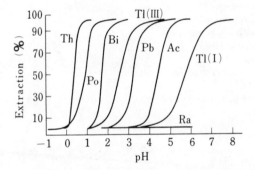

FIGURE 11. Extractability of Ac, Bi, Pb, Po, Ra, Th, and Tl(I)(III) with 0.25 M TTA(7) in benzene as a function of pH. (Reprinted with permission from Hagemann, F., *J. Am. Chem. Soc.*, 72, 768, 1950. Copyright 1980 American Chemical Society.)

Use as Vaporizing Reagent for Metal Ions

Some of metal β-diketonates are so volatile that they can be separated by gas-liquid partition chromatography or by fractional sublimation. The elements that have been investigated by gas chromatography are Al, Be, Co, Cr, Fe, Hf, In, Mg, Mn, Ni, Pb, Rh, Sc, Th, Ti, Tl, U, V, Y, Zr, and rare earths. Some examples are summarized in Table 8. In general practice, the metal chelates are dissolved in volatile organic solvent (0.5 to 5%), such as benzene or chloroform, and 0.1 to several microliters of the solution are injected. The temperature of injection port and detector is maintained at 20 to 40° higher than the column temperature. The detector sensitivity is mostly dependent upon the nature of central metal ion, rather than the ligand.[100]

Table 6
APPLICATION OF β-DIKETONE AS PHOTOMETRIC REAGENTS

β-Diketone	Metal ion[a]	Condition	Metal chelate			Range of determination (ppm)	Ref.
			Ratio	λ_{max} (nm)	$\varepsilon(\times 10^3)$		
(1)	Be	pH 7 $\curvearrowright$ 8, EDTA, citrate, extraction with CHCl$_3$	ML$_2$	295	3.6	0.03 $\curvearrowright$ 0.3	45, 46
	Fe(III)	pH 6 $\curvearrowright$ 8, extraction with butylacetate	ML$_2$	438	—	0.5 $\curvearrowright$ 10	47
	Mo(VI)	6 NH$_2$SO$_4$, extraction with AA + CHCl$_3$ (1:1)	MO$_2$L$_2$	352	1.63	—	48
	UO$_2$$^{2+}$	pH 6.5 $\curvearrowright$ 7.0, extraction with butylacetate	MO$_2$L$_2$	365	—	$\curvearrowright$70	49
	V(III)	pH 2, extraction with AA + CHCl$_3$ (1:1)	ML$_3$	383	0.9	—	50
	V(V)	pH 2 $\curvearrowright$ 7, H$_2$O$_2$, extraction with AA + CHCl$_3$ (1:1)	—	460	0.6	—	50
(2)	Ti(III)	$\simeq$0.05 MHCl, 50% ethanol	—	615	3.0	0.4 $\curvearrowright$ 8	51
	Ti(IV)	$\simeq$0.05 MHCl, 50% ethanol	—	382	8.5	0.4 $\curvearrowright$ 8	51
	UO$_2$$^{2+}$	Extraction with benzene	—	400	2.0	—	M12
(3)	Fe	pH 6 $\curvearrowright$ 7, TOPO, extraction with benzene	—	408	12.0	—	52
	UO$_2$$^{2+}$	HNO$_3$, EDTA, extraction with CCl$_4$	—	410	—	1 $\curvearrowright$ 10	53
	UO$_2$$^{2+}$	pH 6.0 Ca-EDTA, extraction with butylacetate	—	400	16.1	0 $\curvearrowright$ 10	42
	UO$_2$$^{2+}$	Extraction with 10% TBP in iso-octane from 15% solution of Al(NO$_3$)$_3$	—	400 $\curvearrowright$ 415	—	0.5 $\curvearrowright$ 5	M4, 54
	UO$_2$$^{2+}$	Extraction with tetrapropylammonium hydroxide in MIBK from medium of HNO$_3$ + Al(NO$_3$)$_3$	—	400 $\curvearrowright$ 415	—	—	55
(6)	Fe(III)	pH 1.5 $\curvearrowright$ 2.5, extraction with CCl$_4$	ML$_3$	455	—	0.8 $\curvearrowright$ 10	56
(7)	Ce(IV)	1 NH$_2$SO$_4$, extraction with xylene	ML$_3$	440	9.3	$\curvearrowright$10	59, 60
	Cu(II)	pH 3.5 $\curvearrowright$ 4, extraction with benzene	ML$_2$	430	2.2	5 $\curvearrowright$ 80	65, 66
	Cu(II)	pH 5.5, pyridine, extraction with cyclohexane	ML$_2$X$_2$	340	27.6	0.1 $\curvearrowright$ 1	67
	Fe(III)	pH 2 $\curvearrowright$ 3, extraction with benzene	ML$_3$	460	4.9	1 $\curvearrowright$ 10	70
	Fe(III)	2 MHNO$_3$ + 9 M NH$_4$NO$_3$, extraction with xylene	ML$_3$	510	4.9	0 $\curvearrowright$ 5	71
	Fe(III)	pH 1, Aliquat® 336S, extraction with benzene	ML$_3$X$_n$	500	4.64	0 $\curvearrowright$ 12	72
	Ga	pH 4.5 $\curvearrowright$ 6.0, Rhodamine B, extraction with xylene	—	565	1.07	2 $\curvearrowright$ 60	73
	Ga	pH 2.4 $\curvearrowright$ 2.6, TPAC, extraction with CHCl$_3$	ML$_2$X	335 or 380	—	—	74
	Ni	pH 5.5 $\curvearrowright$ 8.0, extraction with acetone + benzene	ML$_2$	420	0.6	7.2 $\curvearrowright$ 64	80
	Ni	pH 5 $\curvearrowright$ 9, Zephiramine®, extraction with 1,2-dichloroethane	ML$_3$X	420	—	—	63

Table 6 (continued)
APPLICATION OF β-DIKETONE AS PHOTOMETRIC REAGENTS

β-Diketone	Metal ion[a]	Condition	Metal chelate Ratio	λ_{max} (nm)	$\varepsilon (\times 10^3)$	Range of determination (ppm)	Ref.
	Pd(II)	pH 4.5 ∿ 8.8, extraction with n-BuOH	ML$_2$	410	1.2	2 ∿ 20	84
	Pd(II)	pH 0.8 ∿ 6.2, extraction with methylpropylketone	ML$_2$	430	42.0	0.6 ∿ 6	85
	Pt(II)	6 NHCl, extraction with n-BuOH	ML$_2$	410	7.9	8 ∿ 60	84
	Rh(III)	pH 5.2 ∿ 6.2, extraction with xylene + acetone (1:1)	ML$_3$	430	39.5	2 ∿ 20	87
	UO$_2^{2+}$	pH 3.5 ∿ 8.0, extraction with benzene	MO$_2$L$_2$	430	2.0	0 ∿ 85	90
	UO$_2^{2+}$	Extraction with 0.05 M TBP in toluene	MO$_2$L$_2$X	395	19	0 ∿ 8	91
	V(V)	pH 2 ∿ 3, extraction with n-BuOH	—	420	3.6	0.8 ∿ 11	92
	V(V)	pH 2 ∿ 3, extraction with butylacetate	—	410	4.5	0 ∿ 10	93

[a] Other metals determined by the photometry with TTA (7) include Au(III),[57] Ca,[58] Co(II)[61] and (III),[62,63] Cr(III),[64] Er,[68,69] Eu,[69] Ho,[65,75] Ir,[76] Mn(II),[77,78] Nb(V),[79] Nd,[68,75] Np(IV),[81] Os(III),[82] Pa(V),[83] Pr(III),[68] Pu(IV),[81] Re(VII),[86] Ru,[88] Sm,[68] Sr,[58] Th(IV),[81] and Ti(IV).[89]

The separation of metal chelates by fractional sublimation[119] or zone melting[120] has been investigated in detail.

Use as NMR Shift Reagent

Some of the lanthanide β-diketonates are known as "NMR shift reagent" which shifts the proton NMR signals of various organic compounds having polar functional groups. The shift is induced by the pseudo contact term due to the interaction between the organic molecule and the coordination unsaturated site of paramagnetic lanthanide ion of the chelate.

The representative shift reagents are summarized in Table 9. Europium chelates are known to shift the proton signal toward the lower magnetic field, and Pr chelates toward the higher magnetic field.[123] These reagents are often very hygroscopic, and once hydrated, they become very difficultly soluble in organic solvents. Therefore, it is advisable to keep the reagents in a dessicator over P$_2$O$_5$.

The requirements for the ligands to be a successful shift reagent are

1. The chelate must have a high adduct formation constant with the observing organic molecule.
2. The shift reagent itself must not give the proton signal in the observing range.
3. The shift reagent must have enough solubility in the organic solvents for NMR measurement.

A great number of papers have been published on the use and the theory of NMR shift reagents, and the details should be consulted with original articles or reviews.[M11,R4]

Other Reagents with Related Structure
Polyfluorinated β-Diketones

Encouraged by the successful separation of metals through their chelates with TTA

Table 7

APPLICATION OF β-DIKETONE AS FLUORIMETRIC REAGENT

β-Diketone	Metal ion	Condition	Metal chelate		Range of determination	Remarks	Ref.
			λ_{exit} (nm)	λ_{emit} (nm)			
(1)	Tb	pH 10 ∼ 11.5, EDTA(X)	Hg lamp	546	16 ppb ∼ 1.6 ppm	MLX	97, 98
(6)	Eu	95% ethanol	312 or 430	614	4 ppb	Pr and Sm interfere	94
	Eu	pH 3, TOPO, extraction with methylcyclohexane	360	615	0.2 ppb ∼ 15 ppm	—	95
	Sm	In ethanol	312	563	30 ppb	Eu and Tb (30-fold) do not interfere	96
	Sm	pH 3, TOPO, extraction with methylcyclohexane	350	565	1.5 ppb ∼ 15 ppm	—	95
	Tb	95% ethanol	312	544	4 ppb	Dy, Er, and Yb interfere	94
	Tb	pH 3, TOPO, extraction with methylcyclohexane	350	550	1.6 ppb ∼ 16 ppm	—	95
(7)	Eu	pH 8.2 ∼ 8.3, EDTA-OH	—	613	0.05 ∼ 76 ppb	Other lanthanoids do not interfere; Al, Be, Cd, Sn, Te, and Th do	97
	Eu	pH 3.6, TOPO, Triton X-100	352	613	1.5 ∼ 760 ppb	Sc and Th(50-fold) do not interfere	170
	Sm	pH 3.6, TOPO, Triton X-100	372	561	1.5 ppb ∼ 1.5 ppm	Sc and Th (50-fold) do not interfere	170

Table 8
GAS CHROMATOGRAPHIC SEPARATION OF METAL β-DIKETONATES

β-Diketone	Metal chelate to be separated (increasing order of retention time)	Column condition		
		Packing material	Temp (°)	Ref.
(1)	Be, Al, Cr	0.5% SE-30/glass beads	170	99
	Sc, Ir, Pd(II), Pt(II)	10% Epon®-1001/Chromosorb W	115	101
(4)	Sc, Yb, Ho, Eu	2% Apiezon-H/Gas-pack F	157	26
	Rare earths (increasing order of ionic radii)	10% Apiezon-N/Gas-pack F	185	26
	Be, Al, Cr, Fe, Ni	0.5% Apiezon-L/Flusin GH	200	169
	Zn, Co(III), Ga, In	0.5% Apiezon-L/Flusin GH	200	169
	Be, Al, Sc, Lu, Eu, Tb, Sm	0.5% Apiezon-L/Flusin GH	200	169
(5)	Be, Al, Cr, Rh	0.5% Daifloil 200/Chromosorb® W	120 ∿ 50	102
	Al, Cr, Rh, Zr	0.5% silicon grease/glass beads or borosilicate glass	135	103
	Be, Al, Ga or Al, Ga, In	0.5% DC silicone 710/glass beads	115 ∿ 20	104
	Ga, Fe, Cu	0.5% Daifloil 200/Chromosorb® W	140	102
	Co(III), Ru(III)	3.8% SE-30/Diatport-S	110	105
	Th, U	Silicon DC-550/Gaschrom CLH	170	106
	trans-RhL₃, *cis*-RhL₃	5% DC silicone/Chromosorb® W	105	103
	trans-CrL₃, *cis*-CrL₃	1% QF-1/Chromosorb® W	130	102
(6)	Be, Al, Cr	7.5% SE-30/firebrick pellet	90	107
	Co(III), Rh(III)	30% SE-30/Chromosorb® W	30 → 100	108
	Zn, Co(II), Ni (as mixed ligand complexes with di-*n*-butylsulfoxide)	5% Dexsil® 300, SE-30 or OV-1/Chromosorb® W-HP	170	109
	Fe(III), Cu(II)	10% SE-52/Chromosorb® W	90	110
(7)	Al	20% Apiezon-L	240 ∿ 50	111
(9)	Al, Cr(III), Fe(III)	15% Apiezon-L/Universal® B	162	112
	Be, Al, Fe(III), In, Th	0.5% XE-60/glass beads	120 ∿ 50	113
	In (III), CO (II), Th	0.5% SE-30/glass beads	170	113
	Ca	25% E-301 silicone rubber/Universal® B	230	29
(9)	Zn (as mixed ligand complexes with TBP or TOPO)	3% OB-101/Chromosorb® W	150 ∿ 240	114
	Sc, Lu, Er, Dy, Sm	5% Silicone grease/Chromosorb® W	150 ∿ 260	35
(11)	Rare earths (increasing order of ionic radii)	10% SE-30/Chromosorb® W or 0.1% Apiezon L/glass beads	171	115
	U(IV), Th(VI)	0.3% DC-QF 1/glass beads	170 ∿ 200	116
	Zr(or Hf)	OV-3, 101 or 17/Chromosorb® W		117

(5) or HFA (6), various kinds of fluorinated β-diketones have been prepared and evaluated as the reagents for gas chromatography of metal chelates. However, very few advantages can be found in their practical applications if one consider the difficulties in the syntheses.

1,1,1,2,2-	Pentafluoro-6,6-dimethyl-3,5-heptanedione[127]
1,1,1,2,2,3,3-	Heptafluoro-7,7-dimethyl-4,6-octanedione[127]
1,1,1,2,2,6,6,6-	Octafluoro-3,5-hexanedione[M11,127]
1,1,1,2,2,3,3,7,7,7-	Decafluoro-4,6-heptanedione[M11,127]
1,1,1,2,2,6,6,7,7,7-	Decafluoro-3,5-heptanedione[128,129]
1,1,1,2,2,3,3,7,7,8,8,8-	Dodecafluoro-4,6-octanedione[130,131]

Table 9
REPRESENTATIVE NMR SHIFT REAGENTS

β-Dike-tone	Shift reagent	Abbreviated name	Formula, mol wt	Physical properties	Ref.
(4)	Tris(dipivaloylmethanato)-europium(III)	Eu-DPM, Eu-THD	$C_{33}H_{57}O_6Eu$, 701.77	Pale yellow to yellow-brown powder, hygroscopic, mp 187 ∿ 9°	124
	Tris(dipivaroylmethanato)-praseodymium(III)	Pr-DPM, Pr-THD	$C_{33}H_{57}O_6Pr$, 690.72	Pale yellow to yellow-brown powder, hygroscopic, mp 219 ∿ 21°	123
(11)	Tris(heptafluorobutanoyl-pivaroylmethanato)-euro-pium(III)	Eu-FOD	$C_{30}H_{30}O_6F_{21}Eu$, 1037.49	Pale yellow to yellow-brown powder, hygroscopic, mp 100 ∿ 200°, decomp. with acid	5, 122
	Tris(heptafluorobutanoyl-pivaroylmethanato)-praseodymium(III)	Pr-FOD	$C_{30}H_{30}O_6F_{21}Pr$, 1026.44	Pale yellow to yellow-brown powder, hygroscopic, mp 180 ∿ 221°, decomp. with acid	125
TFMC[a]	Tris[3-(trifluoromethyl hy-droxymethylene)-*d*-camphor-ato)europium (III)	Eu-TFMC	$C_{36}H_{42}O_6F_9Eu$, 893.67	Pale yellow to ∿ yellow-brown powder, hygroscopic, No sharp mp	126
TFMC[a]	Tris[3-(trifluoromethyl hy-droxymethylene)-*d*-camphor-ato) praseodymium(III)	Pr-TFMC	$C_{36}H_{42}O_6F_9Pr$, 882.62	Pale yellow to ∿ yellow-brown powder, hygroscopic, No sharp mp	126

[a]   3-(Trifluoromethylhydroxymethylene)-*d*-camphor.

1-Phenyl-3-methyl-4-benzoyl-5-pyrazolone (PMBP)

colorless   yellow-green

This reagent (9mp 92°) has been investigated extensively as an extraction reagent mostly by Russian chemists. It is claimed that PMBP has advantages over TTA (5): being more easily synthesized[134] and being more stable; almost insoluble in water ($\sim 10^{-4}$ M at pH 2 to 4), but easily soluble in chloroform (1.78 M), 1,2-dichloroethane (1.43 M), and heptane (0.03 M);[135] pKa (enol OH) = 4.04 (μ = 0.01, 20°);[135] and log D (benzene/water) = 3.66, log D (chloroform/water) = 4.0 (μ = 0.1), log β_2 (ZnL$_2$) = 6.56, and log D (benzene/water) for ZnL$_2$ = 3.57.[172] Extraction ratio of some metal ions with PMBP as a function of pH is illustrated in Figure 12. The optimum conditions for the extraction of metals with PMBP are summarized in Table 10.

The sulfur analogue of PMBP has been investigated recently.[137,172]

β-Isopropyltropolone (IPT)

$C_{10}H_{12}O_2$
mol wt = 164.20

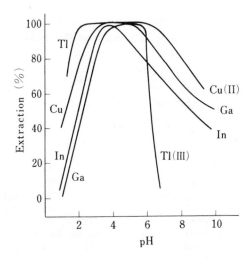

FIGURE 12. Extractability of Cu(II), Ga, In, and Tl(III) with 0.03 M PMBP in xylene as a function of pH. (From Mirza, M. Y., *Talanta*, 25, 685, 1978. With permission.)

In contrast to the parent compound (tropolone), this reagent is more stable, more easily synthesized, and has more favorable extraction properties:[165,166] pKa = 7.04, log D (CHCl$_3$/H$_2$O) = 3.37 (μ = 0.1, 25°).

2,2′-Bipyridine or triphenylphosphineoxide have been used as a synergistic reagent on IPT extraction of Tb[169] or Ga,[168] respectively.

MONOGRAPHS

M1. **Morrison, G. H. and Freiser, H.,** *Solvent Extraction in Analytical Chemistry,* John Wiley & Sons, New York, 1957.

M2. Nippon Bunseki Kagakukai, Ed., *Shin Bunseki Kagaku Koza, Separation Methods by Organic Reagents,* Vol. 5, Kyoritsu, Tokyo, 1959.

M3. **Stary, J.,** *The Solvent Extraction of Metal Chelates,* Pergamon Press, New York, 1964.

M4. **Ainsworth, L. R. and Harrap, K. R.,** 1, 3-Diphenyl propan-1,3-dione, in *Organic Reagents for Metals,* Vol. 2, Johnson, W. C., Ed., Hopkin & Williams, Essex, England, 1964, 75.

M5. **Moshier, R. W. and Sievers, R. E.,** *Gas Chromatography of Metal Chelates,* Pergamon Press, New York, 1964.

M6. **Arakawa, M. and Tanikawa, K.,** *Gas Chromatography of Organometal Compounds,* Gihodo, Tokyo, 1969.

M7. **Guiochon, G. and Pommier, C.,** *Gas Chromatography in Inorganic and Organometallics,* Ann Arbor Science, Ann Arbor, Mich., 1973.

M8. **Welcher, F. J.,** Acetylacetone and Benzoylacetone, in *Organic Analytical Reagents,* Vol. 1, Van Nostrand, New York, 1947, 404, 415.

M9. **Zolotov, Yu. A.,** *Solvent Extraction of Chelate Compounds,* Humphrey Science, Ann Arbor, Mich., 1970.

M10. **Tanaka, M.,** *Chemistry of Solvent Extraction,* Kyoritsu, Tokyo, 1977.

M11. **Sievers, R. E.,** *NMR Shift Reagents,* Academic Press, New York 1973.

M12. **Sandell, E. B. and Onishi, H.,** *Photometric Determination of Traces of Metals,* 4th ed., John Wiley & Sons, New York, 1978.

M13. **Mehrotra, R. C., Bohra, R., and Gaur, D. P.,** *Metal β-Diketonates and Allied Derivatives,* Academic Press, New York, 1977.

M14. **Sekine, T. and Hasegawa, Y.,** *Solvent Extraction Chemistry,* Marcel Dekker, New York, 1977.

Table 10
EXTRACTION CONDITION OF METAL WITH PMBP

Metal ion[a]	Condition	Extraction solvent	Remarks	Ref.
Be	1 ∿ 11 M LiCl	Benzene, cyclohexane, CHCl$_3$, CCl$_4$, or ethylacetate	—	142
Bi	0.1 N HNO$_3$, TBP, or TOPO	Benzene or cyclohexane	—	143
Ca	pH > 7	Benzene or CHCl$_3$	—	132
Cd	pH > 8, TBP	Benzene	ML$_2$, quantitative extraction	144
Ce(III)	pH 3 ∿ 4	CHCl$_3$	ML$_3$	145
Co(II)	pH > 3.5, TBP	CHCl$_3$ and iso-amylalcohol	Quantitative extraction	147
Cr(III)	pH 3 ∿ 6, 70 ∿ 100°	CHCl$_3$	Partial extraction	148
Cu(II)	pH 2 ∿ 10, TBP	Benzene, iso-AmOH, or MIBK	ML$_2$, quantitative extraction	149
Eu	pH 1 ∿ 2	Iso-amyl-, n-butyl-, or benzylalcohol	Quantitative extraction	146
Eu	0.1 ∿ 0.3 N HNO$_3$, TBP, or TOPO	Cyclohexane + CHCl$_3$	ML$_3$(TBP)$_2$ or ML$_3$(TOPO)$_2$	139, 151
Fe(III)	1 ∿ 2 N HCl, 1 ∿ 9 N H$_2$SO$_4$, or 1 ∿ 6.5 N HNO$_3$	Benzene, 1,2-dichloroethane or CHCl$_3$	ML$_3$, (H$_2$L$^+$·FeCl$_4$ in higher conc HCl solution)	152
Ga	pH 3.5 ∿ 5.0	Xylene	Can be separated from In and Tl	136
Ga	TBP, piperidine, or 1,10-phenanthroline	Cyclohexane, CHCl$_3$, or CCl$_4$	—	150
Hf	2 M HClO$_4$, alcohol or propane-1,2-diol	Benzene, toluene, CHCl$_3$, or CCl$_4$	ML$_4$	153
Nb	6 N HCl, SCN$^-$, or catechol	Benzene	λ_{max} = 410 nm, $\varepsilon \simeq 10^4$	156
Nd	pH 3 ∿ 9	Benzene, CHCl$_3$, or CCl$_4$	ML$_3$, log K$_{ex}$ = 5.63	157
Nd	TBP, piperidine, or 1,10-phenanthroline	Cyclohexane, CHCl$_3$, or CCl$_4$		150
Ni	pH 4 ∿ 9, ClO$_4^-$	CHCl$_3$	ML$_2$, quantitative extraction	158
Sr		O-containing solvents	Separated from large amount of Ca	161
Th	0.25 N HCl or HNO$_3$	Benzene	Extracted > 98%, separation from Eu	162
Ti	pH 1.5 ∿ 5	CHCl$_3$		147
Zn	TBP or TOPO	Iso-amylalcohol + benzene	ML$_2$X	164

[a]   Extractions of the following metals have also been investigated using PMBP as an extraction reagent: Am(III),[138-140] (V),[141] Cf,[143] Cm(III),[146] Er,[145,150] Gd,[145] In,[154] Mn,[155] Mo(VI),[147] Pa,[159] Pb,[160] Pu,[133,152,160] Sc,[161] Tl,[136] V(IV)(V).[163]

REVIEWS

R1. Kuroya, H. and Kitagawa, T., Acetylacetone and its metal chelates, *Kagaku (Kyoto)*, 16, 126, 1961.

R2. Komarov, V. A., β-Diketones and their use in analytical chemistry. Gas-chromatographic determination of metals, *J. Anal. Chem. USSR*, 31, 309, 1976.

R3. Uden, P. C. and Henderson, E. E., Determination of metals by gas chromatography of metal complexes, *Analyst*, 102, 889, 1977.

R4. Yamasaki, A., NMR shift reagents, *Kagaku (Kyoto)*, 29, 349, 435, 1974.

R5. Miyamoto, M., Gas chromatography of metal chelates, *Bunseki*, 485, 1978.

R6. Fujinaga, T. and Kuwamoto, T., Gas chromatography of metal chelates — especially on the separation of rare earth elements by use of the ligand vapour gas chromatography, in *Recent Progress of Gas Chromatography*, Kagaku no Ryoiki, extra issue No. 120, Nankodo, Tokyo, 1978.

R7. Akaiwa, H. and Kawamoto, H., Extraction of chelates, *Bunseki*, 374, 1976.

REFERENCES

1. Sprague, J. M., Beckham, L. J., and Adkins, H., *J. Am. Chem. Soc.*, 56, 2665, 1934.
2. Reid, J. C. and Calvin, M., *J. Am. Chem. Soc.*, 72, 2948, 1950.
3. Man, E. H., Swamer, F. W., and Hause, C. R., *J. Am. Chem. Soc.*, 73, 901, 1951.
4. Park, J. D., Brown, H. A., and Lacker, J. R., *J. Am. Chem. Soc.*, 75, 4753, 1953.
5. Springer, C. S., Jr., Meek, D. W., and Sievers, R. E., *Inorg. Chem.*, 6, 1105, 1967.
6. Badoz-Lambling, M. J., *Ann. Chim. (Paris)*, 8, 586, 1953; *Chem. Abstr.*, 48, 10474c, 1954.
7. Calvin, M. and Wilson, K. W., *J. Am. Chem. Soc.*, 67, 2003, 1945.
8. Van Uitert, L. G., Fernelius, W. C., and Douglas, B., *J. Am. Chem. Soc.*, 75, 457, 1953.
9. Sekine, T. and Ihara, N., *Bull. Chem. Soc. Jpn.*, 44, 2942, 1971.
10. Sillén, L. G. and Martell, A. E., *Stability Constants of Metal Complexes* and Suppl. 1, The Chemical Society, London, 1964, 1971.
11. Van Uitert, L. G., Thesis, State College, 1951.
12. Sweet, T. R. and Brengartner, D., *Anal. Chim. Acta*, 52, 173, 1970.
13. Yoshimura, Y. and Suzuki, N., *Talanta*, 25, 489, 1978.
14. Belford, R. L., Martell, A. E., and Calvin, M., *J. Inorg. Nucl. Chem.*, 2, 11, 1956.
15. Berg, E. W. and Truemper, J. T., *J. Phys. Chem.*, 64, 487, 1960.
16. Berg, E. W. and Truemper, J. T., *Anal. Chim. Acta*, 32, 245, 1965.
17. Berg, E. W. and Hartlage, F. R., Jr., *Anal. Chim. Acta*, 33, 173, 1965.
18. Berg, E. W. and Hartlage, F. R., Jr., *Anal. Chim. Acta*, 34, 46, 1966.
19. Berg, E. W. and Acosta, J. J. C., *Anal. Chim. Acta*, 40, 101, 1968.
20. Richardson, M. F., Wagner, W. F., and Sands, D. E., *Inorg. Chem.*, 7, 2495, 1968.
21. Berg, E. W. and Reed, K. P., *Anal. Chim. Acta*, 42, 207, 1968.
22. Schweitzer, G. K., Pullen, B. P., and Fang, Y. H., *Anal. Chim. Acta*, 43, 332, 1968.
23. Belcher, R., Majer, J. R., Rerry, R., and Stephen, W. I., *Anal. Chim. Acta*, 43, 451, 1968.
24. Melia, T. P. and Merrifield, R., *J. Inorg. Nucl. Chem.*, 32, 1489, 2573, 1970.
25. Utsunomiya, K. and Shigematsu, T., *Anal. Chim. Acta*, 58, 411, 1972.
26. Eisentraut, K. J. and Sievers, R. E., *J. Am. Chem. Soc.*, 87, 5254, 1965.
27. Schwarberg, J. E., Sievers, R. E., and Moshier, W., *Anal. Chem.*, 42, 1828, 1970.
28. Swain, H. A., Jr. and Karraker, D. G., *Inorg. Chem.*, 9, 1766, 1970.
29. Belcher, R., Cranley, C. R., Majer, J. R., Stephen, W. I., and Uden, P. C., *Anal. Chim. Acta*, 60, 109, 1972.
30. Berg, E. W. and Herrera, N. M., *Anal. Chim. Acta*, 60, 117, 1972.
31. Chattoraj, S. C., Lynch, C. T., and Mazdiyasni, K. S., *Inorg. Chem.*, 7, 2501, 1968.
32. Richardson, M. F. and Sievers, R. E., *Inorg. Chem.*, 10, 498, 1971.
33. Dilli, S. and Patsalides, E., *Aust. J. Chem.*, 29, 2369, 1976.
34. Chattoraj, S. C., Cupka, A. G., Jr., and Sievers, R. E., *J. Inorg. Nucl. Chem.*, 28, 1937, 1966; *Inorg. Chem.*, 6, 408, 1967.
35. Shigematsu, T., Matsui, M., and Utsunomiya, K., *Bull. Chem. Soc. Jpn.*, 41, 763, 1968.
36. Shigematsu, T., Matsui, M., and Utsunomiya, K., *Bull. Chem. Soc. Jpn.*, 42, 1278, 1969.
37. Springer, C. S., Jr., Meek, D. W., and Sievers, R. E., *Inorg. Chem.*, 6, 1105, 1967.
38. Belcher, R., Majer, J. R., Stephen, W. I., Thomson, I. J., and Uden, P. C., *Anal. Chim. Acta*, 50, 423, 1970.
39. Honjo, T., Imura, H., Shima, S., and Kiba, T., *Anal. Chem.*, 50, 1547, 1978.
40. Dilli, S. and Patsahidas, E., *Aust. J. Chem.*, 29, 2351, 1976.
41. Rosenstreich, J. L. and Goldberg, D. E., *Inorg. Chem.*, 4, 909, 1965.
42. Shigematsu, T., Tabushi, M., Matsui, M., and Munakata, M., *Bull. Chem. Soc. Jpn.*, 41, 1610, 1968.
43. Stary, J. and Hladky, E., *Anal. Chim. Acta*, 28, 227, 1963.
44. Hagemann, F., *J. Am. Chem. Soc.*, 72, 768, 1950.
45. Adam, J. A., Booth, E., and Strickland, J. D. H., *Anal. Chim. Acta*, 6, 462, 1952.
46. Ota, K., *Nippon Kinzoku Gakkaishi*, 28, 338, 1964; *Anal. Abstr.*, 13, 1180, 1966.
47. Tabushi, M., *Bull. Inst. Chem. Res. Kyoto Univ.*, 37, 245, 1959; *Chem. Abstr.*, 54, 7427g, 1960.
48. McKaveny, J. P. and Freiser, H., *Anal. Chem.*, 29, 290, 1957.
49. Tabushi, M., *Bull. Inst. Chem. Res. Kyoto Univ.*, 37, 237, 1959; *Chem. Abstr.*, 54, 7434i, 1960.
50. McKaveny, J. P. and Freiser, H., *Anal. Chem.*, 30, 526, 1968.
51. Opasova, R. G., Savostina, V. M., and Labonov, F. I., *Zh. Anal. Khim.*, 32, 974, 1977.
52. Akaiwa, H. and Kawamoto, H., *Bunseki Kagaku*, 28, 477, 1979.
53. Fremeoux, B. and Catting, G., *Chim. Anal. (Paris)*, 50, 34, 1968; *Chem. Abstr.*, 68, 119086b, 1968.
54. Heunisch, G. W., *Mikrochim. Acta*, 258, 1970.
55. Holzbecher, Z., Divis, L., Kral, M., Sucka, L., and Ulacil, F., *Handbook of Organic Reagents in Inorganic Analysis*, Ellis Horwood, Chichester, England, 1976.

56. Sarghie, I. and Fisel, S., *Rev. Chim. (Bucharest)*, 25, 744, 1974; *Chem. Abstr.*, 82, 164451s, 1975.
57. Rangnekar, A. V. and Khopkar, S. M., *Fresenius Z. Anal. Chem.*, 230, 425, 1967.
58. Poluektov, N. S. and Beltyukova, S. V., *Zh. Anal. Khim.*, 25, 2106, 1970.
59. Onishi, H. and Banks, C. V., *Anal. Chem.*, 35, 1887, 1963.
60. Onishi, H. and Toita, Y., *Anal. Chem.*, 36, 1867, 1964.
61. De, A. K. and Mujumder, S. K., *Anal. Chim. Acta*, 27, 153, 1964.
62. De, A. K. and Rhaman, M. S., *Anal. Chim. Acta*, 34, 233, 1966.
63. Noriki, S., *Anal. Chim. Acta*, 76, 215, 1975.
64. Mujumdar, S. K. and De, A. K., *Anal. Chem.*, 32, 1337, 1960.
65. Khophar, S. M. and De, A. K., *Fresenius Z. Anal. Chem.*, 171, 241, 1959.
66. Akaiwa, H., *Bunseki Kagaku*, 12, 457, 1963.
67. Akaiwa, H., Kawamoto, H., and Abe, M., *Bull. Chem. Soc. Jpn.*, 44, 117, 1971.
68. Mishchenko, V. T., Lauser, R. S., Efryushina, N. P., and Poluektov, N. S., *Zh. Anal. Khim.*, 20, 1073, 1965.
69. Taketatsu, T. and Toriumi, N., *Bull. Chem. Soc. Jpn.*, 41, 1275, 1968.
70. Khopkar, S. M. and De, A. K., *Anal. Chim. Acta*, 22, 223, 1960.
71. Testa, C., *Anal. Chim. Acta*, 25, 525, 1961.
72. Kawamoto, H. and Akaiwa, H., *Bunseki Kagaku*, 23, 495, 1974.
73. Dhond, P. V. and Khopkar, S. M., *Talanta*, 23, 51, 1976.
74. Rahaman, Md. S. and Finston, H. L., *Anal. Chem.*, 40, 1709, 1968.
75. Taketatsu, T. and Banks, C. V., *Anal. Chem.*, 38, 1524, 1966.
76. Rangnekar, A. V. and Khopkar, S. M., *Chemist-Analyst*, 56, 84, 1967.
77. Onishi, H. and Toita, Y., *Talanta*, 11, 1357, 1964.
78. Solanke, K. R. and Khopkar, S. M., *Fresenius Z. Anal. Chem.*, 275, 286, 1975.
79. Savrova, O. D., Gibalo, I. M., and Lobanov, F. I., *Anal. Lett.*, 5, 669, 1972; *Chem. Abstr.*, 78, 11138n, 1972.
80. De, A. K. and Rahaman, M. S., *Anal. Chim. Acta*, 27, 591, 1962.
81. Ramanujam, A., Nadkarni, M. N., Ramakrishna, V. V., and Patil, S. K., *J. Radioanal. Chem.*, 42, 349, 1978.
82. Rangnekar, A. V. and Khopkar, S. M., *Bull. Chem. Soc. Jpn.*, 41, 600, 1968.
83. Myasoedov, B. and Muxart, R., *Zh. Anal. Khim.*, 17, 340, 1962.
84. De, A. K. and Rahman, M. S., *Analyst*, 89, 795, 1964.
85. Rangnekar, A. V. and Khopkar, S. M., *Bull. Chem. Soc. Jpn.*, 38, 1696, 1965.
86. De, A. K. and Rahman, M. S., *Talanta*, 12, 343, 1965.
87. Rangnekar, A. V. and Khopkar, S. M., *Bull. Chem. Soc. Jpn.*, 39, 2169, 1966.
88. Rangnekar, A. V. and Khopkar, S. M., *Mickrochim. Acta*, 272, 1968.
89. De, A. K. and Rahman, M. S., *Anal. Chim. Acta*, 31, 81, 1964.
90. Khopkar, S. M. and De. A. K., *Analyst*, 85, 376, 1960.
91. Obernovi c-Palgoric, I., Gal, I. J., and Vajgand, V., *Anal. Chim. Acta*, 40, 534, 1968.
92. De, A. K. and Rahaman, M. S., *Anal. Chem.*, 35, 1095, 1963.
93. Ikehata, A. and Shimizu, T., *Bull. Chem. Soc. Jpn.*, 38, 1835, 1965.
94. Williams, D. E. and Guyon, J. C., *Anal. Chem.*, 43, 139, 1971.
95. Fisher, R. P. and Winefordner, J. D., *Anal. Chem.*, 43, 454, 1971.
96. Williams, D. E. and Guyon, J. C., *Mikrochim. Acta*, 194, 1972.
97. Tishchenko, M. A., Zheltvai, I. I., Bakshun, I. V., and Poluektov, N. S., *Zh. Anal. Khim.*, 28, 1954, 1973.
98. Tishchenko, M. A., Zheltvai, I. I., and Poluektov, N. S., *Zavod. Lab.*, 39, 670, 1973; *Chem. Abstr.*, 79, 152550g, 1973.
99. Yamakawa, K., Tanikawa, K., and Arakawa, K., *Chem. Pharm. Bull.*, 11, 1405, 1963.
100. Tanikawa, K., *Bunseki Kagaku*, 25, 721, 1976.
101. Karayannis, N. M. and Corwin, A. H., *J. Chromatogr. Sci.*, 8, 251, 1970.
102. Tanikawa, K., Hirano, K., and Arakawa, K., *Chem. Pharm. Bull.*, 15, 915, 1967.
103. Sievers, R. E., Ponder, P. W., Morris, M. L., and Moshier, R. W., *Inorg. Chem.*, 2, 693, 1963.
104. Schwarberg, J. E., Moshier, R. W., and Walsh, J., *Talanta*, 11, 1231, 1964.
105. Veening, H., Bachman, W. E., and Wilkinson, D. M., *J. Gas Chromatogr.*, 5, 248, 1967.
106. Fujinaga, T., Kuwamoto, T., and Murai, B., *Anal. Chim. Acta*, 71, 141, 1974.
107. Hill, R. D. and Gesser, H., *J. Gas Chromatogr.*, 1, 11, 1963.
108. Veening, H., Bachman, W. E., and Wilkinson, D. M., *J. Gas Chromatogr.*, 5, 248, 1967.
109. O'Brien, T. P. and O'Laughlin, J. W., *Talanta*, 23, 805, 1976.
110. Arakawa, M. and Tanigawa, K., *Bunseki Kagaku*, 16, 812, 1967.
111. Dono, T., Ishihara, Y., Saito, K., and Nakazawa, T., *Bunseki Kagaku*, 15, 181, 1966.
112. Belcher, R., Jenkins, C. R., Stephen, W. I., and Uden, P. C., *Talanta*, 17, 455, 1970.

113. Tanikawa, K., Ochi, H., and Arakawa, K., *Bunseki Kagaku,* 19, 1669, 1970.
114. Shigematsu, T., Uchiike, T., Aoki, T., and Matsui, M., *Bull. Inst. Chem. Res. Kyoto Univ.*, 51, 273, 1973; *Chem. Abstr.*, 81, 85500h, 1974.
115. Buchtela, K., Grass, F., and Mueller, G., *J. Chromatogr.*, 103, 141, 1975.
116. Fontaine, R., Santoni, B., Pommier, C., and Guichon, G., *Anal. Chim. Acta*, 62, 337, 1972.
117. Leary, J. J., *James Madison J.*, 35, 89, 1977; *Chem. Abstr.*, 87, 77807x, 1977.
118. Saito, S., *Bunseki Kagaku,* 28, 227, 1979.
119. Yoshida, I., Kobayashi, H., and Ueno, K., *Talanta,* 24, 61, 1977.
120. Ueno, K., Kobayashi, H., and Yoshida, I., *Mem. Fac. Eng. Kyushu Univ.*, 38, 83, 1978; *Chem. Abstr.*, 89, 122189z, 1978.
121. Sanders, J. K. M. and Williams, D. H., *J. Chem. Soc.*, 422, D1970.
122. Sievers, R. E. and Randeau, R. E., U.S. Air Force Systems Command, Aerospace Research Laboratory, Tech. Doc. Rep., ARL 70-0285, 26, 1970; *Chem. Abstr.*, 75, 27926s, 1971.
123. Briggs, J., Frost, G. H., Hart, F. A., Moss, G. P., and Staniforth, M. L., *J. Chem. Soc.*, 749, D1970.
124. Hinckley, C. C., *J. Org. Chem.*, 35, 2834, 1970.
125. Springer, C. P., Jr., Meek, D. N., and Sievers, R. E., *Inorg. Chem.*, 6, 1105, 1967.
126. Whitesides, G. M. and Lewis, D. W., *J. Am. Chem. Soc.*, 92, 6979, 1970.
127. Scribner, W. G. and Sievers, R. E., 5 th, Solvent Extr. Res., Proc. Int. Conf. Solvent Extr. Chem., 1968, 1; *Chem. Abstr.*, 73, 92153k, 1970.
128. Burgett, C. A., Report 1972, IS-T-541, 152; *Chem. Abstr.*, 78, 66504j, 1973.
129. Burgett, C. A. and Fitz, J. S., *Talanta,* 20, 363, 1973.
130. Sieck, R. F. and Banks, C. V., *Anal. Chem.*, 44, 2307, 1972.
131. Mitchell, J. W. and Banks, C. V., *Talanta,* 19, 1157, 1972.
132. Zolotov, Yu. A. and Lambrev, V. G., *Zh. Anal. Khim.*, 20, 659, 1965.
133. Zolotov, Yu. A., Chmutova, M. K., and Palei, P. N., *Zh. Anal. Khim.*, 21, 1217, 1966.
134. Jensen, B. B., *Acta Chem. Scand.*, 13, 1668, 1890, 1959.
135. Sizonenko, N. T. and Zolotov, Yu. A., *Zh. Anal. Khim.*, 24, 1305, 1969; *J. Anal. Chem. USSR,* 24, 1053, 1969.
136. Mirza, M. Y., *Talanta,* 25, 685, 1978.
137. Rao, G. N. and Chouhan, V. S., *Chem. Era,* 14, 395, 1978; *Chem. Abstr.*, 91, 101393q, 1979.
138. Vdovenko, V. M., Kovalskaya, M. P., and Smirnova, E. A., *Sov. Radiochem.*, 15, 320, 1973; *Anal. Abstr.*, 27, 3189, 1974.
139. Kochetkova, N. E., Chmutova, M. K., and Myasoedov, B. F., *Zh. Anal. Khim.*, 27, 678, 1972.
140. Myasoedov, B. F., Kochetkova, N. E., and Chumutova, M. K., *Zh. Anal. Khim.*, 28, 1723, 1973.
141. Myasoedov, B. F. and Molochnikova, N. P., *Radiochem. Radioanal. Lett.*, 18, 33, 1974; *Anal. Abstr.*, 28, 2B82, 1975.
142. Sevastyanov, A. I., Lanskaya, N. G., and Rudenko, N. P., *Sov. Radiochem.*, 15, 260, 1973; *Anal. Abstr.*, 27, 1843, 1974.
143. Chmutova, M. K., Pribylova, G. A., and Myasoedov, B. F., *Zh. Anal. Khim.*, 28, 2340, 1973.
144. Arora, H. C. and Rao, G. N., *Indian J. Chem.*, 11, 488, 1973.
145. Efimov, I. P., Tamilova, L. G., Voronets, L. S., and Peshkova, V. M., *Zh. Anal. Khim.*, 28, 267, 1973.
146. Chumutova, M. K. and Kochetkova, N. E., *Zh. Anal. Khim.*, 24, 1757, 1969.
147. Zolotov, Yu. A., Sizonenko, N. T., Zolotovitskaya, E. S., and Yakovenko, E. I., *Zh. Anal. Khim.*, 24, 20, 1969; *J. Anal. Chem. USSR.*, 24, 15, 1969.
148. Freger, S. V., Lozovich, A. S., and Ovrutskii, M. I., *Zh. Anal. Khim.*, 26, 2380, 1971.
149. Akama, Y., Nakai, T., and Kawamura, F., *Nippon Kaisui Gakkai shi,* 33, 120, 1979.
150. Tomilova, L. G., Efimov, I. P., and Peshkova, V. M., *Zh. Anal. Khim.*, 28, 666, 1973.
151. Chmutova, M. K. and Kochetkova, N. E., *Zh. Anal. Khim.*, 25, 710, 1970.
152. Zolotov, Yu. A., Chmutova, M. K., Palei, P. N., and Kochetkova, N. E., *Zh. Anal. Khim.*, 24, 216, 711, 1969.
153. Hala, J. and Prihoda, J., *Collect. Czech. Chem. Commun.*, 40, 546, 1975.
154. Zolotov, Yu. A. and Gavrilova, L. G., *Zh. Neorg. Khim.*, 14, 2157, 1969; *Chem. Abstr.*, 71, 95543g, 1969.
155. Akama, Y., Ishii, T., Nakai, T., and Kawamura, F., *Bunseki Kagaku,* 28, 196, 1979.
156. Savrova, O. D., Gibalo, I. M., Spiridonova, S. S., and Labanov, F. I., *Zh. Anal. Khim.*, 28, 817, 1973.
157. Roy, A. and Nag, K., *Indian J. Chem.*, 15A, 474, 1977; *Chem. Abstr.*, 87, 157829w, 1977.
158. Joshi, S. N., Enanova, E. K., and Peshkova, V. M., *Indian J. Chem.*, 11, 78, 1973; *Anal. Abstr.*, 25, 3806, 1973.
159. Mysoedov, B. F. and Malochnikova, N. P., *Zh. Anal. Khim.*, 24, 702, 1969.

160. Akama, Y., Nakai, T., and Kawamura, F., *Bunseki Kagaku*, 25, 496, 1976.
161. Zolotov, Yu. A. and Lambrev, V. G., *Radiokhimiya*, 8, 627, 1966.
162. Karalova, Z. K. and Pyzhova, Z. I., *Zh. Anal. Khim.*, 23, 1564, 1968.
163. Shcherbakova, S. A., Melchakova, N. V., and Peshkova, V. M., *Zh. Anal. Khim.*, 31, 318, 1976.
164. Zolotov, Yu. A. and Gavrilova, L. G., *J. Inorg. Nucl. Chem.*, 31, 3613, 1969; *Chem. Abstr.*, 71, 129385z, 1969.
165. Dyrssen, D., *Acta Chem. Scand.*, 15, 1614, 1961.
166. Dyrssen, D., *Trans. R. Inst. Technol. Stockholm*, 188, 1962.
167. Lengyel, T., *Radiochem. Radioanal. Lett.*, 25, 197, 1976.
168. Lengyel, T., *Radiochem. Radioanal. Lett.*, 28, 453, 1977.
169. Okubo, T. and Aoki, F., *Nippon Kagaku Kaishi*, 1681, 1973.
170. Taketatsu, T. and Sato, A., *Anal. Chim. Acta*, 108, 429, 1979.
171. Agrawal, Y. K. and Shukla, J. P., *Indian J. Chem.*, 13, 94, 1975; *Anal. Abstr.*, 29, 5C21, 1975.
172. Navratil, O. and Smola, J., *Collect. Czech. Chem. Commun.*, 36, 3549, 1971.
173. Cheng, K. L., *Anal. Chem.*, 33, 783, 1961.

N-BENZOYL-N-PHENYLHYDROXYLAMINE AND RELATED REAGENTS

Synonyms

Reagents covered in this section are listed in Table 1, together with their synonyms.

Source and Method of Synthesis

All reagents are commercially available. (1) and (2) are prepared by the reaction of phenylhydroxylamine with benzoylchloride[1,2] and cinnamoylchloride,[3,9] respectively. (3) is prepared by the reaction of o-tolylhydroxylamine with benzoylchloride.[4]

Analytical Uses

As these reagents are structurally related with Cupferron (p.71), their reactions with metal ions are similar, but they are superior to Cupferron in many respects, because they are more stable in mineral acid solutions. (1) to (3) are used as the extraction photometric reagents for V(V) and Ti(IV). (3) is most specific for V. They are also used as solvent extraction reagents for the separation of various metal ions, as well as precipitating reagent or gravimetric reagent for Al, Cu(II), Fe(III), Ti(IV), and other elements. A number of compounds structurally related with substituted hydroxylamine have been investigated as analytical reagents and are discussed in detail.[M1,M2,R1]

Properties of Reagents

BPA (1) is colorless needles, mp 121 to 122°, is stable against heat, light, and air, and does not decompose even in strong mineral acid except nitric acid (3 N).[1,2] Almost insoluble in water. Easily soluble in various organic solvents; also soluble in aqueous mineral acids, acetic acid, and aqueous ammonia; solubility in water, 0.04 g/100 ml (25°),[R1] and in 96% ethanol, 10.9 g/100 ml (22°).[66] It is monobasic acid, pKa = 8.2 (μ = 0.1 NaClO$_4$, 25°),[7] 9.13 (20 v/v% dioxane), and 10.20 (40% dioxane, 35°).[5]

CPA (2) is light yellow-green needles, mp 158 to 163°C.[3,9,67] Similar physical properties and solubility behavior to (1); solubility in water, 0.022 g/100 ml (18°), and in absolute ethanol, 0.972 g/100 ml (18°);[9] pKa = 11.1.[9]

BTA (3) is colorless needles, mp 104°,[4] similar physical properties and solubility behavior to (1).

Aqueous solutions of (1) to (3) are colorless and show no absorption band in the visible region. Absorption spectra of (1) in the UV range are shown in Figure 1.[8]

This group of reagents is often used for the extraction of easily hydrolyzed multivalent metal ions from rather concentrated acid solutions. In this connection, distribution of BPA (1) between chloroform and hydrochloric acid was investigated as shown in Figure 2. The distribution ratio of (1) for various systems is summarized in Table 2. No data are available for (2) and (3).

Complexation Reactions and Properties of Complexes

BPA was first synthesized by Bamberger, who noted that it gave colored precipitates

Table 1
N-BENZOYL-*N*-PHENYLHYDROXYLAMINE AND RELATED COMPOUNDS

Number	Reagent	Synonyms	Molecular formula	Mol wt
(1)	*N*-Benzoyl-*N*-phenyl-hydroxylamine	*N*-Phenylbenzohydroxamic acid, BPA, NBPHA, BPHA	$C_{13}H_{11}NO_2$	213.24
(2)	*N*-Cinnamoyl-*N*-phenylhydroxylam-ine	*N*-Phenylcinnamohydrox-amic acid, CPA	$C_{15}H_{13}NO_2$	239.31
(3)	*N*-Benzoyl-*N*-o-tolylhydroxylamine	*N*-o-Tolyl-benzohydroxamic acid, BTA	$C_{14}H_{13}NO_2$	227.26

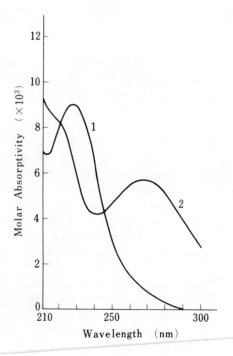

FIGURE 1. Absorption spectra of BPA (1) in chloroform equilibrated with aqueous phase of various pH. (1) pH 1 ~ 5; (2) pH 11.7. (From Bianco, P., Haladjian, J., and Pilard, R., *Anal. Chim. Acta*, 93, 255, 1977. With permission.)

with certain transition group elements.[1] Later, Shome found that this reagent has definite advantages over Cupferron in gravimetry[91] (K_{sp} for GaL$_3$, 1.6×10^{-34}; InL$_3$, 7.2×10^{-31}).[104] The solubility of the metal chelates in organic solvents also opened up the possibility of separation of metal ions by solvent extraction method.

The reaction with metal ion may be written as:

$$M^{m+} + nHL = ML_n^{m-n} + nH^+$$

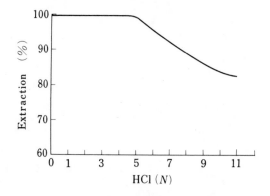

FIGURE 2. Distribution of (1) between chloroform and HCl phases. (From Vita, O. A., Levier, W. A., and Litteral, E., *Anal. Chim. Acta*, 42, 87, 1968. With permission.)

Table 2
DISTRIBUTION RATIO OF
BPA (1)

Phase	D	Ref.
CHCl$_3$/H$_2$O	214	11
CHCl$_3$/1.0 MHClO$_4$	137 ± 5	12
CHCl$_3$/6 MHClO$_4$	24 ± 1	12
C$_6$H$_6$/1.0 MHClO$_4$	23,[a] 37[b]	12

[a] Total concentration of BPA in benzene is 1×10^{-3} M.
[b] Total concentration of BPA in benzene is 6×10^{-3} M.

The optimum conditions for the precipitation reaction of selected metal ions with (1) and (3) are summarized in Table 3. The structure of ML$_2$-type chelate may be shown as below.

Reagents (2) and (3) also behave similarly.

Due to the poor solubility of the metal chelates, very few data are available on the chelate stability constants of these reagents. Reported values are summarized in Table 4.

The precipitates of metal chelates which are listed in Table 3 are more or less soluble in polar or nonpolar solvents. Rather extensive investigations have been done on the extraction behavior of metal ions with (1), but less has been done with (2) and (3). The elements that can be extracted with BPA (1) are summarized in Figure 3.   The

Table 3

OPTIMUM CONDITIONS FOR THE PRECIPITATION OF VARIOUS METAL IONS WITH BPA (1) AND BTA (3)

Reagent	Metal ions	Condition (pH)	Composition of chelate	Drying temp[a] (°)	Separation from	Ref.
(1)	Al	3.6 ~ 6.4	ML_3	100	Be, Co, Mn, Ni, U(VI), Zn	13
	Be	5.5 ~ 6.5	ML_2	110 or I.O.	Al, Fe(III), Ti	14
	Bi	6.0 ~ 6.8	ML_3	110 ~ 115	Al, As(V), Be, Cd, Co, Cu, Fe(III), Hg(II), La, Mn, Mo(VI), Ni, Pb, Pd, Sb(III), Sn(IV), Th, Ti, U(VI), V(V), W(VI)	15
	Cd	5.8 ~ 6.5	ML_2	110	—	103
	Ce(III)	6.5 ~ 7.5	ML_3	I.O.	Th	16
	Co(II)	5.5 ~ 6.5	ML_2	110 ~ 120 or I.O.	Cu	17
	Cu(II)	3.6 ~ 6.0	ML_2	100	As(III) (V), Be, Cd, Co, Hg(II), Mn, Ni, Pb, U(VI), Zn, PO_4^{3-}	13
	Fe(III)	3.0 ~ 5.5	ML_3	100	Co, Mn, Ni, U(VI), Zn	13
	Ga	2.0 ~ 3.0	ML_3	110 ~ 120 or I.O.	Al, Be, Ce(III), Cu, Fe(II), In, Ti, U(VI), Zn	18, 19
	Ge	≥2 N HClO₄	ML_4	I.O.	—	20
	Hg(II)	3.6 ~ 6.0	ML_2	<105	Ag, As(V), Bi, Cd, Co, In, Mo(VI), Ni, Pb, Sb, Sn(IV), Tl(I), W(VI), Zn	21
	In	4.3 ~ 8.0	ML_3	110 ~ 120 or I.O.	Ca, Cu, Fe(III), Ni, Sn, Zn	18
	La(III)	6.4 ~ 7.2 (50°)	ML_3	110 ~ 150 or I.O.	Al, Ce(IV), Cu, Fe(II) (III), Ga, In, Ni, Sc, Th, U(VI), Zn	22
	Mg	8.0 ~ 9.0	ML_2	I.O.	Al, Be, Co, Cu, Fe(III), Ni, Th, Zn	23
	Mo(VI)	0.01 ~ 2.5 N HCl (<70°), EDTA	MO_2L_2	110 ~ 150 or I.O.	Al, As, Be, Bi, Cd, Co, Cu, Cr, Fe, Ga, In, Mn, Ni, Pb, U, Zn, rare earths	24, 92

Element	Conditions	Composition	Drying temp.	Interferences	Ref.
Nb(V)	$<0.0 (H_2SO_4)$	ML_3	I.O.	Al, As(III) (V), Be, Ce(III) (IV), Cd, Co, Cr(III) (VI), Fe(III), Hg(II), Mg, Mn, Ni, Th, U(VI), Zn, PO_4^{3-}	25, 26
Nb(V)	3.5 ∼ 3.6, tartarate, EDTA	ML_3	110	Al, As(III) (V), Ba, Be, Bi, Ca, Cd, Ce(III) (IV), Co, Cr(III) (VI), Cu, Fe(III), Hg(II), Mg, Mn, Ni, Pb, Sb(III), Sn(IV), Sr, Ta, Th, U(VI), W(VI), Zn, rare earths, PO_4^{3-}	27—32
Sb(III)	1.5 ∼ 3 NHCl	ML_3	105 ∼ 110	Sb(IV) is not pptd	89
Sc(III)	5.4	ML_3	I.O.	Zn, rare earths	33
Sn(IV)	0.1 ∼ 0.5 NHCl	ML_2Cl_2	110 or I.O.	Cu, Pb, Zn	34
Ta(V)	1.0	—	I.O.	Nb, Ti, Zr	35
Ta(V)	1.0 ∼ 1.5	—	I.O.	Al, As(III) (V), Be, Ce(III) (IV), Cd, Co, Cu, Cr(III) (VI), Fe(III), Hg(II), Mg, Mn, Ni, Th, U(VI), Zn, PO_4^{3-}	25, 26, 35
Th	4.0 ∼ 8.5	ML_4	105 ∼ 20 or I.O.	Al, Ce(III), Fe(III), Ga, In, La, Ti, U(VI), V(V), Zr, rare earths	16, 36, 37
Ti(IV)	0.12 ∼ 0.5 NHCl, H_2O_2, EDTA	—	I.O.	Cd, Ce(III), Fe(III), Mo(VI), Th, V(V) Zr (bauxite)	38
Ti(IV)	0.02 ∼ 0.5 NH_2SO_4, tartarate EDTA (65°)	MOL_2	110	Al, Ce(III), Co, Cr(III), Cu, Fe(III), Mn, Ni, Th, U(VI), V(IV), Zn, PO_4^{3-}	39
U(VI)	5.2 ∼ 5.6, Mg-EDTA	MO_2L_2	110 or I.O.	Al, Bi, Ce(III), Fe(III), Mo(VI), Pb, Th, Ti, Zr	40
W(VI)	0.05 ∼ 1.0 NHCl or 0.5 ∼ 1.0 NHCl	MO_2L_2	115 or I.O.	Fe(III), Mo(VI), Ti, U(VI), V(V)	41
W(VI)	3 ∼ 7 NHCl or 9 ∼ 12 NH_2SO_4	MO_2L_2	115 or I.O.	Al, As, Be, Bi, Cd, Co, Cu, Cr, Fe, Ga, Hg, In, Mn, Ni, Pb, U, Zn, rare earths	92
Zr	2.4 NHCl or 3.6 N H_2SO_4, H_2O_2, tartarate	—	I.O.	Al, Cr, Fe(III), Nb, Ta, Ti, V(V), rare earths	42
Zr	0.5 ∼ 0.6 NHCl, H_2O_2,	—	I.O.	Al, Bi, Cd, Ce(III), Cr(III), Cu, Dy, Ga, La, Mg, Mn, Nd, Ni, Pb, Sm, Th, U(VI), Yt, Zn	43

Table 3 (continued)

OPTIMUM CONDITIONS FOR THE PRECIPITATION OF VARIOUS METAL IONS WITH BPA (1) AND BTA (3)

Reagent	Metal ions	Condition (pH)	Composition of chelate	Drying temp[a] (°)	Separation from	Ref.
(3)	Zr	0.5 ~ 0.6 NH_2SO_4	ML_4			44
	Be	6.3 ~ 8.3	ML_2	110	Al, Ba, Ca, Cd, Co, Cu, Hg(II), Mg, Mn(II), Ni, Th(IV), Zn	108
	Bi	6.2 ~ 7.3, tartarate	ML_3	110 ~	As, Mo(VI), Sb, W(VI)	95
	Ga	1.8 ~ 5.5	ML_3	110 ~ 120	In	96
	Hg(II)	3.0 ~ 3.2		102 ~ 110	Ag, As, Ba, Ca, Cd, I.a, Mg, Mn(II), Sb, Sr, Tl(I), U(VI), Zn	97
	In	4.5 ~ 8.5	ML_3	110 ~ 120	Ga	96

a I.O., ignition to oxide.

Table 4
CHELATE STABILITY CONSTANTS OF BPA (1)

Metal ion	$\log K_{ML}$	$\log K_{ML_2}$	$\log K_{ML_3}$	$\log K_{ML_4}$	Condition		Ref.
					Temp	μ	
Fe(III)	5.28	—	—	—	—	—	45
Ga	9.2	β_2 18.0	β_3 25.3	—	25	3(NaClO$_4$)	8
Hf	13.7	13.2	12.3	12.0	25	2 (NaClO$_4$)	12
In	9.2	β_2 18.4	β_3 26.3	—	25	0.1 (NaClO$_4$)	7
Th	—	β_2 7.67	—	—	25	0.1 (NaClO$_4$)	46
Os(VIII)[a]	—	—	—	β_4 17.6	Room temp	—	106

[a]   Value for (3).

pH values at which the metal ion can be extracted quantitatively are also included in the figure.

The extractions are often conducted from strongly acid solution to attain the selectivity of metal ions and to avoid the hydrolysis of multivalent metal ions. The extraction from hydrochloric acid was found to be superior to sulfuric acid. Figures 4 to 9 illustrate the extraction characteristics of elements in Group IV through VIII in the (1)-chloroform-hydrochloric acid system.[10]

Extraction constants and distribution ratio of metal chelates with (1) and optimum conditions for the extraction of some important elements with (1) are summarized in Tables 5[47] and 6,[49,50] respectively. No quantitative data is available on (2) and (3).

Metal ions which form colored precipitates with these reagents show a single absorption peak in the visible range when they are extracted into an organic solvent. Vanadium(V) forms two types of chelates with (1), according to the acidity of aqueous phase, as illustrated in Figure 10. Titanium(IV) behaves similarly to V. The spectral characteristics of the analytically important metal chelates with (1), (2) and (3) are summarized in Table 7.

Purification and Purity of Reagents

(1) to (3) can be easily purified by recrystallization from appropriate solvents: (1) from hot water, benzene, or acetic acid; (2) from ethanol; and (3) from aqueous ethanol.

(1) to (3) are well-characterized crystalline materials, and their purities can easily be checked by observing their melting points.

Analytical Applications
Use as Precipitating and Gravimetric Reagent

As described in Table 3, BPA (1) has been widely used as the gravimetric reagent for a large number of elements. Their separations from other metal ions can be attained under the conditions controlled by pH and masking agents, such as EDTA or tartaric acid. The optimal conditions for the precipitation of various metal ions with (1) are summarized in Table 3. The precipitates with (1) are often directly weighable, but in some cases, they have to be ignited to oxides before weighing. The most important uses of (1) are the determination and separation from each other of Nb and Ta, the direct determination of Nb in the presence of Ta and other ions, the separation of Nb, Ta, and Ti from each other, and the determination of Zr in the presence of Nb, Ta, Ti, and V.[M1]

Similar application for the determination of Nb and Ta and their separation from other ions are reported on (2) and (3).

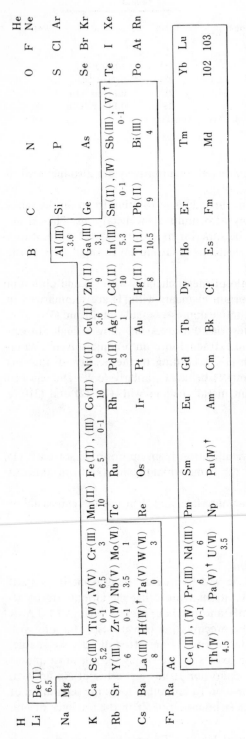

The number under each element indicates the pH at which the element can be extracted quantitatively. Where two numbers occur, they refer to the two oxidation states of the element taken in the same order.

† Extracted almost completely from high acid concentrations.

FIGURE 3.   Elements extracted by BPA (I). (From Shendrikar, A. D., *Talanta*, 16, 51, 1969. With permission.)

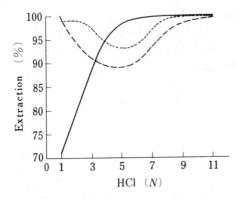

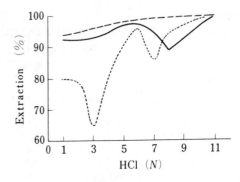

FIGURE 4. Extraction of group IV-B elements with BPA (1); Ti,——; Hf, ————; Zr, ·····
····· . (From Vita, O. A., Levier, W. A., and Litteral, E., *Anal. Chim. Acta,* 42, 87, 1968. With permission.)

FIGURE 5. Extraction of group V-B elements with BPA (1); V, ——; Nb,——; Ta, ·····; V(IV); no extraction.

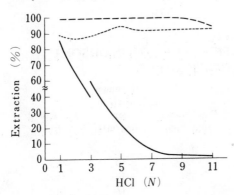

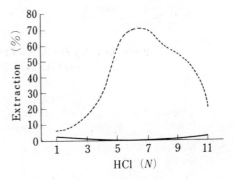

FIGURE 6. Extraction of group VI-B elements with BPA (1); Cr(VI), ——; Mo(VI), ————; W, ·······; Cr(III), no extraction.

FIGURE 7. Extraction of group VII-B elements with BPA (1); Mn,——; Te, ———.

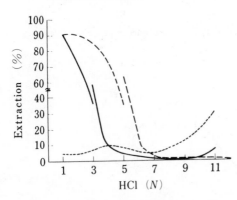

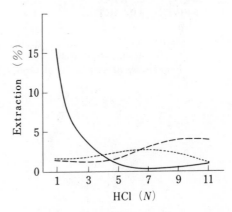

FIGURE 8. Extraction of group IV-A elements with BPA (1); Sb(III), ——; Sb(V), ·····; Sn(IV), ————; Bi and Pb, no extraction.

FIGURE 9. Extraction of group VIII elements with BPA (1). Fe, ——; Co, ————; Ni, ·····. (From Vita, O. A., Levier, W. A., and Litteral, E., *Anal. Chim. Acta,* 42, 87, 1968. With permission.)

Table 5
K_{ex} AND log K_D OF METAL CHELATES WITH BPA (1) IN A CHLOROFORM-WATER SYSTEM[47]

Chelate	log K_{ex}	log K_D (ML$_n$)	Ref.
CdL$_2$	-12.1	1.73	
CeL$_3$	-13.2	—	
CoL$_2$·HL	1.0	—	
CuL$_2$	-0.7	2.35	
FeL$_3$	5.3	—	
GdL$_3$	-12.6	2.75	
InL$_3$	-1.8	3.03	
LaL$_3$·HL	-13.6 (-14.4)	—	48
ThL$_4$	-0.7	3.45	
TlL	-7.3 (-6.8)	0.30	48
UO$_2$L$_2$	-3.14	—	48
YL$_3$	-12.4	—	
ZnL$_2$	-10.0	1.93	

Table 6
OPTIMUM CONDITIONS FOR THE EXTRACTION OF SOME METAL IONS WITH BPA (1)[49,50]

Metal ion[a]	Acidity	Conc of BPA (1) in CHCl$_3$ (% w/v)	Equilibration time (min)	Maximum metal ion conc (mg/mℓ)	Extraction[b] (%)
Bi(III)	0.1 N HCl	0.6	10	8	99.4
Ga	pH 3.1	1.0	12	10	99.4
In	pH 5.3	1.0	10	10	99.2
Pb	pH 9.0	1.0	10	12	97.4
Sb(III)	1.0 N HCl	0.5	10	10	93.1
Sb(V)	9.4 N HClO$_4$	0.9	15	8	98.3
Sn(IV)	0.8 N HCl	1.0	10	10	94.4
Sn(IV)	4.0 N HCl	1.0	10	10	96.1
Tl(I)	pH 10.5	0.7	8	8	99.7

[a] As(III) and Ge are extracted with chloroform alone from the aqueous phase of 11 N and 8 N HCl acidity, respectively.

[b] For a single extraction with equal volumes of aqueous and chloroform phases.

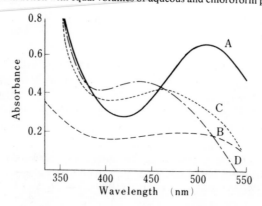

FIGURE 10.   Absorption spectra of V-BPA chelate in chloroform. Concentration of HCl in aqueous phase: (A) 3.6 M; (B) 1.8 M; (C) 1.0 M; (D) 0.01 M. (From Priyadarshini, U. and Tandon, S. G., *Anal. Chem.*, 33, 435, 1961. With permission.)

Table 7

SPECTRAL CHARACTERISTICS OF METAL CHELATES
WITH (1), (2), AND (3)

Reagent	Metal ion	Extracted		Metal chelate			Ref.
		From	Into	Ratio	λ_{max} (nm)	$\varepsilon(\times 10^3)$	
(1)	Ce(IV)	pH 8 ~ 10	CHCl₃	ML₄	460	4.57	52
	Hg(II)	pH 6.1 ~ 7.4	CHCl₃	ML₂	340	2.69	53
	Fe(III)	pH 3	Benzene	ML₃	460 ~ 480	5.38	54, 55
	Ga	pH 3.75	CHCl₃	—	400	—	104
	Nb(V)	9 ~ 12 NHCl	Toluene	ML₂	365	10.0	56
	Nb(V)	7 ~ 8 NHCl + SCN⁻	CHCl₃	ML₂X	360	46.5	57
	Ti(IV)	pH 0.8 ~ 1.4	CHCl₃	M(OH)₂L₂	340	7.5	93
	Ti(IV)	2 NHCl	CHCl₃	ML₄	355	5.2	58
	Ti(IV)	>9.6 NHCl	CHCl₃	ML₂Cl₂	371	6.7	59
	Ti(IV)	6.5 ~ 8.0 NHCl + SCN⁻	CHCl₃	ML₂X	350	16.7	60
	Ti(IV)	pH 0.7 ~ 1.0 phenylfluorone	CHCl₃	—	550	75.0	109
	U(VI)	pH 4	CHCl₃	—	510	—	105
	V(V)	pH 2	CHCl₃	M₂O₃L₄	440	3.6	61
	V(V)	2.8 ~ 4.6 NHCl	Benzene	ML₂	530	4.5	94
	V(V)	3.6 NHCl	CHCl₃	—	510	4.65	51
	V(V)	5 ~ 9 NHCl	CHCl₃	ML₂	530	5.1	62
	Zr(IV)	0.05 ~ 0.1 N H₂SO₄	CHCl₃	—	350	5.2	63
(2)	Fe(III)	pH 1 ~ 7	iso-AmOH	ML₃	{ 360 / 480	15.0 / 3.4	64
	Nb(III)	8 ~ 9 NHCl + SCN⁻	CHCl₃	—	370	—	65
	Ti(IV)	8 ~ 9 NHCl + SCN⁻	CHCl₃	ML₃X	410	—	65, 90
	UO₂²⁺	pH 5.5 ~ 8.5	Ethylacetate, butanol, or iso-amylal-cohol	ML₃	355	16.0	64
	V(V)	2.7 ~ 7.5 NHCl	CHCl₃	—	542	6.30	67
(3)	Os	pH 11 ~ 11.2 NaOH−H₂SO₄ 90°	Aqueous soln	ML₄	465	14.9	106
	V(V)	4 ~ 8 NHCl	CHCl₃	ML₂	510	5.25	68, 69

Use as Extraction Reagent

As described in Figure 3 and Table 8, extensive investigations have been done on the extraction characteristics and on some novel separations of metal ions with BPA (1). The extraction of metal ions with a solution of (1) in immiscible solvent is a function of shaking time, concentration of (1), nature of solvent, type of mineral acid, and the acidity. Solvents used for extraction are chloroform, benzene, and iso-amyl alcohol, of which chloroform is the preferred solvent.

Higher concentration of (1) is needed for the extraction from sulfuric acid than from hydrochloric acid. In the case of nitric acid, the extraction becomes less effective with the increase of acid concentration, although nitrate is preferred for the extraction of some elements.

The extraction behaviors of various metal ions in the BPA (1)-chloroform-HCl system are illustrated in Figures 4 to 9.[10] The typical examples of the separation of various elements by solvent extraction with (1) are summarized in Table 8. The separation factors for various pairs of elements are also listed in Table 9.

Table 8
TYPICAL EXAMPLES OF SEPARATION OF ELEMENTS BY SOLVENT EXTRACTION WITH BPA (1)

Extraction of	From	Condition of extraction			
		Acidity	Solvent	Conc of BPA (1)	Ref.
Al	Uranium fuel	$(NH_4)_2CO_3-(Na\,PO_3)_6$ + thioglycollic acid + KCN + H_2O_2	Benzene	Ca. 0.2%[a]	70
Fe, Ti	Al	0.5 NH_2SO_4	$CHCl_3$	Ca. 1%	71
Nb	Ta	pH 4.5 ~ 5.0 + tartarate	$CHCl_3$	Ca. 2%[a]	72
Nb	Pa, Zr	1 NHCl + 0.05 MF^-	$CHCl_3$	0.2%	73, 74
Nb, Zn	U	10 NHCl	$CHCl_3$	0.5%	75
Pa	Al, Bi, Ce(IV), Cr, La, Mn, Ni, Sn(II)(IV), Th, Ti, U	7 NH_2SO_4	Benzene	0.1 M	76, 77
Pa	Th, U	≥10 NHCl + 0.025 MF^-	$CHCl_3$	1%	78
Pu(IV)	Am, U, Zr, other fission products	3 NHNO₃	$CHCl_3$	0.4 M	79
Sc	La	pH ≥4.5 (HCl)	iso-AmOH	0.5%	80
Sb, Sn	In	10 ~ 12 NH_2SO_4	$CHCl_3$	0.4 M	81
Ta	Pa	1 NHCl + 0.4 MF^-	$CHCl_3$	0.5%	74
Th	La	pH 4.5 (HCl)	iso-AmOH	3%	82
Th, U	La	pH 4.5 (HClO₄)	$CHCl_3$	0.1 M	83
W	Al, As(III), Bi, Ba, Cr(VI), Cu, Fe, Ni, Pb, Sb(III), Ti, U, V(IV)	1 ~ 8 NHCl or 14 ~ 22 NH_2SO_4	$CHCl_3$, benzene, or CCl_4	0.2%	84
Zr, Ti	Sc	2 ~ 5 NHCl	iso-AmOH	3%	80

[a]   Ethanolic solution of (1) is first added before extraction with immiscible solvent.

Use as Photometric Reagent

As a photometric reagent, BPA (1) has been much more extensively investigated than its analogues. The elements that have been studied are listed in Table 7. Of these elements, determinations of V and Ti are of most practical importance, as those highly colored chelates can be extracted selectively from acidic solution.

Determination of Vanadium

Vanadium(V) combines with these reagents over a wide range of acidity, forming the complexes which can be extracted with chloroform. At pH 2.5, the mahogany-red chelate is formed with BPA (1) in the presence of ethanol, but the reaction is not selective. From 5 to 9 N hydrochloric acid, the chloroform extract is purple, and the color system obeys the Beer-Lambert law at 530 nm. With increasing acid concentration or the exposure to daylight, the optical density of the system decreases.

CPA (2) has been recommended as a sensitive reagent for V, but is not specific for V.[67] Later, (1) was claimed to be more selective,[62] but still reacts with Mo, Ti, and Zr to give yellow to orange-colored complexes that interfere with the determination of trace of V. BTA (3), finally, seems to have the necessary sensitivity and selectivity for V.[68,69] In 4 to 8 N hydrochloric acid, (3) is an almost specific reagent for V and can be used in the analysis of rocks and minerals. The violet complex is formed in 6 N hydrochloric acid solution, extracted with carbon tetrachloride or chloroform, and measured at 510 nm. Only interfering ion is Ti which can be masked with fluoride.[69]

Table 9

SEPARATION FACTORS FOR VARIOUS PAIRS OF ELEMENTS WITH (1)

| Separation | | Condition of extraction | | |
Of (major component)	From (minor component)	Acidity	Conc in chloroform (%)	Separation factor[a]
Ga	In	pH 3.1	1	10^4
Ga	Pb	pH 3.1	1	10^4
Ga	Ge	pH 3.1	1	10^3
In	Sn	pH 5.3	1	10^3
In	Pb	pH 5.3	1	10^4
Nb	Ta	1 NHCl + 0.05 NHF	0.2	10^2
Nb	Pa	1 NHCl + 0.05 NHF	0.2	10^3
Nb	Zr[b]	1 NHCl + 0.05 NHF	0.2	10^4
Pa	Sb	Con HCl + 0.025 NHF	1	2×10^2
Nb	Sn	Conc HCl + 0.025 NHF	1	7×10^2
Nb	U	Conc HCl + 0.025 NHF	1	$>4 \times 10^3$
Nb	Th	Conc HCl + 0.025 NHF	1	3×10^4
Pb	Tl	pH 9.0	0.7	6×10^2
Sb(V)	Bi(III) or (V)	9.0 NHClO$_4$	1	7×10^2
Sn(IV)	Sb(V)	0.8 NHCl	1	10^2
Sn(IV)	Bi(III) or (V)	4.0 NHClO$_4$	1	9×10^2
Sn(IV), Sb(V)	Bi(III) or (V)	9.0 NHClO$_4$	1	7×10^2
Sn	Ga	0.8 NHCl or 4.0 NHClO$_4$	1	10^3
Sn	In	0.8 NHCl or 4.0 NHClO$_4$	1	10^3
Sn	Pb	0.8 NHCl or 4.0 NHClO$_4$	1	10^4
Ta	Pa	1 NHCl + 0.4 NHF	0.5	10^3

[a] The ratio of the initial and final amounts of minor component. The aqueous phase was extracted twice with equal volumes of the chloroform phase and then washed with two successive portions of chloroform alone.

[b] One batch extraction is sufficient.

From Shendrickar, A. D., *Talanta*, 16, 51, 1969. With permission.

Procedure for the determination of V in rocks or minerals[69]

Decompose a 100-mg finely powdered sample by treating with sulfuric, nitric, and hydrochloric acids in a platinum crucible in the usual way, removing excess sulfuric acid by heating on a hot plate. Fuse the dry residue in platinum or preferably silica crucible with potassium pyrosulfate and extract the fused melt with 10 mℓ of water containing two drops of 20 N sulfuric acid. Transfer the solution to a separatory funnel and add 0.02 M potassium permanganate solution drop by drop until an excess, giving a pink color for 5 min to oxidize to V(V). The volume of the solution at this stage should be about 20 mℓ. Add 2 mℓ of 0.05 M sulfamic acid solution, 2 mℓ of saturated sodium fluoride solution, and 20 mℓ of concentrated hydrochloric acid. Add by pipette 10 mℓ of 0.02% BTA (3) in carbon tetrachloride or chloroform (ethanol free), stopper and shake the funnel for 30 sec. After phase separation, filter the lower layer through a small wad of cotton-wool into a cell. Measure the absorbance of this solution against a reagent blank at 510 nm.

Determination of Titanium

Titanium(IV), like V(V), forms different types of chelates with (1), depending upon the acidity of aqueous phase. Above pH 1, the color of chelate ($TiOL_2$) is less intense than that obtained at higher acidities and is more sensitive to pH changes.[58] When the hydrochloric acid concentration is increased to 2 N, a more intensely colored TiL_4-type chelate (λ_{max}, 355 nm, in $CHCl_3$) is formed. In strong hydrochloric acid solution, it gives ML_2Cl_2-type chelate (λ_{max}, 380 nm, in $CHCl_3$).[59] An excess of chloride also keeps U(VI) in anionic form unreactive to (1). If thiocyanate ion is added before the extraction of Ti from acidic solution into chloroform, the color intensity increases considerably, probably due to the formation of the ternary complex of ML_2SCN.[60] Similar effect has also been noticed with (2) and (3).

Procedure for the determination of Ti in alloys[59]

To an aliquot of the sample solution (containing not more than 100 μg of Ti) in a separatory funnel, add sufficient amount of concentrated hydro-chloric acid so that the acidity of the final solution is >9.6 N. Add 10 mℓ of 0.1% chloroform solution of BPA (1) to this solution and shake the mixture for 1 to 2 min, filter the lower layer through a dry filter paper, and measure the absorbance at 380 nm against a reagent blank.

Presence of Fe, Sn, W, V, and Zr, up to several milligrams, and of Mo(VI), if reduced by Sn(II), is not harmful. The complex has a maximum absorption at 371 nm, but it is preferable to measure at 380 nm because the absorbance of the reagent at this wavelength is lower.

Miscellaneous Uses

(1) has been used as a metal indicator in the chelatometric titrations of In(III),[87] Fe(III),[86] Tl(III),[87] Th,[87] and V(V).[85,107] Chromatographic behavior of a large number of metal ions on paper impregnated with (1) has been investigated for the separation of mixtures of metal ions.[88]

(1), (2), and (3) also have been employed as a titrant in the amperometric titration of Ga, Hf, Nb, Sc, Ti, and Zr[98,99,100,101] and in the redox titration of Mn.[102]

MONOGRAPHS

M1. Majumdar, A. K., *N-Benzoylphenylhydroxylamine and Its Analoques*, Pergamon Press, Oxford, 1972.

M2. Jablonoki, W. Z. and Skidmore, P. R., N-Benzoyl-N-phenylhydroxylamine, N-cinnamoyl-N-phenyl-hydroxylamine, in *Organic Reagents for Metals*, Vol. 2, Johnson, W. C., Ed., Hopkin & Williams, Chadwell Heath, England, 1964.

REVIEW

R1. Shendrikar, A. D., Substituted hydroxylamines as analytical reagents, *Talanta*, 16, 51, 1969.

REFERENCES

1. Banberger, E., *Ber.,* 52, 1116, 1919.
2. Ryan, D. E. and Lutwich, G. D., *Can. J. Chem.,* 31, 9, 1953.
3. Majumdar, A. K. and Mukherjee, A. K., *Anal. Chim. Acta,* 22, 514, 1960.
4. Majumdar, A. K. and Das, G., *Anal. Chim. Acta,* 31, 147, 1964.
5. Shukla, J. P. and Tandon, S. G., *J. Inorg. Nucl. Chem.,* 33, 1981, 1971.
6. Gupta, K. R. and Tandon, S. G., *Talanta,* 21, 249, 1974.
7. Schweizer, G. K. and Anderson, M. M., *J. Inorg. Nucl. Chem.,* 30, 1051, 1968.
8. Bionco, P., Haladjian, J., and Pilard, R., *Anal. Chim. Acta,* 93, 255, 1977.
9. Zharovskii, F. G. and Sukhomlin, R. I., *Ukr. Khim. Zh.,* 30, 750, 1964; *Chem. Abstr.,* 61, 12602a, 1964.
10. Vita, O. A., Levier, W. A., and Litteral, E., *Anal. Chim. Acta,* 42, 87, 1968.
11. Dyrssen, D., *Acta Chem. Scand.,* 10, 353, 1956.
12. Fouche, K. F., *Talanta,* 15, 1295, 1968.
13. Shome, S. C., *Analyst,* 75, 27, 1950.
14. Das, J. and Shome, S. C., *Anal. Chim. Acta,* 24, 37, 1961.
15. Das, B. and Shome, S. C., *Anal. Chim. Acta,* 40, 338, 1968.
16. Sinha, S. K. and Shome, S. C., *Anal. Chim. Acta,* 21, 415, 1959.
17. Sinha, S. K. and Shome, S. C., *Anal. Chim. Acta,* 21, 459, 1959.
18. Das, H. R. and Shome, S. C., *Anal. Chim. Acta,* 27, 545, 1962.
19. Alimarin, I. P. and Hamid, S. A., *Zh. Anal. Khim.,* 18, 1332, 1963.
20. Alimarin, I. P., Sokolova, I. V., and Smolina, E. V., *Vestn. Mosk. Univ. Khim.,* 23, 67, 1968; *Chem. Abstr.,* 69, 54761s, 1968.
21. Das, B. and Shome, S. C., *Anal. Chim. Acta,* 35, 345, 1966.
22. Das, B. and Shome, S. C., *Anal. Chim. Acta,* 32, 52, 1965.
23. Cardwell, T. J. and Magee, R. J., *Mikrochem. J.,* 13, 467, 1968.
24. Sinha, S. K. and Shome, S. C., *Anal. Chim. Acta,* 24, 33, 1961.
25. Mukherjee, A. K., Ph.D. thesis, Jadavpur University, 1959.
26. Majumdar, A. K. and Mukherjee, A. K., *Fresenius Z. Anal. Chem.,* 189, 339, 1962.
27. Majumdar, A. K. and Mukherjee, A. K., *Naturwissenschaften,* 44, 491, 1957.
28. Majumdar, A. K. and Mukherjee, A. K., *Anal. Chim. Acta,* 19, 23, 1958.
29. Majumdar, A. K. and Mukherjee, A. K., *Anal. Chim. Acta,* 21, 245, 1959.
30. Langmyhr, F. J. and Hongslo, T., *Anal. Chim. Acta,* 22, 301, 1960.
31. Majumdar, A. K. and Pal, B. K., *Anal. Chim. Acta,* 24, 497, 1961.
32. Pal, B. K., Ph.D. thesis, Jadavpur University, 1964.
33. Alimarin, I. P. and Chieh, Y. H., *Talanta,* 8, 317, 1961.
34. Ryan, D. and Lutwick, G. D., *Can. J. Chem.,* 31, 9, 1953.
35. Moshier, R. W. and Schwarberg, J. E., *Anal. Chem.,* 29, 947, 1957.
36. Alimarin, I. P. and Chieh, Y. H., *Vestn. Mosk. Univ. Khim.,* 15, 53, 1960; *Chem. Abstr.,* 55, 241a, 1961.
37. Das, B. and Shome, S. C., *Anal. Chim. Acta,* 33, 462, 1965.
38. Kaimal, V. R. M. and Shome, S. C., *Anal. Chim. Acta,* 27, 298, 1962.
39. Kaimal, V. R. M. and Shome, S. C., *Anal. Chim. Acta,* 29, 286, 1963.
40. Das, J. and Shome, S. C., *Anal. Chim. Acta,* 27, 58, 1962.
41. Kaimal, V. R. M. and Shome, S. C., *Anal. Chim. Acta,* 31, 268, 1964.
42. Alimarin, I. P. and Chieh, Y. H., *Zh. Anal. Khim.,* 14, 574, 1959; *Talanta,* 9, 9, 1962.
43. Ryan, D. E., *Can. J. Chem.,* 38, 2488, 1960; *Chem. Abstr.,* 55, 9161i, 1960.
44. Majumdar, A. K., *N-Benzoylphenylhydroxylamine and Its Analogues,* Pergamon Press, Oxford, 1972, 57.
45. Armor, C. H. and Ryan, D. E., *Can. J. Chem.,* 35, 1454, 1957.
46. Rydberg, J., *Acta Chem. Scand.,* 14, 157, 1960.
47. Riedel, A., *J. Radioanal. Chem.,* 13, 125, 1973.
48. Dyrssen, D., *Sven. Kem. Tidskr.,* 68, 212, 1956; *Chem. Abstr.,* 50, 14147c, 1956.
49. Lyle, S. J. and Shendrikar, A. D., *Anal. Chim. Acta,* 32, 575, 1965.
50. Lyle, S. J. and Shendrikar, A. D., *Anal. Chim. Acta,* 36, 286, 1966.
51. Priyadarshini, U. and Tandon, S. G., *Anal. Chem.,* 33, 435, 1961.
52. Murugaiyan, P. and Das, M. S., *Anal. Chim. Acta,* 48, 155, 1969.
53. Das, B. and Shome, S. C., *Anal. Chim. Acta,* 35, 345, 1966.
54. Chieh, Y. H., Candidate's thesis, Moscow State University, 1960; *Usp. Khim.,* 31, 989, 1962.
55. Ishii, H. and Einaga, H., *Bunseki Kagaku,* 17, 1296, 1968.
56. Villarreal, R. and Barker, S. A., *Anal. Chem.,* 41, 611, 1969.

57. Che-Ming, N. and Chu-Chuan, L., *Hua Hsueh Hsueh Pao,* 30, 540, 1964; *Chem. Abstr.,* 62, 11142g, 1965.
58. Zharovskii, F. G., Shpak, E. A., and Piskunova, E. V., *Ukr. Khim. Zh.,* 28, 1104, 1962; *Chem. Abstr.,* 59, 4544d, 1963.
59. Tanaka, K. and Takagi, N., *Bunseki Kagaku,* 12, 1175, 1963.
60. Afghan, B. K., Marryatt, R. G., and Ryan, D. E., *Anal. Chim. Acta,* 41, 131, 1968.
61. Zharovskii, F. G. and Pilipenko, A. T., *Ukr. Khim. Zh.,* 25, 230, 1959; *Anal. Abstr.,* 7, 3209, 1960.
62. Dyan, D. E., *Analyst,* 85, 569, 1960.
63. Shigematsu, T., Nishikawa, Y., Goda, S., and Hirayama, H., *Bull. Inst. Chem. Res. Kyoto Univ.,* 43, 347, 1965; *Chem. Abstr.,* 65, 2982b, 1966.
64. Zharovskii, F. G. and Sukhomlin, R. I., *Zh. Anal. Khim.,* 21, 59, 1966.
65. Dutt, N. K. and Seshadri, T., *Indian J. Chem.,* 6, 741, 1968.
66. Zharovskii, F. G. and Shpak, E. A., *Ukr. Khim. Zh.,* 25, 800, 1959; *Chem. Abstr.,* 54, 11646g, 1960.
67. Priyadarshini, U. and Tandon, S. G., *Analyst,* 86, 544, 1961.
68. Majumdar, A. K. and Das, G., *Anal. Chim. Acta,* 31, 147, 1964.
69. Jeffery, P. G. and Kerr, G. O., *Analyst,* 92, 763, 1967.
70. Villarreal, R., Krsul, J. R., and Baker, S. A., *Anal. Chem.,* 41, 1420, 1969.
71. Zharovskii, F. G., *Ukr. Khim. Zh.,* 25, 245, 1959; *Anal. Abstr.,* 7, 3254, 1960.
72. Alimarin, I. P., Petrukhin, O. M., and Chieh, Y. H., *Dokl. Acad. Nauk SSSR,* 136, 1073, 1961; *Anal. Abstr.,* 9, 2274, 1962.
73. Lyle, S. J. and Shendrikar, A. D., *Radiochim. Acta,* 3, 90, 1964.
74. Lyle, S. J. and Shendrikar, A. D., *Talanta,* 12, 573, 1965.
75. Vita, O. A., Levier, W. A., and Litteral, E., *Anal. Chim. Acta,* 42, 87, 1968.
76. Palshin, E. S., Myasoedov, B. F., and Novikov, Yu. P., *Zh. Anal. Khim.,* 18, 657, 1963.
77. Myasoedov, B. F., Palshin, E. S., and Palei, P. N., *Zh. Anal. Khim.,* 19, 105, 1964.
78. Lyle, S. J. and Shendrikar, A. D., *Talanta,* 13, 140, 1966.
79. Chmutova, M. K., Petrukhin, O. M., and Zolotov, Yu. A., *Zh. Anal. Khim.,* 18, 588, 1963.
80. Alimarin, I. P. and Chieh, Y. H., *Zavod. Lab.,* 12, 1435, 1959; *Talanta,* 8, 317, 1961.
81. Rakovskii, E. E. and Petrukhin, O. M., *Zh. Anal. Khim.,* 18, 539, 1963.
82. Alimarin, I. P. and Chieh, Y. H., *Vestn. Mosk. Univ. Khim.,* 15, 53, 1960; *Chem. Abstr.,* 55, 241a, 1961.
83. Pyrssen, D., *Acta Chem. Scand.,* 10, 353, 1956.
84. Che-Ming, N., Chung-Fen, C., and Shu-Chuan, L., *Acta Chim. Sinica,* 29, 249, 1963; *Chem. Abstr.,* 62, 5872, 1965.
85. Kaimal, V. R. M. and Shome, S. C., *Anal. Chim. Acta,* 27, 594, 1962.
86. Alimarin, I. P. and Chieh, Y. H., *Vestn. Mosk. Univ. Khim.,* 16, 59, 1961; *Chem. Abstr.,* 56, 10901e, 1962; *Anal. Abstr.,* 8, 4176, 1962.
87. Das, H. R. and Shome, S. C., *Anal. Chim. Acta,* 43, 140, 1968.
88. Fritz, J. S. and Sherma, J., *J. Chromatogr.,* 25, 153, 1966.
89. Shirguppi, M. K. and Haldar, B. C., *J. Indian Chem. Soc.,* 53, 848, 1976; *Chem. Abstr.,* 86, 182468q, 1977.
90. Bag, S. P. and Khastagir, A. K., *J. Indian Chem. Soc.,* 54, 611, 1977; *Anal. Abstr.,* 35, 3B93, 1978.
91. Shome, S. C., *Curr. Sci. India,* 13, 257, 1944.
92. Ostroumov, E. A., Kulumbegashvili, V. A., and Tetrashvili, M. S., *Org. Reagenty Anal. Khim. Tezisy Dokl. Vses. Konf.,* 4th, 2, 25, 1976; *Chem. Abstr.,* 87, 161135j, 1977.
93. Shpak, E. A. and Zulfigarov, O. S., *Org. Reagenty Anal. Khim., Tezisy Dokl. Vses. Konf.,* 4th, 2, 8, 1976; *Chem. Abstr.,* 87, 161130d, 1977.
94. Einaga, H. and Ishii, H., *Bunseki Kagaku,* 17, 836, 1968.
95. Lahiri, S., *J. Indian Chem. Soc.,* 51, 903, 1974; *Anal. Abstr.,* 29, 6B94, 1975.
96. Bag, S. P. and Lahiri, S., *Indian J. Chem.,* 13, 415, 1975; *Anal. Abstr.,* 29, 6B54, 1975.
97. Lahiri, S., *Indian J. Chem.,* 12, 1206, 1974; *Anal. Abstr.,* 29, 2B62, 1975.
98. Gallai, Z. A. and Sheina, N. M., *Usp. Anal. Khim.,* 278, 1974; *Chem. Abstr.,* 83, 37018b, 1975.
99. Pilipenko, A. T., Shpak, E. A., and Samchuk, A. I., *Ukr. Khim. Zh.,* 40, 266, 1974; *Anal. Abstr.,* 27, 2527, 1974.
100. Gallai, Z. A., Sheina, N. M., Svediene, N., and Oliferenko, G. L., *Zh. Anal. Khim.,* 30, 2346, 1975.
101. Gallai, Z. A., Sheina, N. M., Svediene, N., and Oliferenko, G. L., *Zavod. Lab.,* 44, 1180, 1978; *Chem. Abstr.,* 90, 66114e, 1979.
102. Ahmed, M. K. and Rao, C. S., *Talanta,* 25, 708, 1978.
103. Agrawal, Y. K., *Talanta,* 20, 1213, 1973.
104. Zharovskii, F. G., Ostrovskaya, M. S., and Babyuk, L. A., *Ukr. Khim. Zh.,* 39, 73, 1973; *Anal. Abstr.,* 25, 2145, 1973.

105. Shukla, J. P., Agrawal, Y. K., and Bhatt, K., *Sep. Sci.*, 8, 389, 1973; *Anal. Abstr.*, 26, 3174, 1974.
106. Yeole, V. V., Patil, P. S., and Shinde, V. M., *Fresenius Z. Anal. Chem.*, 294, 46, 1979.
107. Gordeeva, M. N. and Ryndina, A. M., *Vestn. Leningr. Univ. Fiz. Khim.*, 16(3), 149, 1972; *Anal. Abstr.*, 25, 148, 1974.
108. Das, M. K. and Chakraborti, D., *Indian J. Chem.*, 12, 773, 1974; *Anal. Abstr.*, 29, 1B46, 1975.
109. Pilipenko, A. T., Shpak, E. A., and Zulfigarov, O. S., *Zh. Anal. Khim.*, 29, 1074, 1974.

POLY(MACROCYCLIC) COMPOUNDS

Synonyms

Poly(macrocyclic) compounds, such as crown polyethers and cryptands, which may be of interest in analytical chemistry are summarized in Table 1, together with their synonyms and physical properties. As the naming of these compounds of IUPAC rule is very cumbersome, the trivial names "crown" and "cryptand" are often used for monocyclic and bicyclic polyether, respectively. In the former case, the total number of atoms in the polyether ring, the "crown", and the number of oxygen atoms in the main ring are indicated. In the latter case, the "cryptand" and the number of oxygen atoms on each bridge are indicated.

Source and Method of Synthesis

All compounds listed in Table 1 are commercially available. The syntheses of this kind of the reagents often necessitate laborious laboratory works, and each reference in the table should be consulted for the detailed synthetic procedure.

Analytical Uses

As complexing agents for the solvent extraction separation and photometric determination of alkaline earth and alkali metal ions.

Properties of Reagents

In general, crown reagents (1) to (6) are slightly soluble in water and organic solvents except methylene chloride, chloroform, pyridine, and formic acid. As an example, the solubilities of (3) in various organic solvents are summarized in Table 2.

The solubility of crown compounds in water decreases with the increase in temperature, as shown in Table 3 for (5), the reason for this being ascribed to breakdown of hydrogen bonds between ether oxygen and hydraated water.

Distribution ratio of (5) between benzene-water is reported to be >170 at 25°.[R2] Cryptand (7) to (9) behave as diacidic base, and the proton dissociation constants of conjugate acids are summarized in Table 7.

Complexation Reaction and Properties of Complexes

Poly(macrocyclic) compounds are known as ligands which form fairly stable stoichiometric complexes with certain cations. Special importance is due to their preference for alkali metal ions which do not form complexes with many of numerous ligands for the transition metal ions.

The reagents (1) to (9) form stable crystalline complexes and solution of the complexes with some or all of the cations of Li, Na, K, Rb, Cs, NH_4, RNH_3, Ag(I), Au(I), Ca, Sr, Ba, Ra, Zn, Cd, Hg(I) (II), La(III), Tl(I), Ce(III), and Pb(II).[1]

Some of them, e.g., (5), also form complexes with Co(II), U(VI), and some other transition metal ions.[8] The saturated ligands (5) and (6) are better complexing agents than the corresponding aromatic compounds (2) and (4).

These reagents have holes of different diameters in the center of macrocycle rings. The uncomplexed cations also have their own ionic radii. The most stable complex is formed with a cation which fits well into the hole of a given ligand. Depending, however, on the relative sizes of the hole and the cation, complexes of ligand / cation ratio of 1:1, 3:2 and 2:1 have also been reported. The structures of these complexes are schematically shown in Figure 1.

The solubility of macrocyclic polyether in organic solvents is not so high as shown in Table 2, but in the presence of inorganic salts, the ligands and complexable salts

Table 1
POLY(MACROCYCLIC) COMPOUNDS AS ANALYTICAL REAGENTS

Trivial name	Structural formula (chemical name)	Molecular formula (mol wt)	Physical properties	Solubility	Ref.
15-Crown-5 (1)	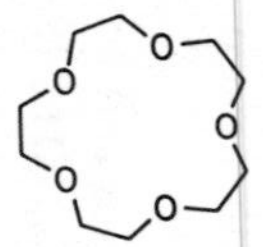 1,4,7,10,13-Pentaoxacyclopentadecane	$C_{10}H_{20}O_5$ (220.27)	Colorless viscous liquid, bp 100 $\sim$ 135°/0.2 Torr 78°/0.05 Torr	Insoluble in water; soluble in alcohol, aromatic hydrocarbon, and petroleum ether	1, 2, 7
18-Crown-6 (2)	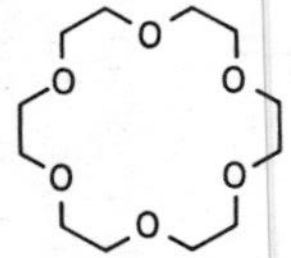 1,4,7,10,13,16-Hexaoxacyclooctadecane	$C_{12}H_{24}O_6$ (264.32)	Colorless crystal, mp 36.5 $\sim$ 38°	Similar to (1)	1-3, 7
Dibenzo-18-crown-6 (3)	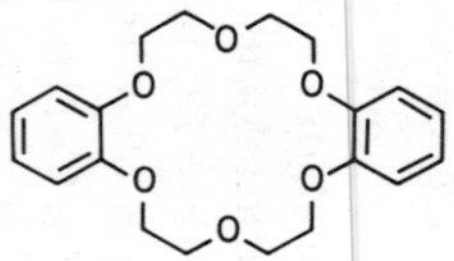 2,3,11,12-Dibenzo-1,4,7,10,13,16-hexaoxacyclooctadeca-2,11-diene	$C_{20}H_{24}O_6$ (360.41)	Colorless fibrous crystal, mp 162 $\sim$ 164°, bp 380 $\sim$ 384°/769 Torr, irritating to skin	Slightly soluble in water and alcohol; soluble in benzene, dioxane, methylene chloride, chloroform, pyridine, and formic acid	1, 4

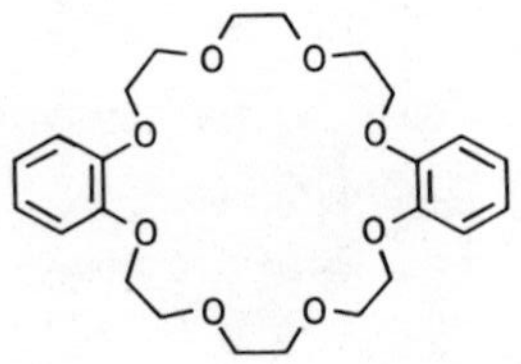 Dibenzo-24-crown-8 (4) 2,3,14,15-Dibenzo-1,4,7,10,14,16,19,22-octaoxacyclotetraeicosane-2,14-diene	$C_{24}H_{32}O_8$ (448.51)	Colorless crystal; mp 113∽114°	Similar to (3)	1, 4
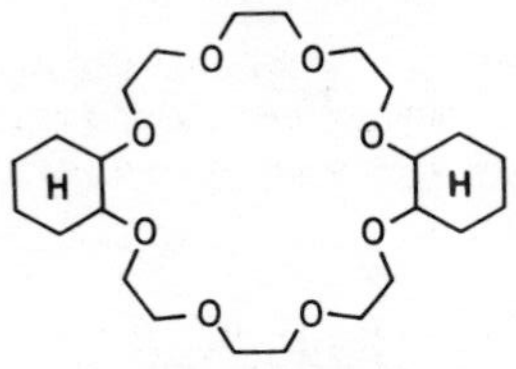 Dicyclohexyl-18-crown-6 (5) 2,3,11,12-Dicyclohexyl-1,4,7,10,13,16-hexaoxacyclooctadecane	$C_{20}H_{36}O_6$ (372.50)	Colorless or pale yellow wax; mixture of isomers of different steric conformations of cyclohexane rings; mp 38∽54°; irritating to skin	Similar to (1)	1, 4
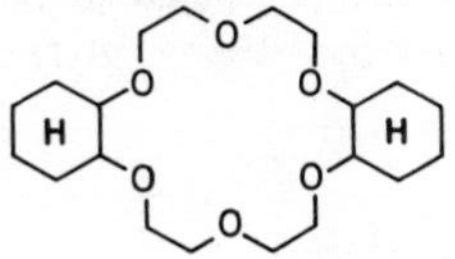 Dicyclohexyl-24-crown-8 (6) 2,3,14,15-Dicyclohexyl-1,4,7,10,13,16,19,22-octaoxacyclotetraeicosane	$C_{24}H_{44}O_8$ (460.61)	Pale yellow viscous liquid; mp <20°	Similar to (1)	1, 4

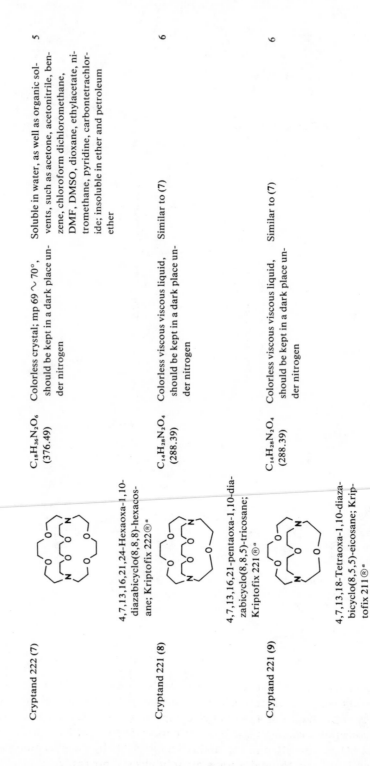

Cryptand 222 (7) 4,7,13,16,21,24-Hexaoxa-1,10-diazabicyclo(8,8,8)-hexacos-ane; Kriptofix 222®ᵃ	$C_{18}H_{36}N_2O_6$ (376.49)	Colorless crystal; mp 69 ∼ 70°, should be kept in a dark place under nitrogen	Soluble in water, as well as organic solvents, such as acetone, acetonitrile, benzene, chloroform dichloromethane, DMF, DMSO, dioxane, ethylacetate, nitromethane, pyridine, carbontetrachloride; insoluble in ether and petroleum ether	5
Cryptand 221 (8) 4,7,13,16,21-pentaoxa-1,10-diazabicyclo(8,8,5)-tricosane; Kriptofix 221®ᵃ	$C_{14}H_{28}N_2O_4$ (288.39)	Colorless viscous viscous liquid, should be kept in a dark place under nitrogen	Similar to (7)	6
Cryptand 221 (9) 4,7,13,18-Tetraoxa-1,10-diazabicyclo(8,5,5)-eicosane; Kriptofix 211®ᵃ	$C_{14}H_{28}N_2O_4$ (288.39)	Colorless viscous viscous liquid, should be kept in a dark place under nitrogen	Similar to (7)	6

ᵃ Kryptofix is a tradename of E. Merck Co. (West Germany) for nitrogen containing macrocyclic polyether ligands.

Table 2
SOLUBILITIES OF DIBENZO-18-CROWN-6 (3) (26 ± 5°)

Solvent	Dielectric constant	Solubility (mg/ml)	Solvent	Dielectric constant	Solubility (mg/ml)
Cyclohexane	2.05	0.24	Ethanol	25	0.32
Carbon-tetrachloride	2.24	1.8	Methanol	33.1	0.36
Benzene	2.28	64.9	Formic acid	—	382
Chloroform	5.05	75.7	DMF	36.7	20.2
Ethyl acetate	6.4	3.6	Acetonitrile	38.8	28.5
Tetrahydrofuran	7	7.9	Nitromethane	39	16.9
1-Butanol	7.8	0.36	DMSO	45	17.3
Pyridine	12.5	43.2	Water	80	0.03
Acetone	21.4	3.3			

From Pedersen, C. J., *J. Am. Chem. Soc.*, 89, 7017, 1967. With permission.

Table 3
SOLUBILITY OF DICYCLOHEXYL-18-CROWN-6 (5) IN WATER

Solvent	Temp	Solubility (mg/ml)
Water	26	13.4
Water	53	8.2
Water	82	3.7
1 N KOH soln	26	331
1 N KCl soln	26	> 346

From Pedersen, C. J., *J. Am. Chem. Soc.*, 89, 7017, 1967. With permission.

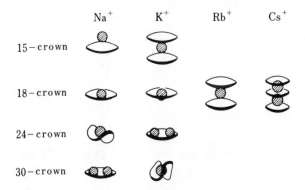

FIGURE 1.   Schematic presentation of polyether complexes. Solid dot stands for metal ions, and the loop stands for polyether.

mutually increase their solubilities considerably. Some examples are given in Tables 4 and 5. Addition of coordinating solvents such as methanol further increases their solubilities as shown in Table 4. Although complexing of cation is an obvious prerequisite for solubilization, the anion also plays an important role. Salts of "hard" anions, such as F^- and SO_4^{2-}, are often not solubilized appreciably by these reagents, while "soft" anions, such as I^-, SCN^-, and picrate, are solubilized more easily.

Table 4
SOLUBILIZATION OF ALKALIMETAL HALIDES IN
ORGANIC SOLVENTS BY ADDITION OF 50 m*M* OF
DICYCLOHEXYL-18-CROWN-6 (*5*)

Solvent	Methanol (m*M*)	Solubility (m*M*)[a]				
		NaCl	NaBr	KCl	KBr	KI
Benzene	0	0.01	1.8	0.03	2.3	9.2
	250	0.48	24	8.7	30	46
Carbon tetra-chloride	0	0.03	2.7	0.6	4.1	0.8
	250	1.1	28	8.8	34	15
Chloroform	0	1.8	37	21	41	43
	250	5.7	41	34	44	44
Methylene chloride	0	1.8	35	17	41	43
	250	5.8	42	33	42	44
Terahydrofuran	0	0.02	1.2	0.1	3.6	45
	250	0.04	5	0.4	13	50

[a] Salt concentration after agitation of the polyether solution with enough salt to give 50 m*M* solution.

From Pedersen, C. J. and Frensdorf, H. K., *Angew. Chem. Int. Ed. Engl.*, 11, 16, 1972. With permission.

Table 5
SOLUBILIZATION OF
DIBENZO-18-CROWN-6 (3) IN
METHANOL BY ADDITION
OF 25 m*M* OF INORGANIC
SALTS AT 30°[11]

Inorganic salt (25 m*M*)	Solubility of (*3*) (m*M*)
No salt	1.1
NaSCN	23.6
KF	24.7
RbSCN	25.6
AgNO₃	22.2
SrCl₂	17.9
BaCl₂	26.6

Reagents such as (3) and (4) have an absorption band at 275 nm (in methanol) due to aromatic rings. Complexing with a cation results in a noticeable change in this band as shown in Figure 2.

Stability constants for the poly(macrocyclic)-cation complexes are straightforward measures of the strength of complexing in solution. They are collected in Tables 6 and 7. The stability constant for each cation goes through a maximum with increasing polyether ring size. These optimum ring sizes are exactly those which provide the closest fit between the cation and the ligand hold, as depicted in Figure 3.

Complex cation formed between crown polyether and metal ions can be extracted with an appropriate anion into immiscible solvent as an ion-pair. The extent of extraction is governed by many factors, such as stability constant of the complex, kind of polyether, anion, and solvent. Such extraction processes are often utilized in analytical

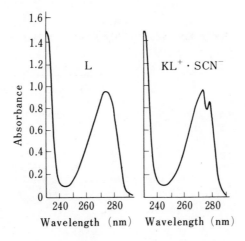

FIGURE 2. UV absorption spectra of dibenzo-18-crown-6 (3) and its KSCN complex (1.8×10^{-4} mol/ℓ in methanol, 1-cm cell). (From Pedersen, C. J. and Frensdorf, H. K., *Angew. Chem. Int. Ed. Engl.*, 11, 16, 1972. With permission.)

Table 6
STABILITY CONSTANT OF SOME CROWN POLYETHERS[12,a]

		$\log K_{ML}$ (25°, $\mu = 10^{-4}$ to 10^{-2})									
Ligand	Solvent[a]	Li	Na	K	Rb	Cs	NH₄	Ag	Ba	Sr	Cs
(2)	W	—	< 0.3	2.06	—	0.8	1.1	1.6	—	—	—
	M	—	4.32	6.10	—	4.62	—	—	—	—	1.30
(3)	W/T	—	—	—	1.35	—	—	—	—	—	—
	M	—	4.36	5.00	—	3.55					2.92
(4) (isomer A)	W	0.6	1.5 ⌒ 1.8	2.18	1.52	1.25	1.4	2.3	3.57	3.24	—
	M	—	4.08	6.01	—	4.61	—	—	—	—	0.59
(4) (isomer B)	W	—	1.2 ⌒ 1.6	1.78	0.87	—	0.8	1.8	3.27	2.64	—
	M	—	3.68	5.38	—	3.49	—	—	—	—	0.9
(4) (mixture of isomers)	M	—	4.05	5.35	—	3.85	—	—	—	—	—
(5)	M	—	—	3.49	—	3.78	—	—	—	—	—
(6)	M	—	—	—	—	1.9	—	—	—	—	—

[a] Further data on (7), (8), and (9) are available in R5.
[b] Key: W, water; M; methanol W/T, 1:1 mixture of water and THF.

chemistry as means of the separation of metal ions, the extractive photometric determination of alkali metals, or as the ion carriers in ion selective electrodes. The extraction constants for some crown polyethers are summarized in Table 8. No data are available for cryptand polyethers at present.

Purification and Recovery of Reagent

Commercially available crown polyethers and cryptand polyethers are usually of high purity for most analytical uses. The polar impurities in crown polyethers may be removed by passing a *n*-heptane solution of crude material through a column of acid washed alumina.[1] The cryptand polyethers are rather expensive and may be recovered by the following procedure.[19]

Table 7

STABILITY CONSTANTS OF SOME CRYPTAND POLYETHERS[13]

| Ligand | solvent[a] | pK₁ | pK₂[b] | log K_{ML}[e] (25°) | | | | | | | |
				Li	Na	K	Rb	Cs	Mg	Ca	Sr
(7)	W	7.28	9.60	<2.0	3.9	5.4	4.35	<2.0	<2.0	4.4	8.0	9.5
	M/W	6.64	9.85	1.8	7.21	9.75	8.40	3.54	<2.0	7.60	11.5	12[d]
	M	—	—	2.6	>8.0	>7.0	>6.0	4.4	—	—	—	—
(8)	W	7.50	10.53	2.50	5.40	3.95	2.55	<2.0	<2.0	6.95	7.35	6.30
	M/W	6.60	10.42	4.18	8.84	7.45	5.80	3.90[e]	<2.0	9.61	10.65	9.70
	M	—	—	>5.0[c]	>8.0	>7.0	>6.0	∿5.0[e]	—	—	—	—
(9)	W	7.85	10.64	5.5	3.2	<2.0	<2.0	<2.0	2.5 ± 0.3	2.50	<2.0	<2.0
	M/W	6.56	11.00	7.58	6.08	2.26	<2.0	<2.0	4.0 ± 0.8	4.34	2.90	<2.0
	M	—	—	>6.0	6.1	2.3	1.9	<2.0	—	—	—	—

[a] Key: W, water; M/W, methanol / water 95/5; M, methanol.
[b] Proton dissociation constants of conjugate acids (H₂L²⁺ and HL⁺).
[c] Total ionic strength: 0.05 in W; 0.01 in M/W; 0.01 in M.
[d] Values are less accurate.
[e] External complexes of ML₂ stoichiometry may be present.

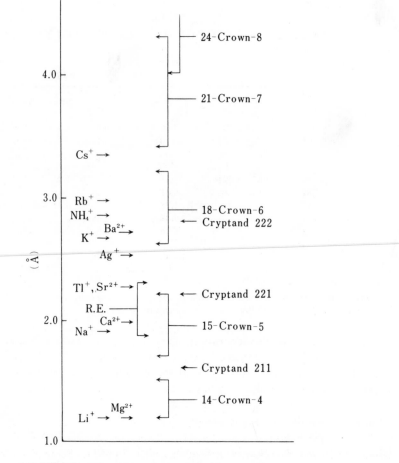

FIGURE 3.   Relationship between cation diameter and ligand hole size. The values are taken from References R2 and M1.

Table 8
EXTRACTION CONSTANTS OF CROWN POLYETHERS

		Distribution		$\log k_{ex}^c$						
Ligand	Anion	between[a]	$\log D_L^b$	Li	Na	K	Rb	Cs	Ca	Ref.
(2)	Picrate	B/W	−1.24	—	3.31	6.00	5.43	4.28	—	15
(3)	Picrate	B/W	2.90	2.9	2.21	4.65	3.75	3.07	—	16
	Picrate	DCM/W	—	—	—	2.85	—	—	—	
(5)	Picrate	DCM/W	—	—	4.18	6.32	—	—	—	17
	Picrate	Hex/W	—	—	—	3.58	—	—	—	
(5)	Picrate	DCM/W	3.60	—	—	6.60	—	—	—	17
(isomer A)	Picrate	Hex/W	0.79	—	—	—	—	—	—	
(5)	Picrate	DCM/W	3.56	—	—	6.04	—	—	—	17
(isomer B)	Picrate	Hex/W	0.85	—	—	—	—	—	—	
(5)	Picrate	DCM/W	—	—	4.20	6.36	—	—	—	18
	2,5-Dinitro-phenol	DCM/W	—	—	—	3.98	—	—	—	
	2,6-Dinitro-phenol	DCM/W	—	—	—	4.26	—	—	—	
	Dipicrylamine	DCM/W	—	—	7.81	8.16	—	—	10.23	

[a] Key: B/W, benzene/water; DCM/W, dichloromethane/water; Hex/W, n-hexane/water.

[b] $D_L = [L]_{org}/[L]_{aq}$.

[c] $K_{ex} = [LMA]_{org}/[L]_{org} \cdot [M^+]_{aq} \cdot [A]_{aq} \cdot$ or $= [LMA_c]_{org}/[L]_{org} \cdot [M^{2+}]_{aq} \cdot [A^-]^2_{aq}$.

Evaporate the solutions of metal cryptates in various solvents to dryness in vacuo at $\sim 10^{-2}$ Torr at room temperature. Dissolve the solids (0.1 to 0.3 g) in 20 mℓ of aqueous 6 N HCl with gentle heating and evaporate the solution again to dryness at $\sim 10^{-2}$ Torr. Dissolve the residue, containing alkali salts and diprotonated cryptand, in 20 mℓ of aqueous 0.1 N HCl. (If the original solutions contain a variety of anions, convert the salts to the chloride form by anion exchange column in Cl⁻ form). Pass the solution through a cation exchange column in H⁺ form (Dowex® 50 × 8, 100 to 200 mesh, 1.4 × 22 cm). Elute metal ions with 150 mℓ of 1.0 N HCl. Finally elute the cryptand with ~ 120 mℓ of 6 N HCl. Dry the cryptand solution in vacuo to obtain the diprotonated ligand (L · 2 HCl).

Analytical Applications

In spite of very interesting complexing properties of poly(macrocycle) compounds, the papers related with analytical applications have appeared only recently. They include the selective separation of metal ions by solvent extraction,[15,16,18,20-23] the uses as an ion carrier in ion selective electrode,[24] the analytical applications of polymers containing polyether functions,[25] and the determination of Hg(II) in HPLC.[26] The interesting application is, however, the extraction photometric determination of alkali metal ions using crown polyethers.

The principle of this method is based on the selective complex formation of crown polyether with alkali metal ions, followed by the solvent extraction of the ion-pair formed with a highly colored anion such as Bromo Cresol Green.[27]

Another approach to the photometric determination of alkali metals is to develop the new chromogenic polyether reagents. These reagents complex alkali metal ion selectively to give a color change, one of successful examples being given below.[28]

4'-Picrylamino-5'-nitro-benzo-18-crown-6 (10)

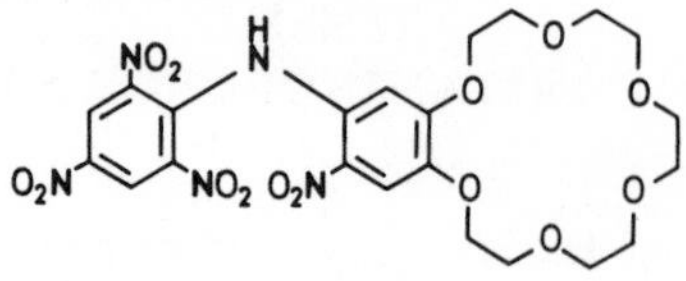

HL

(10)

Dark red needles, mp 160°; very slightly soluble in water, soluble in various organic solvent; and pKa (NH) 10.55 in 50% aqueous dioxane (μ = 0.1 LiCl). The color of HL in chloroform is orange (λ_{max} 425 nm, ε = 1.46 × 10^4) which changes to blood red (λ_{max}, nm, ε = 2.11 × 10^4) upon complex formation at pH $\simeq$ 11 (EDTA-LiOH), according to following equation.

$$HL_{org} + M^+ \rightleftharpoons ML_{org} + H^+$$

Values of log K$_{ex}$ for chloroform / water system are K, −9.19 ; Rb, −9.42 ; Cs, −9.54 (at 25°). Absorption spectra of HL and KL in chloroform are illustrated in Figure 4.[28]

Extraction photometric determination of potassium in serum by ion-pair extraction[27]
Place 0.1 mℓ of serum sample in a 10-mℓ centrifuge tube and add 4.9 mℓ of 1.5% aqueous trichloroacetic acid. Mix well, then centrifuge the mixture for 5 min. Transfer 2 mℓ of the supernatant liquid into another 10 mℓ centrifuge tube and add 0.2 mℓ of 11% aqueous lithium acetate solution, 0.4 mℓ of 0.2 M lithium acetate buffer solution (pH 3.9), 1.4 mℓ of 0.16% Bromo Cresol Green solution (in 20% ethanol), and 4.0 mℓ of 0.3% 18-crown-6 benzene solution. Shake for 5 min with a mechanical shaker and centrifuge for 1 min. Measure the absorbance of the organic phase at 410 nm against a reagent blank.

The calibration curve is linear up to 5 ppm of potassium, and sodium up to 500 ppm does not interfere.

Extraction photometric determination of potassium in Portland cement®️ with a chromogenic crown polyether (10)[23]
Digest a 2-g sample with 10 mℓ concentrated HCl and evaporate to dryness. Dissolve the residue into 20 mℓ water, neutralize it with 0.1 N LiOH, and dilute to 100 mℓ. Filter the solution with dry filter paper and dilute 10-mℓ aliquot to 100 mℓ. Place 2 mℓ sample solution, 2 mℓ 0.2 M EDTA-Li$_4$ solution, and 5 mℓ 1 × 10^{-4} M CHCl$_3$ solution of the reagent (10) in a stoppered test tube, shake for 3 min, and after phase separation, observe the absorbance at 560 nm against a reagent blank. A linear relationship is observed in the range of 40 to 400 ppm of potassium. Tenfold molar excess of sodium does not interfere, and interferences from multivalent metal ions are masked with EDTA.

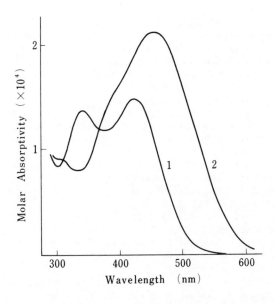

FIGURE 4. Absorption spectra of picrylaminobenzo-crown reagent (1) HL; (2) KL in chloroform. (From Nakamura, H., Takagi, M., and Ueno, K., *Anal. Chem.*, 52, 1668, 1980. With permission.)

MONOGRAPH

M1. Izatt, R. M. and Christensen, J. J., Eds., *Synthetic Multidentate Macrocyclic Compounds*, Academic Press, New York, 1978.

REVIEWS

R1. Truter, M. R. and Pedersen, C. J., Cryptates, *Endeavour*, 30, 142, 1971.

R2. Pedersen, C. J. and Frensdorf, H. K., Macrocyclic polyethers and their complexes, *Angew. Chem. Int. Ed. Engl.*, 11, 16, 1972.

R3. Christensen, J. J., Eatough, D. J., and Izatt, R. M., The synthesis and ion binding of synthetic multidentate macrocyclic compounds, *Chem. Rev.* 74, 351, 1974.

R4. Pedersen, C. J., *Aldrichim. Acta*, 4, (1), 1, 1971.

R5. Kolthoff, I. M., Application of macrocyclic compounds in chemical analysis, *Anal. Chem.*, 51, 1R, 1979.

REFERENCES

1. Pedersen, C. J., *J. Am. Chem. Soc.*, 89, 7017, 1967.

2. Geraldine, J., Ranson, C. J., and Reese, C. B., *Synthesis*, 515, 1976. *Chem. Abstr.*, 85, 177387n, 1976.

3. Dale, J. and Kristiansen, P. O., *Chem. Commun.* 673, 1971.

4. Pedersen, C. J., *Org. Synth.,* 52, 66, 1972.
5. Dietrich, B., Lehn, J. M., and Sauvage, J. P., *Tetrahedron Lett.,* 34, 2885, 1969.
6. Lehn, J. M. and Sauvage, J. P., *Chem. Commun.,* 440, 1971.
7. Johns, G., Ranson, C. J., and Reese, C. B., *Synthesis,* 515, 1976.
8. Su, A. C. L. and Weiher, J. F., *Inorg. Chem.,* 7, 176, 1968.
9. Poonia, N. S., *J. Am. Chem. Soc.,* 96, 1012, 1974.
10. Pedersen, C. J., *J. Am. Chem. Soc.,* 92, 386, 1970.
11. Pedersen, C. J., *J. Am. Chem. Soc.,* 89, 2495, 1967.
12. Christensen, J. J., Eatough, D. J., and Izatt, R. M., *Chem. Rev.,* 74, 351, 1974.
13. Lehn, J. M. and Sauvage, J. P., *J. Am. Chem. Soc.,* 97, 6700, 1975.
14. Lehn, J. M., *Struct. Bonding, (Berlin),* 16, 1, 1973.
15. Iwachido, T., Sadakane, A., and Toei, K., *Bull. Chem. Soc. Jpn.,* 51, 629, 1978.
16. Iwachido, T. and Toei, K., *Bull. Chem. Soc. Jpn.,* 48, 60, 1975.
17. Frensdorf, H. F., *J. Am. Chem. Soc.,* 93, 4684, 1971.
18. Jawaid, M. and Ingman, F., *Talanta,* 25, 91, 1978.
19. Shih, J. S., Liu, L., and Popov, A. I., *J. Inorg. Nucl. Chem.,* 39, 552, 1977.
20. King, R. B. and Heckley, P. P., *J. Am. Chem. Soc.,* 96, 3118, 1974.
21. Kimura, T., Iwashima, K., Ishimori, T., and Hamaguchi, H., *Chem. Lett.,* 562, 1977.
22. Mitchell, J. W. and Shanks, D. L., *Anal. Chem.* 47, 642, 1975.
23. Takeda, Y., Suzuki, S., and Ohyagi, Y., *Anal. Lett.,* 11, 1377, 1978.
24. Moody, G. J. and Thomas, J. D. R., *Selective Ion-Sensitive Electrode,* Merrow, England, 1971.
25. Blasius, E., Jensen, K. P., Adrian, W., Klautke, G., Korscheider, P., Maurer, P. G., Nguyen, B., Tien, T., Sholten, G., and Stockemer, J., *Fresenius Z. Anal. Chem.,* 284, 337, 1977.
26. Mangia, A. and Parolari, G., *Anal. Chim. Acta,* 92, 111, 1977.
27. Sumiyoshi, H., Nakahara, K., and Ueno, K., *Talanta,* 24, 763, 1977.
28. Takagi, M., Nakamura, H., and Ueno, K., *Anal. Lett.,* 10, 1115, 1977; Takagi, M., Nakamura, H., and Ueno, K., *Talanta,* 26, 921, 1979; *Anal. Chem.,* 52, 1668, 1980.

MISCELLANEOUS *o,o*-DONATING REAGENTS

Chromotropic Acid: 1,8-Dihydroxy-3,6-naphthalenedisulfonic acid, disodium salt ($C_{10}H_6O_8S_2Na_2 \cdot 2H_2O$, mol wt = 400.28)

Na$_2$H$_2$L

Pure material is a colorless crystalline powder, but most commercial samples are pink or red-brown due to the air oxidation during the storage, and it is rather difficult to purify the colored sample. However, a slightly colored sample may be satisfactory for the general purposes as an analytical reagent. It does not melt below 300°; easily soluble in water, but not soluble in common organic solvents; pKa_3(OH) = 5.36 and pKa_4(OH) = 15.6 (20°, μ = 0.1).[36]

Chromotropic acid forms colored soluble chelates with Ag, Au, Bi, Cu(II), Fe(III), Hg(II), Ti, UO_2^{2+}, V(IV) (V), W(VI), Zr, and anions,[1] such as BO_3^{3-}, $Cr_2O_7^{2-}$, MnO_4^-, and NbO_3^- and has been known as a photometric reagent for Ti (pH 1.0, ε = 1.6 × 10^4 at 470 nm, 0.5 to 5 ppm)[2,3] and BO_3^{3-} (pH 2.5 to 4.5, EDTA, extraction with 1,2-dichloroethane, ε = 1.4 × 10^4 at 351 nm),)[4] but better reagents are now available. Chromotropic acid is also used as a chromogenic reagent for NH_3 (absorption into $CuCl_2$ solution, 520 nm, 0.05 to 5 ppm)[5] and formaldehyde (570 nm)[6] and its precursors (methanol[7] and EDTA [8]).

Morin: 3,5,7,2′,4′-Pentahydroxyflavone, C.I. 75660 ($C_{15}H_{10}O_7 \cdot 2H_2O$, mol wt = 338.27)

H$_5$L

A colorless or pale yellow crystalline powder, mp 285 to 300°[M1]; almost insoluble in water (0.09% at 100°), but easily soluble in aqueous alkali or common organic solvents except ether and acetic acid. Solid reagent turns to brown upon air oxidation; pKa values for the first to fifth protons are −1, 4.8, 7, 9, and 13, respectively.[9]

Morin is one of the most frequently used polyhydroxyflavones as an analytical reagent.[R1,R3] The most important use is as a fluorimetric reagent. Morin itself shows a weak green fluorescence at pH 4 to 9, but it intensifies significantly upon complexation with metal ions, such as Al (pH 3; λ_{exit}, 440 nm; λ_{emit}, 525 nm);[10] B (dilute HCl, 365 nm, 490 nm),[11] Be (0.04 N NaOH, 460 nm, 540 nm);[9,12] Ga(pH 2.5 to 2.9, 400 nm, 445 nm);[13] Th (0.01 N HCl, 365 nm, 404.7 nm);[14] Zr; Hf (2 N HCl, 450 nm, 502 nm);[15] and rare earths (pH 2.5, 401 nm, 501 nm).[16] Thus, these elements can be determined fluorimetrically. EDTA and DTPA were often used as masking agents.

Morin also forms intensely colored chelates with various metal ions, such as Th (yellow), Ga, In, U, and Zr (red-brown in NH_3 alkali), and can be used as a photometric reagent for these elements,[M1] but the fluorimetry is much more sensitive.

Alizarin Red S: 1,2-Dihydroxyanthraquinone-3-sulfonic acid, Alizarin S, Alizarin Carmine, C.I. Mordant Red 3, C.I. 58005 ($C_{14}H_7O_7$-SNa·H_2O, mol wt = 360.27)

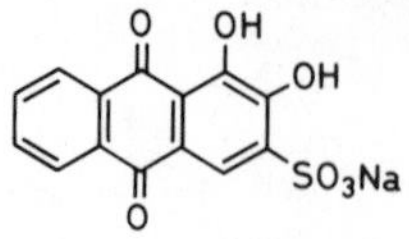

NaH₂L

Commercially available samples are a monosodium, monohydrate, and yellow-brown or yellow-orange crystalline powder; easily soluble in water to give a yellow solution, but almost insoluble in common organic solvents. It is soluble in concentrated H_2SO_4 to give an orange solution; pKa_2 (β-OH) = 5.39 and pKa_3 (α-OH) = 10.72 (25°, μ = 0.5).[17] The aqueous solution shows an absorption maximum at 420 nm (pH <3.5) or at 515 nm (pH >3.5).[23]

It forms colored soluble or insoluble chelates with many metal ions[M1,M2,M4] and has been recommended as a detecting reagent for Al,[18] F⁻,[19] and BO_3^{3-} and as a photometric reagent for Al (pH 4.4 to 4.6, ε = 1.8 × 10⁴ at 490 nm, 0 to 0.8 ppm),[20] Be (pH 5.4 to 5.6, 480 nm, ε = 4.3 × 10³, 0.2 to 4.7 ppm),[21] Zr (pH 3.9 to 4.6, trioctylamine, extraction with toluene, 538 nm),[22] and B (pH 7.7 to 8.2, EDTA, ε = 1250 at 426 nm).[23] Besides these elements, the following metal ions have been determined with Alizarin Red S: In,[17] Mo,[24] Rh,[25] Zn,[26] and rare earths.[27]

Many other hydroxyanthaquinones including quinalizarine (1,2,5,8-tetrahydroxyanthraquinone, $C_{14}H_8O_6$, mol wt = 272.21) have been investigated as the reagents for metal ions, especially for Al, Be, and B. Some details may be found in the reference.[M3]

Stilbazo: Stilbene-4,4′-*bis* (1-azo)-3,4-dihydroxybenzene-2,2′-disulfonic acid, diammonium salt ($C_{26}H_{18}O_{10}N_4S_2$·2(NH_4), mol wt = 646.65

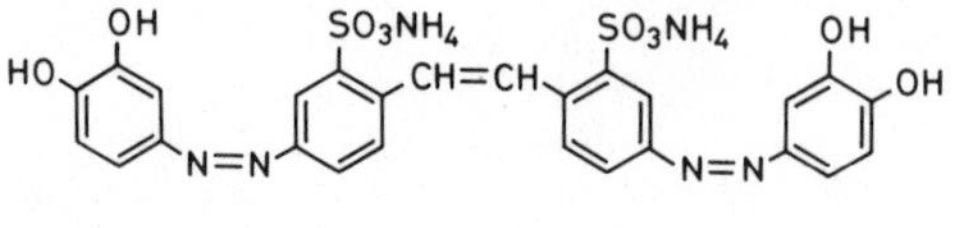

(NH_4)₂H₄L

Commercially available. A dark brown powder which is slightly soluble in water to give a yellow (pH 3 to 7), orange (pH ≃9), or red solution (pH ≃11). Although Stilbazo forms colored complexes with wide variety of metal ions,[M4,R2,28] the analytical application of practical importance is as a photometric reagent for Al in the presence of Fe (pH 5 to 6, ε = 1.95 to 3.46 × 10⁴ at 500 to 520 nm, 0.2 to 1.2 ppm).[M4,29] Interference of Fe (up to 100 ppm) can be masked with ascorbic acid. In the presence of Zephiramine®, a much higher sensitivity can be attained (pH 10, 570 nm, 0.08 to 0.64 ppm Al).[30] Other elements, such as B (pH 8.9 to 9.1, ε = 1340 at 414 nm, 0 to 2 ppm),[31] Ga,[32] In,[33] Mo,[34] and Sn,[35] have also been determined with Stilbazo.

MONOGRAPHS

M1. Sandell, E. B. and Onishi, H., *Photometric Determination of Traces of Metals,* John Wiley & Sons, New York, 1978.

M2. Welcher, F. J., *Organic Analytical Reagents,* Vol. 3, Van Nostrand, Princeton, N.J., 1947, 423.

M3. Johnson, E. A. and Toogood, M. J., Quinizarin, in *Organic Reagents for Metals,* Vol. 1, Johnson, W. C., Ed., Hopkin & Williams, Chadwell Heath, England, 1955, 139.

M4. Budesinsky, B. W. and Curtis, K. E., Aromatic ortho-dihydroxy compounds as reagents for inorganic analysis, in *Chelates in Analytical Chemistry,* Flaschka, H. A. and Barnard, A. J., Jr., Eds., Marcel Dekker, New York, 1976, 194.

REVIEWS

R1. Katyal, M., Flavones as analytical reagents, *Talanta,* 15, 95, 1968; Katyal, M. and Prakash, S., Analytical reactions of hydroxyflavones, *Talanta,* 24, 367, 1977.

R2. Hiiro, K. and Muraki, I., Study on the spectrophotometric determination of boron using organic reagents, Report Govern. Ind. Res. Inst., Osaka, No. 321, 1964.

R3. Nevskaya, E. M. and Nazarenko, V. A., Use of hydroxyflavones in analytical chemistry, *Zh. Anal. Khim.,* 27, 1699, 1972.

REFERENCES

1. Sommer, L., *Publ. Fac. Sci. Univ. Masaryk Brun., C.S.R.,* 398, 397, 1958.

2. Sommer, L., *Fresenius Z. Anal. Chem.,* 163, 412, 1958; *Fresenius Z. Anal. Chem.,* 164, 299, 1958; *Acta Chim. Acad. Sci. Hung.,* 18, 121, 1959; *Collect. Czech. Chem. Commun.,* 27, 2212, 1962.

3. Akiyama, K. and Kobayashi, Y., *Bunseki Kagaku,* 15, 694, 1966.

4. Korenaga, T., Motomizu, S., and Toei, K., *Analyst,* 103, 745, 1978.

5. Someya, T. and Ohsawa, S., *Nippon Kagaku Zasshi,* 744, 1972.

6. Bricker, C. E. and Vail, W. A., *Anal. Chem.,* 22, 720, 1950.

7. Akiya, H. and Sasao, H., *Nippon Yakugaku Zasshi,* 71, 1325, 1951.

8. Shain, Y. and Mayer, A. M., *Israel J. Chem.,* 1, 39, 1963; *Chem. Abstr.,* 60, 3493b, 1964.

9. Fletcher, M. H., *Anal. Chem.,* 37, 550, 1965.

10. Will, F., III, *Anal. Chem.,* 33, 1360, 1961.

11. Murata, A. and Yamanouchi, F., *Nippon Kagaku Zasshi,* 79, 231, 1958.

12. Sill, C. W. and Willis, C. P., *Anal. Chem.,* 31, 598, 1959.

13. Tovarek, J. and Sommer, L., *Scr. Fac. Sci. Nat. Univ. Purkynianae, Brun.,* 7, 17, 1977; *Chem. Abstr.,* 87, 161158u, 1977.

14. Milkey, R. G. and Fletcher, M. H., *J. Am. Chem. Soc.,* 79, 5425, 1957.

15. Cazzotti, R. I., Gomiero, L. A., and Abrao, A., *Relat. Inst. Energ. Atom S. Paulo,* 401, 13, 1976; *Anal. Abstr.,* 32, 6B69, 1976.

16. Chang, T. H., Volkova, O. I., and Getman, T. O., *Visn. Kiiv. Univ. Ser. Fiz. Khim.,* 43, 80, 1971; *Anal. Abstr.,* 24, 709, 1973.

17. Otomo, M. and Tonosaki, K., *Talanta,* 18, 438, 1971.

18. Eegriwe, E., *Fresenius Z. Anal. Chem.,* 76, 440, 1929.

19. Tanaka, Y., Hiratsuka, S., and Tanaka, Y., *Bunseki Kagaku,* 14, 810, 1965.

20. Corbett, J. A. and Guerin, B. D., *Analyst,* 91, 490, 1966.

21. Govil, P. K. and Banerji, S. K., *J. Inst. Chem. Calcutta,* 44, 128, 1972; *Anal. Abstr.,* 25, 1464, 1974.

22. Ishibashi, N., Kohara H., and Fukamachi, K., *Bunseki Kagaku,* 17, 1524, 1968.

23. Hiiro, K., *Nippon Kagaku Zasshi,* 83, 711, 1962.

24. Ishibashi, N., Kohara, H., and Abe, K., *Bunseki Kagaku,* 17, 154, 1968.

25. Saxena, K. K. and Agarwala, B. V., *Indian J. Chem.,* 14A, 634, 1976; *Chem. Abstr.,* 86, 132958d, 1977.

26. Govil, P. K. and Banerji, S. K., *Bull. Inst. Chem. Acad. Sci.*, 24, 81, 1977; *Anal. Abstr.*, 35, 2B29, 1978.
27. Akhmedli, M. K., Ayubova, A. M., and Babaeva, T. R., *Azerb. Khim. Zh.*, 121, 1972; *Anal. Abstr.*, 27, 2502, 1973.
28. Iwasaki, I. and Ohmori, T., *Kogyo Yosui*, 30, 1963.
29. Kuznetsov, V. I., Karanovich, G. G., and Drapkina, D. A., *Zavod. Lab.*, 16, 787, 1950.
30. Ohmori, T., *Kogyo Yosui*, 32, 1978.
31. Hiiro, K., *Bunseki Kagaku*, 10, 1281, 1961.
32. Yampolskii, M. Z., *Ich. Zap. Kursh. Gos. Ped. Inst.*, 67, 1958; *Chem. Abstr.*, 53, 19692a, 1959.
33. Yampolskii, M. Z., *Tr. Kom. Anal. Khim. Akad. Nauk SSSR*, 11, 261, 1960; *Chem. Abstr.*, 55, 10198b, 1961.
34. Busev, A. I. and Chang, F., *Vestn. Mosk. Univ. Khim. Ser.* 16, 55, 1961; *Chem. Abstr.*, 56, 10903h, 1962.
35. Ishiashi, M., Yamamoto, Y., and Todoroki, R., *Bunseki Kagaku*, 10, 1272, 1961.
36. Heller, J. and Schwarzenbach, G., *Helv. Chim. Acta*, 34, 1876, 1951.

O,N-DONATING CHELATING REAGENTS

o,o'-DIHYDROXYARYLAZO COMPOUNDS

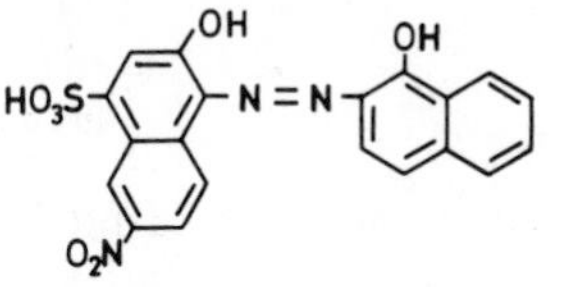

Synonyms

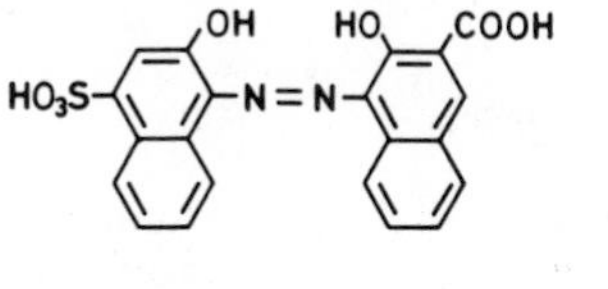

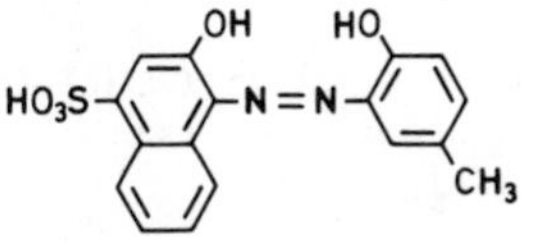

Azo compounds covered in this section are listed in Table 1, together with their synonyms and molecular formula.

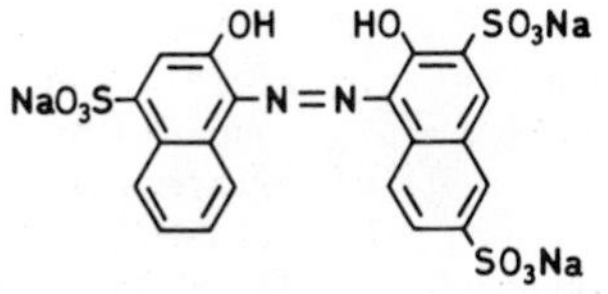

Source and Method of Synthesis

All compounds are commercially available. These azo dyes are synthesized by the standard procedure of azo coupling reaction. Namely, (1) from diazotized 1-amino-6-nitro-2-naphthol-4-sulfonic acid and 1-naphthol,[1,2] (2), (3), and (4) from diazotized 1-amino-2-naphthol-4-sulfonic acid and 2-hydroxynaphthoic acid,[3] p-cresol,[4,7] and 2-naphthol-3,6-disulfonic acid, respectively.

Analytical Uses

Azo dyes of this class are widely accepted as a metal indicator in the chelatometry, especially for the EDTA tritration of alkaline earth metals. They are sometime used as a photometric reagent for Ca and Mg in aqueous solution or after extraction into immiscible solvent.

Properties of Reagents

Eriochrome® Black T (1) is usually supplied as a monosodium salt, is dark violet powder with faint metallic sheen, and is easily soluble in water and alcohol, but insoluble in common organic solvents. Aqueous solution is red at pH <6, blue at pH 7 to 11, and orange at pH >12. In an alkaline solution, it is easily oxidized to become colorless, pKa (H_2L^-) = 6.3, and pKa (HL^{2-}) = 11.55 ($\mu = 0.08 \sim 0.008$, 18 $\sim$ 20°).[3]

Calcon carboxylic acid (2) is usually supplied as a free acid, is dark violet powder, and is slightly soluble in water and alcohol to give a pink solution at pH < 8. It turns to blue at pH 10 to 13 and faint pink in strong alkali,[3] pKa (H_2L^{2-}) = 9.26, and pKa (HL^{3-}) = 13.67 ($\mu = 0.1$ KCl, 24°).[6]

Calgamite® (3) is usually supplied as a free acid, is dark violet powder, and is easily soluble in water to give a bright red solution at pH < 7. It turns to blue at pH 9.1 to 11.4 and reddish-orange at pH > 13.[4] Aqueous solution is most stable among the azo dyes listed in Table 1, pKa (H_2L^-) = 7.92, and pKa (HL^{2-}) = 12.50 ($\mu = 0.1$, KNO$_3$, 25°).[8,9]

Hydroxynaphthol Blue (4) is usually supplied as a trisodium salt, is dark violet hygroscopic powder, and is easily soluble in water and aqueous alcohol to give a red-violet solution at pH < 6. It turns to blue at pH 7 to 12 and pink at pH > 13, pKa (H_2L^{3-}) = 6.44, and pKa (HL^{4-}) = 12.91 ($\mu = 0.1$, KCl, 24°),[6,10]

Absorption spectra of the aqueous solution of these dyes at different stages of deprotonation are illustrated in Figures 1 to 4.

Table 1
o,o'-DIHYDROXYARYLAZO COMPOUNDS

Reagent	Name	Synonyms	Molecular formula and mol wt
(1)	Eriochrome® Black T	1-(1-Hydroxy-2-naphthylazo)-6-nitro-2-naphthol-4-sulfonic acid, monosodium salt; Solochrome Black T, Mordant Black 11, C.I. 14645, Erio T, EBT, BT	$C_{20}H_{12}O_7N_3S$ Na, 461.38
(2)	Calcon carboxylic acid	2-Hydroxy-1-(2-hydroxy-4-sulfo-1-naphthylazo)-3-naphtoic acid, Patton and Reeder's dye, Cal Red, HSN, NANA, HHSNNA, NN	$C_{21}H_{14}N_2O_7S$, 438.41
(3)	Calmagite®	1-(1-Hydroxy-4-methyl-2-phenylazo)-2-naphthol-4-sulfonic acid	$C_{17}H_{14}N_2O_5S$, 358.37
(4)	Hydroxynaphthol Blue	1-(2-Hydroxy-4-sulfo-1-naphthylazo)-2-naphthol-3,6-disulfonic acid, trisodium salt, HNB	$C_{20}H_{11}N_2O_{11}S_3Na_3$, 620.46

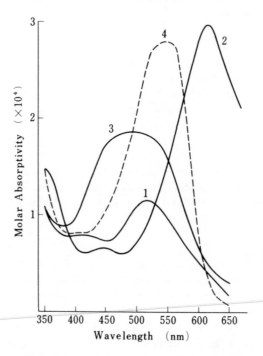

FIGURE 1. Absorption spectra of Eriochrome® Black T(1) and Mg chelate in water. (1) pH 5 (H_2L^-); (2) pH 8 (HL^{2-}); (3) pH 13 (L^{3-}); (4) pH 10 (MgL^-). (From Schwarzenbach, G. and Flaschka, H., *Die Komplexometrische Titration*, 2nd ed., Ferdinard Enke, Stuttgart, 1965. With permission.)

Complexation Reaction and Properties of Complexes

In the presence of metal ions other than alkali metals, the blue species of dyes turn to reddish in the aqueous solution, and such color reactions are utilized in the photometry of metal ions or in the chelatometry as metal indicators.

The color reactions of (1) with various metal ions were investigated in some detail and are summarized in Table 2. The structural change of the reagent accompanied by the color reaction may be schematically shown as below:

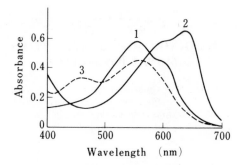

FIGURE 2. Absorption spectra of Calcon carboxylic acid (2) and Ca chelate in water. (1) pH 3.7 (H_2L^{2-}) ; (2) pH 9.9 (HL^{3-}); (3) pH 13 (CaL^{2-}). Dye, 2.4×10^{-5} M. (From Itoh, A. and Ueno, K., *Analyst*, 95, 583, 1970. With permission.)

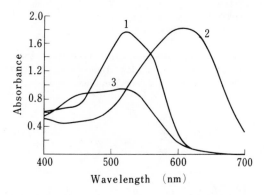

FIGURE 3. Absorption spectra of Calmagite® (3) in water. (1) pH 2.14 (H_2L^-); (2) pH 10.29 (HL^{2-}); (3) pH 13.8 (L^{3-}). Dye, 9.1×10^{-5} M. (Reprinted with permission from Lindstrom, F., *Anal. Chem.*, 32, 1124, 1960. Copyright 1960 American Chemical Society.)

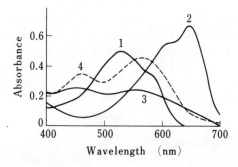

FIGURE 4. Absorption spectra of Hydroxynaphthol Blue (4) and Ca chelate in water. (1) pH 3.3 (H_2L^{3-}); (2) pH 9.8 (HL^{4-}); (3) pH 13.8 (L^{5-}); (4) pH 13 (CaL^{3-}). Dye, 2.4×10^{-5} M. (From Itoh, A. and Ueno, K., *Analyst*, 95, 583, 1970. With permission.)

Table 2
COLOR REACTION OF
ERIOCHROME® BLACK T WITH
METAL ION AT pH 9

Metal ion	Color	Metal ion	Color
None	Blue	Mg	Red
Al	Reddish	Mn(II)	Red
Bi	Gray - green	Nd	Red
Ca	Red	Ni	Red
Cd	Red	Os(VIII)	Pink
Ce(III)	Gray - green	Pt(IV)	Gray - green
Co(II)	Reddish	Th	Red
Cr(III)	Gray - green	U(VI)	Gray - green
Cu(II)	Reddish	V(V)	Light red
Fe(III)	Brown ppt	W(VI)	Blue
Hg(II)	Gray - blue	Zn	Red
La	Deep pink		

Reprinted with permission from Lott, P. F., Cheng, K. L., and Kwan, C. H., *Anal. Chem.*, 32, 1702, 1960. Copyright 1960 American Chemical Society.

blue   red

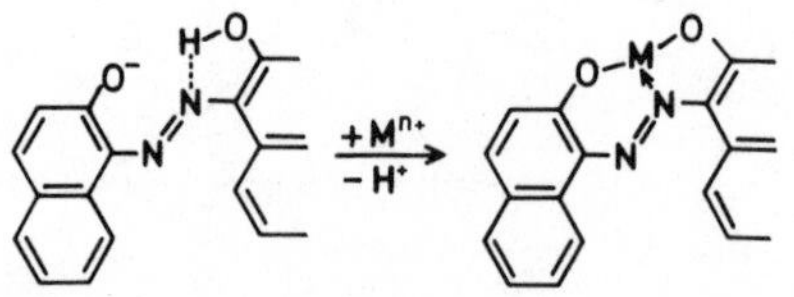

Stability constants of the dyes with several metal ions have been determined mostly by spectrophotometric method. Those values are summarized in Table 3. The conditional stability constants (log K'_{ML}) of (1), (3), and (4) at pH 10 are also reported as follows: (1), Ca (3.8), Mg (5.4); (3), Ca (3.67), Mg (5.69);[107] (4), Be (3.63), Mg (3.43), Ca (2.82), Sr (2.05), Ba (1.75), La (3.57), Pr (4.03), Nd (4.13), Sm (4.09), Eu (4.10), Ga (3.76), Tb (3.77), Dy (3.71), Ho (3.68), Er (3.60), Yb (3.50),[11], UO_2^{2+} (4.10 or 3.99).[13]

Absorption spectra of some representative metal chelates are illustrated in Figures 1, 2, 4, and 5. The colored chelates can be extracted into polar organic solvents such as *n*-amyl alcohol for the subsequent photometry. As example, the absorption spectra of Eriochrome® Black T (1) and its Mg chelate in *n*-amyl alcohol are shown in Figure 6. The anionic chelates can also be extracted with the long chain quaternery ammonium ion such as Zephiramine®[15] or Aliquat® 336 S[16] into organic solvent like chloroform or 1,2-dichloroethane.

Purification and Purity of Reagents

Commercial samples are pure enough for use as metal indicators in the chelatometry. However, they contain a substantial amount of inorganic salts and have to be purified for use in the physicochemical studies.

(1) is purified by recrystallization of dimethylammonium salt from dimethylformamide.[8,17] (2) is purified through its *p*-toluidinium salt. Namely, dye in warm 20% aqueous methanol is treated with *p*-toluidine to precipitate the toluidinium salt after

Table 3
CHELATE STABILITY CONSTANTS OF o,o'-DIHYDROXYARYLAZO DYES

	log K_{ML}					log K_{ML_2}			
Reagent	Ca	Ba	Mg	Mn(II)	Zn	Mn(II)	Zn	Condition	Ref.
(1)	5.4	3.0	7.0	9.6	12.9	8.0	7.1	—	M2
(2)	5.85	—	—	—	—	—	—	$24°, \mu = 0.1$ (KCl)	6
(3)	6.1	—	8.1	—	12.52	—	7.71	$\mu = 0.1$ (KCl)	4, 9
(4)	6.11	—	7.11	—	—	—	—	$24°, \mu = 0.1$ (KCl)	6, 10

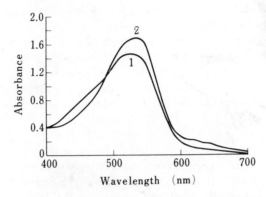

FIGURE 5. Absorption spectra of Ca and Mg Calmagite® chelate at pH 10; Dye, 3.5×10^{-5} M. (1) Ca in excess; (2) Mg in excess. (Reprinted with permission from Lindstrom, F., *Anal. Chem.*, 32, 1124, 1960. Copyright 1960 American Chemical Society.)

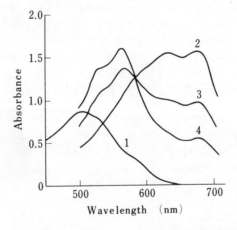

FIGURE 6. Absorption spectra of Eriochrome® Black T and Mg chelate in n-amyl alcohol. (1) In n-amyl alcohol; (2) in n-amyl alcohol shaken with pH 11.5 solution not containing Mg; (3) shaken with pH 11.5 solution containing 0.1 ppm Mg; (4) shaken with pH 11.5 solution containing 0.2 ppm Mg. Reference, water. (From Nishimura, M. and Nakaya, S., *Bunseki Kagaku*, 16, 463, 1967. With permission.)

cooling which is eventually recrystallized from hot water.[6] **(3)** may be purified by extracting the crude sample with anhydrous ether.[4] **(4)** is purified by chromatography on cellulose powder column with propanol-ethyl acetate-water (5 + 5 + 4) as an eluent according to the following procedure. The crude sample which had been treated with hot ethanol to remove ethanol soluble impurity is dissolved in 20% aqueous methanol and is subjected to chromatography. Of the three zones, the upper red-violet zone is eluted to obtain pure dye which may be precipitated as a monosodium salt trihydrate by adding concentrated HCl to the concentrated eluant.[6]

Analytical Applications
Use as a Photometric Reagent

Azo dyes of this class are widely used as photometric reagents for various metal ions. Although most of photometric procedures are carried out in the aqueous system, the colored chelates may be extracted into higher alcohol like iso-amyl alcohol. Alternatively, they can be extracted into chlorinated solvents, such as chloroform or 1,2-dichloroethane, as an ion-pair with the long chain quaternary ammonium ion. Higher sensitivity and selectivity are claimed for the solvent extraction procedures. Examples of the application of the azo dyes as a photometric reagent are summarized in Table 4.

Use as a Metal Indicator

These azo dyes have been used as metal indicators in the chelatometry. The dissociation constants of the last proton of **(2)** and **(4)** are so high that the color change with Ca is observed at higher pH region (12 to 13), where Mg precipitates as $Mg(OH)_2$. Accordingly, they are recommended as the selective calcium indicator in the presence of Mg. The conditions of titrations with these azo dyes are as follows: titrations with Eriochrome® Black T **(1)** (from red to blue at the end point), Ca, Mg,[M2, 27,28] Cd, Zn[36] (pH 10, NH_3), Ba, Sr (pH 10, NH_3, Mg-EDTA),[35] Hg(II), [37] In,[39]Pb,[37] rare earths[42] (pH 8 to 10, NH_3, tartarate, 100°), Co(II), Ni(pH 10, NH_3, Mn-EDTA),[38] Mn(II) (pH 8 to 10, ascorbic acid, 80°),[40] Sc(pH 7.5 to 8, NH_3, malic acid, 100°),[41] VO^{2+} (pH 10, NH_3, Mn-EDTA, ascorbic acid);[36] titration with Calcon carboxylic acid **(2)** (from red to blue), Ca (pH 12 to 13, KOH);[10] titration with Calmagite® **(3)** (from red to blue), Ca, Mg(pH 10, NH_3);[4,30] titration with Hydroxynaphthol Blue **(4)** (from red to blue), Ca (pH 10 to 13, KOH) and Mg (pH 10, NH_3).[43]

The recommended formulations for the indicator solution are

1. **(1)**, A methanol solution containing 0.5 g dye and 4.5 g hydroxylamine hydrochloride or a solution of 0.2 g dye in 15 m*l* triethanolamine and 5 m*l* of methanol
2. **(2)**, a dry powder diluted with solid KNO_3 or KCl (1:100 or 1:200)
3. **(3)**, a 0.5% aqueous solution
4. **(4)**, a dry powder diluted with solid NaCl or KCl (1:150) or a 0.01 to 0.05% aqueous glycerol (1:1) solution

Other Reagents with Related Structure
Gallion

(5-Amino-3-(3-chloro-2-hydroxy-5-nitrophenylazo)-4-hydroxynaphthalene-2,7-disulfonic Acid, Disodium Salt) and Lumogallion (4-Chloro-6-(2,4-dihydroxynaphthylazo)-1-hydroxybenzene-2-sulfonic Acid, Monosodium Salt)

Gallion (H_4L)
$C_{16}H_9N_4O_{10}S_2ClNa_2$
mol wt = 562.82

Table 4
APPLICATION OF o,o'-DIHYDROXYARYLAZO-DYES AS PHOTOMETRIC REAGENTS

Azo-dyes	Metal ion[a]	Condition	Chelate Ratio	λ_{max} (nm)	ε (×10⁴)	Range of determination (ppm)	Interferences and remarks	Ref.
(1)	Cd	pH ≃ 6, 1,10-phenanthroline, extraction with CHCl₃	ML₂X₂	522	2.2	0.22 ∿ 5.6	In Devarda alloys	18
	Co(II)	TPAC, extraction with CHCl₃	—	580	5.0	0.6 ∿ 18	EDTA, DDTC interfere	46
	Co(II)	pH 5.5 ∿ 8.2, SCN⁻, extraction with benzene + Aliquat® 336 S	—	587	6.62	0.1 ∿ 0.8	Mn, Ni, VO₃⁻ interfere	19
	Mg	pH 10.9 ∿ 11.3, KCN, triethanolamine	—	525 ∿ 30	—	0.1 ∿ 1	—	21—24
	Mg	pH 9.5 ∿ 11.2, extraction with AmOH or BuOH	—	545	2.0	0.5 ∿ 5	Ca, Cd, Ce, Fe, Zn interfere	14, 25
	Mg	pH 11.2 ∿ 12.2, Zephiramine®, extraction with dichloroethane	—	690[b]	—	0.03 ∿ 0.1	Ca(tenfold) does not interfere	15, 26
	Zn	pH ≃ 6, 1,10-phenanthroline, extraction with CHCl₃	ML₂X₂	522	2.3	0.64 ∿ 3.7	In Devarda alloys	18
(2)	Rare earths	pH 6.35 ∿ 6.8, diphenyl guanidine, extraction with iso-AmOH	ML₂X₂	550 ∿ 600	2.6 ∿ 3.5	0.1 ∿ 15	—	44
	UO₂²⁺	Alkali	ML₂	570	1.36	14 ∿ 60	After ion-exchange separation	45
(3)	Al	pH 8.2 ∿ 8.8, KCN, EDTA, extraction, with Aliquat® 336 S in CHCl₃	ML₃	570	4.2	0.1 ∿ 0.5	Cd, Cu, Ni, Pb, Zn (60-fold) do not interfere	29
	Co(II)	pH 4.2, extraction with Aliquat® 336 S in CHCl₃	—	580	0.8	—	—	10
	Mg	pH 11.2 ∿ 12.2, Zephiramine®, extraction with dichloroethane	—	690	6.3	2 ∿ 7	Ca (tenfold) does not interfere	15
	Mo(VI)	pH 11, extraction with Aliquat® 336 S in CHCl₃	—	575	0.6	—	—	16

Table 4 (continued)
APPLICATION OF o,o'-DIHYDROXYARYLAZO-DYES AS PHOTOMETRIC REAGENTS

Azo-dyes	Metal ion[a]	Condition	Chelate			Range of determination (ppm)	Interferences and remarks	Ref.
			Ratio	λ_{max} (nm)	ε ($\times 10^4$)			
(4)	Alkaline earths	pH 6, EDTA	—	650	—	1 $\sim$ 600	—	32—33
	UO_2^{2+}	pH 4.0	ML	530	0.41	0.3 $\sim$	—	13
	Rare earths	pH 6 $\sim$ 10, EDTA	ML	650	—	1 $\sim$ 300	Alkaline earths, Be, UO_2^{2+} interfere	32—33

a   Other metals investigated with (1) are Cu(II),[98] Er, Gd, La, Nd,[20] Yb, Ni,[98] Th[12] and with (3) and (1) are Cu(II).[34]

b   At this wavelength, the difference of absorbances between chelate and dye is maximum.

Lumogallion (H_4L)
$C_{12}H_8N_2O_6SC1Na$
mol wt = 365.71

Gallion is used as a photometric reagent for Ga (ML, pH 6, 560 nm, $\varepsilon = 1.54 \times 10^4$), Mn, Pb, and Zn.[99] The formation of a ternary complex of Ga-Gallion-Oxine (MLX, pH 3, 640 nm, $\varepsilon = 2.3 \times 10^4$ in butanol) is also utilized for the determination of Ga.[47]

Lumogallion (C.I. Mordant Red 72, red-brown powder) is also recommended as a photometric reagent for Al (pH 5, 576 nm)[48] or pH 3.3, 558 nm, in presence of Antarox® CO 890,[56] Ga (pH 3, 580 nm)[48] or pH 3.1 to 4.3, 553 nm in presence of Antarox® CO 890),[100] In (pH 2.8 to 6.5, 500 nm),[9] Mo(VI) (510 nm, $\varepsilon = 1.13 \times 10^4$),[50] Nb (pH 5.6 to 7.6, 546 nm),[51] Sc (pH 2, 500 nm),[52] Sn(IV) (pH 1, 510 nm),[53] and W (pH 2.6 to 3.5, 555 nm)[54] or as a fluorimetric reagent for these metals. The extraction photometric determination of Fe, Sc, and Y through their ternary complexes with Lumogallion-diethylether[55] and of Nb through Lumogallion-BPA ternary complex[57] are reported. Anionic chelates of Lumogallion with Cu, Fe, Mo(VI), Pd, V(IV), and W(VI) can also be extracted with Aliquat® 336 S in chloroform.[16]

SPADNS

(SPANS, 3-(4-Sulfophenylazo)-4,5-dihydroxy-2,7-naphthalenedisulfonic Acid, Trisodium Salt)

Na_3H_2L (H_5L)
$C_{16}H_9N_2O_{11}S_3Na_3$
mol wt = 570.40

Red powder. SPADNS is suggested as a photometric reagent for Al (pH 4 to 5, 590 nm),[58] Ca (pH 10, 570 nm),[59], Ba (pH 10 to 11, 570 nm),[103] Mg (pH 8, 570 nm),[59] Pd (pH 2.5 to 4.5, 550 nm),[60] Th (pH 2.6 to 3.6, 580 nm),[101] Zr (0.5 N HCl, 580 nm),[102] and rare earths (pH 5 to 7.5, 540 nm).[62] The colored anionic chelates may be extracted as an ion-pair with diantipyrylmethane for the photometric determination of Sc (ML_2X_2, pH 5.5 to 8.5, 580 nm).[61,104] Fluoride can also be determined indirectly through the decoloration of Zr-chelate (0.7 N HCl, 570 nm).[63,64]

Berryllon II

(2-(3,6-Disulfo-8-hydroxynaphthylazo)-1,8-dihydroxynaphthalene-3,6-disulfonic Acid, Tetrasodium Salt, DSNADNS R-8)

Na_4H_3L (H_7L)
$C_{20}H_{10}O_{15}N_2S_4Na_4$
mol wt = 738.50

Black-violet powder.[65] The pKa values for the last five dissociation steps are 2.39, 4.43, 5.71, 7.01, and 10.88 (20°).[66] It has been used as a photometric reagent for Be (pH 12 to 13.2, 600 nm),[67] Th (pH 2, 620 nm),[68] rare earths (pH 7.6, 610 nm),[70] and B (concentrated H_2SO_4, 650 nm).[71] Indirect determination of Zr was also proposed by precipitating ZrL from strong acid solution, followed by the photometry of freed dye after dissolution of precipitate in NH_3-triethanolamine.[69]

Calcichrome

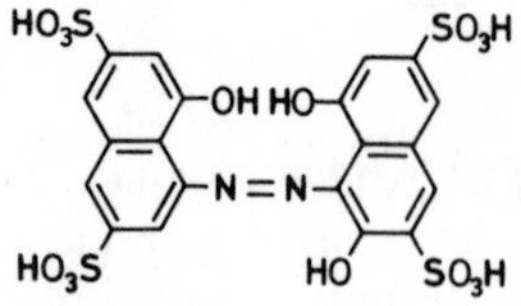

(A)

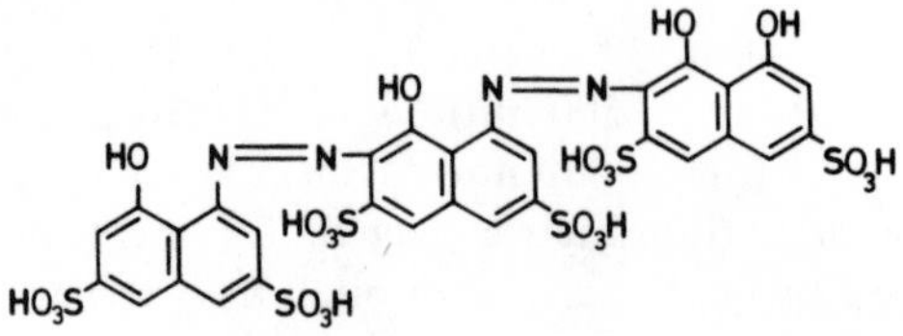

(B)

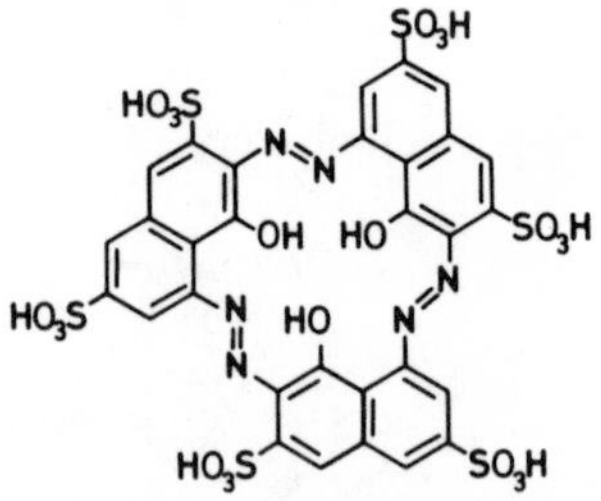

(C)

Monoazo structure (A),[72] bisazo structure (B) (so-called Calcion IREA),[73] or cyclic-trisazo structure (C) (so called Calcichrome)[74] have been assigned to this reagent, and the structural assignment is the subject of discussions by many authors[75-77] ($pKa_7 = 7.2$, $pKa_8 = 11.5$, log $K_{CuL} = 26.1$).[78] Calcichrome has been used as a photometric reagent for Al (pH 6, 317 nm),[79] Co(II) (pH 12, 595 nm),[80] Cu(II) (pH 3 to 10, 535 nm),[81] Fe(III) (pH 6, 555 nm),[82] Mg (pH 11.3, 535 nm),[83] Mn(II) (pH 8 to 12, 590 nm),[84] Ni (pH 13, 625 nm),[85] Ti(IV) (pH 4, 530 nm),[86] and V(IV) (pH 5.3, 634 nm). However, the most practical application is the determination of Ca in the presence of large amount of Ba and Sr (pH 12 to 13, λ_{max}, 615 nm; $\varepsilon = 7.6 \times 10^3$, 0.1 to 4 ppm).[88,89]

Magon

(1-   Azo-2-hydroxy-3-(2,4-dimethylcarboxyanilido)-naphthalene-1-(2-hydroxyben-zene) (Xylidyl Blue® II, Xylyazo Violet I, Mann dye) and Magon Sulfate (Xylidyl Blue® I, Xylylazo Violet II, Bohuon Reagent)

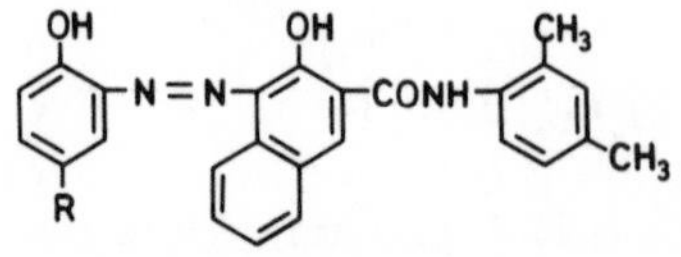

Magon® (R = H, H_2L, $C_{25}H_{21}N_3O_3$, mol wt = 411.46) is a red powder, pKa_1 (phenol) = 8.99, pKa_2 (naphthol) = 14.01, log β_2 (for MgL_2) = 9.78 (50% ethanol).[90]

It has been used as a photometric reagent for Mg (pH 9, 505 to 510 nm in 70 to 80% ethanol)[91,92] and as a detection reagent for Mg on ion-exchange resin beads (detection limit 0.007 μg, dilution limit 1:5 × 10⁶).[93]

Magon sulfate (R = SO_3H, H_3L, supplied as monosodium salt, $C_{25}H_{20}N_3O_6SNa$, mol wt = 513.50) is a dark violet powder. The reagent and metal chelates are only slightly soluble in water. It is recommended as a photometric reagent for Mg (pH 9, MgL, 510 or 540 nm, $\varepsilon = 4.9 \times 10^4$) in aqueous solution[94] and in biological samples.[95]

The reagent and its Mg chelate become more soluble in water and λ_{max} of HL^{2-} shifts from 555 nm to 615 nm if in the presence of 1.6% Triton® X-100 (nonionic surfactant). The dual-wavelength photometry was applied directly to this system (pH 11.2, 515 and 620 nm, 0.004 to 0.12 ppm Mg)[96] or after preconcentration.[97]

Magneson

A group of *o*-hydroxyazo compounds named as Magneson (Magneson I,[M2] Magneson II,[M2] Magneson IREA,[105] and Lumomagneson[106]) have been used as a reagent for Mg. A sharp color change is observed when free dye is adsorbed on the precipitate of $Mg(OH)_2$. Although the reaction is useful for the detection and the photometric determination of Mg, better reagents are now available.

MONOGRAPHS

M1. **Diehl, H.**, Calcein, Calmagite and *o,o'*-Dihydroxyazobenzene: Titrimetric, Colorimetric and Fluorimetric Reagents for Calcium and Magnesium, G.F. Smith Chemical Co., Columbus, Ohio, 1964.

M2. **Ainsworth, L. R. and Yardley, J. T.**, Eriochrome Black T, HSN Indicator: Patton and Reeder's Indicator, L. R. Ainsworth, "Calmagite", and I.N.C. Johnson "Magneson I and II" L. R. Ainsworth and J. T. Yardley, "Calcichrome", in *Organic Reagents for Metals*, Vol. 2, Johnson, W. C., Ed., Hopkin & Williams, Essex, England, 1964.

REFERENCES

1. **Hargenback, J.**, British Patent 15418/1904, 15982/1904; U.S. Patent 790363/1904; German Patent 164655 and 169683.
2. **Ruggli, P. and Knapp, F.**, *Helv. Chim. Acta*, 12, 1034, 1929 and 13, 748, 756, 1930.
3. **Patton, J. and Reeder, W.**, *Anal. Chem.*, 28, 1026, 1956.
4. **Lindstrom, F. and Diehl, H.**, *Anal. Chem.*, 32, 1123, 1960.
5. **Schwarzenback, G. and Biedermann, W.**, *Helv. Chim. Acta*, 31, 678, 1948.
6. **Itoh, A. and Ueno, K.**, *Analyst*, 95, 583, 1970.
7. **Lindstrom, F. and Isaac, R.**, *Talanta*, 13, 1003, 1966.
8. **Lindstrom, F. and Womble, A. E.**, *Talanta*, 20, 589, 1973.
9. **Nakagawa, G., Wada, H., and Fujita, Y.**, *Bull. Chem. Soc. Jpn.*, 46, 489, 1973; *Anal. Chem., Essays in Memory of Anders Ringbom*, 281, 1977.
10. **Itoh, A. and Ueno, K.**, *Bunseki Kagaku*, 19, 393, 1970.
11. **Brittain, H. G.**, *Anal. Chim. Acta*, 96, 165, 1978.
12. **Lott, P. F., Cheng, K. L., and Kwan, C. H.**, *Anal. Chem.*, 32, 1702, 1960.
13. **Brittain, H. G.**, *Anal. Lett.*, 10, 263, 1977.
14. **Nishimura, M. and Nakaya, S.**, *Bunseki Kagaku*, 16, 463, 1967.
15. **Fukamachi, K., Kohara, H., and Ishibashi, N.**, *Bunseki Kagaku*, 19, 1529, 1970.
16. **Woodward, C. and Freiser, H.**, *Talanta*, 20, 417, 1973.
17. **Diehl, H. and Lindstrom, F.**, *Anal. Chem.*, 31, 414, 1959.

18. Shestidesyatnaya, N. L., Kotelyanskaya, L. I., and Semenyuk, V. V., *Zh. Anal. Khim.*, 33, 303, 1978; *J. Chem. Soc. USSR*, 33, 236, 1978.
19. Rokugawa, Y., Sugawara, M., and Kambara, T., *Bull. Chem. Soc. Jpn.*, 52, 2303, 1979.
20. Takano, T., *Bunseki Kagaku*, 16, 27, 1967.
21. Harvey, A. E., Jr., Komarmy, J. M., and Wyatt, G. M., *Anal. Chem.*, 25, 498, 1953.
22. Berlin, R. D., *Anal. Bichem.*, 14, 135, 1966.
23. Kolicka, M. and Kubica, M., *Chem. Anal. (Warsaw)*, 12, 841, 1315, 1967; *Anal. Abstr.*, 15, 6007, 1968; 16, 675, 1969.
24. Nosticzius, A., *Magy. Kem. Foly.*, 74, 612, 1968; *Anal. Abstr.*, 18, 2258, 1970; *Chem. Abstr.*, 71,63932w, 1969.
25. Zolotov, Yu. A. and Begreev, V. V., *Zh. Anal. Khim.*, 22, 1423, 1967.
26. Pyatnitskii, I. V., Pinaeva, S. G., and Pospelova, N. V., *Zh. Anal. Khim.*, 30, 2316, 1975.
27. Takama, H. and Ueno, K., *Bunseki Kagaku*, 12, 1194, 1963.
28. Miyake, S. and Sakamoto, I., *Bunseki Kagaku*, 16, 960, 1967.
29. Woodward, C. and Freiser, H., *Talanta*, 15, 321, 1968.
30. Wharton, H. W. and Chapman, L. R., *Anal. Chem.*, 36, 1079, 1964.
31. Brittain, H. C., *Anal. Chim. Acta*, 106, 401, 1979.
32. Brittain, H. C., *Anal. Chem.*, 49, 969, 1977.
33. Brittain, H. C., *Anal. Chim. Acta*, 96, 165, 1978.
34. Sugawara, M., Rokugawa, Y., and Kambara, T., *Bull. Chem. Soc. Jpn.*, 50, 3206, 1977.
35. Manns, T. J., Reschovsky, M. U., and Certa, A. J., *Anal. Chem.*, 24, 908, 1952.
36. Biedermann, W. and Schwarzenbach, G., *Chimia*, 2, 56, 1948.
37. Flaschka, H., *Mikrochem. Mikrochim. Acta*, 39, 38, 315, 1952.
38. Kinnunen, J. and Wennerstrand, B., *Chemist-Analyst*, 44, 33, 1955.
39. Flaschka, H. and Amin, A. M., *Fresenius Z. Anal. Chem.*, 140, 6, 1953.
40. Flaschka, H., *Chemist-Analyst*, 42, 56, 1953.
41. Wünsh, L., *Collect. Czech. Chem. Commun.*, 20, 1107, 1955.
42. Brunisholz, G. and Cahen, R., *Helv. Chim. Acta*, 39, 2136, 1956.
43. Goettsch, R. W., *J. Pharm. Sci.*, 54, 317, 1965.
44. Akhmedli, M. K., Granovskaya, P. B., and Neimatova, R. A., *Zh. Anal. Khim.*, 28, 278, 1973.
45. Mavrodin, M., *Rev. Roum. Chim.*, 17, 1199, 1972; *Anal. Abstr.*, 24, 715, 1973; *Chem. Abstr.*, 77, 159764p, 1972.
46. Bianchi, G. and Gerlo, C. E., *Afinidad*, 35, 30, 1978; *Chem. Abstr.*, 90, 114450k, 1979.
47. Akhmedli, M. K., Glushchenko, E. L., and Kyazimova, A. K., *Azerb. Khim. Zh.*, 102, 1971; *Anal. Abstr.*, 24, 65, 1973.
48. Nishikawa, Y., Hiraki, K., Morishige, K., and Shigematsu, T., *Bunseki Kagaku*, 16, 692, 1967.
49. Salikhov, V. D. and Yampolsky, M. Z., *Zh. Anal. Khim.*, 22, 998, 1967.
50. Busev, A. I. and Chang, F., *Zh. Neorg. Khim.*, 6, 1308, 1961; *Chem. Abstr.*, 56, 10903d, 1962.
51. Pilipenko, A. T., Volkova, A. I., and Zhebentyaev, A. I., *Zh. Anal. Khim.*, 26, 2048, 1971.
52. Akhmedli, M. K. and Gambrov, D. G., *Zh. Anal. Khim.*, 22, 276, 1967.
53. Marchenko, P. V. and Obolonchik, N. V., *Zh. Anal. Khim.*, 22, 725, 1967.
54. Pilipenko, A. T., Zhebentyaev, A. I., and Volkova, A. I., *Ukr. Khim. Zh.*, 41, 1087, 1975; *Anal. Abstr.*, 30, 4B121, 1976.
55. Alimarin, I. P., Gibalo, I. M., and Pigaga, A. K., *Dokl. Akad. Nauk. SSSR*, 186, 1323, 1969; *Chem. Abstr.*, 71, 85120c, 1969.
56. Ishibashi, N. and Kina, K., *Anal. Lett.*, 5, 637, 1972.
57. Patratti, Yu. V. and Pilipenko, A. T., *Ukr. Khim. Zh.*, 44, 752, 1978; *Chem. Abstr.*, 89, 190347b, 1978.
58. Banerjee, G., *Fresenius Z. Anal. Chem.*, 271, 284, 1974.
59. Banerjee, G. and Kanji, S. K., *Fresenius Z. Anal. Chem.*, 282, 47, 1976.
60. Saxema, K. K. and Dey, A. K., *Indian J. Appl. Chem.*, 32, 255, 1969.
61. Ganago, L. I. and Alinovskaya, L. A., *Zh. Anal. Khim.*, 27, 261, 1972.
62. Zelinski, S. and Radecka, V., *Zh. Anal. Khim.*, 31, 1910, 1976.
63. Bellack, E. and Schouboe, P. J., *Anal. Chem.*, 30, 2032, 1958.
64. Fleury, S., *Anal. Chim. Acta*, 37, 232, 1967.
65. Lukin, A. K. and Zavorkhina, G. B., *Zh. Anal. Khim.*, 11, 393, 1956.
66. Adamovich, L. P. and Miraya, A. P., *Zh. Anal. Khim.*, 18, 292, 1963.
67. Karanovich, G. G., *Zh. Anal. Khim.*, 11, 400, 1956.
68. Datta, S. K., *Fresenius Z. Anal. Chem.*, 167, 105, 1959.
69. Datta, S. K. and Saha, S. N., *Fresenius Z. Anal. Chem.*, 184, 177, 1961.
70. Bagbanly, I. L. and Rustamov, N. Kh., *Azerb. Khim. Zh.*, 163, 1972; *Anal. Abstr.*, 26, 1462, 1974; *Dokl. Akad. Nauk. Azerb. SSR*, 28, 42, 1972; *Anal. Abstr.*, 26, 67, 1974.

71. Hiiro, K., *Bunseki Kagaku,* 11, 337, 1962.
72. Stead, C. V., *J. Chem. Soc.,* 693, 1971.
73. Lukin, A. M., Smirnova, K. A., and Zavarikhina, G. B., *Zh. Anal. Khim.,* 18, 444, 1963; *J. Anal. Chem. USSR,* 18, 389, 1963; *Metody Anal. Khim. Reakt. Prep.,* 12, 57, 1966; *Chem. Abstr.,* 67, 17533g, 1967.
74. Close, R. A. and West, T. S., *Talanta,* 5, 221, 1960; *Analyst,* 87, 630, 1962.
75. Bezdekova, A. and Budesinsky, B., *Collection Czech. Chem. Commun.,* 30, 811, 1965.
76. Mendes Bezerra, A. A. and Stephen, W. I., *Analyst,* 94, 1117, 1969.
77. Shibata, S., *Bunseki Kagaku,* 21, 842, 1972.
78. Ishii, H. and Einaga, H., *Bull. Chem. Soc. Jpn.,* 39, 1154, 1966.
79. Ishii, H. and Einaga, H., *Bull. Chem. Soc. Jpn.,* 39, 1721, 1966.
80. Ishii, H. and Einaga, H., *Bunseki Kagaku,* 16, 328, 1967.
81. Ishii, H. and Einaga, H., *Bull. Chem. Soc. Jpn.,* 38, 1416, 1965.
82. Ishii, H. and Einaga, H., *Nippon Kagaku Zasshi,* 87, 440, 1966; *Bunseki Kagaku,* 15, 577, 1966.
83. Ishii, H., *Bull. Chem. Soc. Jpn.,* 40, 1531, 1967.
84. Ishii, H. and Einaga, H., *Bunseki Kagaku,* 15, 1124, 1966.
85. Ishii, H. and Einaga, H., *Bunseki Kagaku,* 16, 332, 1967.
86. Ishii, H. and Einaga, H., *Bunseki Kagaku,* 15, 821, 1966.
87. Ishii, H. and Einaga, H., *Bunseki Kagaku,* 19, 371, 1970.
88. Herrero Lancina, M. and West, T. S., *Anal. Chem.,* 35, 2131, 1963.
89. Pakalns, P. and Florence, T. M., *Anal. Chim. Acta,* 30, 353, 1964.
90. Svoboda, V. and Chromy, V., *Anal. Chim. Acta,* 54, 121, 1971.
91. Mann, C. K. and Yoe, J. H., *Anal. Chim. Acta,* 16, 155, 1957.
92. Abbey, S. and Maxwell, J. A., *Anal. Chim. Acta,* 27, 233, 2962.
93. Kato, K., Ichikawa, T., and Kakihana, H., *Nippon Kagaku Zasshi,* 87, 718, 1966.
94. Mann, C. K. and Yoe, J. H., *Anal. Chem.,* 28, 202, 1956.
95. Bohuon, C., *Clin. Chim. Acta,* 7, 811, 1962.
96. Watanabe, H. and Tanaka, H., *Bunseki Kagaku,* 26, 635, 1977.
97. Tanaka, H. and Watanabe, H., *Bunseki Kagaku,* 27, 189, 1978.
98. Kaneniwa, N., *Yakugaku Zasshi,* 76, 136, 1956.
99. Dedkov, Yu. M., Khokhlov, L. M., and Salikhov, V. D., *Zh. Anal. Khim.,* 26, 2350, 1971.
100. Kina, K. and Ishibashi, N., *Microchem. J.,* 19, 26, 1974.
101. Banerjee, G., *Anal. Chim. Acta,* 16, 56, 1957; Cooper, J. A. and Vernon, M. J., *Anal. Chim. Acta,* 23, 351, 1960.
102. Banerjee, G., *Anal. Chim. Acta,* 16, 62, 1957.
103. Banerjee, G. and Bhattacharjee, A. K., *J. Indian Chem. Soc.,* 55, 617, 1978; *Anal. Abstr.,* 37, 2B63, 1979.
104. Ganago, L. I. and Alinovskaya, L. A., *Zh. Anal. Khim.,* 34, 111, 1979.
105. Smirnov, K. A., Lukin, A. M., and Uysokova, N. N., *Tr. Vses. Nauchno-Issled. Inst. Khim. Reakt. Osobo Chist. Khim. Veshchestv,* 51, 14, 1969.
106. Kurbatova, I. I., *Zavodsk. Lab.,* 32, 1064, 1966.
107. Ringbom, A., *Complexation in Analytical Chemistry,* John Wiley & Sons, New York, 1963, Table A-7.

ARESENAZO I AND MONOAZO DERIVATIVES OF PHENYLARSONIC ACID

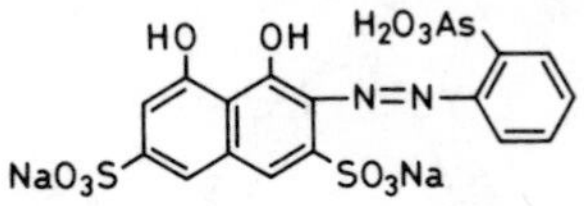

$C_{16}H_{11}N_2O_{11}S_2Na_2As$
mol wt = 592.29

Na₂H₄L

Synonyms

2-(4,5-Dihydroxy-2,7-disulfo-3-naphthylazo)phenylarsonic acid, disodium salt, Neothorin, Neothoron, Uranon

Source and Method of Synthesis

Commercially available as a disodium salt. Synthesized by the coupling of diazotized 2-aminophenylarsonic acid with chromotropic acid.[1,2]

Analytical Uses

As a photometric reagent for Al, Be, In, Th, Zr, rare earths, and actinoid elements; also used as a metal indicator in the chelatometric titration of Ca, Mg, rare earths, and Th and as an indicator in the precipitation titration of SO_4^{2-} with Ba.

Properties of Reagent

Disodium salt is a dark red crystalline powder (see Figure 1), is easily soluble in water, to give an orange-red solution (λ_{max}, 500 nm at pH 1 to 8), and is insoluble in most organic solvents, $pKa_1 = 0.6$; $pKa_2 = 0.8$; $pKa_3 = 3.5$; $pKa_4 = 8.2$; $pKa_5 = 11.6$; and $pKa_6 = 15$ ($\mu = 0.1, 20°$).[3]

Complexation Reaction and Properties of Complexes

Arsenazo I reacts with wide variety of metal ions to form colored soluble chelates in a range of pH 1 to 8, but the reactions which occur in fairly acidic solution are of practical importance because of the higher selectivity of the reaction. The color reactions occurring in hydrochloric acid solution are Nb(III), Ti(III), Zr (violet), Sn(IV), U, rare earths (orange-red), Th (blue-violet), and Ta(III) (red-violet).[4]

The reagent usually forms 1:1 chelate such as shown below.[5]

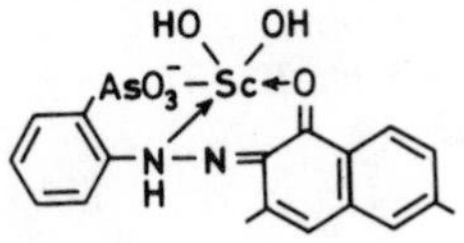

Chelate stability constants for alkaline earth ions and La were reported as follows: log K_{ML} for Mg, 5.58; Ca, 5.09; Sr, 4.41; and Ba, 4.15 ($\mu = 0.1$ KNO₃, 25°)[6] and log K_{MH2L} for La, 28.8.[7]

Absorption spectra of metal chelates in aqueous solution are pH dependent. The spectral curves on a La-Arsenazo I system at different pH values are illustrated in Figure 2 as an example.[8] The molar absorptivities of the chelates are in a order of 10^4, and such color reactions can be utilized for the photometric determination of traces of elements.

These metal chelates are negatively charged and can be extracted as an ion-pair with a large cation such as diphenylguanidinium or long chain alkylammonium ion.

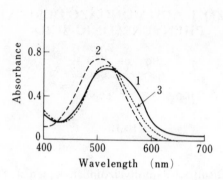

FIGURE 1.   Absorption spectra of Arsenazo I at various acidities. (1) pH 15.80; (2) pH 6.45; (3) $H°$ −3.42. Dye concentration 1.85×10^{-5} M. (From Budesinsky, B., *Talanta*, 16, 1277, 1969. With permission.)

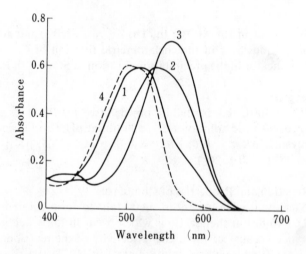

FIGURE 2.   Absorption spectra of Arsenazo I and La-Arsenazo I chelate in aqueous solution of various pH. (1) La-Arsenazo I, 2.25×10^{-5} M, pH 4; (2) La-Arsenazo I, 2.25×10^{-5} M, pH 5; (3) La-Arsenazo I, 2.25×10^{-5} M, pH 9; (4) Arsenazo I, 2.25×10^{-5} M, pH 8. (From Fritz, J. S., Richard, M. J., and Lane, W. J., *Anal. Chem.*, 30, 1776, 1958. With permission.)

Purification and Purity of Reagent

Arsenazo I from commercial supply sources are often impure, although they can be satisfactorily used for the general purposes like as a metal indicator or as a chromogenic reagent for the metal determination. However, they have to be purified for use in the physicochemical work. The reagent can be purified by slowly dropping a saturated aqueous solution of the reagent into an equal volume of concentrated hydrochloric acid to precipitate orange crystals which is eventually filtered, washed with acetonitrile, and dried at 100° for 2 hr.[8]

Purity of the reagent may be determined by measuring the absorbance of aqueous solution of known regent concentration at pH 8.0 (λ_{max}, 500 nm; $\varepsilon = 2.67 \times 10^4$).[8,9]

Analytical Applications

Use as a Photometric Reagent

Arsenazo I has been used in the direct photometry of metal ions in aqueous solution

Table 1
APPLICATION OF ARSENAZO I AS A PHOTOMETRIC REAGENT

Metal ion	Condition	Metal chelate λ_{max} (nm)	ε ($\times 10^3$)	Range of determination (ppm)	Interference	Ref.
B	Conc H_2SO_4, 35°, 15 min	635	6.4	0 ∿ 1	K,Na,NH_4 do not interfere	10
Be	pH 11 ∿ 12	580	10.0	0 ∿ 0.15	—	11
Ca	pH 9.6, GEDTA	580	—	4 ∿	Sum of Ca + Mg in serum	12
Ce(III)	0.02 NHCl, $NH_2OH \cdot HCl$	253	7.6	—	La,Fe(equi. mol), Th(100-fold) do not interfere	13, 14
Ga	pH 3 ∿ 5.5	550	—	0.4 ∿ 5.6	Ce(IV),Cu,Fe,Tl(III), UO_2^{2+} interfere	15
In	pH 5.6 ∿ 6.2	580	58.2	2 ∿ 10	Cu,Fe(III),Pb, U(VI),Zn,PO_4^{3-}, tartarate interfere	16
La	pH 7.0	570	27.5	1 ∿ 2	Fe(III),U(VI) must be absent	8, 17
Pd(II)	pH 2	565	11.5	1.0 ∿ 10.6	Cr(III), Cu(II),Fe, La,Ru,Y,oxalate, citrate,EDTA, CyDTA interfere	18
Pd(II)	pH 12	585	14.5	0.64 ∿ 6.4		
Ru(III)	pH 6.0	580	—	1 ∿ 7	Various metals, citrate, oxalate, EDTA interfere	19
Sc	pH 6.1 ∿ 6.3	570	17.0	∿ 2.0	After extraction separation of interfering metals with TTA or oxine	20
Ta	pH 1 ∿ 2	580	—	0.01 ∿ 0.8	Ti does not interfere	21
Th	pH 1.0	546 (600)[a]	23.0	0.5 ∿ 5	Ce(IV), F^-,SO_4^{2-}, interfere, F^- can be masked with Al^{3+}	22
Ti	pH 2.5 ∿ 3.0	580	—	0.1 ∿ 2.5		23
U(VI)	pH 3 ∿ 7, EDTA	600	21.2	0.2 ∿ 1.6	Al,Fe(III),Th,Zn do not interfere	24
Y	pH 6.0	560	43.8	0.2 ∿ 5.7	—	25

[a]   Observing wavelength.

and in the extraction photometry of an ion-pair with diphenylguanidinium ion. However, as Arsenazo III-type reagents are more sensitive, Arsenazo I reagents are now being replaced by the latter reagents. The typical examples with Arsenazo I are summarized in Table 1. Practically important application of Arsenazo I may be the determination of Th in various samples, although interferences with Nb, Sn, Ta, Ti, and Zr are claimed.

Use as a Metal Indicator

Arsenazo I is recommended as a metal indicator in the chelatometric titration of Ca, Mg (pH 10), Th (pH 1.3 to 3.0), and rare earths (pH 5.5 to 6.5). The color change at the end point of direct titration is from red-violet to orange or orange-red.[26]

It was also used as an indicator in the precipitation titration of SO_4^{2-} with a standard solution of $BaCl_2$ (color change, orange to violet),[27] but dimethylsulfonazo-III is known to be a better indicator.[28]

Other Reagents Derived from Phenylarsonic Acid
Thorin

2-(2-Hydroxy-3, 6-disulfo-1-naphtylazo)benzenearsonic Acid, Disodium Salt; Thoron, Thoronol, Naphtharson, Naphthazarin, APANS, APNS

$$C_{16}H_{11}N_2O_{10}S_2Na_2As$$
$$\text{mol wt} = 576.29$$
$$(\text{as disodium salt})$$

Na_2H_3L

The materials from commercial supply sources are 1 to 5 sodium salt, orange-red (Na salt) to rust-red (Na_5 salt) crystalline powder, which is easily soluble in water, but insoluble in common organic solvents. The aqueous solution is orange-yellow in acidic range (λ_{max}, 510 nm; $\varepsilon = 1.4 \times 10^4$ at pH 1) and orange-red in alkaline range (λ_{max}, 490 nm; $\varepsilon = 1.4 \times 10^4$ at pH 9 to 10).[29] It is red in concentrated sulfuric acid, $pKa_3 = 3.7$, $pKa_4 = 8.4$ (-AsO(OH)$_2$), and $pKa_5 = 11.8$ (-OH).[30]

It forms colored soluble chelates in 0.1 N acid solution with Bi (0.02 to 0.04 N $HClO_4$, λ_{max}, 535 nm; $\varepsilon = 9.5 \times 10^3$),[36] Th (pH 0.5 to 1.5, ML_2; λ_{max}, 545 nm),[31] U(VI) (pH 3 to 4; ML_2; λ_{max}, 545 nm),[32] Zr and Hf (in HCl; λ_{max}, 555 nm),[33] and Pu(IV) (in 0.05 to 0.25 N HCl; λ_{max}, 540 nm).[34] The reactions are fairly selective for these metals. Thorin is also used for the photometric determination of other metals at higher pH region, although the reactions are not so selective: Be (pH 10; λ_{max}, 522 nm),[35] Li (0.4% NaOH, 70% acetone, $\varepsilon = 3.5 \times 10^3$ at 486 nm),[37] Rh(III) (pH 4.0; heat for 70 min; λ_{max}, 550 nm),[38] Ru(III) (pH 4.0; λ_{max}, 510 nm),[38] and for indirect determination of fluoride via Zr chelate.[39]

Thorin can be used as a metal indicator in the chelatometric titration of Bi[40] and Th[41] (pH 1 to 3), color change at the end-point being from red to yellow, and as an indicator in the titration of U (VI) (pH 2.5) using oxalic acid.[42]

Rezarson

2,2′,4′-Trihydroxy-3-arsono-5-chloroazobenzene.

$$C_{12}H_{10}N_2O_6AsCl$$
$$\text{mol wt} = 388.60$$

H_5L

Rezarson from soluble colored 1:1 chelates with various metal ions, including Cu, Ga, Ge, Ni, Pd, and Sn, in a medium of acidic aqueous 50 to 90% methanol.[43] It was suggested as a photometric reagent for Ge[44] and Sn (HCl, 510 nm).[45]

MONOGRAPH

M1. Johnson, E. A., Newman, E. J., and Titshall, M., Thorin, in *Organic Reagents for Metals*, Vol. 2, Johnson, W. C., Ed., Hopkin & Williams, Essex, England, 1964.

REVIEWS

R1. Johnston, M. B., Barnard, A. J., Jr., and Broad, W. C., Thorin: an interesting chromogenic agent and chelatochromic indicator, *Rev. Univ. Ind. Santander,* 2, 137, 1960.
R2. Savvin, S. B., Reagents of the arsenazo-thorin group, *Usp. Khim.,* 32, 195, 1963; *Chem. Abstr.,* 58, 13099h, 1963.
R3. Toei, K., Neothorine, *Kagaku (Kyoto) Zokan,* 14, 996, 1959.

REFERENCES

1. Kuznetsov, V. I., *C. R. Acad. Sci. USSR,* 31, 898, 1941; *Chem. Abstr.,* 37, 845, 1943.
2. Emi, K., Toei, K., and Furukawa, K., *Nippon Kagaku Zasshi,* 79, 681, 1958.
3. Budesinsky, B., *Talanta,* 16, 1277, 1969.
4. Kuznetsov, V. I., *J. Gen. Chem. USSR,* 14, 914, 1944; *Chem. Abstr.,* 39, 4561, 1945.
5. Kuzin, E. L., Savvin, S. B., and Grikov, L. A., *Zh. Anal. Khim.,* 23, 490, 1968; *J. Anal. Chem. USSR,* 23, 418, 1968.
6. Nakashima, S., Miyata, H. and Toei, K., *Bull. Chem. Soc. Jpn.,* 41, 2632, 1968.
7. Budesinsky, B., *Fresenius Z. Anal. Chem.,* 207, 105, 1965.
8. Fritz, J. S., Richard, M. J., and Lane, W. J., *Anal. Chem.,* 30, 1776, 1958.
9. Shibata, S., Goto, K., and Ishiguro, Y., *Anal. Chim. Acta,* 62, 305, 1972.
10. Hiiro, K., *Bunseki Kagaku,* 11, 223, 1962.
11. Shibata, S., Takeuchi, F., and Matsumae, T., *Bull. Chem. Soc. Jpn.,* 31, 888, 1958.
12. Lamkin, E. G. and Williams, M. B., *Anal. Chem.,* 37, 1029, 1965.
13. Stewart, D. C. and Kato, D., *Anal. Chem.,* 30, 164, 1958.
14. Hiiro, K., Russell, D. S., and Berman, S. S., *Anal. Chim. Acta,* 37, 209, 1967.
15. Joshi, A. P. and Munshi, K. N., *Microchem. J.,* 18, 277, 1973.
16. Matsumae, T., *Bunseki Kagaku,* 8, 167, 1959.
17. Banks, C. V., Thompson, J. A., and O'Laughlin, J. W., *Anal. Chem.,* 30, 1792, 1958.
18. Khalifa, H. and Issa, Y. M., *Microchem. J.,* 20, 287, 1975.
19. Khalifa, H. and Issa, Y. M., *Fresenius Z. Anal. Chem.,* 274, 126, 1975.
20. Onishi, H. and Banks, C. V., *Anal. Chim. Acta,* 29, 240, 1963.
21. Nikitina, E. I., *Zh. Anal. Khim.,* 8, 72, 1958.
22. Ishibashi, M. and Higashi, S., *Bunseki Kagaku,* 4, 14, 1955; *Bunseki Kagaku* 4, 5, 135, 1956.
23. Nikitina, E. I., *Zh. Anal. Khim.,* 14, 431, 1959.
24. Shibata, S. and Matsumae, T., *Bull. Chem. Soc. Jpn.,* 32, 279, 1959.
25. Pande, S. P. and Munshi, K. N., *Curr. Sci.,* 41, 330, 1972; *Anal. Abstr.,* 24, 704, 1973.
26. Fritz, J. S., Oliver, F. T., and Pietrzyk, D. J., *Anal. Chem.,* 30, 1111, 1958.
27. Fritz, J. S., Yamamura, S. S., and Richard, M. J., *Anal. Chem.,* 29, 158, 1957.
28. Budesinsky, B. and Krumlova, L., *Anal. Chim. Acta,* 39, 375, 1967.
29. Michaylov, V. A., *Zh. Anal. Khim.,* 16, 141, 1961; *J. Anal. Chem. USSR,* 16, 145, 1961.
30. Margerum, D. W., Byrd, C. H., Beed, S. A., and Banks, C. V., *Anal. Chem.,* 25, 1219, 1953.
31. Banks, C. V. and Byrd, C. H., *Anal. Chem.,* 25, 416, 1953.
32. Michaylov, V. A., *Zh. Anal. Khim.,* 16, 141, 1961.
33. Horton, A. H., *Anal. Chem.,* 25, 1331, 1953.
34. Healy, T. V. and Brown, P. E., *Atomic Energy Research Estab.,* AERE C/R, 1287, 8 (1957); *Anal. Abstr.,* 6, 1724, 1959.

35. Keil, R., *Fresenius Z. Anal. Chem.*, 262, 273, 1972.
36. Mottola, H. A., *Anal. Chim. Acta*, 27, 136, 1962.
37. Thomason, P. F., *Anal. Chem.*, 28, 1527, 1956.
38. Shrivastava, S. C., Munshi, K. N., and Dey, A. K., *Microchem. J.*, 14, 37, 1969.
39. Tanaka, Y., Saito, H., Nishimura, K., and Nakashima, M., *Bunseki Kagaku*, 26, 824, 1977.
40. Rady, G. and Erdey, L., *Fresenius Z. Anal. Chem.*, 152, 253, 1956.
41. Ford, J. J. and Fritz, J. S., U.S. Atomic Energy Commission, Rep. ISC-520, 1954, 47; *Chem. Abstr.*, 50, 7656, 1956.
42. Nadkarni, M. N., Ramanujam, A., Venkateson, M., Gopalakrishman, V., and Kazi, J. A., *Radiochem. Radioanal. Lett.*, 36, 139, 1978.
43. Lukin, A. M., Kaslina, N. A., Fadeeva, V. I., and Petrova, G. S., *Vestn. Mosk. Gos. Univ. Ser. Khim.*, 247, 1972; 77, *Chem. Abstr.*, 69658c, 1972.
44. Shcherbov, D. P., Plotnikova, R. N., and Astafeva, I. N., *Zavod. Lab.*, 36, 528, 1970; *Anal. Abstr.*, 20, 2974, 1971.
45. Kaslina, N. A., Lukin, A. M., Fadeeva, V. I., and Petrova, G. S., *Zavod. Lab.*, 44, 520, 1978; *Anal. Abstr.*, 35, 5B39, 1978.

ARSENAZO III AND BISAZO DERIVATIVES OF CHROMOTROPIC ACID

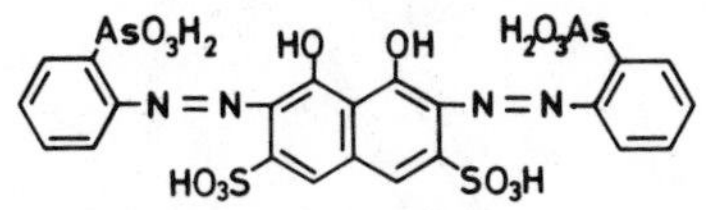

$C_{22}H_{18}N_4O_{14}S_2As_2$
mol wt = 777.37

H_8L

Synonym

2,7-*Bis*(2-arsonophenylazo)-1,8-dihydroxy-3,6-naphthalene disulfonic acid

Source and Method of Synthesis

Commercially available as free acid or disodium salt. It is synthesized by the coupling of diazotized 2-aminophenylarsonic acid with chromotropic acid in HCl medium.[1] Addition of $CaCl_2$ improves the yield.

Analytical Uses

As a highly sensitive photometric reagent in fairly acidic solutions for Th, U, and other hydrolyzable multivalent metal ions, and also as a metal indicator in the chelatometric titration. Arsenazo III is the first reagent of a series of bisazo derivatives of chromotropic acid which are known to be highly sensitive chromogenic reagent for metals.[2]

Properties of Reagent

Free base and mono- or disodium salt which are commercially available are a dark red-violet powder. Slightly soluble in water or in dilute acid, more soluble in aqueous alkali, and less soluble in common organic solvents. Solubilities in 9 N HCl containing 20% organic solvent are 0.27 g/ℓ (ethanol), 0.92 (iso-pentanol), and 0.75 (MIBK)[3]

The aqueous solution is pink or crimson red in a range of 10 N HCl to pH 3 and violet or blue at pH above 4. It is green in concentrated sulfuric acid due to the protonation on the azo groups. Absorption spectra of Arsenazo III in aqueous solution at various pH are illustrated in Figure 1.

The proton dissociation constants of the reagent have been observed by many workers, however, the values in earlier papers may need reinvestigations because of the impure nature of samples. A set of reliable values is shown in Table 1, together with the spectral characteristics of each ionic species.

Complexation Reactions and Properties of Complexes

Arsenazo III reacts with wide variety of metal ions in aqueous solution of weakly acidic to alkaline range, but reacts with only a limited kind of metals in strong mineral acid, forming highly colored ($\varepsilon \simeq 10^5$) soluble chelates. Reacting metals are Hf, Th, U, Zr, rare earths, and actinoid elements to which Arsenazo III has been recommended as a highly sensitive and selective photometric reagent.

The metal to ligand ratio of these chelates may vary from 1:1 to 1:3 or 2:2, depending upon the nature of metal ion, mixing ratio of metal to reagent, and solution condition.[R1] Structures of 1:1 and 2:2 chelates are shown below.[5]

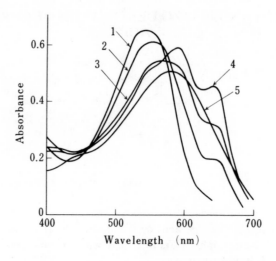

FIGURE 1. Absorption spectra of Arsenazo III in water at various pH; dye concentration, 1.84×10^{-5} M. (1) pH = 3; (2) pH = 6.5; (3) pH = 8.5; (4) pH = 10.6; (5) pH = 13.05. (From Savvin, S. B., *Arsenazo-III as Organic Reagent*, Atomizdat, Moscow, 1971. With permission.)

Table 1
PROTON DISSOCIATION CONSTANTS AND SPECTRAL CHARACTERISTICS OF IONIC SPECIES OF ARSENAZO III ($\mu = 0.1$, 20°)

n	pKa_n $\mu = 0.1$	Ionic species	λ_{max} (nm)	ε ($\times 10^4$)	Color
		$H_{10}L^{2+}$	670	5.10	Green
1	−2.7				
		H_9L^+	—	—	—
2	−2.7				
		H_8L	—	—	—
3	0.6				
		H_7L^-	—	—	—
4	0.8				
		H_6L^{2-}	—	—	—
5	1.6				
		H_5L^{3-}	—	—	—
6	3.4				
		H_4L^{4-}	535	4.05	Red
7	6.3				
		H_3L^{5-}	560	3.78	Wine Red
8	9.1				
		H_2L^{6-}	600	3.98	Violet
9	12.0				
		HL^{7-}	585	3.18	Blue
10	15.1				
		L^{8-}	600	4.52	Blue

(From Budesinsky, B., *Talanta*, 16, 1277, 1969. With permission.)

The alkaline earth metals form a colored chelate in weakly basic solution, and their conditional stability constants are reported to be log K'_{ML} for Ca 5.4,[6] Sr 5.3, and Ba 5.2 (at pH 9, $\mu = 0.06$).[7]

The reactions of Arsenazo III with various elements and the spectral characteristics of the metal chelates are summarized in Table 2. Absorption spectra of metal-Aresenazo III chelates, as illustrated in Figure 2, are characterized with two maxima which are considered to be due to the presence of two weakly interacting chromophoric systems in the chelate. Most of these chelates are anionic nature and can be extracted into butanol as an ion-pair with such cations as diphenylguanidinium ion for the subsequent photometry.[24,25]

Purification and Purity of Reagent

Most of commercial samples are contaminated with monoazo derivatives,[27] starting materials and other by-products. Although it may be partially purified through the reprecipitation of dye from the aqueous alkaline solution by acidification with HCl, the best way to get a pure sample is as follows.[28]

Dissolve 1 to 2 g crude sample into 15 to 25 mℓ of 5% aqueous ammonia and filter. Add 10 mℓ HCl (1:1) to the filtrate to precipitate the dye. Repeat the procedure once more and dissolve 0.5 g solid dye into 7 mℓ of a mixture of n-propanol-concentrated NH$_4$OH-water (1:1:1) at 50°. After cooling, filter the solution and treat the filtrate on a column of microcrystalline cellulose (ϕ 25 mm, h 120 mm) using a mixture of n-propanol-concentrated NH$_4$OH-water(3:1:1) as an eluent at a flow rate of 1.5 mℓ/min. Collect the fraction of blue band and concentrate it to 10 to 15 mℓ by evaporation below 80° and add 10 mℓ concentrated HCl to precipitate pure Arsenazo III. Wash it with ethanol and air dry.

The purity of Arsenazo III can easily be checked by running a paper chromatography using 1 N HCl as a solvent; Arsenazo III (red-violet spot, R$_f$ ∿10.25), Arsenazo I (red spot, R$_f$ ∿0.35), and unknown (brown spot, R$_f$ ∿0.05).[28] Paper electrophoresis is another means of checking purity.[29]

Analytical Applications

Use as a Photometric Reagent

As summarized in Table 2, the color reaction of Arsenazo III is most selective for Th, U, Zr, lanthanoids, and actinoids in fairly acid solution, and the molar absorptivities of the colored complexes are in the order of 10^5. For these reasons, Arsenazo III became of most practical importance for the photometric determination of such ele-

Table 2

APPLICATION OF ARSENAZO III AS A PHOTOMETRIC REAGENT

Metal ion	Condition	Metal chelate			Range of determination (ppm)	Remarks	Ref.
		Ratio	λ_{max} (nm)	ε ($\times 10^4$)			
Ac(III)	pH 3.0 ~ 3.2	ML$_2$	660	3.8	0.025 ~ 1.5	—	8
Am(V)	pH 5 ~ 5.2	ML	650	2.8	0.1 ~ 1.75	—	9
Al	pH 3.5	ML	600	1.98	0.01 ~ 0.6	Ag,Fe(III),Ti(III),F⁻, oxalate, tartarate interfere	10
Ba	pH 5.3	ML	640	0.51	—	—	M4
Be	pH 6.0 ~ 6.5, EDTA,triethanolamine	ML	580	1.8	~0.4	—	11
Ca	pH 8 ~ 9	ML	600 or 650	2.0 or 2.8	0.03 ~ 1.5	In magnesite	12
Cd	pH 10	ML	600	2.6	0.24 ~ 2.38	Zn can be masked with pyridinium nitrate	13
Cm(III)	pH 3	—	650	—	0.01 ~ 2	Am can be determined simultaneously	14
Dy	pH 2.9	M$_2$L$_2$	650	6.31	—	—	15
Gd	pH 2.9	M$_2$L$_2$	650	5.47	—	—	15
La	pH 2.9	M$_2$L$_2$	650	4.5	—	—	16
Li	0.15 ~ 0.3 N alkali	—	600	0.13	0 ~ 4	—	17
Np(V)	pH 4.4 ~ 7.2	ML	650	7.0	—	EDTA,citrate,tartarate interfere	18
Pa(V)	7 ~ 8 N H$_2$SO$_4$, extraction with iso-AmOH	—	680	2.2	0.3 ~	Mn(IV),Mo(VI),Ti(IV) give negative and Fe(III),Sn(IV) give positive error	19
Pd(II)	pH 3.4 ~ 5.9	ML	630	1.5	1.2 ~ 3	Pt(IV)(fourfold), Rh(III), Au(III),Ru(III),Os(III) (threefold), Ir(IV)(twofold) do not interfere	20
Sc	pH 2.2	ML	675	1.9	—	—	21
Th	8 N HClO$_4$	ML$_2$	660	8.6	—	—	16

Th	Extraction with Aliquat® 336 S-Toluene, then strip with 0.1 NHCl.	—	—	12.0	0~0.4	—	22
UO_2^{2+}	pH 1.1~3.4, Zephiramine®, extraction with $CHCl_3$	ML	655	6.2	0~2	Dy,Fe(III),Th,Zr,F⁻ interfere	23
UO_2^{2+}	pH 3~0.2 NHCl, diphenylguanidine, extraction with butanol	ML_2X_n	660	$\simeq 10^5$	0.2~10	Al,Fe,bivalent cations, PO_4^{3-}, F⁻, SO_4^{2-} do not interfere	24
Y	pH 6.3~6.7, diphenylguanidine, extraction with butanol	ML_2X_8	670	5	0.1~4.7	Ca,Cr,Fe,Ga,Mn can be masked with EDTA + triethanolamine	25
Zn	pH 10, I⁻	ML	590	4.28	0.14~1.26	Cd can be masked with NH_4F	13

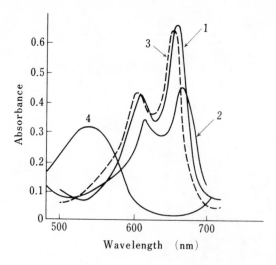

FIGURE 2.   Absorption spectra of some metal che-
lates of Arsenazo III (1) UO_2^{2+} (pH 1.5); (2) Zr (9 *M*
HC1); (3) La (pH 3); (4) free dye (pH 2∿3, 0.5 × 10⁻⁵
M); 10-mm cell, against water. (From Savvin, S. B.,
Talanta, 11, 10, 1964. With permission.)

ments. The reactions with Hf, Nb, Pa, Pu, Th, U, and Zr occur even in 3 *M* or stronger
HC1 solution, and U(VI) in 5 to 6 *M* $HClO_4$ solution, where most of common elements
do not react. With decreasing acidity, the number of reacting metal ions increases.

Determination of uranium in ores using Arsenazo III[30] — Place an aliquot
(1 to 5 m*l*) of the sample solution containing 5 to 50 µg of uranium into a
separatory funnel and add 25 m*l* of 60% aqueous ammonium nitrate solution which is 0.25% in EDTA (disodium salt). Adjust the solution to pH
2.5 to 3.0 with 7 *N* aqueous ammonia and 2 *N* nitric acid. Shake for 2 min
with 15 m*l* of a 20% solution of tributylphosphate in carbon tetrachloride.
Separate the phases and filter the organic phase through a filter paper.
Shake the aqueous phase once more with 10 m*l* of the same extraction
solution and filter the organic phase. Add 15.0 m*l* of 0.006% aqueous Arsenazo III solution to the combined organic phases and shake vigorously
for 1 min. Allow phase separation and measure the absorbance of the
aqueous phase in a 2-cm cell at 655 nm against a reagent blank. The devia-
tions were within ±0.5%. The following ions (in milligrams) do not interfere: Na(500); Cr(III) (VI) (250); Zn, Co(II), Ni, Cu(II) (200); Ca, A1(125);
Mn(II), Be 25, Th 7.5, Fe(III), Mo(VI), W(VI) (5); V(V) (3.75); Zr(2.5);
lanthanoids(0.25); acetate(500); C1⁻ SO_4^{2-} (250); PO_4^{3-}, citrate (25); and F⁻
(10).

Use as a Metal Indicator
   Arsenazo III is recommended as an indicator in the precipitation titration of SO_4^{2-}
with the standard barium chloride solution. The color change (blue → red) becomes
more sensitive in 50% iso-propanol with the titrant containing Pb. The impure dye
containing monoazo derivatives gives inferior color change.
   Arsenazo III is also used as a metal indicator in the chelatometric titration of Bi
(pH 1 to 2),[31] Ca (pH 8.1),[32] and rare earths (pH 5 to 6).[33]

Other Bisazo Derivatives of Chromatropic Acid

Chlorophosphonazo III

2,7-*Bis* (4-chloro-2-phosphonophenylazo)-1,8-dihydroxynaphthalene-3,6-disulfonic acid,

$C_{22}H_{14}N_4O_{14}S_2P_2Cl_2$,
mol wt = 755.34

H_8L

Usually supplied as disodium salt, is dark violet crystalline powder, and is easily soluble in water to give a violet solution above pH 4. It turns to blue in strongly alkaline solution and to green in mineral acid (HCl or H_2SO_4), but decomposes in concentrated HNO_3; pKa values for $H_{10}L^{2+}$ are −1.1, −1.1, 0.6, 0.8, 1.5, 2.5, 5.5, 7.2, 12.2, and 15.1 (μ = 0.1, 20°).[4] Absorption spectra of aqueous solution at various pH are illustrated in Figure 3. It reacts with Th, U, rare earths, and actinoid elements to form blue-violet soluble chelates in more acidic solution than in the case of Arsenazo III.

The composition of the chelates is usually 1:1, but in the case of rare earth chelates, two types of chelate are formed; 1:1 mononuclear chelate with the elements of lower atomic number and 1:1 polynuclear chelates with the elements of higher atomic number.

Spectral characteristics of the chelates with selected metal ions are summarized in Table 3, and the absorption spectra of Ca and Mg chelates are illustrated in Figure 4. The molar absorptivities of Ca and Mg chelates are very much pH dependent as shown in Figure 5. Hence, Ca can be determined in the presence of Mg at pH 2 to 6 and sum of Ca and Mg at pH 7.[34] Ba and Sr behave similarly to Ca.[44] The chelates with lanthanoids,[43] Sc,[45] Th,[39] U,[41] and Zr[42] can be extracted with butanol or 3-methyl-1-butanol, showing increased apparent molar absorptivities than in the aqueous system. EDTA and other complexanes are often used as a masking agent for interfering elements in the determination of U(VI)[48] and lanthanoids.[49]

The reagents commercially available are usually impure, and various methods of purification have been reported in the literature.[46,47]

Analogues of Chlorophosphonazo III have also been investigated as analytical reagents.[50-52]

Sulfonazo III (**1**), Dimethylsulfonazo III (**2**), and Dinitrosulfonazo III (**3**).

H_6L

- Sulfonazo III (**1**) (X = H), 3,6-*bis* (2′-sulfophenylazo)-4,5-dihydroxynaphthalene-2,7-disulfonic acid, $C_{22}H_{14}N_4O_{14}S_4$, mol wt = 686.61.
- Dimethylsulfonazo III (**2**) (X = CH₃), 3,6-*bis* (4′-methyl-2′-sulfophenylazo)-4,5-dihydroxynaphthalene-2,7-disulfonic acid, $C_{24}H_{18}N_4O_{14}S_4$, mol wt = 714.66.
- Dinitrosulfonazo III (**3**) (X = NO₂), 3,6-*bis* (4′-nitro-2′-sulfophenylazo)-4,5-dihydroxynaphthalene-2,7-disulfonic acid, $C_{22}H_{12}N_6O_{18}S_4$, mol wt = 776.61

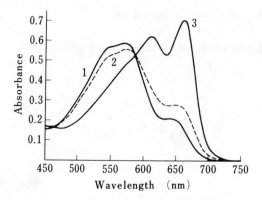

FIGURE 3.   Absorption spectra of Chlorophosphon-
azo-III in water at various pH; dye concentration, 1.6 ×
10^{-5} M. (1) pH 2.2; (2) pH 7; (3) pH 10. (Reprinted with
permission from Ferguson, J. W., Richard, J. J.,
O'Laughlin, J. W., and Banks, C. V., *Anal. Chem.*, 36,
796, 1964. Copyright 1964 American Chemical Society.)

Table 3
SPECTRAL CHARACTERISTICS OF CHLOROPHOSPHONAZO-III CHELATE[M4]

	Metal chelate				
Metal ion	Condition	Ratio[a]	λ_{max} (nm)	ε (× 10⁴)	Ref.
None	pH 2 ∿ 6	—	560 ∿ 5	1.30	
Ba[b]	pH 7.1	ML	660	4.20	
Ca	pH 7.5	ML	660	3.76	
Mg	pH 7.5	ML	660	2.66	
Np(V)	pH 2 ∿ 5	ML	670	6.22	35
Pa(V)	0.1 NHCl, 0.01 MNH₄F	ML	630	3.08	36
Pu(IV)	0.5 ∿ 2 NHCl	—	690	3.7	37
Sc[c]	pH 2 ∿ 3	ML	690	1.25	38
Sr	pH 7.5	ML	660	4.10	
Th	pH 1 ∿ 2	ML₂	690	4.31	38
Th[c]	pH 2 ∿ 3, extraction with 3-methyl-1-butanol	—	670	12.2	39
Ti(III)[c]	pH 1 ∿ 2	ML	690	1.08	38
U(VI)	pH 1.0	ML	670	7.86	40
U(IV)	pH 0 ∿ 1.5, extraction with 3-methyl-1-butanol or BuOH + PhCH₂OH (4:1)	ML₂	670 ∿ 3	12.1	41, 60
Y	pH 2.0	ML₂	665	4.20	
Zr[c]	2 NHCl	ML₂	690	3.3	38
Zr	2 NHCl extraction with 3-methyl-1-butanol	ML₂	675	21.0	42
Rare earths	pH 2.0	ML₂	665	1.65 ∿ 4.31	
Rare earths	pH 1.1 ∿ 1.5, extraction with butanol	ML₂	668	1.5 ∿ 2.3	43

[a]   Indicate metal ligand ratio only, neglecting the protonation state.
[b]   Stability constant for Ba chelate: log K_{ML},7.3; log K_{MHL},22.9; log K_{MH_2L},34.5; log K_{MH_3}L,40.8.
[c]   Conditional stability constants: log K_{ScL},6.9(pH 2 ∿ 3); log β_{ThL_2},13.9(1 ∿ 2 NHCl); log K_{TiL},5.7(pH 1 ∿ 2); log β_{ZrL_2},13.4(2 NHCl).[35]

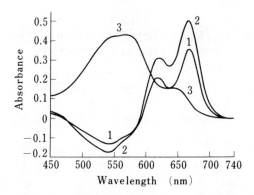

FIGURE 4. Absorption spectra of Ca- and Mg-Chlorophosphonazo III chelates in water; dye concentration, 1.2×10^{-5} M, pH 7.0. (1) Ca, 5 μg; (2) Mg, 5 μg; (3) dye only (Reprinted with permission from Ferguson, J. W., Richard, J. J., O'Laughlin, J. W., and Banks, C. V., *Anal. Chem.*, 36, 796, 1964. Copyright 1964 American Chemical Society.)

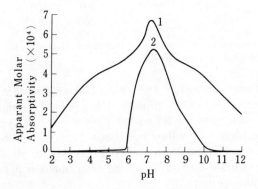

FIGURE 5. pH dependencies of apparent molar absorptivity of Ca and Mg-Chlorophosphonazo III chelates. (1) Ca-chelate; (2) Mg-chelate. Apparent molar absorptivity was calculated from the difference in absorbances of the sample and a reagent blank at 669 nm. (Reprinted with permission from Ferguson, J. W., Richard, J. J., O'Laughlin, J. W., and Banks, C. V., *Anal. Chem.*, 36, 796, 1964. Copyright 1964 American Chemical Society.)

These reagents are usually supplied as disodium salt (Na_2H_4L) or tetrasodium salt (Na_4H_2L). **(1)** and **(2)** are a dark violet crystalline powder, and **(3)** is a dark green powder (decomposes 222°).[M5] All are easily soluble in water and aqueous alcohol, but insoluble in common organic solvents. Dilute aqueous solutions of **(1)**, **(2)**, or **(3)** are red-violet in acid and neutral ranges and turns to blue in strong alkali. The pKa values for **(1)** are −0.3, −0.3, 0.6, 0.8, 2.4, 2.8, 11.6, and 14.4 (μ = 0.1, 20°).[4] **(2)** and **(3)** seem to behave similarly.

These reagents are characterized by the reactivity ≈ with alkaline earth ions, forming the colored chelate (blue-violet) in acidic range (pH ≈ 2.4). The spectral characteristics of **(1)** and its metal chelates are summarized in Table 4. One of the important applications of these reagents is their use as an indicator in the precipitation titration of SO_4^{2-}

Table 4

SPECTRAL CHARACTERISTICS OF SULFONAZO III AND ITS METAL CHELATES[a]

Metal ion	Condition	Ratio	λ_{max} (nm)	ε (× 10^4)	Ref.
H_8L^{2+}	$H° -2$		670	5.10	
H_4L^{2-}	pH ∿1		580	3.98	
H_2L^{4-}	pH 4 ∿ 8		580	4.19	
HL^{5-}	pH ∿13		580	2.93	
L^{6-}	pH ∿16		610	4.41	
Ba	pH 2.4	ML	638	1.73[a]	
Ca	pH 2.4		638	0.53[a]	
Ce(III)		ML_3X_n[b]	660	3.66	53
Cu(II)	pH 2.4		638	0.91[a]	
La(III)		ML_3X_n[b]	660	3.7	53
Mg	pH 2.4		638	0.01	
Nd		ML_3X_n[b]	660	2.88	53
Pd(II)	pH 2.4		638	0.48[a]	
Pr		ML_3X_n[b]	660	3.0	53
Sr	pH 2.4		638	0.13	

[a] Values observed in aqueous acetone (acetone to water = 2:3).
[b] In the presence of antipyrine (X).

with a standard barium solution.[54] The color change becomes more sensitive in aqueous organic solvents (50 to 80% ethanol or acetone).[55] (2) and (3) are claimed to be the most sensitive indicators for Ba among the various azo dyes[M3] and are recommended for the determination of sulfate in various inorganic and organic samples.[56,57]

These reagents can also be used as photometric reagents for alkaline earth and some other elements. For example, Ba and Sr can be determined at pH 6 (λ_{max}, 643 nm; in 35 to 45% acetone; 0.4 to 4 ppm).[58,59]

MONOGRAPHS

M1. **Savvin, S. B.**, *Arsenazo III, Methods of Photometric Determining Rare Earths and The Actinide Elements* (in Russian), Atomizdat, Moscow, 1966.

M2. **Savvin, S. B.**, *Arsenazo III as Organic Reagent* (in Russian), Atomizdat, Moscow, 1971.

M3. **Savvin, S. B., Akimova, T. G., and Dedkova, V. T.**, *Organic Reagents for Determination of Ba^{2+} and SO$_4^{2-}$* (in Russian), Izdatelistvo, Moscow, 1971.

M4. **Budesinsky, B.**, Monoazo and *Bis* (arylazo) derivatives of chromotropic acid as photometric reagents, in *Chelates in Analytical Chemistry*, Vol. 2, Flaschka, H. A. and Barnard, A. J., Jr., Eds., Marcel Dekker, New York, 1969, 1.

M5. **Sandell, E. B. and Onishi, H.**, *Photometric Determination of Traces of Metals*, 4th ed., John Wiley & Sons, New York, 1978.

REVIEWS

R1. **Savvin, S. B.**, Analytical use of Arsenazo III. Determination of thorium, zirconium, uranium and rare earth elements, *Talanta*, 8, 673, 1961.

R2. **Fujinaga, T. and Kuwabara, K.**, Arsenazo III and its analogues, *Kagaku (Kyoto)*, 19, 1047, 1964.

R3. **Imamura, T.**, Synthesis and applications of bisazochromotropic acid analogues, *Bunseki Kagaku*, 18, 1273, 1968; Application of azodyes for analytical chemistry, *Bunseki Kagaku*, 21, 566, 1972.

REFERENCES

1. **Zaikovsky, F. V. and Ivanova, V. N.**, *Zh. Anal. Khim.*, 18, 1030, 1963.
2. **Savvin, S. B.**, *Dokl. Akad. Nauk SSR*, 127, 1231, 1959; *Chem. Abstr.*, 55, 4249b, 1961.
3. **Nemodruk, A. A. and Boganova, A. N.**, *Zh. Anal. Khim.*, 31, 854, 1976; *J. Anal. Chem. USSR*, 31, 698, 1976.
4. **Budesinsky, B.**, *Talanta*, 16, 1277, 1969.
5. **Savvin, S. B. and Kuzin, E. L.**, *Zh. Anal. Khim.*, 22, 1059, 1967.
6. **Michaylova, V. and Kuleva, N.**, *God. Sofii. Univ. Khim. Fak.*, 67, 206, 1972; *Chem. Abstr.*, 86, 128292q, 1977.
7. **Michaylova, V. and Popnikolova, M.**, *God. Sofii. Univ. Khim.*, 67, 301, (1972); *Chem. Abstr.*, 86, 128293r, 1977.
8. **Karalova, Z. K., Myasoedov, B. F., and Rodinova, L. M.**, *Zh. Anal. Khim.*, 28, 942, 1973.
9. **Myasoedov, B. F., Milyukova, M. S., and Ryzhova, L. V.**, *Radiochem. Radioanal. Lett.*, 11, 39, 1972.
10. **Michaylova, V.**, *Acta. Chim. (Budapest)*, 76, 221, 1973; *Chem Abstr.*, 79, 13207d, 1973.
11. **Talipov, Sh. T., Khadeeva, L. A., and Popova, R.**, *Izv. Vyssh. Uchebn. Zaved. Khim. Khim. Tecknol.*, 14, 343, 1971; *Anal. Abstr.*, 23, 89, 1972.
12. **Michaylova, V. and Ilkova, P.**, *Anal. Chim. Acta*, 53, 194, 1971.
13. **Michaylova, V. and Yurukova, L.**, *Anal. Chim. Acta*, 68, 73, 1974.
14. **Michaylova, M. S., Myasoedov, B. F., and Ryzhova, L. V.**, *Zh. Anal. Khim.*, 27, 1769, 1972.
15. **Budesinsky, B.**, *Collect. Czech. Chem. Commun.*, 28, 2902, 1963.
16. **Budesinsky, B., Haas, K., and Vrzalova, D.**, *Collect. Czech. Chem. Commun.*, 30, 2373, 1965.
17. **Lazarev, A. I. and Lazareva, V. I.**, *Zh. Anal. Khim.*, 23, 36, 1968.
18. **Chudinov, E. G. and Yakovlev, G. N.**, *Radiokhimiya*, 4, 505, 1963; *Anal Abstr.*, 10, 4669, 1963.
19. **Palshin, E. S., Myasoedov, B. F., and Palei, P. N.**, *Zh. Anal. Khim.*, 17, 471, 1962.
20. **Sen Gupta, J. G.**, *Anal. Chem.*, 39, 18, 1967.
21. **Ryabchikov, D. I., Savvin, S. B., and Dedkov, Yu. M.**, *Zh. Anal. Khim.* 19, 1210, 1964.
22. **Petrov, H. G. and Strehlow, C. D.**, *Anal. Chem.*, 39, 265, 1967.
23. **Sekine, K.**, *Mikrochim. Acta*, 2, 559, 1976.

24. Kuznetsov, V. I. and Savvin, S. B., *Radiokhimiya*, 2, 682, 1960; *Chem. Abstr.* 55, 15224d, 1961; *Anal. Abstr.*, 8, 4613, 1961.
25. Akhimedli, M. K., Granovskaya, P. B., and Melikova, E. G., *Zh. Anal. Khim.*, 28, 1304, 1973.
26. Savvin, S. B., *Talanta*, 11, 10, 1964.
27. Shibata, S., Goto, K., and Ishiguro, Y., *Anal. Chim. Acta*, 62, 305, 1972.
28. Borak, J., Slovak, Z., and Fischer, J., *Talanta*, 17, 215, 1970.
29. Akimova, T. G., Dekova, V. P., and Savvin, S. B., *Zh. Anal. Khim.*, 32, 1269, 1977; *J. Anal. Chem. USSR*, 32, 1003, 1977.
30. Palei, P. N., Nemodruk, A. A., and Davydov, A. V., *Radiokhimiya*, 3, 181, 1961.
31. Perez Bustamante, J. A., Alvarez Jimenez, D., and Burriel Marti, I., *Anal. Chim. Acta*, 50, 354, 1970.
32. Michaylova, V. and Kouleva, N., *Dokl. Bolz. Akad. Nauk.*, 25, 949, 1972; *Chem. Abstr.*, 78, 37613p, 1973; *Talanta*, 20, 453, 1973.
33. Marsh, W. W., Jr. and Myers, G., Jr., *Anal. Chim. Acta*, 43, 511, 1968.
34. Fergason, J. W., Richard, J. J., O'Laughlin, J. W., and Banks, C. V., *Anal. Chem.*, 36, 796, 1964.
35. Chudinov, E. G., *Zh. Anal. Khim.*, 20, 805, 1965.
36. Chudinov, E. G. and Yakovlev, G. N., *Radiokhimiya*, 4, 605, 1962; *Chem. Abstr.*, 60, 9902g, 1964.
37. Yamamoto, T., *Mikrochim. Acta*, 871, 1974.
38. Fadeeva, V. I. and Alimarin, I. P., *Zh. Anal. Khim.*, 17, 1020, 1962.
39. Yamamoto, T., *Anal. Chim. Acta*, 63, 65, 1973.
40. Nemodruk, A. A., Novikov, Yu. P., Lukin, A. M., and Kalinina, I. D., *Zh. Anal. Khim.*, 16, 180, 1961.
41. Yamamoto, T., *Anal. Chim. Acta*, 65, 329, 1973.
42. Yamamoto, T., Muto, H., and Kato, Y., *Bunseki Kagaku*, 26, 515, 1977.
43. Taketatsu, T., Kaneko, M., and Kohno, N., *Talanta*, 21, 87, 1974.
44. Lukin, A. M., Smirnova, K. A., and Chernysheva, T. V., *Zh. Anal. Khim.*, 21, 1300, 1966, *Tr. Vses. Nauchoo Issled. Inst. Reaktiv. Osobo Chist. Khim. Veshchestv.* 170, 1971; *Chem. Abstr.*, 78, 37505e, 1973.
45. Fadeeva, V. I. and Kuchinskaya, O. I., *Vestn. Mosk. gos. Univ. Ser. Khim.*, 67, 1967; *Anal. Abstr.*, 15, 1905, 1968.
46. Ishiguro, Y., Goto, K., Shibata, S., Nakashima, R., Sasaki, S., Furukawa, M., and Kamata, E., *Rep. Govern. Ind. Res. Inst. Nagoya (Japan)*, 24, 337, 1975; *J. Chromatogr.*, 115, 660, 1975.
47. Zenki, M., *Anal. Chim. Acta*, 93, 323, 1977.
48. Strelow, F. W. E. and Van der Wart, T. N., *Talanta*, 26, 537, 1979.
49. Taketatsu, T. and Sato, A., *Anal. Chim. Acta*, 93, 327, 1977.
50. Budesinsky, B. and Menclova, B., *Talanta*, 14, 688, 1967.
51. Budesinsky, B., Haas, K., and Bendekova, A., *Collect. Czech. Chem. Commun.*, 32, 1528, 1967.
52. Lukyanov, V. F., Duderova, E. P., Barabanova, T. E., Novak, E. F., and Polyakova, I. A., *Zh. Anal. Khim.*, 32, 674, 1977; *J. Anal. Chem. USSR*, 32, 541, 1977.
53. Kharzeeva, S. E. and Maltseva, V. S., *Zh. Anal. Khim.*, 34, 1022, 1979.
54. Budesinsky, B. and Krumlova, L., *Anal. Chim. Acta*, 39, 375, 1967.
55. Archer, E. E., White, D. C., and Mackison, R., *Analyst*, 96, 879, 1971.
56. Budesinsky, B. and Vrzalova, D., *Chemist-Analyst*, 55, 110, 1966.
57. Budesinsky, B., *Microchim. J.*, 20, 360, 1975.
58. Budesinsky, B. W., Vrzalova, D., and Bedekova, A., *Acta Chim. Acad. Sci. Hung.*, 53, 37, 1967; *Chem. Abstr.*, 68, 45935d, 1968.
59. Kemp, P. J. and Williams, M. B., *Anal. Chem.*, 45, 124, 1973.
60. Guertler, O., Holzapfel, H., and Chu-Xuan-Auh, *Z. Chem.*, 11, 389, 1971; *Chem. Abstr.*, 76, 10036z, 1972.

SULFARSAZENE

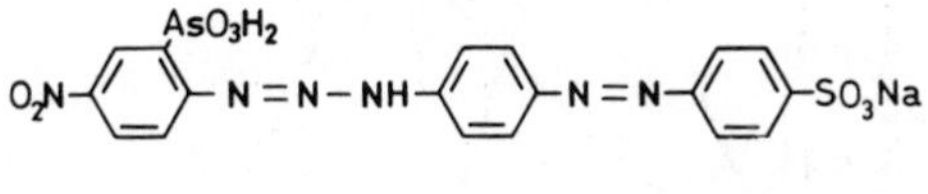

$$O_2N-\langle\ \rangle(AsO_3H_2)-N=N-NH-\langle\ \rangle-N=N-\langle\ \rangle-SO_3Na$$

$C_{18}H_{14}O_8N_6SNaAs$
mol wt = 572.32

NaH_2L

Synonyms

4″-Nitrobenzene-1″,4-diazoamino-1,1′-azobenzene-2″-arsono-4′-sulfonic acid, monosodium salt, 5-nitro-2-[3-(4-p-sulfophenylazo-phenyl)-1-triazene]-benzenearsonic acid, monosodium salt, Plumbon, Plumbon S, Plumbon IREA.

Source and Method of Synthesis

Commercially available. It is prepared by the coupling of diazotized 2-amino-4-nitrobenzenearsonic acid with 4-aminoazobenzene-4′-sulfonic acid.[1]

Analytical Uses

As a selective detection and photometric reagent for Pb and as a metal indicator in the chelatometric titration of Cd, Pb, Ni, and Zn.

Properties of Reagent

Commercially available samples are usually monosodium dihydrate salt ($NaH_2L \cdot 2H_2O$),[1] are a red-brown crystalline powder, and are soluble in water and dilute aqueous alkali, but insoluble in common organic solvents. Aqueous solution is yellow in acidic and weak alkaline range, but turns to violet in strong alkaline range. It decomposes in concentrated mineral acids. Absorption spectrum in aqueous solution (pH $\simeq$ 9) is illustrated in Figure 1.

Complexation Reactions and Properties of Complexes

Although Sulfarsazene had originally been introduced as a highly selective chromogen for Pb, it is known to form colored chelates with other elements, such as Cd, Co, Cu, La, Mn(II), Ni, U, and Zn.[1] The yellow solution of the reagent turns to orange-red upon addition of Pb at pH 8 to 10. The structure of the resulting Pb-chelate is proposed as shown below,[2] and the absorption spectrum is illustrated in Figure 1. The stability constants of the chelates are reported on the following metals; log K_{ML} for Cd, 9.8; Ni, 8.1; Pb, 10.5; Zn, 10.8[3] and the conditional constant log K'_{ML} for Cu, 5.09, (at pH 7).[7]

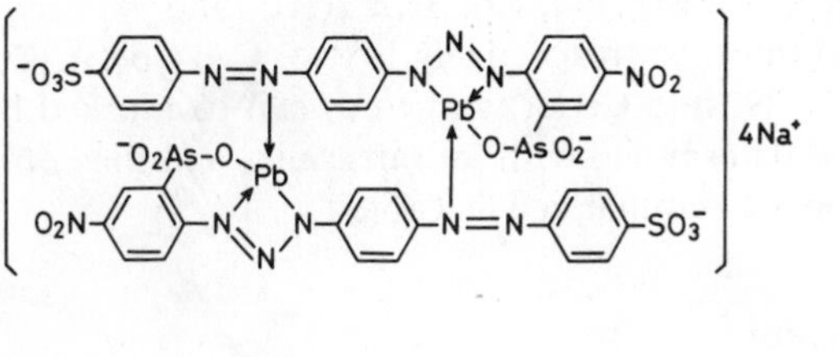

Analytical Applications

Use as a Photometric Reagent

Sulfarsazene has been used as a highly selective photometric reagent for Pb. Each 25 μg of Cr, Te, Y; 10 μg of Al, Be, Sn(IV); 5 μg of Th; and 2 μg of Sc do not interfere in the determination of 1 to 10 μg of Pb. Serious interferences are observed for Fe(III), Mo(VI), Ti, V, and Zr. Up to 50 μg of following elements do not interfere; As, Ba,

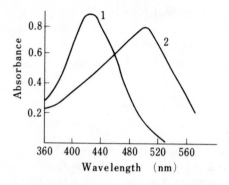

FIGURE 1.   Absorption spectra of Sulfarsazene and its metal chelate; dye concentration, 2
× 10⁻⁵ *M*, pH 9 (0.05 *M* Na₂B₄O₇). (1) Dye; (2)
Pb-chelate. (From Lukin, A. M. and Petrova, G.
S., *Zh. Anal. Khim.*, 15, 295, 1960. With permission.)

Table 1
APPLICATIONS OF SULFARSAZENE AS A PHOTOMETRIC REAGENT[R1]

Metal ion[a]	Condition	Metal chelate				Range of determination (ppm)	Remarks	Ref.
		Ratio	λ_{max} (nm)	ε (× 10⁴)				
Cd	pH 8 ∿ 9.8	M₂L₂	510	6.1		0.02 ∿ 3	—	
Co(II)	pH 8 ∿ 9.8	—	533	6.2		0.01 ∿ 2	—	6
Cu(II)	pH 7	M₂L₂	500	4.0		0.12 ∿ 1.3	Zn,Ni, EDTA interfere	7
Hg(II)	pH 9.5 ∿ 10.5	M₂L₂	500 ∿ 20	4.0		0.05 ∿ 1	Fe can be masked with HIDA[b]	8
Ni	pH 8.6 ∿ 9.6	—	530	6.2		0.01 ∿ 2		
Pb	pH 8 ∿ 10	M₂L₂	500	4.5		0.05 ∿ 9	After extraction with dithizone	1, 9
Zn	pH 8 ∿ 9.2	M₂L₂	505	4.6		0.1 ∿ 1.6	After chromatographic separation	10,11

[a]   Other metals determined with sulfarsazene are Ag,[5] Mn(II), and rare earths.[12]
[b]   HIDA: *N*(2 Hydroxyethyl)-iminodiacetic acid.

Bi, Ca, Cs, Ga, Ge, K, Li, Mg, Na, Rb, Sr, Tl(III), and W. Zn and Cu can be masked
by ferrocyanide and thiourea, respectively.[1] Also, Cd, Co(II), Cu, Hg(II), Ni, and Zn
can be masked by KCN, and Cr(III) and V(V) can be masked by H₂O₂.[4] Some other
elements can be also determined with Sulfarsazene, and the conditions for the photometric determination are summarized in Table 1.

Use as a Metal Indicator
Sulfarsazene has been recommended as a metal indicator in the chelatometric titration of Cd, Hg, Ni, Pb, and Zn at pH 9.5 to 10 (borate buffer).[3,8,13] The color change
at the end point is from red to yellow.

It can also be used as an adsorption indicator in place of Titan Yellow in the precipitation titration of excess tetraphenylborate with Zephiramine® on the indirect volumetric determination of potassium.[14]

Other Reagents with Related Structures

Cadion [(4-*p*-nitrophenyltriazeno)azobenzene, 4-nitrodiazoaminoazobenzene, Cadion A]; $C_{18}H_{14}N_6O_2$, mol wt = 346.35

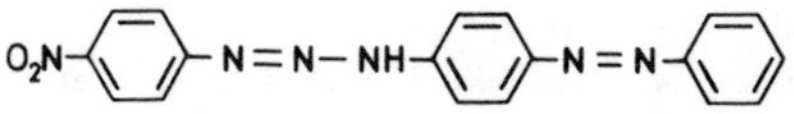

$$O_2N-\langle\rangle-N=N-NH-\langle\rangle-N=N-\langle\rangle$$

Yellow-orange crystalline powder, mp 192° (decomposes at 197°), is easily soluble in common organic solvents, and is also readily soluble in alcoholic alkali to give a violet solution. As the solution gives a bright pink precipitate with Cd and blue one with Mg, it is recommended as a detection reagent for these metals.

It is also used as a photometric reagent for Cd (KOH, tartarate, polyvinylpyrolidone, 560 nm)[15] or (pH 11.5 to 13.0, Triton®-X 100, dual-wavelength photometry at 477 and 560 nm, 8 to 80 ppb)[16] and Hg(II) (extraction with benzene or CHCl$_3$ from acetate buffer, 418 nm, 0.1 to 4 ppm).[17]

Other analogues of Cadion, such as Cadion 2B[1-(4-nitro-1-naphthyl)-3-(4-phenylazophenyl)triazene],[18] Cadion 3B (benzenediazoaminobenzene-4-azo-4′-nitrobenzene), and Cadion IREA (5-nitro-2-*3-[4-(4-sulfophenylazo)phenyl]triazeno* benzenesulfonic acid)[19] have been investigated as the reagent for Cd.

MONOGRAPH

M1. **Welcher, F. J.**, Organic Analytical Reagents, Vol. 3, Von Nostrand, Princeton, N.J., 1947.

REVIEW

R1. **Petrova, G. S., Yagodnitsyn, M. A., and Lukin, A. M.**, Sulfarsazene and its use in analysis, *Zavod. Lab.*, 36, 776, 1970; *Chem. Abstr.*, 73, 105089w, 1970.

REFERENCES

1. **Lukin, A. M. and Petrova, G. S.**, *Zh. Obshch. Khim.*, 31, 1254, 1961; *Chem. Abstr.*, 55, 24623c, 1961; *Zh. Anal. Khim.*, 15, 295, 1960.
2. **Lukin, A. M., Vainshtein, Yu. I., Dyatlova, N. M., and Petrova, G. G.**, *Zh. Anal. Khim.*, 17, 212, 1962.
3. **Partashnikova, M. Z. and Shafran, I. G.**, *Zh. Anal. Khim.*, 20, 313, 1965.
4. **Yamashige, T., Ohmoto, Y., and Shigetomi, Y.**, *Bunseki Kagaku*, 28, 464, 1979.
5. **Stolyarov, K. P. and Firyulina, V. V.**, *Zh. Anal. Khim.*, 24, 1494, 1969.
6. **Bashirov, E. A. and Ayukova, A. M.**, *Uch. Zap. Azerb. Univ. Ser. Khim. Nauk*, 28, 1971; *Chem. Abstr.*, 78, 37535q, 1973.
7. **Bashirov, E. A. and Abudullaeva, T. E.**, *Uch. Zap. Azerb. Gos. Univ. Ser. Khim. Nauk*, 53, 1970; *Chem. Abstr.*, 76, 20897v, 1972.

8. Lukin, A. M., Smirnova, K. A., and Petrova, G. S., *Tr. Vses. Nauchno Issled. Inst. Khim. Reaktiv. Osobo Chist. Khim. Veshchestv.*, 29, 290, 1966; *Chem Abstr.*, 68, 18242u, 1968.

9. Markova, A. I., *Zh. Anal. Khim.*, 17, 952, 1962.

10. Kamaeva, L. V., Podchainova, V. N., Fedorova, N. D., and Ozorina, I. A., *Tr. Vses. Nauchno Issled. Inst. Stand. Obraztov*, 7, 88, 1971; *Anal. Abstr.*, 24, 63, 1973.

11. Ackermann, G. and Kothe, J., *Talanta*, 26, 693, 1979.

12. Koroleva, G. N., Poluektov, N. S., and Kirillov, A. I., *Zh. Anal. Khim.*, 32, 2357, 1977.

13. Petrova, G. S., *Zavod. Lab.*, 26, 1162, 1960; *Chem. Abstr.*, 61, 6362f, 1964.

14. Muto, G., Wada, Y., and Doi, A., *Jpn. Soc. Anal. Chem.*, *19th, Annual Meeting*, 1C29, B266, 1970.

15. Holybecher, Z., Divis, L., Kral, M., Sucha, L., and Ulacil, F., *Handbook of Organic Reagents in Inorganic Analysis*, Ellis Horwood, Chichester, England, 1976.

16. Watanabe, H. and Ohmori, H., *Talanta*, 26, 959, 1979.

17. Popa, G., Danet, A. F., and Popescu, M., *Talanta*, 25, 546, 1978.

18. Dao, H. K., Jen, S. H., and Wang, N. C., *Acta Chim. Sinica*, 29, 344, 1963; *Anal. Abstr.*, 12, 52, 1965.

19. Shestidesyatnaya, N. L., Milyaeva, N. M., and Voronich, O. G., *Zh. Anal. Khim.*, 34, 94, 1979.

AZOAZOXY BN

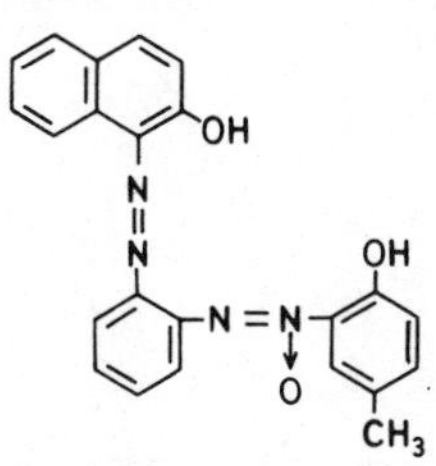

$C_{23}H_{17}N_4O_3$
mol wt = 397.41

H₂L

Synonyms

2-[(2″-Hydroxy-1-naphthyl-1″-azo)-2′-phenylazo]-4-methylphenol, Azoazohydroxy BN, AABN, NAHM

Source and Method of Synthesis

Commercially available only in USSR. It can be prepared by the coupling of diazotized 2-amino-2′-hydroxy-5′-methylazobenzene with 2-naphthol.[1]

Analytical Uses

As an extraction reagent for Ca and other alkaline earth metals.

Properties of Reagent

Dark red crystalline powder, mp 229 to 230°[1] and is practically insoluble in water, but slightly soluble in aqueous alkali (2×10^{-7} M in 0.05 M NaOH; 6.5×10^{-6} M in 0.5 M NaOH) to give an orange or orange-red solution. It is also slightly soluble in chloroform (6×10^{-3} M), TBP (4.6×10^{-4} M) and other polar solvents.[M1] $pKa_1 = 10.5$, $pKa_2 = 13.3$ ($\mu = 1$), log K_D(CCl₄/H₂O) = 5.42, log K_D (CCl₄-TBP(4:1)/H₂O) = 5.60 ($\mu = 1$).[2]

Complexation Reactions and Properties of Complexes

The alkaline solution of Azoazoxy BN forms red soluble chelates with alkaline earth metals, Co, Cu, and Zn. The structure of Ca-chelate is proposed to be such as shown below.

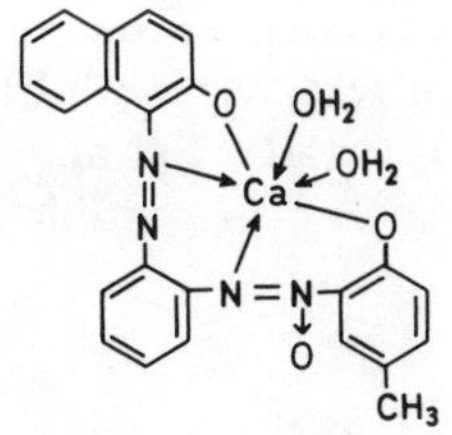

By replacing the remaining coordinated water of the central metal ion with the uncharged ligand, such as alkylamines (butylamine, hexylamine), higher alcohols (hexylalcohol, nonylalcohol), ethers (butylcellosolve), ketones, or alkylphosphates (TBP), the chelate can be extracted into an organic solvent (cyclohexane, carbon tetrachloride or chloroform) quantitatively. Of these ligands, TBP is the most preferred one. The extraction constants of such mixed ligand complexes can be expressed by

$$K_{ex} = \frac{[CaL \cdot 2TBP]_{org} \cdot [H^+]^2}{[Ca^{2+}] \cdot [H_2L]_{org} \cdot [TBP]_{org}^2}$$

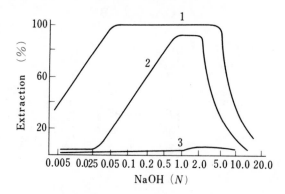

FIGURE 1.   Dependencies of percent extraction of alkaline earth metals with Azoazoxy BN on the alkalinity of
aqueous phase. (1) Ca; (2) Sr; (3) Ba. (From Gorbenko, E.
P. and Lapitskaya, E. V., *Zh. Anal. Khim.*, 23, 1139, 1968.
With permission.)

The values of log K_{ex} are -20.0 (CCl$_4$-TBP/H$_2$O), -18.4 (cyclohexane-TBP/H$_2$O), and -22.14 (CHCl$_3$/H$_2$O).[3]

Although the extracted species are colored (Ca, 490 to 510 nm; Zn, 450, 495 nm; Co(II), 600 to 650 nm; and Cu, 580 nm), the molar absorptivities are not so high ($\varepsilon = 5 \times 10^3$ for CaL) that they are generally not of practical importance for the photometric determination of traces of elements. It is a common practice to determine the extracted metal ions by other means after back extraction of the metal ion with an acid (0.01 to 0.5 N HCl).

Analytical Applications

Use as a Solvent Extraction Reagent

Extraction of each member of alkaline earth ions with Azoazoxy BN is very much dependent upon the alkalinity of aqueous phase. When microgram amounts of alkaline earth ions are extracted with a 0.01 to 0.04% solution of the reagent in CCl$_4$-TPB (4:1)[5] from an equal volume of 0.005 to 20 N NaOH solution, the values of percent extraction vary with alkalinity of aqueous phase, as illustrated in Figure 1. Thus, it is possible to extract Ca from 0.05 N NaOH solution, then Sr from 1.0 to 2.0 N NaOH solution, leaving Ba in the aqueous phase.[4] As Mg is not extracted from 0.8 N NaOH solution, it is possible to separate Ca from Mg. The elements in the organic phase can be back extracted with 0.01 to 0.1 N HCl. In the extraction of Ca from 0.05 N NaOH, the following elements do not interfere: Mg (with EDTA); Al, 0.5 g; Be, 5 g; Pb, 1.5 g; Zn, 2g; Sn, 3 g (with NH$_2$OH); Cr(III), 2 g; and Ti (with glucose).[R1] Separation of Zn from Al, Ba, Be, Cu, and Sn is also possible by extraction from 0.2 to 0.4 N NaOH solution.[6,7]

Use as a Photometric Reagent

Direct extraction photometry of metals with Azoazoxy BN is not advantageous because of the lower molar absorptivity, although a few papers have been found on Ca,[8] Cu,[9] and Co[10]

Other Reagents with Related Structure

Of the several members of Azoazoxy BN analogues, Azoazoxy AN (8-acetamido-1-[2-(2-hydroxy-5-methylphenylazoxy)phenylazo]-2-naphthol)[11] is reported to be selective for Ca and Cu, and Azoazoxy FMP (Azoazoxy PMP, 4-[2-(2-hydroxy-5-methylphenylazoxy)phenylazo]-3-methyl-1-phenyl-5-pyrazolin-5-one[12-14] is reported to be selective for Sr and Y.

MONOGRAPH

M1. Sandell, E. B. and Onishi, H., *Photometric Determination of Traces of Metals,* 4th ed., John Wiley & Sons, New York, 1978.

REVIEW

R1. Kodama, K. and Asai, T., Azoazoxy compounds, *Bunseki Kagaku,* 21, 584, 1972.

REFERENCES

1. Dziomko, V. M. and Dunevskaya, K. A., *Zh. Obshch. Khim.,* 30, 628, 1960; *Chem. Abstr.,* 54, 24565i, 1960; *Zh. Obshch. Khim.,* 30, 3708, 1960; *Chem. Abstr.,* 55, 19837h, 1961, *Zh. Obshch. Khim.,* 31, 68, 1961; *Chem. Abstr.,* 55, 23394d, 1961; *Zh. Obshch. Khim.,* 31, 3712, 1961; *Chem. Abstr.,* 57, 9755b, 1962; *Zh. Obshch. Khim.,* 31, 3385, 1961; *Chem. Abstr.,* 57, 2112c, 1962.
2. Gorbenko, F. P. and Sachko, V. V., *Zh. Anal. Khim.,* 24, 15, 1969.
3. Gorbenko, F. P. and Sachko, V. V., *Zh. Anal. Khim.,* 25, 1884, 1970.
4. Gorbenko, F. P. and Lapitskaya, E. V., *Zh. Anal. Khim.,* 23, 1139, 1968.
5. Gorbenko, E. P. and Lapitskaya, E. V., *Anal. Khim. Ekstr. Protsessy,* 44, 1970; *Chem. Abstr.,* 74, 44456h, 1971.
6. Gorbenko, F. P. and Degtyarenko, L. I., *Poluch. Anal. Veshcheotv Osobai Chist. Mater. Vses. Konf., Gorky, USSR,* 279, 1963; *Chem. Abstr.,* 67, 50117x, 1967.
7. Gorbenko, F. P., Dziomko, V., Lapitskaya, E. V., and Velshtein, E. I., *Metody Anal. Khim. Reaktiv. Prep.,* 84, 1971; *Chem. Abstr.,* 77, 121792h, 1972.
8. Gorbenko, F. P. and Sacho, V. V., *Ukr. Khim. Zh.,* 30, 402, 1964; *Chem. Abstr.,* 61, 3666d, 1964.
9. Minczewski, J. and Wieteska, E., *Chem. Anal. (Warsaw),* 9, 365, 1964; *Chem. Abstr.,* 61, 6379h, 1964.
10. Wieteska, E. and Kamela, M., *Chim. Anal., (Warsaw),* 17, 85, 1972; *Chem. Abstr.,* 77, 10358b, 1972.
11. Gorbenko, F. P. and Enaleva, L. Ya., *Tr. Vses, Nauchno Issled. Inst. Khim. Reaktivov. Osobo Chist. Khim. Veshchestv.,* 42, 1966,; *Anal. Abstr.,* 15, 3123, 1968.
12. Gorbenko, F. P. and Nadezhda, A. A., *Zh. Anal. Khim.,* 24, 671, 1969.
13. Nadezhda, A. A., Gorbenko, F. P., Dunaevskaya, K. A., and Kostrinskaya, L. A., *Tr. Vses. Nauchno Issled. Inst. Khim. Reaktivov. Osobo Chist. Khim. Veshchestv.,* 200, 1970; *Anal. Abstr.,* 23, 104, 1972.
14. Nadezhda, A. A., Gorbenko, F. P., Dunaevskaya, K. A., and Zelinskaya, M. I., *Tr. Vses. Nauchno Issled. Inst. Khim. Reaktivov. Osobo Chist. Khim. Veshchestv.,* 204, 1970; *Anal. Abstr.,* 23, 105, 1972.

PYRIDYLAZONAPHTHAOL

$C_{15}H_{11}N_3O$
mol wt = 249.27

HL

Synonyms

1-(2-Pyridylazo)-2-naphthaol, α-Pyridylazo-β-naphthaol, PAN, β-PAN, o-β-PAN

Source and Method of Synthesis

Commercially available. It is prepared by the coupling of diazotized 2-aminopyridine with 2-naphthol in dry methanol.[1,2]

Analytical Uses

It is used as a photometric reagent (with or without extraction) for wide variety of metal ions and as a metal indicator in the chelatometric titration. As the reagent itself is rather unselective, the selectivity is attained by choosing pH and masking agents.

Properties of Reagent

Orange-yellow needles, mp 140 to $142°$[3,6] (commercial samples often melt at $138°$ or lower); sublimes at 130 to $140°$ (2×10^{-2} Torr).[4,6] PAN is very stable in solid state and can be kept for many years in an amber bottle and is almost insoluble in water, slightly soluble in strong acid, soluble in aqueous alkali, and in various organic solvents (yellow solution). The proton dissociation scheme can be written as below.

$$H_2L^+ \xrightleftharpoons[\quad]{pKa_1 (N^+H)=2.9} HL \xrightleftharpoons[\quad]{pKa_2 (OH) = 11.6} L^- \quad (\mu = 0.1\ NaClO_4,\ \text{room temperature})[6]$$

Pale Green	Yellow	Red
pH <2.5	pH 2.5 to 12	pH >12

The pKa values in 20% dioxane are reported as $pKa_1 = 1.90$ and $pKa_2 = 12.2$.[5] log K_D (CHCl$_3$/H$_2$O) = 5.4, log K_D (CCl$_4$/H$_2$O) = 4.0.[3]

Absorption spectra of PAN in water at various pH are illustrated in Figure 1.

Complexation Reaction and Properties of Complexes

PAN (HL to yellow) reacts with wide variety of metal ions to form colored water-insoluble chelates.[53] Metal ions that form colored and/or insoluble chelates in aqueous media are indicated in Figure 2.

The metal-PAN chelates are red, except those of Pd(II), Co(III), and Rh(III) (green), Fe(III) (dull red), and V(V) and Tl(III) (purple). Some of the typical absorption spectra are illustrated in Figure 3. The spectral characteristics of some of metal-PAN chelates are summarized in Table 3.

Some of metal-PAN chelates exhibit fluorescence under UV light. The fluorescent color varies with the metal and solvents used. A strong fluorescence is observed on Al-PAN λ_{ex}, 355, 550 ± 10 nm; λ_{em} 570 to 590 nm) and Co(III)-PAN (λ_{ex}, 320, 380 nm; λ_{em}, 436 nm).[7-9]

PAN behaves as a terdentate ligand to form MLX-type mixed ligand chelate (X: unidentate ligand) with metal ions of coordination number 4, and ML$_2$-type normal

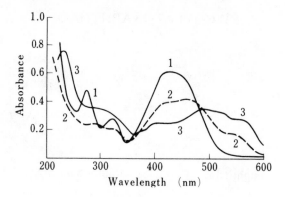

FIGURE 1.   Absorption spectra of PAN in water at various pH; PAN, 6.6×10^{-6} M, $\mu = 0.10$, 50-mm cell. (1) pH 1.0; (2) pH 4.0; (3) pH 9.8. (Reprinted with permission from Pease, B. F. and Williams, M. B., *Anal. Chem.*, 31, 1044, 1959. Copyright 1959 American Chemical Society.)

Period	A								B							
	1	2	3	4	5	6	7	8	1	2	3	4	5	6	7	8
2	Li	Be									B	C	N	O	F	Ne
3	Na	Mg									Al	Si	P	S	Cl	Ar
4	K	Ca	Sc *	Ti 5	V 1.5~4.5	Cr 3.2~3.7	Mn 7~10	Fe 1.6~5 Co 3~6 Ni 4~10	Cu 3~7	Zn 8~9.5	Ga 3.6~6	Ge	As	Se	Br	Kr
5	Rb	Sr	Y 8~10	Zr 1N HNO₃	Nb 2	Mo	Tc	Ru 4.5~5.2 Rh 1.4~5.6 Pd 2~5	Ag *	Cd 8.7~10	In 1.5~7	Sn	Sb 0.25~0.5N H₂SO₄	Te	I	Xe
6	Cs	Ba	La	Hf 4	Ta	W	Re	Os 3.5~5.2 Ir 4.6~5.9 Pt 3~5	Au	Hg 6.0~11.5	Tl	Pb	Bi 0.01~0.1N HNO₃	Po	At	Rn
7	Fr	Ra	Ac													

Lanthanide Series 8~10	La	Ce	Pr	Nd	Pm	Sm	Eu	Gd	Tb	Dy	Ho	Er	Tm	Yb	Lu
Actinide Series	Ac	Th	Pa	U 9~10	Np	Pu	Am	Cm	Bk	Cf					

* In alkaline range

FIGURE 2.   Metal ions that reacts with PAN in aqueous media to form chlored and/or insoluble chelates. Figures under each element indicate the pH range where PAN gives color reaction. (From Shibata, S., *Chelate in Analytical Chemistry*, Vol. 4, Falschka, H. A. and Barnard, A. J., Jr., Eds., Marcel Dekker, New York, 1972.)

chelate with those of coordination number 6. Structures of both types are shown as (1) and (2), respectively.[10,11] In some cases, PAN behaves as bidentate ligand to form ML_2-type chelate with the metal ions of coordination number 4.

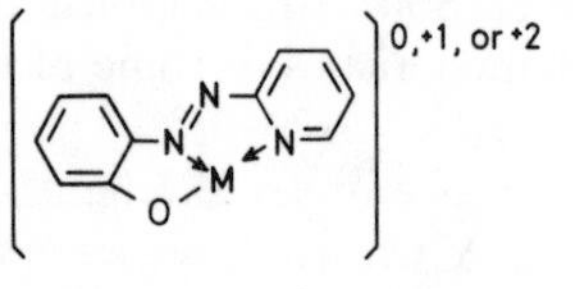

(1)

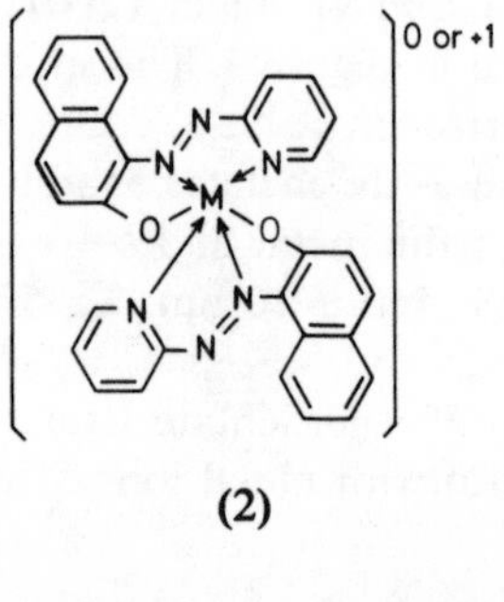

(2)

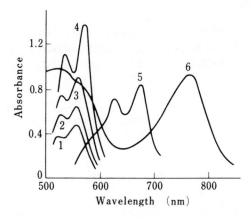

FIGURE 3. Absorption spectra of metal - PAN chelates against a reagent blank. (1) Zn in chloroform; (2) Mn(II) in ether; (3) In; (4) Ni; (5) Pd(II); (6) Fe(III) in chloroform. (From Shibata, S., *Chelate in Analytical Chemistry,* Vol. 4, Falschka, H. A. and Barnard, A. J., Jr., Eds., Marcel Dekker, New York, 1972.

Table 1
STABILITY CONSTANT OF METAL-PAN CHELATES

Metal ion	$\log K_{ML}$	$\log K_{ML_2}$	Condition	Ref.
Al	12.9	—	50% ethanol, $\mu = 0.06, 25°$	12
Co(II)	>12	—	50% dioxane, 25°	13
Cu(II)	15.6	≈8.4	5% dioxane for $\log K_{ML}$, 50% dioxane for $\log K_{ML_2}$ 15°, $\mu = 0.1$, NaClO$_4$	14
Eu	12.4	11.4	$\log K_{ML_3} = 10.4$ $\log K_{ML_4} = 9.5$ $\mu = 0.05$ ClO$_4^-$	15
Ga	15.1	—	50% ethanol, $\mu = 0.06, 25°$	12
Ho	12.8	11.6	$\log K_{ML_3} = 10.4$ $\log K_{ML_4} = 9.3$ $\mu = 0.05$ ClO$_4^-$	15
In	13.1	—	50% ethanol, $\mu = 0.06, 25°$	12
Mn(II)	8.5	7.9	50% dioxane, 25°	13
Ni	12.7	12.6	50% dioxane, 25°	13
Zn	11.2	10.5	50% dioxane, 25°	13

The chelate stability constants were observed, in many cases, in aqueous organic solvent media. The reported values are summarized in Table 1.

The coordination-saturated uncharged chelates, such as CdL$_2$, MnL$_2$, and ZnL$_2$, are almost insoluble in water, but can be extracted into immiscible solvents like chloroform (log K_{ex} for ZnL$_2$, −5.04; CdL$_2$, −8.98),[16], carbon tetrachloride (log K_{ex} for ZnL$_2$, −4.7; MnL$_2$, −10.5),[3], iso-amylalcohol, or dichloroethane.

The charged chelates are slightly soluble in water, and some of them can also be extracted into organic solvent as a mixed ligand chelate if appropriate anionic ligand (acetate, chloride, bromide, perchlorate, or nitrate) is present. For example, Cu(II) is extracted as CuLX (X = acetate, Cl$^-$, Br$^-$, SCN$^-$, and OH$^-$) at pH 8, but is extracted as CuL$_2$ at pH 10.[16]

The compositions of the extracted species are not known for all metal ions, and

Table 2
COLOR REACTIONS OF METALS
WITH PAN IN AQUEOUS MEDIA IN
THE PRESENCE OF MASKING
AGENTS

Metal ion	KCN	Citrate	Fluoride	EDTA
Ag		+		
Bi	+	−	−	−
Cd	−	+	+	−
Ce(III)	+			−
Co(II)	−	+	+	−
Cu(II)	−	+		−
Fe(III)	−	−	−	−
Ga		−	−	−
Hg(III)	−	+	+	−
In	+	+	+	−
La	+			−
Mn(II)	+	+	+	−
Ni	−	+	+	−
Pb	+	−	−	−
Pd(II)	−			
Pt(IV)	−			
UO_2^{2+}	+		−	+
Y(III)	+	+	−	−
Zn	−	+	+	−

Note: +, indicates that reaction occurs with PAN;
−, indicates that the metal ion is masked.

From Shibata, S., *Chelate in Analytical Chemistry,* Vol. 3, Flaschka, H. A. and Barnard, A. J., Jr., Eds., Marcel Dekker, New York, 1972. With permission.

information on the extractability of the PAN chelates of a tri- or quadrivalent metal is rather complicated due to the possible formation of charged chelates. Extraction of In(III)-PAN chelate into chloroform was investigated in detail.[17] When In is extracted with chloroform solution of PAN, the degree of extraction increased with the shaking time, although the extraction rate is slow. On the contrary, when PAN was added to the aqueous phase prior to the chloroform extraction, the degree of extraction was high at the beginning, falling down with the increase of shaking time. The extraction percentage was also dependent upon the buffer, being 90 to 92% in acetate buffer (pH 6), 13% in phthalate buffer, and ∿ 0% in citrate buffer.

The selection of a suitable solvent is important for an effective extraction of metal-PAN chelates. Chloroform is most widely used. The extraction is often incomplete with carbon tetrachloride, ether, and benzene.

Insoluble metal-PAN chelates may be solubilized either by adding water-miscible organic solvent (ethanol, DMF, or dioxane) or by using nonionic surfactant (Triton® X-100),[18] thus enabling the photometry of trace elements without extraction.

PAN is a nonselective photometric reagent, but the selectivity may be improved by the selection of pH and the combined use of masking agents. Some of the color reactions of PAN in the presence of various masking agents are summarized in Table 2.

Purification and Purity of Reagent

PAN is well-defined stable organic molecules and can be easily purified by repeated recrystallization from methanol until melting point reaches 141°. The sublimation in vacuum is also effective method for a small quantity (130 to 140° at 1 to 0.02 Torr).[4]

Purity of PAN can be checked by TLC method on a silica-gel plate, using a mixed solvent (petroleum ether 10 vol + ether 10 vol + ethanol 1 vol); R_f = 0.68.[19]

Analytical Applications
Use as a Photometric Reagent

As PAN is a nonselective reagent, metal ions have to be separated in advance or the selectivity of PAN has to be improved by the selections of extraction condition and masking agents. Table 3 summarizes the application of PAN as a photometric reagent.

A 0.1 to 0.2% methanol solution is used for this purpose. The solution is stable for many weeks, if stored in an amber bottle.

Use as a Metal Indicator in Chelatometry

Metal ions that can be titrated with EDTA using PAN indicator are as follows: Bi (pH 1),[52] Ga (pH 2 to 2.6),[55] In (pH 2.5 to 3.5),[56] Ni,[57] Tl(III), (pH 4 to 5),[58] Cu (pH 4 to 9),[58] Cd,[53] Ce(III),[54] Zn (pH 5 to 6),[57] and UO_2^{2+} (pH 4.4 to 4.6).[59] The color change at the end point of direct titration is from red-violet to yellow. As the color change is often slow at room temperature, the titration at higher temperature (50 to 80°) or in the presence of miscible organic solvent is recommended. The color change is most sensitive for Cu(II), so that many kinds of metal ions can be determined by the back titration with a standard Cu(II) solution and PAN indicator.[M2] A 0.1 to 0.01% alcohol solution is recommended for the indicator use.

A mixture of Cu-EDTA and PAN is a very useful indicator in the EDTA titration of metal ions that do not give a sharp color reaction with PAN.[M1,60] The color change at the end point is similar to that of PAN. Metal ions and the optimum pH of titration are Al(3), Co(II), Ni (>3), Ga, Hg(II) (3 to 3.5), In, Th (2 to 3.5), V(IV) (3.5), Cd, Zn, Fe(III), Pb (4.5) and Ca, Mg, and Mn(II) (10). A 50% aqueous dioxane containing 1.3 g of Cu-EDTA and 0.1 g PAN in 100 mℓ is recommended for this purpose.

Other Uses

PAN is used as a detecting reagent for metal ions in solution or on ion-exchange resin beads. The latter is very sensitive test using a drop of test solution soaked in the resin beads.[61] Cation exchange paper treated with PAN is also used for the separation and detection of metal ions (detection limit, 0.001 ppm).[62]

Other Reagents with Related Structure
4-(2-Pyridylazo)-1-naphthol (p-PAN)[63,65]

HL

Brick-red crystals, mp 110 to 115°; pKa_1 (N⁺H) = 2.95, pKa_2 (OH) = 11.23 in methanol-water (1:4);[64] insoluble in water and dilute aqueous alkali; easily soluble in organic solvents. It is yellow-green in acidic solution (pH <2), yellow at pH 2 to 11, and red at pH >12. Although hydroxyl group in para position can not participate in the chelate formation, p-PAN forms insoluble chelates with wide variety of metal ions, and the color is generally deeper than that of corresponding PAN chelates.

Table 3

APPLICATION OF PAN AS A PHOTOMETRIC REAGENT

Metal ion	Condition	Metal chelate			Extraction solvent	Range of determination (ppm)	Remarks	Ref.
		Ratio	λ_{max} (nm)	$\varepsilon(\times 10^4)$				
Cd	pH 8.7~10	ML_2	555	4.9	$CHCl_3$	~2.5	—	M1,11
Co(III)	pH 3~6	ML_2	640	2.5	$CHCl_3$	0.01~2.4	Co(II) is oxidized to Co(III) upon complex formation	M1,11
Co(III)	pH 5, Triton® X-100 + dodecylbenzenesulfonate, EDTA	—	620	1.9	Aqueous	0.4~3.2	—	22
Cu(II)	pH 3.2, 20% dioxane	ML	550	—	—	—	ML_2 at pH 5.0	16, 24
Hf	pH 4.0, 40% methanol	ML_2	545	3.86	Aqueous	0.2~3.6	Zr(sixfold) does not interfere	27
Mn(II)	pH 8.8~9.6, NH_3, KCN	ML_2	562	4.8	$CHCl_3$	0.2~1.2	—	30
Mn(II)	pH 9.2, KCN, Triton® X-100	ML_2	562	4.4	Aqueous	~10	Al,Ca,Co,Fe,Mg,Ni,Pb do not interfere	31
Ni	pH 4~10, NH_3, KCN	ML_2	570	5.0	Benzene, $CHCl_3$, or CCl_4	0~1.5	After decomposing $Ni(CN)_4$ by heating with H_2O_2	34
Os(VIII)	pH 3.5~5.2	ML	550	2.8	$CHCl_3$	~9.2	Ru(III),EDTA, thiourea, $S_2O_3^{2-}$, SCN^- interfere	35
Tl(III)	pH 4~6, Cl^-, Br^-, I^-, SCN^-, NO_3^-, or acetate	ML_2X	560~570	3.2	$CHCl_3$	1~9	Al,Bi,La,Sn,Ti interfere	45
Zn	pH 8.0~9.5, Triton® X-100	ML_2	555	5.6	Aqueous	0.2~2	Cd,Co,Fe,Mn,Ni interfere	18, 47, 48
Zn	Tartrate, I^-, $S_2O_3^{2-}$ ascorbic acid, CN^-, then HCHO	ML_2	550	—	$CHCl_3$	—	Ag,Cu,Hg,Pb do not interfere	49
Zr	1 M HNO_3, Al^{3+}, pyridine	—	555	3.2	$CHCl_3$	0.4~2.6	After extraction with dibutylphosphate	50
Rare earths	pH 8~9.5	ML_2	530	6~7	ether or $CHCl_3$	~2	Red chelate except Pr which forms yellow one	11, 51

Note: Other metal ions which have been determined with PAN are Ag,[20] Bi,[21] Cr(III),[23] Fe(III),[M1,25] Ga,[11,26] Hg(II),[11,28] In,[11,26] Ir,[29] Nb(V),[33] Pd (II),[36-38] Pt(II),[37,39] Rh(II),[37] (III),[29] Ru,[40] SB(III),[41] Ti(IV),[42] U(VI),[45,79,80] and V(V).[46]

2-(2-Pyridylazo)-1-naphthol (α-PAN)[66]

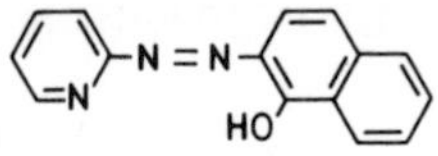

HL

α-PAN is another member of the PAN family, and its melting point is different by different workers (128° and 102 to 104°).[M1, R3] It sublimes at 122°.[66] It shows similar chelating behavior to that of PAN. α-PAN has been used as a reagent for extraction photometric determination of Bi[67] and Zn[68] and as a metal indicator for Tl.[69] α-PAN sulfonic acid was used as a chromogenic reagent for Cu[70] and Ni.[71]

2-(5-Bromo-2-pyridylazo)-5-dimethylaminophenol (5-Br-DMPAP)

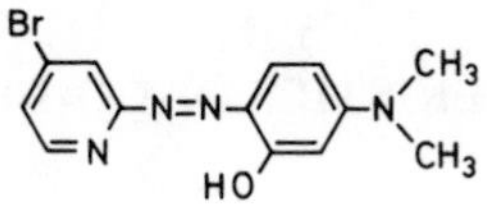

$C_{13}H_{13}N_4OBr$
mol wt = 464.22

HL (or H_3L^{2+})

One of the most sensitive reagents among the pyridylazo dyes;[R5] dark red needles, mp 190° (with sublimation). Aqueous solution is yellow in neutral range (λ_{max}, 444 nm; $\varepsilon = 4.7 \times 10^4$); pKa_1 (pyridine), <1, pKa_2 (amino) = 3.4, and pKa_3 (o-OH) = 11.6 (40% methanol, $\mu = 0.1$, KCl, 25°),[R5] It turns to red-purple upon complex formation with metal ions (Al, Cd, Co(II), Cu(II), Fe(III), La(III), Mg, Mn(II), Ni, Pb, and Zn). Spectral characteristics of some metal chelates are summarized in Table 4. The reagent is now commercially available, but can be synthesized according to Shibat a.[72]

Similarly, 2-(5-bromo-2-pyridylazo)-5-diethylaminophenol (5-Br-DEPAP) was used as a chromogenic reagent for Co,[81] Ga,[R3] U,[74,75] and Zn,[76] and 5-Cl-DEPAP for Zn.[32]

4-(5-Chloro-2-pyridylazo)-1,3-diaminobenzene (5-Cl-PADAB)

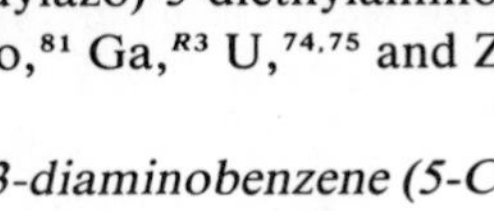

$C_{11}H_{10}N_5Cl$
mol wt = 247.69

L (or H_3L^{3+})

Highly sensitive and selective chromogenic reagent for Co(II);[R4,77] commercially available; red-brown needles, mp 250° (sublimes above 200°); slightly soluble in water, but easily soluble in organic solvents. Aqueous solution is yellow in acidic range and yellow-orange in neutral or alkaline range; pKa_2 (o-N⁺H₃) = 1.3 and pKa_3 (p-N⁺H₃) = 5.4 ($\mu = 0.2$, KCl, 20°). It forms colored chelates with Fe(II), (III), Cu(I), (II), Hg(II), Ni, Zn, and Co(II) at pH 4 to 7. Most of the other common metals give no color reaction at pH 4 to 7 and 8 to 10. The Co chelate (CoL_2^{2+}) once formed in weakly acidic solution does not dissociate, even in concentrated sulfuric acid. Thus, the reagent is highly specific for Co(II). Spectral characteristics of the metal chelates are summarized in Table 5. The protonated species ($CoL_2H_2^{4+}$) shows the highest absorbance among the Co-PADAB chelates. The bromo analogue (5-Br-PADAB) can also be used similarly.[78]

Table 4
SPECTRAL CHARACTERISTICS OF 5-Br-DMPAP CHELATES IN 50% ETHANOL[R5]

| Metal ion | Condition | Ratio | Chelates | | Ref. |
			λ_{max} (nm)	$\varepsilon(\times 10^4)$	
Cd	pH 8 $\sim$ 10.5	ML$_2$	525 $\sim$ 555	14.1[a]	73
Co(II)	pH 3.0 $\sim$ 10.0	ML$_2$	588	8.8	
Cu(II)	pH 2.0 $\sim$ 6.0	ML	558	6.1	
Cu(II)	pH 9.0 $\sim$ 10.0	ML$_2$	555	10.0	
Ni	pH 4.5 $\sim$ 10.0	ML$_2$	558	12.8	
Zn	pH 7.5 $\sim$ 10.0	ML$_2$	552	13.3	

[a]   Extracted into 3-methyl-1-butanol.

Table 5
SPECTRAL CHARACTERISTICS OF CO-5-CL-PADAB CHELATE[R5]

$$CoL_2H_n^{(2+n)+} \xrightleftharpoons{pKa_1 = 5.1^a} CoL_2H_2^{4+} \xrightleftharpoons{pKa_2 = 0.9} CoL_2H^{3+} \xrightleftharpoons{pKa_3 = 4.8} CoL_2^{2+}$$

λ_{max} (nm)	$\epsilon (\times 10^4)$	λ_{max} (nm)	$\epsilon (\times 10^4)$	λ_{max} (nm)	$\epsilon (\times 10^4)$	λ_{max} (nm)	$\epsilon (\times 10^4)$
404	3.0	530	8.7	520	6.2	480	4.9
424	3.1	568	11.4	558	5.3	504	5.0
448	2.7	—	—	—	—	—	—

[a]   Expressed in term of Hammett's acidity function in sulfuric acid.

MONOGRAPHS

M1. **Shibata, S.**, 2-Pyridylazo compounds, in *Chelate in Analytical Chemistry,* Vol. 4, Flaschka, H. A. and Barnard, A. J., Jr., Eds., Marcel Dekker, New York, 1972, 1.

M2. **Ueno, K.**, *Chelatometric Titration,* 2nd ed., Nankodo, Tokyo, 1972.

M3. **Pribil, R.**, *Analytical Application of EDTA and Related Compounds,* Pergamon Press, New York, 1972, 267.

REVIEWS

R1. **Danzuka, T.**, Analytical applications of pyridylazonaphthol and related compounds, *Kagaku (Kyoto),* 16, 686, 1961.

R2. **Busev, A. I. and Ivanov, V. M.**, The use of pyridylazo compounds in analytical chemistry, *Zh. Anal. Khim.,* 19, 1238, 1964.

R3. **Anderson, R. G., and Nickless, G.**, Heterocyclicazo dyestuffs in analytical chemistry, *Analyst,* 92, 207, 1967.

R4. Shibata, S., 1-(2-pyridylazo)-2-naphthol (PAN) and its analogues as organic analytical reagents, *Bunseki Kagaku*, 21, 551, 1972.

R5. Shibata, S. and Furukawa, M., Preparation of new chromogenic reagents which have molar absorptivities of the order of 10^5 for different metals, *Bunseki Kagaku*, 23, 1412, 1974.

R6. Wada, H., Applications of PAN and closely related compounds as a complexometric indicator, *Bunseki Kagaku* 21, 543, 1972.

R7. Ivanov, V. M., Use of heterocyclicazo compounds in analytical chemistry, *Zh. Anal. Khim.*, 31, 993, 1976; *J. Anal. Chem. USSR*, 31, 810, 1976.

REFERENCES

1. Chichibabin, A. E. and Rjasanzew, M., *Zh. Russ. Fiz. Khim. Ova.*, 47, 1582, 1915; *Chem. Zentralbl.*, 3, 299, 1916.
2. Chichibabin, A. E., *Zh. Russ. Fiz. Khim. Ova.*, 50, 513, 1918; *Chem. Zentralbl. III, 1022, 1923*.
3. Betteridge, D., Fernando, Q. and Freiser, H., *Anal. Chem.*, 35, 294, 1963.
4. Honjo, T., Imura, H., Shima, S., and Kiba, T. *Anal. Chem.*, 50, 1547, 1978.
5. Pease, B. F. and Williams, M. B., *Anal. Chem.*, 31, 1044, 1959.
6. Nakagawa, G. and Wada, H., *Nippon Kagaku Zasshi*, 84, 639, 1963.
7. Haworth, D. T., Starshak, R. J., and Surak, J. G., *J. Chem. Educ.*, 41, 436, 1964.
8. Surak, J. G., Herman, M. F., and Haworth, D. T., *Anal. Chem.*, 38, 428, 1965.
9. Schenk, G. H., Dilloway, K. P., and Coulter, J. S., *Anal. Chem.*, 41, 510, 1969.
10. Ooi, S., Carter, D., and Fernando, Q., *Chem. Commun.*, 1301, 1967.
11. Shibata, S., *Anal. Chim. Acta*, 25, 348, 1961.
12. Toropova, V. F., Budnikov, G. K., Maistrenko, V. N., and Nagorskaya, I. V., *Zh. Obshch. Khim.*, 43, 2126, 1973; *Chem. Abstr.*, 80, 41499q, 1974.
13. Corsini, A., Yih, M. L., Fernando, Q., and Freiser, H., *Anal. Chem.*, 34, 1090, 1962.
14. Wada, H. and Nakagawa, G., *Nippon Kagaku Zasshi*, 85, 549, 1964.
15. Navratil, O., *Collect. Czech. Chem. Commun.*, 31, 2492, 1966.
16. Galik, A., *Talanta*, 16, 201, 1969.
17. Zolotov, Yu. A., Seryakova, I. V., and Berobyeva, G. A., *Talanta*, 14, 737, 1962.
18. Watanabe, H. and Sakai, Y., *Bunseki Kagaku*, 23, 396, 1974.
19. Pollard, F. H., Nickless, G., Samelson, T. J., and Anderson, R. G., *J. Chromatogr.*, 16, 231, 1964.
20. Eshwar, M. C. and Subrahmanyam, B. *Zh. Anal. Khim.*, 31, 2319, 1976.
21. Subrahmanyam, B. and Eshwar, M. C., *Chem. Anal. (Warsaw)*, 21, 873, 1976; *Chem. Abstr.*, 86, 100336h, 1977.
22. Watanabe, H., *Talanta*, 21, 295, 1974.
23. Subrahmanyam, B. and Eshwar, M. C., *Bull. Chem. Soc. Jpn.*, 49, 347, 1976.
24. Kitagawa, T., *Bunseki Kagaku*, 8, 594, 1959.
25. Shibata, S., Goto, K., and Nakashima, R., *Anal. Chim. Acta*, 46, 146, 1969.
26. Cheng, K. L. and Goydish, B. L., *Anal. Chim. Acta*, 34, 154, 1966.
27. Subrahmanyam, B. and Eshwar, M. C., *Mikrochim. Acta*, 585, 1976.
28. Ho, L. S., Kuo, C. N., Shih, C. S., and Chiang, W., *Hua Hsueh Tung Pao*, 250, 1965; *Chem. Abstr.*, 63, 17135f, 1965.
29. Stokeley, J. R. and Jacobs, W. D., *Anal. Chem.*, 35, 149, 1963.
30. Donaldson, E. M. and Inman, W. R., *Talanta*, 13, 489, 1966.
31. Goto, K., Taguchi, S., Fukue, Y., Ohta, K., and Watanabe, H., *Talanta*, 24, 752, 1977; *Mizu Shori Gijutsu (Tokyo)*, 19, 749, 1978; *Chem. Abstr.*, 90, 33484x, 1979.
32. Furukawa, M., *Nagoya Kogyo Gijutsu Shikensho Hokoku*, 22, 233, 1978; *Anal. Abstr.*, 37, 2B64, 1979.
33. Gagliardi, E. and Wolf, E., *Mikrochim. Acta*, 104, 1967.
34. Dono, T., Nakagawa, G., and Wada, H., *Nippon Kagaku Zasshi*, 82, 590, 1961; *Nippon Kagaku Zasshi*, 84, 636, 1963.
35. Pandeya, K. B., Singh, R. P., and Bhoon, Y. K., *Ann. Chim. (Rome)*, 65, 735, 1975; *Chem. Abstr.*, 86, 50178k, 1977.
36. Dono, T., Nakagawa, G., and Hayashi, H., *Nippon Kagaku Zasshi*, 81, 1703, 1960.
37. Ivanov, V. M., Figurovskaya, V. N., and Busev, A. I., *Zavod. Lab.*, 38, 1311, 1972; *Anal. Abstr.*, 24, 3489, 1973; *Zavod. Lab.*, 39, 132, 1973; *Anal. Abstr.*, 25, 1582, 1973.

38. Busev, A. I. and Kiseleva, L. V., *Vestn. Mosk. Univ. Ser. Mat. Mekh. Astron. Fiz. Khim.*, 13, 179, 1958; *Chem. Abstr.*, 53, 11105f, 1959.
39. Ivanov, V. M., Figurovskaya, V. N., and Busev, A. I., *Zavod. Lab.*, 39, 270, 1973; *Ind. Lab.*, 39, 371, 1973; *Anal. Abstr.*, 27, 1240, 1974.
40. Kodama, K., *Nagoya-shi Kogyo Kenkyusho Kenkyu Hokoku (Japan)*, 42, 48, 1969; *Chem. Abstr.*, 72, 74429e, 1970.
41. Ho, L. S., Ching, S. S., and Ching, L. H., *Hua Hsueh Tung Pao*, 56,1965; *Chem. Abstr.*, 63, 15535d, 1965.
42. Puschel, R. and Lassner, E., *Mikrochim. Acta*, 977, 1967.
43. Nazarenko, V. A., Biryuk, E. A., and Ravitskaya, R. V., *Zh. Anal. Khim.*, 33, 2362, 1978.
44. Busev, A. I. and Tipsova, V. G., *Zh. Anal. Khim.*, 15, 573, 1960.
45. Shibata, S., *Anal. Chim. Acta*, 22, 479, 1960.
46. Staten, F. W. and Huffman, E. W. D., *Anal. Chem.*, 31, 2003, 1959.
47. Watanabe, H. and Tanaka, H., *Talanta*, 25, 585, 1978.
48. Watanabe, H., Yamaguchi, N., and Tanaka, H., *Bunseki Kagaku*, 28, 366, 1979.
49. Flaschka, H. and Weiss, R. H., *Mikrochem. J.*, 15, 653, 1970.
50. Rolf, R. F., *Anal. Chem.*, 33, 125, 1961.
51. Shibata, S., *Anal. Chim. Acta*, 28, 388, 1963.
52. Busev, A. I., *Zh. Anal. Khim.*, 12, 386, 1957.
53. Cheng, K. L. and Bray, R. H., *Anal. Chem.*, 27, 782, 1958.
54. Konkin, V. D. and Zhikareva, V. I., *Sb. Tr. Ukr. Nauchn. Issled. Inst. Metal.*, 444, 1964; *Chem. Abstr.*, 62, 5887h, 1965.
55. Busev, A. I. and Skrebkova, L. M., *Izv. Sib. Otd. Akad. Nauk SSR*, 57, 1962; *Chem. Abstr.*, 58, 5032g, 1963.
56. Cheng, K. L., *Anal. Chem.*, 27, 1582, 1955.
57. Cheng, K. L., *Anal. Chem.*; 30, 243, 1958.
58. Busev, A. I. and Tiptsova, V. G., *Zh. Anal. Khim.*, 16, 275, 1961.
59. Lassner, E. and Sharf, R. S., *Fresenius Z. Anal. Chem.*, 164, 398, 1958.
60. Flaschka, H. and Abdine, H., *Chemist - Analyst*, 45, 2, 58, 1956.
61. Fujimoto, M. and Iwamoto, T., *Mikrochim. Acta*, 655, 1963.
62. Watanabe, H., *Nippon Kagaku Zasshi*, 86, 513, 1965.
63. Betteridge, D., Todd, P. K., Fernando, Q., and Freiser, H., *Anal. Chem.*, 35, 729, 1963.
64. Gusev, S. I. and Nikolaeva, E. M., *Zh. Anal. Khim.*, 21, 166, 1966.
65. Anderson, R. G. and Nickless, G., *Analyst*, 93, 13, 1968.
66. Betteridge, D. and John, D., *Analyst*, 98, 377, 390, 1973.
67. Shurlava, L. M., *Tr. Permsk. Gos. Med. Inst.*, 108, 27, 1972; *Anal. Abstr.*, 26, 830, 1974.
68. Nikolaeva, E. M., *Tr. Permsk. Gos. Med. Inst.*, 108, 17, 1972; *Anal. Abstr.*, 26, 50, 1974.
69. Gusev, S. I. and Kurepa, G. A., *Tr. Permsk. Gos. Med. Inst.* 108, 21, 1972; *Anal. Abstr.*, 26, 65, 1974.
70. Koblikova, V., Kuban, V., and Sommer, L., *Collect. Czech. Chem. Commun.*, 43, 2711, 1978.
71. Wada, K., Yotsuyanagi, T., and Aomura, K., *Nippon Kagaku Kaishi*, 131, 1978.
72. Shibata, S., Furukawa, M., and Toei, K., *Anal. Chim. Acta*, 66, 397, 1973.
73. Shibata, S., Kamata, E., and Nakashima, R., *Anal. Chim. Acta*, 82, 169, 1976.
74. Pakalns, P., *Anal. Chim. Acta*, 69, 211, 1974.
75. Johnson, D. A. and Florence, T. M., *Talanta*, 22, 253, 1975.
76. Gusev, S. I., Nikolaev, E. M., and Pirozhkova, E. A., *Zh. Anal. Khim.*, 26, 1740, 1971.
77. Shibata, S., Furukawa, M., Ishiguro, Y., and Sasaki, S., *Anal. Chim. Acta*, 55, 231, 1971 and 71, 85, 1974.
78. Kiss, E., *Anal. Chim. Acta*, 66, 385, 1973.
79. Cheng, K. L., *Anal. Chem.*, 30, 1027, 1958.
80. Cheng, K. L., *Talanta*, 9, 736, 1962.
81. Zbiral, J. and Sommer, L., *Fresenius Z. Anal. Chem.*, 306, 129, 1981.

4-(2-PYRIDYLAZO)-RESORCINOL

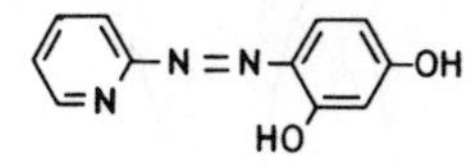

$$C_{11}H_9N_3O_2$$
mol wt = 215.21

H₂L

Synonym
PAR

Source and Method of Synthesis
Commercially available as free dye (H₂L), monosodium salt (NaHL · H₂O), or disodium salt (Na₂L·2H₂O). It is synthesized by the coupling of diazotized 2-aminopyridine with resorcinol in alcohol.[1,2]

Analytical Use
Highly sensitive photometric reagent for wide variety of metal ions.

Properties of Reagent
Free acid (H₂L) is an orange-red to brown amorphous powder. It is slightly soluble in water (5 mg/100 mℓ at 10°) and alcohol, but is more soluble in acidic or alkaline solution. Disodium salt is a slightly hygroscopic brown powder which is easily soluble in water.[1]

The proton dissociation scheme can be represented as below:[3]

$$H_3L^+ \xrightleftharpoons{pKa_1} H_2L \xrightleftharpoons{pKa_2} HL^- \xrightleftharpoons{pKa_3} L^{2-}$$

Yellow	Yellow	Orange	Red
pH <2.5	pH 3 ~ 5.5	pH 6 ~ 12.5	pH >12.5
λ_{max}, 395 nm	λ_{max}, 383 nm	λ_{max}, 415 nm	λ_{max}, 485 nm[54]
$\epsilon = 1.55 \times 10^4$	$\epsilon = 1.57 \times 10^4$	$\epsilon = 2.59 \times 10^4$	$\epsilon = 1.73 \times 10^4$

and the dissociation constants are pKa_1 (N⁺H) = 3.1, pKa_2 (p-OH) = 5.6 and pKa_3 (o-OH) = 11.9 (μ = 0.2, water).[4,50] The absorption spectra of the last two deprotonated species of PAR are illustrated in Figure 1. Two additional protonated species H₄L²⁺ and H₅L³⁺ can exist in 50 and 90% sulfuric acid solution, respectively.[4]

Complexation Reactions and Properties of Complexes
PAR behaves as a terdentate or bidentate ligand to form soluble or insoluble colored chelates with a wide variety of metal ions. Most PAR chelates are red or red-violet, and, in some cases, the color changes with pH. In the case of soluble chelate, photometric determination can be done in aqueous solution.

The coordination structure of the metal chelates with a univalent PAR anion (HL⁻) may be schematically illustrated as below.[R7]

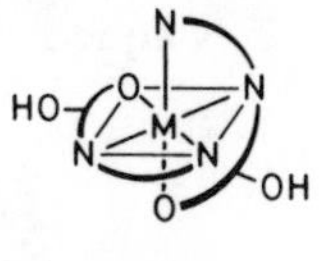

(1) 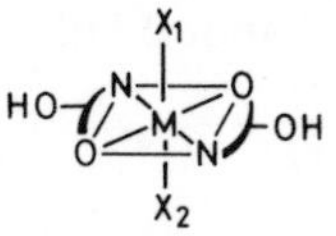   (2) 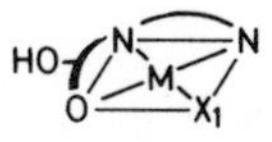   (3) 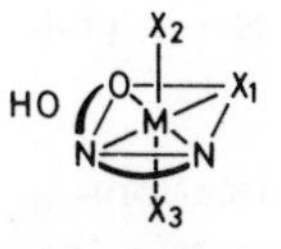   (4)

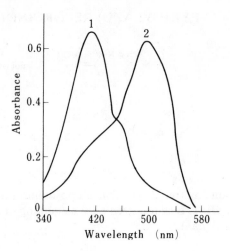

FIGURE 1.   Absorption spectra of PAR in water at various pH; PAR, 2.0×10^{-5} M. (1) pH 7.25 (HL$^-$); (2) pH 12.25 (L^{2-}). (From Shibata, S., *Chelates in Analytical Chemistry*, Vol. 4, Flaschka, H. and Barnard, A. J., Jr., Eds., Marcel Dekker, New York, 1972. With permission).

In the case of a divalent metal ion, the chelate is an uncharged coordination-saturated one (1) which is the only extractable chelate species. The rest of the chelates are charged and/or coordination-unsaturated chelates, all of them being extractable into organic solvent only in the presence of auxiliary ligand or suitable pairing cation. Such examples are summarized in Table 1.

The phenolic OH group of the ligands in these chelates may deprotonate at the higher pH region to give a chelate with higher negative charge or with higher ligand to metal ratio. At the same time, the deprotonation of *p*-OH group results in a substantial increase of molar absorptivity of the chelate, attaining a higher sensitivity in the photometry. These negatively charged chelate can be extracted into immiscible solvent as an ion pair in the presence of lipophilic bulky cation. The examples are also summarized in Table 1. The absorption spectra of the typical MHL and ML$_2$-type chelates in water are illustrated in Figure 2.

Although PAR is nonselective ligand like PAN, the selectivity may be improved by using masking agents and choosing the condition of solvent extraction because the extractability of the chelates may vary to a great extent with metal ion, pH, and the nature of the auxiliary ligand or pairing cation. The equilibria existing in such system are rather complex and are illustrated in Figure 3 on a system of Ni-PAR-Zephiramine® as an example.

As metal-PAR chelates are often slightly soluble in water, the chelate stability constants were measured in aqueous phase in many cases as summarized in Table 2.

Purification and Purity of Reagent

Sodium salt can be purified by recrystallization from 1:1 ethanol-water. Pure sample should give one spot on a TLC analysis. R$_f$ value is 0.7 on a silica-gel plate, eluted with a mixed solvent (*n*-butanol 6 vol. + ethanol 2 vol. + 2 N aqueous ammonia 2 vol.).[40]

Analytical Applications

Use as a Photometric Reagent

Numerous papers have appeared on the use of PAR for the photometric determina-

Table 1

EXTRACTION BEHAVIORS AND SPECTRAL CHARACTERISTICS OF MTAL-PAR CHELATES

Metal ion	Condition	Chelate			Solvent	Ref.
		Ratio	λ_{max} (nm)	$\varepsilon (\times 10^4)$		
Bi(III)	pH 25 ~ 3.5, HNO$_3$,TBP	M(HL)	530	0.92	Benzene or xylene	7
Bi(III)	pH 3.0 ~ 6.0, ClO$_4^-$, antipyrine(X)	M(HL)$_2$X$_2$ClO$_4$	520	2.9	CHCl$_3$-iso-BuOH(1:1)	8
Co(III)	pH 5.8 ~ 10, zephiramine(X)	ML$_2$X	520 ~ 530	5.9 ~ 6.2	CHCl$_3$	23, 24
Co(III)	pH 4 ~ 10, pH$_4$P, or Ph$_4$As(X)	ML$_2$X	518	6.0	CHCl$_3$	25
Cr(III)	pH 5 ~ 5.5, Zephiramine®(X)	M(HL)$_2$LX	540	4.8	CHCl$_3$	9, 26
Ga	pH 2.4 ~ 7.4, Zephiramine®(X)	ML$_2$X	513	10.7	CHCl$_3$	12, 28, 29
In	pH 5 ~ 5.6, acetate, antipyrine(X)	M(HL)X$_2$2AcO	530	—	CHCl$_3$	5
Ni	pH 8 ~ 9.5, Zephiramine®(X)	ML$_2$X$_2$	500	8.1	CHCl$_3$	13, 14, 23
Pb	pH 8.0 ~ 8.8, Aliquat®336 S, I$_2$	—	514	—	n-Octanol	31, 51
Sn(II)	pH 5.5 ~ 6.1, KI	1:1	515	4.85	Toluene	18
U(VI)	pH 6.8 ~ 10.7, tridodecylethylammonium bromide	MO$_2$LX	550	4.25	CHCl$_3$	33
V(V)	pH 4.6 ~ 5.1, crystal violet(X)	MO$_2$LX	585	11.0	Benzene-MIBK (3:2) + 15% EtOH	34—36
V(V)	pH 4.5 ~ 6, quinine(X)	MO$_2$(HL)X	560	3.7	CHCl$_3$ –iso-BuOH (2:1)	20—22
Zn	pH 6 ~ 7, diphenylguanidine(X)	ML$_2$X$_3$	515	6.73	CHCl$_3$	37
Zn	pH 9.7, Zephiramine®	ML$_2$X$_2$	505	9.21	CHCl$_3$	38

Note: Other metal ions extracted and determined as PAR complex are Al,[5] Au(III),[6] Cu(II),[27] Fe(II),[R7,10,23] Hf,[11] Nb(V)[30] Pd(II),[15,16,52] Sb(III),[17] Ti(IV),[19] Tl(III),[5] and V(III)[32]

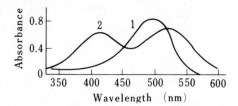

FIGURE 2. Absorption spectra of PAR chelates in water. (1) CuHL (Cu, 3.05×10^{-2} M; PAR, 4.35×10^{-5} M; pH 2.3); (2) ZnL₂ (Zb, 7.53×10^{-5} M; PAR, 2.60×10^{-5} M, pH 8.67); 10-mm cell; reference, water. (From Shibata, S., *Chelates in Analytical Chemistry*, Vol. 4, Flaschka, H. and Barnard, A. J., Jr., Eds., Marcel Dekker, New York, 1972. With permission.)

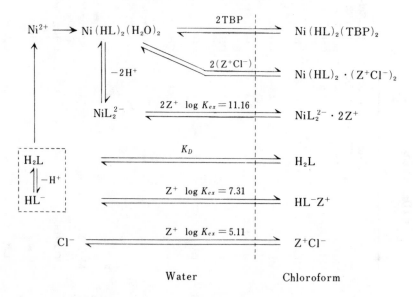

FIGURE 3. Equilibria in a system of Ni-PAR-Zephiramine®. Z^+Cl^- stands for tetradecyldimetylbenzylammonium chloride (or zephiramine). (From Hoshino, H., Yotsuyanagi, T., and Aomura, K., *Anal. Chim. Acta*, 83, 317, 1976. With permission.)

tion of traces of elements. The determinations may be carried out either in aqueous solution or after extraction into immiscible solvent. They are, more or less, based on the principles summarized in Table 1, and unusual high sensitivities can be attained by utilizing the high molar absorptivity of metal chelates with L^{2-} PAR anion.

Due to poor selectivity of PAR, the preseparation by solvent extraction or the combined use of masking agent is necessary. EDTA and CyDTA are useful masking agents in PAR photometry. The selective masking behavior of EDTA and CyDTA against metal-PAR reaction is summarized as shown in Figure 4. The elements in area A are masked with EDTA or CyDTA at any temperature. The elements in area B cannot be masked with these reagents at any temperature. The elements in area C are masked only above 80°.

Colorless organic cations can also be determined by the ion-pair extraction of anionic PAR or PAR chelate.[53]

A 0.1 to 0.01% aqueous solution of PAR (sodium salt) is recommended for the photometric uses and is quite stable for many months if protected from light.

Table 2
STABILITY CONSTANTS OF METAL-PAR CHELATES[39]

Metal ion	log K_{ML}	log K_{ML_2}	Condition	Ref.
Al	11.5	—	$18 \sim 22°, \mu = 0.1(NaClO_4)$	
Bi	17.2(MHL)	—	$18 \sim 22°, \mu = 0.1(NaClO_4)$	
Cd	10.5(MHL)	—	$18 \sim 22°, \mu = 0.1(NaClO_4)$	
Co(III)	10.0	7.1	$25°, \mu = 0.1$	
Cu(II)	14.8	9.1	$25°, \mu = 0.1$	
Fe(III)	4.37	—	$25°, \mu = 0.1(KNO_3)$	50
Ga	14.6(MHL)	β_2 30.3	$18 \sim 22°, \mu = 0.2(NaClO_4)$	
In	9.3	—	$25°$	
Ir(III)	—	β_2 25.4	50% dioxane	41
La	9.2	—	$18 \sim 22°, \mu = 0.1(NaClO_4)$	
Mn(II)	9.7(MHL)	$9.2[M(HL)_2]$	$25°, \mu = 0.005$, 50% dioxane	
Nb(V)	4.3	—	$25°$	
Ni	13.2(MHL)	$12.8[M(HL)_2]$	$25°, \mu = \sim 0.005$, 50% dioxane	
Pb	11.9(MHL)	—	$18 \sim 22°, \mu = 0.1(NaClO_4)$	
Sc(III)	12.8(MHL)	—	$18 \sim 22°$	
Ta(V)	4.5	—	$25°$	
Th	7.17($M^{4+} + 4HL \rightleftharpoons ML_4 + 4H^+$)	—	—	55
Tl(III)	9.9	—	$25°$	
UO_2^{2+}	12.5	8.4	$25°, \mu = 0.1$	
VO_2^+	16.10		$25°, \mu = 0.1(NaCl)$	22
Zn	10.5	6.6	$25°, \mu = 0.1$	

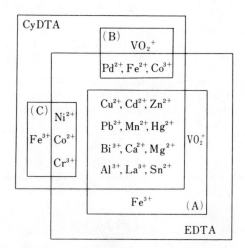

FIGURE 4. Masking behavior of EDTA and CyDTA for metal-PAR reaction. (From Yotsu-yanagi, T. and Hoshino, H., *Bunseki*, 743, 1976. With permission.)

Use as a Metal Indicator[M2,R3,R5]

PAR is used as an indicator in the EDTA titration of the following metal ions in the pH range indicated in parentheses. The color change at the end point of titration is from red-violet to yellow and is claimed to be sharper than that of PAN; Al (3, at 100°), Bi(1 to 2), Ca(11.5),[42] Cu(II) (5 to 9),[43] Fe(III) (2 to 3),[44] Ga, In (2 to 2.5, at 80°),[45,46] Hg(II) (3 to 6), Mn(II) (9), Ni (5 at 90°), Pb, Zn, Cd (5 to 6 or 9 to 10), and rare earths (6 to 7).[42]

Similar to Cu-EDTA-PAN system, a mixture of Cu-EDTA and PAR is useful as the metal indicator in the titration of metal ions other than listed above.[47,48]

PAR has been also used as an indicator in the precipitation titration of the following anions with a lead nitrate solution: MoO_4^{2-} (pH 5.5 to 6.5), PO_4^{3-} (6.6 to 7.3), WO_4^{2-} (5.9 to 6.8), and SO_4^{2-} (5.5 to 6.2).[49]

MONOGRAPHS

M1. **Shibata, S.**, 2-Pyridylazo compounds in analytical chemistry, in *Chelates in Analytical Chemistry,* Vol. 4, Flaschka, H. A. and Barnard, A. J., Jr., Eds., Marcel Dekker, New York, 1972, 232.
M2. **Pribil, R.**, *Analytical Applications of EDTA and Related Compounds,* Pergamon Press, New York, 1972.

REVIEWS

R1. **Danzuka, T.**, Analytical applications of pyridylazonaphthol and related compounds, *Kagaku (Kyoto),* 16, 686, 1961.
R2. **Busev, A. I. and Ivanov, V. M.**, The use of pyridylazo compounds in analytical chemistry, *Zh. Anal. Khim.,* 19, 1238, 1964.
R3. **Anderson, R. G. and Nickless, G.**, Heterocyclicazo dyestuffs in analytical chemistry, *Analyst,* 92, 207, 1967.
R4. **Shibata, S.**, 1-(2-Pyridylazo)-2naphthol (PAN) and its analogs in organic analytical reagents, *Bunseki Kagaku,* 21, 551, 1972.
R5. **Wada, H.**, Applications of PAN and closely related compounds as a complexometric indicator, *Bunseki Kagaku,* 21, 543, 1972.
R6. **Ivanov, V. M.**, Use of heterocyclicazo compounds in analytical chemistry, *Zh. Anal. Khim.,* 31, 993, 1976; *J. Anal. Chem. USSR,* 31, 810, 1976.
R7. **Yotsuyanagi, T. and Hoshino, H.**, New application of PAR, *Bunseki,* 743, 1976.

REFERENCES

1. **Pollard, F. H., Hanson, P., and Geary, W. J.**, *Anal. Chim. Acta,* 20, 26, 1959.
2. **Lindstrom, F. and Womble, A. E.**, *Talanta,* 20, 589, 1973.
3. **Geary, W. J., Nickless, G., and Pollard, F. H.**, *Anal. Chim. Acta,* 26, 575, 1962.
4. **Hnilickova, M. and Sommer, L.**, *Collect. Czech. Chem. Commun.,* 26, 2189, 1961 and 29, 1424, 1964.
5. **Biryuk, E. A. and Ravitskaya, R. V.**, *Zh. Anal. Khim.,* 26, 735, 1971, 28, 1500, 1973.
6. **Nagarkar, S. G. and Eshwar, M. C.**, *Anal. Chim. Acta,* 71, 461, 1974.
7. **Tomioka, H. and Terashima, K.**, *Bunseki Kagaku,* 16, 698, 1967.
8. **Nevskaya, E. M., Shelikhina, E. I., and Antionovich, V. P.**, *Zh. Anal. Khim.,* 30, 1560, 1975.
9. **Yotsuyanagi, T., Takeda, Y., Yamashita, R., and Aomura, K.**, *Anal. Chim. Acta,* 67, 297, 1973.
10. **Biryuk, E. A. and Ravitskaya, R. V.**, *Zh. Anal. Khim.,* 31, 1327, 1976; *J. Anal. Chem. USSR,* 31, 1085, 1976.
11. **Subrahmanyam, B. and Eshwar, M. C.**, *Anal. Chem.,* 47, 1692, 1975.
12. **Pyatniskii, I. V. and Kolomiets, L. L.**, *Zh. Anal. Chim.,* 31, 1562, 1976.
13. **Nonova, D. and Likhareva, N. L.**, *Dokl. Bolg. Akad. Nauk,* 27, 815, 1974; *Chem. Abstr.,* 81, 114142a, 1974.

14. Hoshino, H., Yotsuyanagi, T., and Aomura, K., *Anal. Chim. Acta,* 83, 317, 1976.
15. Busev, A. I. and Ivanova, I. M., *Zh. Anal. Khim.,* 19, 232, 1964.
16. Yotsuyanagi, T., Hoshino, H., and Aomura, K., *Anal. Chim. Acta,* 71, 349, 1974.
17. Talipov, Sh. T.,Dzhiyanbaeva, R. Kh., and Abdisheva, A. V., *Zavod. Lab.,* 37, 387, 1971; *Chem. Abstr.,* 75, 44511v, 1971.
18. Kasiura, K. and Olesiak, K., *Chem. Anal. (Warsaw),* 14, 139, 1969; *Chem. Abstr.,* 70, 111405a, 1969; *Anal. Abstr.,* 18, 3871, 1970.
19. Quy, H. V., Lobanov, F. I., and Gibalo, I. M., *Vestn. Mosk. Univ. Khim.,* 14, 460, 1973; *Chem. Abstr.,* 80, 7655c, 1974.
20. Pogranichnaya, R. M., Reznik, B. E., Nerubashchenko, V. V., Zezyanova, A. G., and Tsevina, A. V., *Zh. Anal. Khim.,* 30, 180, 1975.
21. Babko, A. K., Volkava, A. I., and Getman, T. E., *Zh. Neorg. Khim.,* 11, 374, 1966; *Chem. Abstr.,* 64, 13736g, 1966.
22. Tajika, M., Hoshino, H., Yotsuyanagi, T., and Aomura, K., *Nippon Kagaku Kaishi,* 85, 1979.
23. Yotsuyanagi, T., Yamashita, R., and Aomura, K., *Bunseki Kagaku,* 19, 981, 1970; *Bunseki Kagaku,* 20, 1283, 1971.
24. Okochi, H., *Bunseki Kagaku,* 21, 51, 1972.
25. Siroki, M., Maric, Li., Stefanac, Z., and Marak, M. J., *Anal. Chim. Acta,* 75, 101, 1975.
26. Yotsuyanagi, T., Takeda, Y., Yamashita, R., and Aomura, K., *Anal. Chim. Acta,* 67, 297, 1973.
27. Siroki, M. and Herak, M. J., *Fresenius Z. Anal. Chem.,* 278, 285, 1976.
28. Harimaya, S., Oji, H., and Odajima, K., *Tetsu To Hagane (Japan),* 60, 1869, 1974; *Chem. Abstr.,* 82, 51036s, 1975.
29. Siroki, M. and Herak, M. H., *Anal. Chim. Acta,* 87, 193, 1976; *J. Inorg. Nucl. Chem.,* 39, 127, 1977.
30. Siroki, M., Djordjevic, C., Maric, L., and Herak, M. J., *J. Less-Common Metals,* 23, 228, 1971; *Anal. Chem.,* 43, 1375, 1971 and 48, 55, 1976.
31. Zelaskowski, C. A., U.S. Patent 3934976; *Chem. Abstr.,* 84, 167189g, 1976.
32. Hoshino, H., Yamashita, R., Yotsuyanagi, T., and Aomura, K., *Nippon Kagaku Kaishi,* 495, 1974.
33. Shijo, Y. and Sakai, K., *Bull. Chem. Soc. Jpn.,* 51, 2574, 1978.
34. Mincewski, J., Chwastowska, J., and Mai, P. T. H., *Analyst,* 100, 708, 1975.
35. Yotsuyanagi, T., Yamashita, Y., and Aomura, K., *Bunseki Kagaku,* 19, 981, 1970.
36. Siroki, M. and Djordjevic, C., *Anal. Chim. Acta,* 57, 301, 1971.
37. Mamuliya, S. G., Ryatniskii, I. V., and Grigelashvili, K. I., *Ukr. Khim. Zh.,* 44, 410, 1978; *Anal. Abstr.,* 35, 4B46, 1978.
38. Nonova, D., Nenov, V., and Likareva, N., *Talanta,* 23, 679, 1976.
39. Sillen, L. G. and Martell, A. E., *Stability Constants of Metal-Complexes* and Suppl. 1, The Chemical Society, London, England (1964, 1971).
40. Pollard, F. H., Nickless, G., Samuelson, T. J., and Anderson, R. G., *J. Chromatogr.,* 16, 231, 1964.
41. Shusupova, T. I., and Ivanov, V. M., *Koord. Khim.,* 21, 1659, 1976; *Chem. Abstr.,* 86, 111755v, 1977.
42. Wehber, P., *Fresenius Z. Anal. Chem.,* 166, 186, 1959.
43. Wada, H. and Nakagawa, G., *Nippon Kagaku Zasshi,* 85, 549, 1969.
44. Iwamoto, T., *Bunseki Kagaku,* 10, 190, 1961.
45. Busev, A. I. and Skrebkova, L. M., *Izu. Sib. Otd. Akad. Nauk SSSR,* 57, 1962; *Chem. Abstr.,* 58, 5032g, 1963.
46. Busev, A. I. and Karaev, N. A., *Nauchn. Dokl. Vyssh. Skh. Khim. Khim. Tekhnol.,* 299, 1959; *Chem. Abstr.,* 53, 18747c, 1959.
47. Nonova, D. and Likhareva, N., *Talanta,* 23, 439, 1976.
48. Nonova, D. and Likhareva, N., *God. Sofi. Univ. Khim. Fak.,* 67, 589, 1972; *Chem. Abstr.,* 87, 47643y, 1977.
49. Sammer, L. and Janoscova, L., *Collect. Czech. Chem. Commun.,* 39, 101, 1974.
50. Russeva, E., Kuban, V., and Sommer, L., *Collect. Czech. Chem. Commun.,* 44, 374, 1979.
51. Kleckova, Z., Langova, M., and Havel, J., *Collect. Czech. Chem. Commun.,* 43, 3163, 1978.
52. Mizuno, K., Uwano, A., and Miyatani, G., *Bunseki Kagaku,* 25, 113, 1976; *Bull. Chem. Soc. Jpn.,* 49, 2479, 1976.
53. Siroki, M. and Maric, Li, *Anal. Chim. Acta,* 79, 265, 1975, *Fresenius Z. Anal. Chem.,* 276, 371, 1975.
54. Geary, W. J., Nickless, G., and Pollard, F. H., *Anal. Chim. Acta,* 26, 575, 1962.
55. Busev, A. I. and Ivanov, V. M., *Izv. Vyssh. Uchebn. Zaved. Khim. Khim. Technol.,* 4, 914, 1961; *Chem. Abstr.,* 56, 14922a, 1962.

THIAZOLYLAZOPHENOLS AND THIAZOLYLAZONAPHTHOLS

Synonyms

Thiazolylazophenols and thiazolylazonaphthols covered in this section are listed in Table 1, together with their synonyms, structural formulas, and other information.

Source and Method of Synthesis

All are commercially available. They are prepared by the coupling of diazotized 2-aminothiazol with the corresponding coupling components.[1,71,78,82]

Analytical Uses

These reagents behave similarly to PAN or PAR, but with slight differences in reactivity toward metal ions. They are used as a photometric reagent and as a metal indicator. TAR is the most widely used reagent of this group.

Physical Properties of Reagents

TAR (1) is an orange crystalline powder, mp 200 to 202° (with decomposition)[78] and is slightly soluble in water, but is easily soluble in organic solvents. The color change in the aqueous solution and the acid dissociation constants are [R4.2-4]

$$H_3L \underset{\longleftarrow}{\overset{pKa (N^+H) = 0.96}{\longrightarrow}} H_2L \underset{\longleftarrow}{\overset{pKa(p-OH) = 6.23}{\longrightarrow}} HL^- \underset{\longleftarrow}{\overset{pKa(o-OH) = 9.44}{\longrightarrow}} L^{2-} \quad \text{(in water, } \mu = 0.1 \text{ NaClO}_4\text{)}$$

Red	Yellow	Yellow-Orange	Red-Violet
λ_{max}, 484 nm	λ_{max}, 439 nm	λ_{max}, 481 nm	λ_{max}, 510 nm
$\epsilon = 2.9 \times 10^4$	$\epsilon = 2.0 \times 10^4$	$\epsilon = 2.9 \times 10^4$	$\epsilon = 3.45 \times 10^4$
(in 20% dioxane)	(in 20% dioxane)	(in 20% dioxane)	(in water)

Distribution of H_2L between 1 M (H,Na)Cl and benzene is reported as log K_D = 1.79.[5]

TAN (2) is orange-yellow needles, mp 139 to 141°[71,82] and is slightly soluble in water, but is easily soluble in organic solvents. The color change in an aqueous solution associated with the proton dissociation are shown below.[6,7,38]

$$H_2L^+ \underset{\longleftarrow}{\overset{pKa (N^+H) = 2.37}{\longrightarrow}} HL \underset{\longleftarrow}{\overset{pKa (OH) = 8.71}{\longrightarrow}} L^- \quad \text{(in water, } \mu \rightarrow 0)$$

Pale Yellow	Yellow	Red
λ_{max}, 440 nm	λ_{max}, 485 nm	λ_{max}, 535 nm
$\epsilon = 3.4 \times 10^4$	$\epsilon = 3.6 \times 10^4$	$\epsilon = 3.9 \times 10^4$ (in 10% methanol)

TAC (3) is a yellow-brown crystalline powder, mp 129 to 130°[78,82] and is slightly soluble in water, but is easily soluble in mineral acid or alkali hydroxide solution and in common organic solvents. The color change of an aqueous solution and the acid dissociation equilibria are[8,9,52,82]

$$H_2L^+ \underset{\longleftarrow}{\overset{pKa (N^+H) <0.5}{\longrightarrow}} HL \underset{\longleftarrow}{\overset{pKa (OH) = 8.34}{\longrightarrow}} L^- \quad \text{(in 20% dioxane, 25°, } \mu = 0.1)$$

Red-Orange	Yellow	Red-Purple
λ_{max}, 381 nm	λ_{max}, 373 nm	λ_{max}, 544 nm
$\epsilon = 1.7 \times 10^4$	$\epsilon = 1.3 \times 10^4$	$\epsilon = 1.6 \times 10^4$ (in 20% dioxane)

Table 1
THIAZOLYLAZO REAGENTS

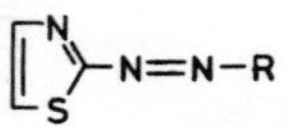

Reagent	Name	Synonym	R	Molecular formula	Mol wt
(1)	TAR H₂L	4-(2-Thiazolylazo)-resorcinol	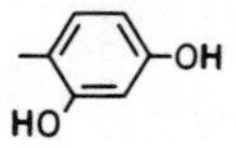	$C_9H_7N_3O_2S$	221.23
(2)	TAN HL	1-(2-Thiazolylazo)-2-naphthol *o*-TAN, *β*-TAN	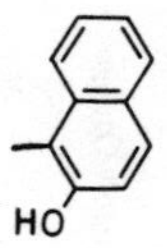	$C_{13}H_7N_3OS$	255.28
(3)	TAC[a] HL	2-(2-Thiazolylazo)- *p*-cresol, 2-(2-Thiazolylazo)-4-methylphenol	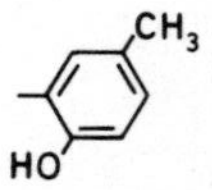	$C_{10}H_9N_3OS$	219.26
(4)	TAM[a] HL	2-(2-Thiazolylazo)-5-dimethylaminophenol	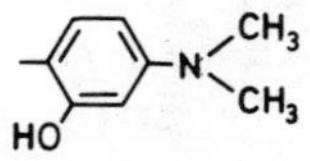	$C_{11}H_{12}N_4OS$	248.30

[a] There is some confusion regarding the naming of TAM and TAC. In some cases, TAM stands for thiazolylazomethoxyphenol, and TAC for thiazolylazochlorophenol.

TAM (**4**) is a red crystalline powder, mp 216°[74] and is slightly soluble in water, but is easily soluble in mineral acid or alkali hydroxide solution and in common organic solvents. The color change of aqueous solution and the acid dissociation equilibria are[10]

$$H_2L^+ \xrightleftharpoons[]{pKa\ (N^+H)\ =\ 3.13} HL \xrightleftharpoons[]{pKa\ (OH)\ =\ 8.65} L^- \qquad (\mu = 0.1,\ NaClO_4)$$

Red	Orange-Yellow	Red
λ_{max}, 540 nm	λ_{max}, 500 nm	λ_{max}, 526 nm
$\epsilon = 4.8 \times 10^4$	$\epsilon = 3.36 \times 10^4$	$\epsilon = 3.75 \times 10^4$ (in aqueous solution)

Complexation Reactions and Properties of Complexes

Reagents of this class behave as a bi- or terdentate ligand, giving colored (red or red-violet) chelates with many metal ions. In acidic and slightly acidic solutions, metal ions form ML-type chelates or mixture of ML and ML_2-type chelates. In alkaline solution, the equilibrium shifts towards ML_2-type chelate. In addition to these normal chelates, some metal ions form protonated chelates (MHL) with TAR in acidic solutions, and an increasing amount of $M(HL)_2$ is formed when the pH is raised. The ease of formation of the protonated chelate with TAR may be due to the behavior of TAR which tends to coordinate to metal ion with p-OH group of the ligand undissociated. The coordination structure of metal-TAN chelate was investigated by X-ray crystallography, and the structure of Fe-TAN chelate can be shown as below schematically.[11]

As to the absorption spectra of metal chelates, the reagents of this group appear to be similar in character to pyridylazo derivatives. The spectral characteristics of these chelates are summarized in Tables 4 to 7. As an example, the absorption spectra of Cu-TAR chelate are illustrated in Figure 1.

Chelate stability constants of these reagents are observed on a limited number of metal ions, and they are summarized in Tables 2 and 3. The proton dissociation constants (pKa_1 and pKa_2) of the protonated TAR chelates ($M(HL)_2$) are Cu(II), 4.34; --, Mn(II), 7.88, 9.38; Ni, 6.84, 8.55; and Zn 7.12, 8.74 (in 25% dioxane, 25°).[79] Due to the lower basicity of the thiazole nitrogen, substitution of the pyridine ring with a thiazole ring should result in a shift of the complexation reaction towards the more acidic region. Therefore, the thiazolylazo reagents form metal chelates in more acidic solution.

Similar to PAN and PAR chelates, the coordination saturated uncharged chelate can be extracted into an inert solvent, such as chloroform or benzene. The charged or coordination unsaturated chelates are somewhat soluble in water, but can be extracted into a donating solvent, such as alcohol or ketone, or into an inert solvent in the presence of an auxiliary ligand or pairing cation. The optimum conditions for the extraction may be found in Tables 4 to 7. The extraction constants for the normal uncharged TAR chelates in a water-iso-amylalcohol system are reported as follows: log K_{ex} ($Cd(HL)_2$), −11.0, ($Zn(HL)_2$) −10.4, ($Ni(HL)_2$) − 5.5, and ($Co(HL)_3$) 0.3 (at 20°).[5] The similar values for TAC chelates in a water-benzene system are log K_{ex}(CoL_2), −6.18, (ZnL_2) −8.99, and (NiL_2) −4.97 (at 30°).[14]

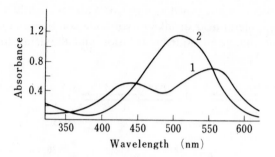

FIGURE 1.   Absorption spectra of Cu - TAR(1) che-
late. (1) pH 2.06; (2) pH 6.11. Dye, 3.71×10^{-5} *M*; Cu,
2.94×10^{-3} *M*. (From Hnilichova, M. and Sommer, L.,
Talanta, 13, 667, 1966. With permission.)

Table 2
CHELATE STABILITY CONSTANTS OF TAR (1)

Metal ion	Chelate log K_{ML}	log K_{ML_2}	Condition	Ref.
Au(III)	12	—	25°, 50% CH_3OH	80
Ba	< 3	—	25°, 50% CH_3OH	80
Bi	13.11	—	$\mu = 0.1\ ClO_4^-$	2
Ca	3.5	—	25°, 50% CH_3OH	80
Cd	6.96	—	$\mu = 0.1\ ClO_4^-$	2
Co(II)	12.05	11.23	25°, 50% dioxane	79
Cr(III)	10	—	25°, 50% CH_3OH	80
Cu(II)	11.56	—	$\mu = 0.1\ ClO_4^-$	2
Cu(II)	12.3	9.9	25°, 50% CH_3OH	80
Fe(II)	—	β_2 21.6	25°, 50% CH_3OH	80
Ga	12	—	25°, 50% CH_3OH	80
In	10.8	—	25°, 50% CH_3OH	80
Ir(III)	—	β_2 29.46	50% DMF	26
Mg	< 3	—	25°, 50% CH_3OH	80
Mn(II)	9.43	8.6	50% dioxane	79
Nb	9.5	—	25°, 50% CH_3OH	80
Ni	12.94	11.82	25°, 50% dioxane	79
Pb	8.34	—	$\mu = 0.1\ ClO_4^-$	2
Pt(II)	12	—	25°, 50% CH_3OH	80
Rh(III)	12	—	25°, 50% CH_3OH	80
Ti²⁺	13	—	25°, 50% CH_3OH	80
Tl(I)	< 3	—	25°, 50% CH_3OH	80
Tl(III)	13.41	—		25
Sc	10.4	9.9	25°, 50% CH_3OH	80
Sr	⩽ 3	—	25°, 50% CH_3OH	80
UO_2^{2+}	10.9	9.7	25°, 50% CH_3OH	80
V(V)	10.22	—	25°, 30% C_2H_5OH	36
Zn	7.19	—	$\mu = 0.1\ ClO_4^-$	2
Zn	11.08	10.10	50% dioxane	79
ZrO^{2+}	13	—	25°, 50% CH_3OH	80
Rare earths	9.56 ∿ 9.81	9.20 ∿ 9.51	log K_{ML_3} 8.84 ∿ 9.21 log K_{ML_4} 6.48 ∿ 8.91	13

Purification and Purity of Reagents

TAR can be purified by dissolving the dye in alkali hydroxide solution, followed by
extraction with diethyl ether and subsequent reprecipitation with dilute hydrochloric

Table 3
CHELATE STABILITY CONSTANTS OF TIAZOLYLAZO REAGENTS

Metal ion	TAN (2)			TAC (3)			Ref.
	$\log K_{ML}$	$\log K_{ML_2}$	Condition	$\log K_{ML}$	$\log K_{ML_2}$	Condition	
Cd	6.59	5.61	50% dioxane	6.03	5.60	50% dioxane	12
Co(II)	8.70	8.24	50% dioxane	7.25	7.45	50% dioxane	
Cu(II)	10.92	11.60	$\mu = 0.1\ ClO_4^-$	9.19	5.60	50% dioxane	81
Fe(II)	—	—	—	8.00	9.51	50% dioxane	
Hg(II)	—	—	—	6.11	6.11	25°, 10% dioxane	52
Ir(III)	—	—	—	—	β_2 13.70	30% DMF	26
Mn(II)	—	—	—	—	β_2 7.6	25°, 50% CH_3 OH	80
Ni	9.55	9.45	25°, 50% dioxane	7.93	7.33	25°, 20% dioxane	R3,78
Zn	12.29	12.29	$\mu = 0.1\ ClO_4$	—	—	—	81

[a] Values for TAM (4) chelates: $\log K_{ML}$ and $\log K_{ML_2}$ for Cu(II), 6.4 and 5.38 ($\mu = 0.1$, ClO_4^-),[57] and for Ni, 8.22 and 7.62(25°, 20% dioxane),[78] respectively.

acid. The rest of the reagents of this group can be purified similarly with a slight modification.

Purity of the sample is checked by a thin-layer chromatography on a silica gel plate using a mixed solvent of petroleum ether-diethyl ether-ethanol (10:10:1) as a mobile phase. R_f values are 0.50 for TAR, 0.70 for TAN, and 0.80 for TAC.[15]

Alternatively, their purities can be assayed by the photometric titration with a standard copper (II) solution.[16]

Analytical Applications
Use as a Photometric Reagent

Due to a poor selectivity of the reagents and poor solubility of the metal chelates, the photometric methods are applied, in most cases, after solvent extraction in the presence of masking agent. Insoluble chelate may be solubilized with an aid of nonionic surfactant, enabling the photometry in aqueous solution.

In the case of TAR chelates, the molar absorptivities are generally in the range of 2 to 5×10^5, attaining the very high sensitivity in the photometric determination.

The representative examples on the uses of these reagents are summarized in Tables 4 to 7. They are also used as a spraying reagent in TLC analysis of Co, Cu, and Ni.[64]

Use as a Metal Indicator[R3,65]

TAR is recommended as a metal indicator in the EDTA titration of the following metal ions; Cu(II) (pH 3 to 8)[65], Co(II) (pH 4 to 8), Ni (pH 4 to 6), Tl(III) (pH 1.2 to 5.5),[66] and rare earths. A mixed indicator system (Cu-EDTA and TAR) is recommended in the titration of Cd, Fe, Hg, In, Mn, Pb, Sc, and Zn (pH 3 to 6.5),[67] Ga (pH 3.5 to 4),[68] and Ni (pH 4.0 to 6.5).[69] The color change at the end point is from red-violet to yellow.

TAC is recommended for the titration of Ca (pH 12 to 12.5),[70] Hg(II) (pH 7.0 to 8.3),[52] and Ni (pH 6 to 8).[9,71,72] Especially, TAC is claimed to be the best known for the Ni titration, with respect to the sharpness of color change (blue to yellow).[78] TAN[73,74] and TAM[75] can be used similarly.

TAR[81] and TAC[82] are also used as an indicator in the complexometry of anions using a lead nitrate[76] and mercuric nitrate solution,[77] respectively.

Other Reagents with Related Structure

Beside the above-mentioned reagents, a large number of heterocyclic azo dyes have been synthesized and evaluated as the organic reagent. They are reviewed in various articles.[R1-R5]

Table 4

APPLICATION OF TAR (1) AS A PHOTOMETRIC REAGENT

Metal ion[a]	Condition	Ratio	λ_{max} (nm)	$\varepsilon(\times 10^4)$	Extraction solvent	Range of determination and remarks (ppm)	Ref.
Bi	pH 1.7	M(HL)	540	2.24	Aqueous	3.8 ∿ 8.3	2
Cd	pH 4.7 ∿ 6.1	M(HL)	500	3.06	Aqueous		2
Cd	pH 4.5 ∿ 5.0, Cetylpyridinium chloride, or Neocuproine(X)	ML$_2$X$_2$	570	—	iso-AmOH	0.06∿	18
Co(II)	pH 6.2 ∿ 8.6	ML$_2$	510	6.0	Aqueous	0.2 ∿ 2.0	19
Co(II)	pH 6.7 ∿ 10.0, EDTA, Zephiramine®		550	3.42	CHCl$_3$	Quantitative extraction	20
Cu(II)	pH ≃ 6	ML	510	3.11	Aqueous	Bi,Co,Fe(III), Ni give positive error	2
Hf	Acidic, 20% ethanol	ML$_4$	540	5.8	Aqueous	0.2 ∿ 3.4	24
Mo(VI)	NH$_2$OH, heating	ML	530	2.94	Aqueous	In Cu ores	27
Nb	pH 5 ∿ 6, EDTA or CyDTA, tartarate	ML	540	2.85	Aqueous	1 ∿ 10, Seven- to ten- fold V, excess Mo,Ti,Th and U interfere	53
Ni	pH 7.5 ∿ 8.0, 30% ethanol	M(HL)$_2$	540	3.7	Aqueous	—	3
Ni	NH$_3$ alkali		610	3.14	CHCl$_3$	0.1 ∿ 2, Co, interferes	28
Ni	pH 7.2 ∿ 9.1, Zephiramine®(X)	ML$_2$X$_2$	550	3.51	CHCl$_3$	log K$_{ex}$ = 11.25	39
Os(IV)	pH 5.7 ∿ 7.2	M(HL)	550	1.55	Aqueous	0.6 ∿ 9.5	29
Os(VIII)	pH 7.5 ∿ 9.0	ML	550	2.22	Aqueous	0.4 ∿ 6.5	29
Sc	pH 7.0 ∿ 8.5	ML$_3$	540	5.1	Aqueous	0.12 ∿ 1.6, borate, tartarate, phosphate interfere	33
UO$_2^{2+}$	pH 7.5 ∿ 7.8, triethanolamine	MO$_2$L	540 ∿ 545	3.3	Aqueous	0.4 ∿ 4.8, oxalate, carbonate interfere	35
Zn	pH 7 ∿ 8,	ML or M(HL)	500	3.57	Aqueous		2
Zn	pH 7.4 ∿ 8.4, acetone	—	530	—	Aqueous	0 ∿ 1.4	37
Zn	pH 8.2 ∿ 10.1, Zephiramine®(X)	ML$_2$X$_2$	550	3.03	CHCl$_3$	log K$_{ex}$ = 12.96	39
Zr	Acidic, 20% methanol	ML$_4$	550	6.4	Aqueous	0.01 ∿ 1.2	24

[a]   Other metals determined with TAR are Au(III),[17] Co,[40] Cr(III),[21] Fe(III),[4,22,23] Ga, In,[25] Ir,[26] La,[22] Pb,[2] Pd(II),[30] Pr,[31] Rh(III),[32] Ru(IV),[29] Th,[22,34] T1(III),[25] and V(V).[36]

Table 5
APPLICATION OF TAN (2) AS A PHOTOMETRIC REAGENT

| Metal ion[a] | Condition | Metal chelate | | | Extraction solvent | Range of determination and remarks (ppm) | Ref. |
		Ratio	λ_{max} (nm)	ε ($\times 10^4$)			
Cd	pH 4.5 $\sim$ 5.0	ML$_2$	570	3.7	Benzene	0.05 $\sim$ 2.0	38, 40
Co(III)	pH 2.5 $\sim$ 6.0	ML$_3$	619	2.7	CHCl$_3$	0.5 $\sim$ 2.0	38
Cu(II)	pH 8 $\sim$ 9	ML$_2$	580	3.2	CHCl$_3$	0.5 $\sim$ 2.0	38, 41
Mn(II)	pH 9 $\sim$ 11	ML$_2$	570	4.1	CHCl$_3$	0.5 $\sim$ 2.0	38
Ni	pH 4 $\sim$ 10	ML$_2$	595	4.0	CHCl$_3$	0.5 $\sim$ 2.0	38, 43
Ni	pH 6.4 $\sim$ 7.0, Triton® X 100	ML$_2$	595	3.9	Aqueous	0 $\sim$ 1.4, Zn, Cd, Co, are masked with N-(di-thiocarboxyl) glycine + pyrophosphate	44
Tl(III)	pH 1.5 $\sim$ 3.5	ML$_2$	580	3.5	Aqueous	Bi must be absent	49
UO$_2^{2+}$	pH 6 $\sim$ 9	MO$_2$L$_2$	585 $\sim$ 590	2.4	CHCl$_3$ or MIBK	0 $\sim$ 6.5	38, 61
Zn	pH 6 $\sim$ 8	ML$_2$	580	5.0	CHCl$_3$	0.5 $\sim$ 2.0	38

[a] Other metals determined with TAN are Fe(III),[13,38] Ga,[38] Hg(II),[42] In,[38] Os,[45] Pd(II),[46,47] and Rh(III).[48]

Table 6
APPLICATION OF TAC (3) AS A PHOTOMETRIC REAGENT

| Metal ion[a] | Condition | Metal chelate | | | Extraction solvent | Range of determination and remarks (ppm) | Ref. |
		Ratio	λ_{max} (nm)	ε($\times 10^4$)			
Bi	pH 6.0 $\sim$ 7.0, Zephiramine®(X)	ML$_4$X	596	3.44	Aqueous	$\sim$3, Co,Cu,Fe, Ni, and Th give positive,Cr(III), Nb,Sn, and Ta give negative errors	50
Cd	pH 8.7 $\sim$ 10.2	ML$_2$	600	3.67	CHCl$_3$	—	51
Co(II)	pH 5.8 $\sim$ 6.9 with MES buffer[b]	ML$_2$	590	2.63	Benzene	99% extraction	12
Cu(II)	pH 7.8 $\sim$ 8.2, 50% ethanol, TBP	ML$_2$	610	2.87	Benzene	—	8, 82
Fe(II)	pH 4.8 $\sim$ 10.0	ML$_2$	762	1.37	CHCl$_3$	$\sim$ 5, log K$_{cy}$ 54 = $^-$2.70	54
Ni	7.0	ML$_2$	617	3.0	Benzene	—	14
Zn	pH 9.18, CyDTA	ML$_2$	600	3.36	CHCl$_3$	Al,Cr,Fe do not interfere; Hg and Ni can be separated by prior extraction at pH 7.1	51
Rare earths	—	ML$_2$	601 $\sim$ 618	2 $\sim$ 2.5	Aqueous	—	55

[a] Other metals, such as Hg(II)[52] and Ir,[26] can also be determined with TAC.
[b] MES, 2-(N-Morpholino)-ethanesulfonic acid.

Table 7

APPLICATION OF TAM (4) AS A PHOTOMETRIC REAGENT

Metal ion	Condition	Metal Chelate			Extraction solvent	Range of determination and remarks (ppm)	Ref.
		Ratio	λ_{max} (nm)	$\varepsilon(\times 10^4)$			
Bi	pH 4.3 ∿ 5.0 diphenylguianidine	MLXn	585	4.6	Aqueous	∿1.8,Ce,Co,Fe,Ni, Ti,V, and Zr interfere	56
Cu(II)	pH 1.5 ∿ 4.8	ML	590	2.62	Aqueous		57
Cu(II)	pH 10 ∿ 13	ML₂	570	3.94	CHCl₃	0.02 ∿ 0.8,Cd,Co, Ni,Pb,and Zn interfere	57
Nb(V)	pH 3.2 ∿ 3.8, triethanolamine	HL	605	4.8	Aqueous	0 ∿ 1.2, Al,Cu,Fe, Ti, and Zr interfere	58
Ni	pH 7.8 ∿ 8.2, Triton® X-100, *N*-(dithiocarboxyl)-glycine,tartarate, pyrophosphate	—	560	6.5	Aqueous	0.1 ∿ 1.2,Cd,Cu, Fe,Mn, and Zn do not interfere	59
Th	pH 4.3 ∿ 4.8, cetyl-pyridinium chloride	ML₃	570	8.6	Aqueous	Co,Fe,Ni,V, and Zr give positive, and Mo,Ta,W,EDTA, and F⁻ give negative error	
Ti(IV)	pH 4.2 ∿ 5.2, diphenylguanidine	ML₂	583		Benzylalcohol	∿1.2, Co,Fe,Ta, and Zr interfere	60
UO₂²⁺	pH 4.5 ∿ 7.5, pyridine or TBP	MO₂L₂	575	≃5	CHCl₃ or MIBK	0 ∿ 6	61, 62
Zr	pH 3.7 ∿ 4.5, Zephiramine®	ML₄	595	10.5	Aqueous	∿0.4, Cu,Hf,Ta, and W interfere	63

REVIEWS

R1. Anderson, R. G. and Nickless, G., Heterocyclicazo dyestuffs in analytical chemistry, *Analyst,* 92, 207, 1967.

R2. Busev, A. I., Ivanov, V. M., and Krysina, L. S., Thiazolazo compounds in inorganic analysis, *Sovrem. Metody Anal. Mater.,* 135, 1969; *Chem. Abstr.,* 73, 136940w, 1970.

R3. Wada, H., Applications of PAN and closely related compounds as a complexometric indicator, *Bunseki Kagaku,* 21, 543, 1972.

R4. Hovind, H. R., Thiazolylazo dyes and their applications in analytical chemistry, *Analyst,* 100, 769, 1975.

R5. Ivanov, V. I., Use of heterocyclicazo compounds in analytical chemistry, *Zh. Anal. Khim.,* 31, 933, 1976; *J. Anal. Chem. USSR,* 31, 810, 1976.

REFERENCES

1. Traumann, V., *Ann.*, 249, 31, 1888.
2. Hnilichova, M. and Sommer, L., *Talanta*, 13, 667, 1966.
3. Mushran, S. P. and Sommer, L., *Collect. Czech. Chem. Commun.*, 34, 3693, 1967.
4. Russeva, E., Kuban, V., and Sommer, L., *Collect. Czech. Chem. Commun.*, 44, 374, 1979.
5. Navratil, O., *Collect. Czech. Chem. Commun.*, 41, 2682, 1976.
6. Ivanov, V. M., Busev, A. I., and Ershova, N. S., *Zh. Anal. Khim.*, 28, 214, 1973.
7. Ershova, N. S., Ivanov, V. M., and Busev, A. I., *Zh. Anal. Khim.*, 28, 2220, 1973.
8. Sommer, L., Langova, M., and Kuban, V., *Collect. Czech. Chem. Commun.*, 41, 1317, 1976.
9. Wada, H., Nakagawa, O., and Takada, H., *Bull. Chem. Soc. Jpn.*, 50, 2101, 1977.
10. Minczewski, J. and Kasiura, K., *Chem. Anal. (Warsaw)*, 10, 21, 1965; *Chem. Abstr.*, 63, 17109h, 1965.
11. Kurahashi, M., Kawase, A., Hirotsu, K., Fukuyo, M., and Shimada, A., *Bull. Chem. Soc. Jpn.*, 45, 1940, 1972; *Bull. Chem. Soc. Jpn.*, 49, 1419, 1976.
12. Kawase, A., *Bunseki Kagaku*, 13, 553, 1964.
13. Navratil, O., *Collect. Czech. Chem. Commun.*, 31, 2492, 1966.
14. Fujiwara, J. and Kawase, A., *Bunseki Kagaku*, 21, 1191, 1972 and 25, 484, 1976.
15. Pollard, F. H., Nickless, G., Samuelson, T. J., and Anderson, R. G., *J. Chromatogr.*, 16, 231, 1964.
16. Pease, B. F. and Williams, M. B., *Anal. Chem.*, 31, 1044, 1959.
17. Subrahmanyam, B. and Eshwar, M. C., *Anal. Chim. Acta*, 82, 435, 1976.
18. Frei, R. W. and Navratil, O., *Anal. Chim. Acta*, 52, 221, 1970; *Can. J. Chem.*, 49, 173, 1971.
19. Busev, A. I., Nemtseva, Zh. I., and Ivanov, V. M., *Zh. Anal. Khim.*, 24, 1376, 1969.
20. Ueda, K., *Bull. Chem. Soc. Jpn.*, 52, 1215, 1979.
21. Subrahmanyam, B. and Eshwar, M. C., *Mikrochim. Acta*, 579, 1976.
22. Minczewski, J., Grzegrzolka, E., and Kasiura, K., *Chem. Anal. (Warsaw)*, 13, 601, 1968; *Chem. Abstr.*, 70, 25293f, 1969.
23. Biryuk, E. A. and Ravitskaya, R. V., *Zh. Anal. Khim.*, 31, 1327, 1976; *J. Anal. Chem. USSR*, 31, 1085, 1976.
24. Subrahmanyam, B. and Eshwar, M. C., *Indian J. Chem.*, 15A, 66, 1977; *Chem. Abstr.*, 87, 177070v, 1977.
25. Hnilichova, M. and Sommer, L., *Talanta*, 16, 83, 1969.
26. Shurupova, T. I., Ivanov, V. M., and Busev, A. I., *Zh. Anal. Khim.*, 31, 2162, 1976; *J. Anal. Chem. USSR*, 31, 1581, 1976.
27. Szczygielska, M. and Kasiura, K., *Chem. Anal. (Warsaw)*, 18, 799, 1973; *Chem. Abstr.*, 79, 77920; 1974.
28. Gusev, S. I., Zhvakina, M., Kozhevnikova, I. A., and Maltseva, L. S., *Zavod, Lab.*, 42, 19, 1976; *Chem. Abstr.*, 85, 103316s, 1976.
29. Ivanov, V. M., Busev, A. I., Popova, L. V., and Bogdanovich, L. I., *Zh. Anal. Khim.*, 24, 1064, 1969.
30. Lin, H. K., Chen, K. Y., and Chen, Y. F., *Hua Hseuh Tung Pao*, 365, 1966; *Chem. Abstr.*, 66, 25818y, 1967.
31. Vosta, J. and Havel, J., *Collect. Czech. Chem. Commun.*, 42, 2871, 1977.
32. Busev, A. I., Ivanov, V. M., and Gresl, V. G., Anal. Lett., 1, 577, 1968; *Zh. Anal. Khim.*, 23, 1570, 1968.
33. Shimizu, T. and Momo, E., *Anal. Chim. Acta*, 52, 146, 1970.
34. Sakai, T. and Tonosaki, K., *Bull. Chem. Soc. Jpn.*, 42, 2718, 1969.
35. Sommer, L. and Ivanov, V. M., *Talanta*, 14, 171, 1967.
36. Langova, M., Klabenesova, I., Kasiura, K., and Sommer, L., *Collect. Czech. Chem. Commun.*, 41, 2386, 1976.
37. Marshall, B. S., Telford, I., and Wood, R., *Analyst*, 96, 569, 1971.
38. Nakagawa, G. and Wada, H., *Nippon Kagaku Zasshi*, 83, 1185, 1962.
39. Ueda, K., *Anal. Lett.*, All, 1009, 1978.
40. Navratil, O. and Frei, R. W., *Anal. Chim. Acta*, 52, 221, 1970.
41. Goyal, S. S., Misra, G. J., and Tandon, J. P., *Bull. Acad. Pol. Sci. Ser. Sci. Chim.*, 18, 425, 1970; *Chem. Abstr.*, 74, 71270b, 1971.
42. Kolosova, I. V., *Tr. Permsk. Gos. Med. Inst.*, 99, 446, 1970; *Chem. Abstr.*, 78, 37601h, 1973.
43. Wada, H. and Nakagawa, G., *Anal. Lett.*, 1, 687, 1968.
44. Watanabe, H., Matsunaga, H., Miura, J., and Ishii, H., *Bunseki Kagaku*, 25, 35, 677, 808, 1976; *Bunseki Kagaku*, 26, 252, 1977.
45. Busev, A. I., Ivanov, V. M., and Bogdanovich, L. I., *Zh. Anal. Khim.*, 24, 1273, 1969.

46. Busev, A. I., Ivanov, V. M., and Krysina, L. S., *Vestn. Mosk. Gos. Univ. Ser. Khim.*, 23, 80, 1968; *Anal. Abstr.*, 16, 1888, 1969.
47. Gusev, S. I., Gluhkova, I. N., and Ketova, L. A., *Izv. Vyssh. Uchebn. Zaved. Khim. Khim. Tekhnol.*, 19, 1210, 1976; *Anal. Abstr.*, 32, 3B196, 1976.
48. Ivanov, V. M., Busev, A. I., Gresl, V. G., and Zagruzina, A. N., *Zh. Anal. Khim.*, 26, 1553, 1971.
49. Gusev, S. I., Kurepa, G. A., and Shevaldina, I. M., *Zh. Anal. Khim.*, 29, 1535, 1974.
50. Tsurumi, C., Furuya, K., and Kamata, H., *Nippon Kagaku Kaisi*, 1469, 1977.
51. Haraguchi, K. and Ito, S., *Bunseki Kagaku*, 24, 405, 1975.
52. Kai, J., *Anal. Chim. Acta*, 44, 129, 242, 1969.
53. Patrovsky, V., *Talanta*, 12, 971, 1965.
54. Veda, K., Sakamoto, S., and Yamamoto, Y., *Nippon Kagaku Kaishi*, 1111, 1981.
55. Kai, F., Sadakane, Y., Yokoi, H., and Aburada, H., *J. Inorg. Nucl. Chem.*, 35, 2128, 1973.
56. Tsurumi, C. and Furuya, K., *Bunseki Kagaku*, 24, 566, 1975.
57. Minczewski, J. and Kasiura, K., *Chem. Anal. (Warsaw)*, 10, 719, 1965; *Chem. Abstr.*, 64, 5726b, 1966; *Anal. Abstr.*, 14, 541, 1967.
58. Tsurumi, C., Mitsuhashi, H., Furuya, K., and Fuzimura, K., *Buseki Kagaku*, 23, 143, 1974.
59. Ishii, H. and Watanabe, H., *Bunseki Kagaku*, 26, 86, 1977.
60. Tsurumi, C. and Furuya, K., *Bunseki Kagaku*, 26, 149, 1977.
61. Sorensen, E., *Acta Chem. Scand.*, 14, 965, 1960.
62. Kasiura, K. and Minczewski, J., *Mukleonika*, 11, 399, 1966; *Chem. Abstr.*, 65, 19308d, 1966.
63. Tsurumi, C., *Bunseki Kagaku*, 26, 260, 1977.
64. Frei, R. W. and Miketukova, V., *Mikrochim. Acta*, 290, 1971.
65. Wada, H. and Nakagawa, G., *Bunseki Kagaku*, 14, 28, 1965.
66. Chang, C. H., Yin, C. C., and Yu, J. C., *Hua Hsueh Tung Pao*, 34, 1966; *Chem. Abstr.*, 66, 8087j, 1967.
67. Yamada, H., Maeda, T., and Kohima, I., *Anal. Chim. Acta*, 72, 426, 1974.
68. Yamada, H., Kojima, I., and Tanaka, M., *Anal. Chim. Acta*, 52, 35, 1970.
69. Kojima, I., *Anal. Chim. Acta*, 57, 460, 1971.
70. Douglas, W. R., *Anal. Chim. Acta*, 33, 567, 1965.
71. Kaneniwa, E., *Kanazawa Daigaku Yakugakubu Kenkyu Nempo*, 9, 27, 1959; *Chem. Abstr.*, 54, 4541g, 1960.
72. Kasuya, H. and Watanuki, K., *Bunseki Kagaku*, 19, 845, 1970.
73. Boni, I. E. and Hemmeler, A., *Chimica (Milan)*, 34, 443, 445, 1958; *Chem. Abstr.*, 53, 11114f, 1959.
74. Havir, J. and Vrestal, J., *Chem. Listy*, 60, 64, 1966; *Anal. Abstr.*, 14, 2433, 1967.
75. Jensen, B. S., *Acta Chem. Scand.*, 14, 927, 1960.
76. Sommer, L. and Janoscova, J., *Collect. Czech. Chem. Commun.*, 39, 101, 1974.
77. Ciba, J., Langova, M., and Kubickrova, L., *Collect. Czech. Chem. Commun.*, 38, 3405, 1973.
78. Nakagawa, G. and Wada, H., *Nippon Kagaku Zasshi*, 85, 202, 1964.
79. Stanley, R. W. and Cheney, G. E., *Talanta*, 13, 1619, 1966.
80. Nickless, G., Pollard, F. H., and Samuelson, T. J., *Anal. Chim. Acta*, 39, 37, 1967.
81. Navratil, O., *Collect. Czech. Chem. Commun.*, 29, 2490, 1964.
82. Kawase, A., *Bunseki Kagaku*, 11, 621, 628, 1962; *Bunseki Kagaku*, 12, 810, 817, 904, 1963; *Bunseki Kagaku*, 13, 553, 1964.
83. Tsurumi, T., Furuya, K., and Kamada, H., *Bunseki Kagaku*, 28, 754, 1979.

EDTA AND OTHER COMPLEXANES

Synonyms

Reagents covered in this section are listed in Table 1, together with their synonyms.

Source and Method of Synthesis

All commercially available and prepared by the carboxymethylation of corresponding polyamine, either by the alkaline condensation of haloacetic acid[M7,R1,1] or by the modified Strecker reaction with KCN and formaldehyde.[2]

Analytical Uses

EDTA is almost exclusively used as a titrant in the chelatometric titration. EDTA and other complexanes are also used as masking agents in various fields of analytical chemistry.

Properties of Reagents

Free acid (H_4L) of EDTA is a white crystalline powder and is almost insoluble in water (approximately 0.09 g/ℓ at pH 1.6)[3] and in common organic solvents. It is soluble in mineral acids or in aqueous alkalis. The samples obtained as an analytical reagent "EDTA" are usually a dihydrate, disodium salt of EDTA ($Na_2H_2L \cdot 2H_2O$), which can readily be purified by the recrystallization from water (11.1 g/100 g at 20°, 27.0 g/100 g at 98°)[4] and becomes 99.5% pure after drying at 80°. Drying at 100° or above results in the partial loss of hydrated water.

The complexane-type reagents, in a form of free acid, are white crystalline powder and very slightly soluble in water and common organic solvents and become more soluble when the alkalinity of aqueous solution is increased. Due to the amphoteric nature of the reagents, they are also soluble in mineral acid. The stepwise acid dissociation constants of the reagents are summarized in Table 2. Like other amino acids, the complexanes exist as zwitterion.

Complexation Reactions and Properties of Complexes

The complexane-type reagents form fairly stable and soluble chelates with a wide range of metal ions, with the exception of alkali and some other univalent metal ions. Another feature is that the chelating ability is strongly pH dependent. The metal to ligand ratio is 1:1 with the exception of TTHA and NTA. These become the basis for the use of complexanes, especially EDTA, as the titrant in the chelatometric titration, and as the masking agent in various fields of analytical chemistry. The structure of metal-EDTA chelate may be shown as below for the metal ions of coordination number of 4 (a) and (b), respectively.[5]

(a)　　　　(b)

The chelate stability constants of these reagents are summarized in Table 2.

EDTA is the most widely used chelating agent because of the relatively lower price of the reagent, as it is being manufactured in a large scale for the industrial uses.

Table 1
EDTA AND OTHER COMPLEXANES OF ANALYTICAL IMPORTANCE

Reagent	Complexane[a]	Chemical formula	Synonyms	Free acid	Molecular formula, mol wt	Remarks
(1)	EDTA	$HOOCCH_2$ $\diagdown$ NCH_2CH_2N $\diagup$ CH_2COOH / $HOOCCH_2$ … CH_2COOH	Ethylenediamine-N,N,N',N'-tetraacetic acid, ethylenedinitrilo tetraacetic acid, Komplexon II, Sequestrene II, Trilon B	H_4L	$C_{10}H_{16}N_2O_8$, 292.25	Decomposes at 240°
(2)	CyDTA	CH_2COOH / N $\diagdown$ CH_2COOH on cyclohexane ring (H) N $\diagdown$ CH_2COOH / CH_2COOH	Trans-1,2-cyclohexylene dinitrilo tetraacetic acid, 1,2-cyclohexanediamine-N,N,N',N'-tetraacetic acid, CDTA, DCTA	$H_4L \cdot H_2O$	$C_{14}H_{22}N_2O_8 \cdot H_2O$, 364.35	—
(3)	GEDTA	$HOOCH_2C$ $\diagdown$ $NCH_2CH_2OCH_2CH_2OCH_2CH_2N$ $\diagup$ CH_2COOH / $HOOCH_2C$ … CH_2COOH	Ethyleneglycol bis(2-aminoethylether)-N,N,N',N'-tetraacetic acid, glycoletherdiamine tetraacetic acid, GIEDTA, EGTA	H_4L	$C_{14}H_{24}N_2O_{10}$, 380.35	Slightly more soluble in organic solvent(DMF) than EDTA
(4)	DTPA	$HOOCH_2C$ $\diagdown$ $NCH_2CH_2NCH_2CH_2N$ $\diagup$ CH_2COOH / $HOOCH_2C$ — $HOOCH_2C$ — CH_2COOH	Diethylene triamine-N,N,N',N'',N''-pentaacetic acid, DETPA	H_5L	$C_{14}H_{23}N_3O_{10}$, 393.35	—
(5)	TTHA	$HOOCH_2C$ $\diagdown$ $NCH_2CH_2NCH_2CH_2NCH_2CH_2N$ $\diagup$ CH_2COOH / $HOOCH_2C$ — $HOOCH_2C$ — CH_2COOH … CH_2COOH	Triethylene tetramine-N,N,N',N',N'',N''-hexaacetic acid,	H_6L	$C_{18}H_{30}N_4O_{12}$, 494.49	Decomposes at 226°
(6)	HEDTA	$HOCH_2CH_2$ $\diagdown$ NCH_2CH_2N $\diagup$ CH_2COOH / $HOOCCH_2$ … CH_2COOH	N-Hydroxy ethylethylene-diamine-N,N',N'-triacetic acid, EDTA-OH	H_3L	$C_{10}H_{18}N_2O_7$, 278.26	Decomposes at 226°
(7)	NTA	CH_2COOH / N—CH_2COOH \ CH_2COOH	Nitrilotriacetic acid, ammonia-triacetic acid, Komplexon I, Trilon A	H_3L	$C_6H_9O_6N$, 191.11	Decomposes at 247°; solubility in water (0.13 g/100 g, 5°)

[a] "Complexane" is a recommended name by IUPAC Division of Analytical Chemistry for EDTA and other aminopolycarboxylic acids of related structure.

Table 2
ACID DISASSOCIATION CONSTANTS AND CHELATE STABILITY CONSTANTS OF SELECTED COMPLEXANES[6-8]

	EDTA (1)[a]	CyDTA (2)	GEDTA (3)	DTPA (4)	TTHA (5)[b]	HEDTA (6)	NTA (7)	Ref.
pKa_1[c]	1.99	2.43	2.00	2.08	2.42	2.51	1.89	
pKa_2	2.67	3.52	2.65	2.41	2.95	5.31	2.49	
pKa_3	6.16	6.12	8.85	4.26	4.16	9.86	9.73	
pKa_4	10.26	11.70	9.46	8.60	6.16	—	—	
pKa_5	—	—	—	10.55	9.40	—	—	
pKa_6	—	—	—	10.19				

				log K_{ML}[d]				
Ag	7.32	8.15	6.88	8.70	8.67 (13.89)	6.71	5.16	
Al	16.13	18.63	13.90	18.4	19.7 (28.9)	12.43	9.5	
Ba	7.76	8.64	8.41	8.63	8.22 (11.63)	5.54	4.83	
Bi	27.9	31.2	23.8	29.7	—	21.8	—	
Ca	10.96	12.50	11.00	10.74	10.06 (14.16)	8.14	6.41	
Cd	16.46	19.23	16.70	19.31	18.65 (26.85)	13.6	9.54	
Ce(III)	15.98	16.76	15.70	20.50	—	14.11	10.83	
Co(II)	16.31	18.92	12.50	18.4	17.1 (28.8)	14.4	10.38	
Cr(III)	23.40	—	2.54	—	—	—	>10	
Cu(II)	18.80	21.30	17.8	21.53	19.2 (32.6)	17.55	12.96	
Dy	18.30	19.69	17.42	22.82	23.29	15.30	11.74	9
Er	18.38	20.20	17.40	22.74	23.19 (26.92)	15.42	12.03	
Eu	17.35	18.77	17.10	22.30	23.85	15.35	11.52	9
Fe(II)	14.33	16.27	11.92	16.50	17.1	12.2	8.84	
Fe(III)	25.1	28.05	20.5	28.6	26.8 (40.5)	19.8	15.87	
Ga	20.27	22.91	—	23.0	—	16.9	13.6	
Gd	17.0	18.80	16.94	22.46	23.83	15.22	11.54	9
Hf	29.5	—	—	35.40	19.08	—	20.34	
Hg(II)	21.8	24.30	23.12	27.0	26.8 (39.1)	20.1	14.6	
Ho	18.05	19.89	17.38	22.78	23.58	15.32	11.90	9
In	24.95	28.74	—	29.0	—	17.16	16.9	
La	15.50	16.75	15.79	19.48	22.22 (25.62)	13.46	10.36	
Li	2.79	4.13	1.17	—	—	—	2.51	
Lu	19.65	20.91	17.81	22.44	—	15.88	12.49	
Mg	8.69	10.32	5.21	9.3	8.10 (14.38)	7.0	5.46	
Mn(II)	14.04	16.78	12.3	15.50	14.65 (21.19)	10.7	7.44	
Na	1.66	2.70	1.38	—	—	—	2.15	
Nd	16.61	17.69	16.28	21.60	22.82 (26.75)	14.86	11.26	
Ni	18.62	19.4	13.55	20.32	18.1 (32.4)	17.0	11.54	
Pb	18.04	19.68	14.71	18.80	17.1 (28.1)	15.5	11.39	
Pr	16.40	17.23	16.05	21.07	23.45	14.61	11.07	9
Sc	23.1	25.4	—	—	—	—	12.7	
Sm	16.7	18.63	16.88	22.34	23.81	15.28	11.53	9
Sn(II)	18.3	—	23.85	—	—	—	—	10
Sr	8.63	10.54	8.50	9.68	9.26 (12.70)	6.92	4.98	
Tb	17.81	19.30	17.27	22.71	23.61	15.32	11.59	9
Th	23.2	29.25	—	28.78	31.9	18.5	12.4	
TiO^{2+}	17.3	19.9	—	—	—	—	12.3	
Tl(III)	22.5	38.3	—	48.0	—	—	18	
Tm	18.62	20.46	17.48	22.72	—	15.59	12.2	
UO$_2^{2+}$	25.6	26.9[e]	—	—	—	—	9.56	3
VO^{2+}	18.77	19.40	—	—	—	—	—	
Y	18.09	19.41	16.82	22.05	—	14.65	11.48	
Yb	18.88	20.80	17.78	22.62	23.58	15.88	12.40	9

Table 2 (continued)
ACID DISASSOCIATION CONSTANTS AND CHELATE STABILITY CONSTANTS OF SELECTED COMPLEXANES[6-8]

	EDTA (1)[a]	CyDTA (2)	GEDTA (3)	DTPA (4)	TTHA (5)[b]	HEDTA (6)	NTA (7)	Ref.
Zn	16.50	18.67	14.5	18.75	16.65 (28.7)	14.5	10.67	
Zr	29.9	20.74	—	36.9	19.74	—	20.8	

[a] Values for other metals; log K_{ML} for Be(9.27), Cs(0.15), K(0.96), Nb(40.78), Pd(II) (18.5) and Sb(III) (24.8).

[b] Values in parentheses indicate log β_{M_2L}

[c] pKa values were observed in a medium of $\mu = 0.1$ (KCl or KNO₃) at 20°, with an exception for TTHA(at 25°).

[d] log K_{ML} values were observed, in most cases, in a medium of $\mu = 0.1$(KCl, KNO₃ or NaClO₄) at 20 ∿ 25°.

[e] Values for U(IV)

Complexanes other than EDTA and NTA are more expensive, and their uses are limited only in analytical chemistry and other research purposes.

CyDTA forms more stable metal chelates than does EDTA, hence CyDTA has been recommended as a masking agent where the metal ion cannot be masked effectively with EDTA.[11]

GEDTA is known to be more selective for Ca than Mg (Δ log $K_{Ca-Mg} = 5.79$) and for Cd than Zn (Δ log $K_{Cd-Zn} = 2.2$) in comparison with the chelating behavior of EDTA (Δ log $K_{Ca-Mg} = 1.9$, Δ log $K_{Cd-Zn} \simeq 0$). Accordingly, GEDTA has been used as a selective titrant for Ca[12] and Cd.[13]

DTPA and TTHA behave as octadentate and decadentate ligands, forming more stable chelates than does EDTA, with the metal ions of coordination number 8 or higher, and the resulting chelates are more resistant to hydrolysis. In the case of TTHA, it also forms M₂L-type chelates with Cu(II), Sn(II) (pH 3 to 4), Al, Cd, Co(II), Ga, Hg(II), Ni, Pb, Ti(with H₂O₂), Zn (pH 5 to 6), Mg, and Zn (pH 9 to 10).[14,15,18]

In constrast to the Fe(III)-EDTA chelate, Fe(III)-HEDTA chelate is resistant to hydrolysis in a caustic alkali solution, so that it is recommended as a masking agent for Fe(III) in a strongly alkaine solution.

NTA behaves as a quadridentate ligand, forming less stable chelates than EDTA does. In the presence of excess NTA, it forms ML-type as well as ML₂-type chelates. NTA is used as a masking agent where the higher masking effect with EDTA or CyDTA is not desirable.

Purification and Purity of Reagents

Free acid of complexanes can be purified by dissolving it in an aqueous sodium hydroxide solution, followed by the precipitation of the free acid with an addition of dilute sulfuric acid at pH 1 to 2. The white precipitate was filtered and washed with cold water until the filtrate becomes free from sulfate.

Disodium EDTA (Na₂H₂L·2H₂O) may be purified by recrystallization from hot water. The following procedures is also recommended to obtain the pure sample.[16]

Dissolve 10 g of crude sample into 100 mℓ of water and to this solution add ethanol slowly until it becomes slightly turbid. After filtering the solution, add an equal volume of ethanol to the clear filtrate to precipitate the pure salt. Filter the salt, wash with acetone, then with ether. After drying the salt in the air overnight, dry it at 80° for 4 hr.

The purity of complexanes can generally be determined by titrating the standard Zn (Xylenol Orange at pH 5) or Cu(II) (PAN or PAR at pH 4 to 5) solution with the solution of complexane of known concentration.

Analytical Applications

Use as a Titrant in Chelatometry

The stoichiometry in the chelatometry may be written as below:

$$M^{n+} + H_2 L^{2-} \rightleftharpoons ML^{n-4} + 2H^+$$

EDTA is the most widely used complexane in the chelatometry, and a standard solution of 0.01 M EDTA (3.72 g $Na_2H_2L \cdot 2H_2O/\ell$) is recommended for the general purposes. Detailed procedures for the determiation of each element can be found in various monographs on chelatometry.[M1,M2] However, Figure 1 may be useful in finding out the suitable condition for the titration of the respective metal ion.[17] The proper selection of the pH range for the titration and of the metal indicator is essential to get the successful result. Inorganic anions and some organic compounds can also be determined indirectly by the chelatometry.[R2]

Use as a Masking Agent

The range of applications of complexanes as masking agents in analytical reactions is so broad that it may not be appropriate to discuss each example in details. However, Table 3 summarizes some examples of the uses of EDTA as a masking agent. The differences in the masking properties between EDTA and CyDTA against the complexation reaction with cyanide are shown in Table 4 as another example. The detailed treatise on the use of complexanes as a masking agent can be found in some other monographs.[M3,M5]

MONOGRAPHS

M1. Crook, L. R. and Yardley, J. T., Sequestric acid and sequestric acid disodium salt, in *Organic Reagents for Metals*, Vol. 1, Johonson, W. C., Eds., Hopkin & Williams, Chadwell Heath, England, 1955, 157.

M2. Schwarzenbach, G. and Flaschka, H., *Die Komplexometrische Titration*, 2nd ed., Ferdinand Enke Verlag, Stuttgart, 1965.

M3. Perrin, D. D., *Masking and Demasking of Chemical Reactions*, John Wiley, & Sons, New York, 1970.

M4. Ueno, K., *Chelatometric Titration*, 2nd ed., Nankodo, Tokyo, 1972.

M5. Pribil, R., *Analytical Applications of EDTA and Related Compounds*, Pergamon Press, Oxford, 1972.

M6. Bermejo-Martinez, F., "EDTA and aminopoly-carboxylic acids as chromogenic agents, in *Chelates in Analytical Chemistry*, Vol. 5, Flaschka, H. A. and Barnard, A. J., Jr., Eds., Marcel Dekker, New York, 1976, 1.

M7. Ueno, K., Murase, I., Imamura, T., and Ando, T., *Chemistry of Complexane, EDTA*, Nankodo, Tokyo, 1977.

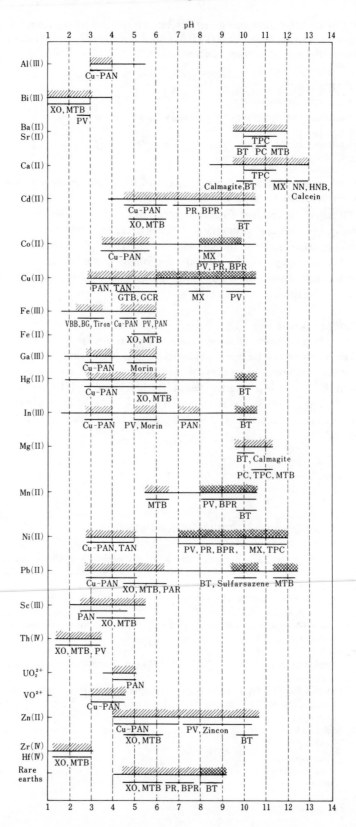

FIGURE 1. The pH range and metal indicator recommended for the titration of common metal ions. ⧼, pH range where metal ion can be titrated; ⧻, pH range where suitable metal indicator is available; −, pH range where auxiliary complexing agent must be used to prevent the hydrolysis of metal ions.

Abbreviations for the Metal Indicator

BG:	Bindschedlar's Green	PAN:	Pyridylazonaphthol
BPR:	Bromopyrogallol Red	PAR:	Pyridylazorescorcinol
BT:	Eriochrome® Black T	PC:	Phthalein Complexone
Cu-PAN:	Cu-EDTA + PAN mixture	PR:	Pyrogallol Red
GCR:	Glycine cresol Red	PV:	Pyrochatecol Violet
GTB:	Glycinethymol Blue	TAN:	Thiazolylazonaphthol
HNB:	Hydroxynaphthol Blue	TAR:	Thiazolylazoresorcinol
MTB:	Methylthymol Blue	TPC:	Thymolphthalein Complexone
MX:	Murexide	VBB:	Variamine Blue B
NN:	Calcon carboxylic acid (Patton and Reeder's dye)	XO:	Xylenol Orange

(Reproduced with slight modifications from Ueno, K., *J. Chem. Educ.,* 12, 432, 1965. With permission.)

Table 3
SOME EXAMPLES OF THE USE OF EDTA AS A MASKING AGENT[M3]

Reaction masked	Application
Fe,Co, and Ni with sodium diethyl-dithiocarbamate	Extraction photometry of Cu and Bi
Fe(III), V(V) with Tiron	Photometric determination of Ti
Fe,Co,Cu,Ni, and Cr with 2-nitroso-1-naphthol	Photometric determination of Pd
Al,Cd,Cr,Cu,Ga,Hg,In,Pb,V,Zr, and rare earths with PAR	Photometric determination of Ni (pH 8.6 ∿ 10)
Bi,Cd, and Cr with Beryllon IV	Photometric determination of Be(pH 7 ∿ 8)
Fe(II),Ce,Co,La,Mg,Mo,Nd, and Pb with Xylenol Orange	Photometric determination of Al(pH 3 ∿ 3.8)
Mo(V) with 8-hydroxyquinoline	Extraction of U, W(VI) (pH 2∿3)

Table 4
REACTION OF KCN WITH EDTA AND CyDTA-METAL CHELATES[11]

Metal ion	Reactivity of KCN with		Metal ion	Reactivity of KCN with	
	EDTA chelate	CyDTA chelate		EDTA chelate	CyDTA chelate
Hg(II)	Reacts instantaneously	Reacts instantaneously	Co(II)	Reacts	Reacts slowly
Cu(II)	Reacts instantaneously[a]	Reacts instanteously[a]	Mn(II)	No reaction	No reaction
Pb	No reaction	No reaction	Co(III)	No reaction	No reaction
Bi	No reaction	No reaction	Zn	Reacts quickly	Reacts fairly quickly
Cd	Reacts instantaneously	Almost no reaction	Fe(II)	Reacts	Reacts slowly
Ni	Reacts	No reaction	Fe(III)	Reacts	No reaction

[a] In the presence of $NH_2OH \cdot HCl$ or H_2O_2

REVIEWS

R1. Ueno, K., Recent development on the synthesis of complexane type chelating agents, *Yuki Gosei Kagaku Kyokai Shi.*, 23, 127, 1965.

R2. Yurist, I. M. and Talmud, M. M., Complexometric determination of inorganic anions, *Zh. Anal. Khim.*, 31, 1984, 1976; *J. Anal. Chem. USSR*, 31, 1441, 1976.

R3. Ueno, K., Absorption spectral characteristics of complexane and its application to absorption photometry, *Kagaku (Kyoto)*, 15, (10), 1, No. 12, 55 (1960).

REFERENCES

1. I. G. Farben Industrie, German Patent, 638071, 1936.
2. Smith, R., Bullock, J. L., Bersworth, F. C., and Martell, A. E., *J. Org. Chem.*, 14, 355, 1949.
3. Klygin, A. E., Smirnova, I. D., Nikolskaya, N. A., *Zh. Neorg. Khim.*, 4, 2766, 1959; *Chem. Abstr.*, 54, 20609i, 1960.
4. Takei, S., *Bunseki Kagaku*, 22, 137, 1973.
5. Martell, A. E., *J. Chem. Ed.*, 29, 270, 1952.
6. Sillen, L. G. and Martell, A. E., *Stability Constants of Metal Complexes*, and supplement, The Chemical Society, London, 1964, 1971.
7. Moeller, T., *The Chemistry of the Lanthanides*, Reinhold, London, 1963.
8. Harju, L. and Ringdom, A., *Anal. Chim. Acta*, 49, 205, 221, 1970; *Anal. Chem Acta*, 50, 475, 1970.
9. Masuda, Y., Nakamori, T., and Sekido, E., *Nippon Kagaku Kaishi*, 190, 199, 204, 1978.
10. Battari, E. and Rufolo, A., *Ric. Sci.*, 38, 735, 1968; *Chem. Abstr.*, 71, 117044z, 1969.
11. Pribil, R., *Collect. Czech. Chem. Commun.*, 20, 162, 1955.
12. Sadek, F. S., Schmid, R. W., and Reilley, C. N., *Talanta*, 2, 38, 1959.
13. Pribil, R. and Veseley, V., *Chemist-Analyst*, 55, 4, 1966.
14. Pribil, R. and Vladimir, V., *Talanta*, 9, 939, 1962.
15. Itoh, A. and Ueno, K., *Bunseki Kagaku*, 17, 327, 1968.
16. Blaedel, W. J. and Knight, H. T., *Anal. Chem.*, 26, 741, 1954.
17. Ueno, K., *J. Chem. Educ.*, 42, 432, 1965.
18. Moya and Cheng, K. L., *Anal. Chem.*, 19.

PHTHALEIN COMPLEXONE (PC) AND THYMOLPHTHALEIN COMPLEXONE (TPC)

H_6L

Synonyms

PC: 3,3'-*Bis*[N,N-di(carboxymethyl)-aminomethyl]-*o*-cresolphthalein, *o*-Cresolphthalein-3,3'-bismethyliminodiacetic acid, Metal Phthalein (R_1 = CH_3, R_2 = H, $C_{32}H_{32}N_{32}N_2O_{12} \cdot H_2O$, mol wt 654.63)

TPC: (R_1 = CH $(CH_3)_2$, R_2 = CH_3, $C_{38}H_{44}N_2O_{12}$, mol wt 720.77) 3,3'-*Bis*[N,N-di(carboxymethyl)-aminomethyl]-thymolphthalein, Thymolphthalexone

Source and Method of Synthesis

Commercially available. PC and TPC can be prepared by the Mannich condensation of formaldehyde and iminodiacetic acid with *o*-cresolphthalein and thymolphthalein, respectively.[1]

Analytical Uses

PC was first introduced as a metal indicator in the chelatometry of alkaline earth metal ions, but is now also being widely used as a photometric reagent for Ca. TPC has been used in a similar manner.

Properties of Reagents

PC[2]

Free acid ($H_6L \cdot H_2O$) is a colorless or pale yellow crystalline powder, mp 186°, is insoluble in water, but easily soluble in aqueous ammonium acetate or alkali solution and in common organic solvents. The aqueous solution is unstable, and the reagent decomposes within a week.

Its sodium salt is also commercially available. It is easily soluble in water, but insoluble in organic solvents.

Aqueous solution of PC is colorless at pH <6, faint pink (fairly pink if PC is contaminated with *o*-cresolphthalein) at pH 7 to 10, and violet-red at pH >11; pKa_1 = 2.2, pKa_2 = 2.9 (−COOH); pKa_3 = 7.0, pKa_4 = 7.8 (−OH); pKa_5 = 11.4, and pKa_6 = 12.0 (N^+H) (μ = 0.1, 20°)[1] Absorption spectra of PC at various pH are illustrated in Figure 1.

TPC[2]

Usually supplied as a free acid which is a white crystalline powder. TPC shows similar solubility in water and organic solvents to that of PC; pKa_3 = 7.0, pKa_4 = 8.0, pKa_5 = 10.8, and Ka_6 = 13.0 (μ = 0.4, 25°).[4] The aqueous solution is colorless at pH <6, faint blue (or almost colorless on diluted solution) at pH 7 to 11.5, and deep blue at pH >12. Absorption spectra of TPC are illustrated in Figure 2.

Complexation Reactions and Properties of Complexes

Almost colorless species of PC in a weekly alkaline range forms violet-red soluble

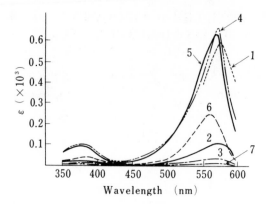

FIGURE 1. Absorption spectra of PC and its metal chelates. (1) ----- L^{6-}; (2) ——— HL^{5-}; (3) –·– H_2L^{4-}; (4) –·––– Ba_2L^{2-}; (5) ——— Ca_2L^{2-}; (6) ––– Mg_2L^{2-}; (7) –·–·– Zn_2L^{2-}. (From Anderegg, G., Flaschka, H., Sallmann, R., and Schwarzenbach, G., *Helv. Chim. Acta*, 37, 113, 1954. With permission.)

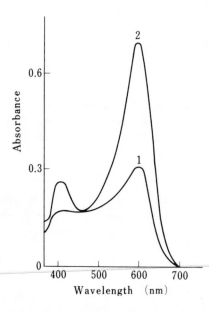

FIGURE 2. Absorption spectra of TPC and Ca-chelate at pH 10 to 12. (1) Dye, 8×10^{-5} *M*; (2) Ca-TPC chelate, Ca 2×10^{-5} *M*. (From Bezdekova, A., and Budesinsky, B., *Collect. Czech. Chem. Commun.*, 30, 818, 1965. With permission.)

chelates with alkaline earth metals, Cd, Mn(II), and Ni at pH 9. However, as the reagent itself is red at pH >11, the sharp color change upon complexation can only be observed at pH 10 to 11.

PC also forms stable soluble chelates with Co(II), Cu(II), Pb, and Zn, but they are all colorless.[2]

TPC behaves similarly, but the color change upon complexation is from colorless to deep blue.

Table 1

CHELATE STABILITY CONSTANT OF

PC[1] ($\mu = 0.1$, 20°)

Metal ion	PC chelate		
	log K_{ML}	log K_{M_2L}	log $K_{MHL}{}^a$
Ba	6.2	3.0	4.8
Ca	7.8	5.0	6.4
Mg	8.9	5.2	7.5
Zn	15.1	9.8	13.8

^{a} $K_{MHL} = [MHL] / [M] [HL].$

The compositions of the chelates were investigated in solution and were found to be ML^{4-}, MHL^{3-}, and M_2L^{2-}, depending upon the solution pH and reacting ratio of metal to ligand.[1] The chelate stability constants are available only for PC, as summarized in Table 1. Absorption spectra of some representative metal chelates of PC and TPC are illustrated in Figures 1 and 2, respectively.

Purification and Purity of Reagents

The reagents of lower grade often show fairly high absorption on the blank solution at pH $\simeq$ 10. This blank is due to the contamination with o-cresolphthalein or thymolphthalein which is one of the starting materials. Such contaminants can be removed by dissolving the crude reagent into an aqueous solution of threefold molar excess of sodium acetate, followed by the fractional precipitation of PC or TPC by adding hydrochloric acid in small portions to the clear filtrate. The fractions of pure material should give one spot on the paper chromatogram (eluted by ethanol-water-phenol (6:3:1) and developed by NaOH).[1]

Analytical Applications

Use as Metal Indicators

PC and TPC have been used as metal indicators in the chelatometry of alkaline earth metal ions (at pH 10 to 11 with NH_3). With the introduction of these indicators, it became first possible to titrate Ba and Sr with EDTA directly.[1] The color change at the end point is from red-violet to faint pink (PC) or from deep blue to almost colorless (TPC) at pH 11 (NH_3). The residual color of PC at the end point can be reduced to almost colorless by adding 0.5 to 1 volume of ethanol (or methanol) to the titrating solution. If the pH of solution is higher than 11, the reagent itself (HL^{5-} and L^{6-}) shows a substantial blank coloration, resulting in a poor color change.

TPC is also recommended as an indicator in the chelatometric titration of Ca and Mg (pH 10.5 to 11 with NH_3), where hydroxy-arylazo indicators such as Eriochrome® Black T or Patton-Reeder's dye do not function properly by the presence of traces of oxidizing species (Fe(III) or Mn(III)). TPC is more resistant to oxidation than the azo dye indicators, so that Ca and Mg can be titrated successfully with TPC, while Fe(II) and Mn(III) being masked with triethanolamine and KCN[5].

Other cations, such as Ba[2] Ca,[6] Mg,[7] Mn(II) (with reducing agent),[8] Sr,[2] (pH 10 with NH_3), and Cd (TPC only, pH 11 with NH_3)[9] can be titrated similarly.

The color change of Hg-PC chelate (from red-violet of HgL^{4-} to colorless of $HgL(CN)^{5-}$) can be used to detect the end point in the titration of CN^- with Hg-EDTA.[10]

A 0.1% methanol solution of PC is recommended as an indicator. In the case of TPC, a dry powder diluted with KNO_3 (1:100) is recommended.

Use as Photometric Reagents

PC and TPC are highly sensitive and selective photometric reagents for alkaline earth metals, as they form highly colored chelates (for PC, CaL^{4-}, pH 10.1 to 10.7, λ_{max}, 575 nm, $\varepsilon = 6.5 \times 10^4$, $\curvearrowright 1$ ppm) with alkaline earth metal ions only.[11] An aqueous solution containing 50 mg of PC, 21 mℓ of 37% HCl, 2.5 g of 8-hydroxyquinoline, and 1 mℓ of 30% Brij® 35 in 1000 mℓ is recommended as a photometric reagent for Ca in serum.[3]

Some other elements have been determined with PC: Ba (pH 11.3, 575 nm, $\curvearrowright 5$ ppm),[11] Hg (pH 9.6 to 10.3, 583 nm, $\varepsilon = 5.26 \times 10^4$, 0.1 to 4 ppm),[12] La (with cetyl-pyridinium bromide, 617 nm, $\curvearrowright 2$ ppm),[13] Mg (pH 10, 570 nm, 5 to 30 ppm),[14] and Sr(pH 11.2, 575 nm, $\varepsilon = 3.2 \times 10^4$, $\curvearrowright 3$ ppm).[11]

TPC was also investigated as a photometric reagent for Ca, Mg, Ba, Sr (pH 11.6, 600 nm, $\varepsilon = 4.06 \times 10^4$, 2.79×10^4, 4.25×10^4, and 3.45×10^4, respectively),[15,16] Ni (pH 7.7, 610 nm),[17] rare earths (pH 9.5 to 11.0, 592 to 594 nm, 0.02 to 0.2 ppm),[18] and V(V) (pH 2.5, 600 nm).[19]

MONOGRAPHS

M1. **Pribil, R.,** *Analytical Applications of EDTA and Related Compounds,* Pergamon Press, New York, 1972.

M2. **Ueno, K.,** *Chelatometric Titration,* Nankodo, Tokyo, 1972.

M3. **Ainsworth, L. R. and Yardley, J. T.,** Metalphthalein in *Organic Reagents for Metals,* Vol. 2, Johnson, W. C., Ed., Hopkin & Williams, Chadwell Heath, England, 1965, 237.

REFERENCES

1. **Anderegg, G., Flaschka, H., Sallmann, R., and Schwarzenbach, G.,** *Helv. Chim. Acta,* 37, 113, 1954.

2. **Korbl, J. and Pribil, R.,** *Chem. Ind.,* 233, 1957; *Collect. Czech. Chem. Commun.,* 23, 1213, 1958.

3. **Stavropoulos, W. S., Thiegs, B. J., and Mack, R. F.,** U.S. Patent 3938954; *Chem. Abstr.,* 84, 132321z, 1976.

4. **Hulanicki, A. and Glab, S.,** *Talanta,* 26, 423, 1979.

5. **Pribil, R., Korbl, J., Krisil, B., and Uobora, J.,** *Collect. Czech. Chem. Commun.,* 19, 58, 465, 1162, 1954; *Collect. Czech. Chem. Commun.,* 24, 1799, 1959.

6. **Pribil, R. and Adam, J.,** *Talanta,* 24, 177 (1977).

7. **Nina, K.,** Bull. Goven. For. Experi, Stat., Report (Tokyo), 125, (1960), 128 (1961).

8. **Belcher, R., Leonard, M. A., and West, T. S.,** *Chem. Ind.,* 128, 1958.

9. **Pribil, R.,** *Talanta,* 13, 1223, 1966.

10. **Nomura, T., Takeuchi, K., and Komatsu, S.,** *Nippon Kagaku Zasshi,* 89, 291, 1968.

11. **Uesugi, K., Tabushi, M., Murakami, T., and Shigematsu, T.,** *Bunseki Kagaku,* 13, 440, 1032, 1964.

12. **Komatsu, S. and Nomura, T.,** Nippon Kagaku Zasshi, (88, 542, 1967.

13. **Poluektov, N. S., Lauer, R. S., and Ovchar, L. A.,** *Zh. Anal. Khim.,* 27, 1956, 1972.

14. **Morimoto, Y., Suzuki, S., and Azami, J.,** *Bunseki Kagaku,* 10, 886, 1961.

15. **Bezdekova, A. and Budesinsky, B.,** *Collect. Czech. Chem. Commun.,* 30, 818, 1965.

16. **Hirata, H. and Arai, M.,** *Bunseki Kagaku,* 16, 820, 1967.

17. **Kotoucek, M. and Divisova, D.,** *Acta Univ. Palacki. Olomuc., Fac. Rerum. Nat.,* 49, 165, 1976; *Anal. Abstr.,* 34, 2B112, 1978.

18. **Prajsnar, D.,** *Chem. Anal. (Warsaw),* 7, 861, 1962; *Chem. Abstr.,* 58, 11a, 1963.

19. **Cherkesov, A. I., Smirnov, A. N., and Kazakov, B. I.,** *Zavod. Lab.,* 41, 933, 1975; *Anal. Abstr.,* 30, 5B198, 1976.

CALCEIN

$$C_{30}H_{26}O_{13}N_2$$
mol wt = 622.54

H₆L

Synonyms

Bis [N,N'-di(carboxymethyl)aminomethyl]fluorescein, 2,6-Dihydroxy-*bis*[N,N'-di(carboxymethyl)aminomethyl]fluoran, Fluorexone, Calcein W, Fluorescein Complexone

Source and Method of Synthesis

Commercially available as free acid or disodium salt and prepared by the Mannich condensation of fluorescein with formaldehyde and iminodiacetic acid.[1-3]

Analytical Uses

Metal fluorescent indicator in the chelatometric titration; fluorimetric reagent for the determination of trace of Ca.

Properties of Reagent

Free acid is an orange powder and melts at 300° with decomposition.[3] Disodium salt is a red-brown powder and melts at 185° with decomposition. The solid sample is quite stable in the dark place, but an aqueous solution of sodium salt is not so stable at room temperature. Free acid is slightly soluble in water, but easily soluble in aqueous alkali to give an orange-red solution.

The following proton dissociation constants were obtained by the poteniometric titration in 30% aqueous methanol at room temperature.[4]

$$H_6L \xrightleftharpoons{pKa_1 = 2.1} H_5L^- \xrightleftharpoons{pKa_2 = 2.9} H_4L^{2-} \xrightleftharpoons{pKa_3 = 4.2} H_3L^{3-} \xrightleftharpoons{pKa_4 = 5.5} H_2L^{4-} \xrightleftharpoons{pKa_5 = 10.8} HL^{5-} \xrightleftharpoons{pKa_6 = 11.7} L^{6-}$$

—COOH   —OH   ⟩N⁺H

The absorption spectra of Calcein in an aqueous solution at different pH are shown in Figure 1. A yellow-green fluorescence is also observed in the neutral solution as shown in Figure 2, but it weakens either in alkaline or acidic range as illustrated in Figure 3. The degree of quenching in a strongly alkaline solution is dependent upon the kind of alkali used as shown in Figure 4, KOH being most effective in quenching the fluorescence.[8] The residual fluorescence also depends upon the viscosity and temperature of the solution. For example, the addition of glycerol increases the residual fluorescence.[7]

Complexation Reactions and Properties of Complexes

When alkaline earth metal ion, especially Ca, is added to the strongly alkaline solution of Calcein, an intense yellow-green fluorescence appears which eventually disappears upon an addition of EDTA.

In acidic solution, the fluorescence appears with an addition of Mg, Al, or Zn, but is quenched with Cu(II), Ni, or Co(II).[7] At the same time, the absorption spectra of

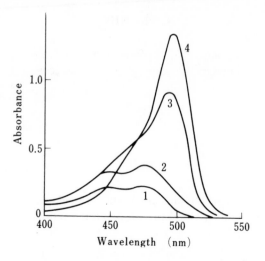

FIGURE 1.   Absorption spectra of Calcein at different
pH values; (1) pH 1.9; (2) pH 4.1; (3) pH 5.5; (4) pH
12; Calcein, 2×10^{-5} *M*. (From Miyahara, T., *Bunseki
Kagaku*, 26, 615, 1977. With permission.)

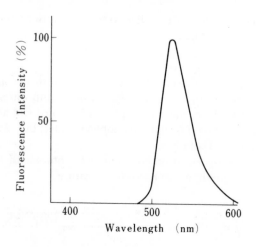

FIGURE 2.   Fluorescence spectrum of Calcein,
1×10^{-6} *M* aqueous solution. (From Iritani, N.,
Miyahara, T., and Takahashi, I., *Bunseki Ka-
gaku*, 17, 1075, 1968. With permission.)

Calcein changes with the complex formation. Figure 5 illustrates the absorption spectra
of Cu(II)-Calcein system.[5] The intensity and λ_{max} for the emmision spectra of some of
the metal Calcein chelates are summarized in Table 1.

Equilibrium constants for the following reactions were determined in aqueous solution at room temperature ($\mu = 0.1$).[5]

$$
\begin{array}{llll}
\text{Cu}^{2+} & + \text{H}_4\text{L}^{2-} & + \text{H}_3\text{L}^{3-} & \rightleftharpoons \text{Cu(H}_4\text{L)(H}_3\text{L)}^{3-} & \log K = 10.43 \\
\text{Cu}^{2+} & + \text{H}_2\text{L}^{4-} & & \rightleftharpoons \text{Cu(H}_2\text{L)}^{2-} & \log K = 8.27 \\
2\text{Cu}^{2+} & + \text{L}^{6-} & & \rightleftharpoons \text{Cu}_2\text{L}^{2-} & \log K = 28.97
\end{array}
$$

Approximate values for the following equilibrium were also determined by fluorimetry in the aqueous solution.[7]

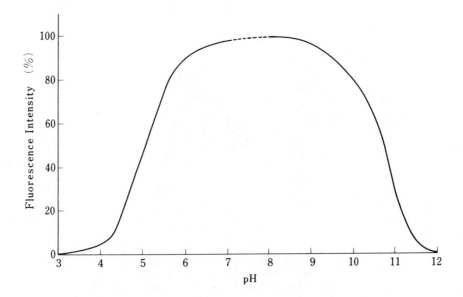

FIGURE 3. Variation of fluorescence intensity of calcein with pH. (Reprinted with permission from Wallach, D. F. H. and Steck, T. L., *Anal. Chem.*, 35, 1035, 1963. Copyright 1963 American Chemical Society.)

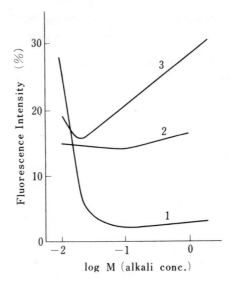

FIGURE 4. Dependencies of fluorescence intensity on alkali concentration. Calcein : 0.6 × 10^{-5} M with excess calcium. (1) KOH; (2) LiOH; (3) NaOH. (From Körbl, J., Vydra, F., Pribil, R., *Talanta*, 1, 281, 1958. With permission.)

$$M^{2+} + ML^{4-} \xrightleftharpoons{K} M_2 L^{2-}$$

log K for Ca, 6.63; Ba, 5.57; Sr, 6.86, and Mg, 7.90.

Purification and Purity of the Reagent

Free acid can be purified by recrystallization from 50% aqueous methanol.[4] Alternatively, it can be purified by reprecipitation as follows:

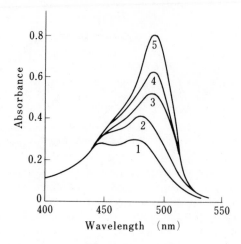

FIGURE 5.   Absorption spectra of Cu(II)-Cal-
cein and Calcein at pH 3.5. (1) Calcein 2×10^{-5}
M; (2) 1 + Cu(II) 10^{-5} M; (3) 1 + Cu(II) 2×10^{-5}
M; (4) 1 + Cu(II) 4×10^{-5} M; (5) 1 + Cu(II) $2 \times$
10^{-3} M. (From T. Miyahara, *Bunseki Kagaku,*
26, 615 (1977). With permission.)

Table 1
FLUORESCENCE CHARACTERISTICS OF
METAL-CALCEIN CHELATES[7]

Metal ion	Fluorescence intensity[a] (arbitrary units)	λ_{emis}(nm)
	At pH 7.4	
None	991	495
Al	730	470
Co(II)	2	495
Cu(II)	10	495
Mg	992	490
Mn(II)	80	495
Ni	4	495
Zn	988	487
	At pH 12	
None	10	500
Ba	735	495
Ca	756	495
Sr	756	495

[a]   Excitation at 495 nm.

Dissolve a 300-mg sample into a minimum amount of 0.1 N NaOH, add
50 mℓ of 10 to 20% aqueous methanol, and filter. To the filtrate, add 1 N
HC1 to adjust pH to 2.5. After keeping it in a refrigerator overnight, filter
the precipitate on a Number 4 glass filter and wash well with methanol.[3,4]

The commercial products are considered to be a mixture of various isomers because
the di(carboxymethyl)aminomethyl groups may be introduced on the any two of 2,4,5,
and 7 position of fluorescein (1).[3,4,7]

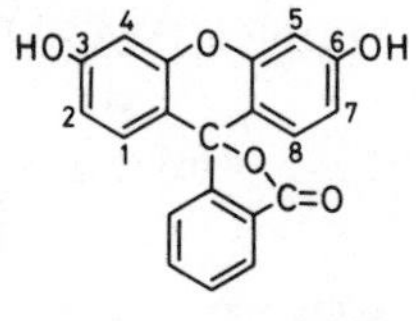

The separation of these isomers is not so easy, and the isomeric mixtures are still satisfactory for the practical uses. The only interfering contaminant in the commercial samples may be fluorescein which can be determined by the following procedure.[9]

> To 0.1 mℓ of 0.1% aqueous Calcein solution, add 10 mℓ of 1 N KOH solution and 1 mℓ of 0.01 M EDTA solution, then dilute it with water to 100 mℓ. Intensity of green fluorescence, if any, should not exceed that of the equal volume of the reference solution containing 6 μg of fluorescein sodium.

The free acid can be assayed by the potentiomatric titration. Two clear inflections are observed at the second and fourth equivalence points on the titration of free acid (H_6L) with 0.1 N KOH in 30% aqueous methanol.[4]

Analytical Applications
Use as a Metallofluorescent Indicator
Calcein is used as a metallofluorescent indicator in the chelatometric titration of Ca (especially in the presence of phosphate).[10-12] A yellow-green fluorescence disappears at the end point of the direct titration. For the metal indicator use, it is a common practice to dilute the solid Calcein with solid K_2SO_4 (1:100 or 1:200) to make a finely powdered mixture. This preparation can be stored at least for a year when protected from light.

The use of Ca-Calcein complex, instead of free Calcein, was proposed to improve the stability of the aqueous solution. The concentration of Ca in the indicator solution is so small that it does not cause any appreciable error in the calcium titration.[13] The combined use of various screening dyes were also proposed to improve the end point color change. Dyes recommended are phenolphthalein,[14,15] acrydine,[16] thymolphthalein,[17,18] murexide,[19] phenolphthalein complexone,[20] and thymolphthalein complexone.[21]

Use as a Fluorimetric Reagent for Nontransition Metals
When nontransition metal ions are complexed with Calcein in the pH region where the fluorescence of free calcein is quenched, a yellow-green fluorescence is observed. The reaction is so sensitive that as low as 0.08 ppm of Ca, for example, can be determined.[22]

An aqueous solution of the sodium salt (1 to 0.5%) is used for this purpose. It is stable for at least a month when kept at 4° in the dark.[23] The solution should be kept in a plastic bottle to avoid the metal contaminations.

MONOGRAPH

M1. Diehl, H., Calcein, Calmagite® and *o,o′*-dihydroxyazobenzen, G. F. Smith Chemical Company, Colombus, Ohio, 1963.

REFERENCES

1. Diehl, H. and Ellingboe, J., *Anal. Chem.,* 28, 882, 1956.
2. Körbl, J., Vydra, F., and Pribil, R., *Talanta,* 1, 138, 1958.
3. Wallach, D. F. H., Surgenor, D. M., Soderberg, J., and Delano, E., *Anal. Chem.,* 31, 456, 1959.
4. Iritani, N. and Miyahara, T., *Bunseki Kagaku,* 22, 174, 1973.
5. Miyahara, T., *Bunseki Kagaku,* 26, 615, 1977.
6. Iritani, N., Miyahara, T., and Takahashi, I., *Bunseki Kagaku,* 17, 1075, 1968.
7. Wallach, D. F. H. and Steck, T. L., *Anal. Chem.,* 35, 1035, 1963.
8. Korbl, J., Vydra, F., and Pribil, R., *Talanta,* 1, 281, 1958.
9. Purity Specification of Organic Reagents, Dojindo Laboratories Ltd. Kumamoto, Japan.
10. Yalman, R. G. and Bruegemann, W., *Anal. Chem.,* 31, 1230, 1959.
11. Austin, J. H. and Klett, C. A., *Chemist-Analyst,* 55, 11, 1966.
12. Kratochvil, B. and Jeremy, P. G., *Talanta,* 24, 126, 1977.
13. Hoyle, W. C. and Diehl, H., *Talanta,* 19, 206, 1972.
14. Svoboda, V., Chromy, V., Körbl, J., and Dorazil, L., *Talanta,* 8, 249, 1961.
15. Sedlacek, B. and Dusek, P., *Z. Libensm. Untersuch. Forsch.,* 129, 333, 1966; *Chem. Abstr.,* 65, 14333e, 1966.
16. Kirkbright, G. F. and Stephen, W. I., *Anal. Chim. Acta,* 27, 294, 1962.
17. Roche, M., *Sucr. Fr.,* 108, 2, 1967; *Chem. Abstr.,* 67, 7687w, 1967.
18. Tucker, B. M., *Analyst,* 82, 284, 1957.
19. Baumann, E. W., *Arkiv. Hig. Rada Toksikol.,* 18, 155, 1967; *Chem. Abstr.,* 69, 80, 12a, 1968.
20. Svoboda, V., Chromy, V., Körbl, J., and Dorazil, L., *Talanta,* 8, 249, 1961.
21. Tulyupa, F. M. and Kurskaya, K. V., *Novye Metody Analiza na Met. i Metalloobrabatyvayushchikh Zavodokh, Sov. Ner. Khoz.Pridneprovsk.Ekon.Admin.Raiona,* 108, 1964; *Chem. Abstr.,* 62, 8388g, 1965.
22. Körbl, J. and Vydra, F., *Chem.Listy.,* 51, 1457, 1957; *Collect. Czech. Chem. Commun.,* 23, 622, 1958.
23. Olsen, R. L., Diehl, H. D., Collins, P. F., and Ellestad, R. B., *Talanta,* 7, 187, 1961.

XYLENOL ORANGE AND METHYL THYMOL BLUE

Synonyms

The reagents covered in this section are listed in Table 1, together with their synonyms.

Source and Methods of Synthesis

Commercially available. Xylenol Orange is prepared by the Mannich condensation of iminodiacetic acid, and Cresol Red with formaldehyde.[1] Methyl Thymol Blue is prepared from Thymol Blue in a similar fashion.[2]

Analytical Uses

As metal indicators for Bi, Cd, Hg, La, Mn(II), Pb, Sc, Th, Zn, Zr, and rare earths in the chelatometric titration. Also as photometric reagents for various metal ions, especially for Bi, Fe(III), Hf, Nb, Th, and Zr.

Properties of Reagents

Xylenol Orange(XO)

Commercial samples are di- or trisodium salt, dark red hygroscopic crystalline powder. Free acid can be obtained as dihydrate (red) which melts at 286° (with decomposition).[3] Sodium salt is very easily soluble in water and slightly soluble in alcohol and other organic solvents. Commercial samples are usually a mixture of XO and semi-XO in varying ratio (up to 17% of semi-XO).[3,5]

Acid dissociation constants of XO reported by various workers are not in good agreement, probably because the impure nature of samples investigated. The values obtained on the purified samples are shown in Table 2.

The aqueous solution of XO is yellow at pH <6 and red-violet at pH >6. The absorption spectra of XO in aqueous solution in this pH range are illustrated in Figure 1 (H_5L^- or H_4L^{2-}, λ_{max}, 434 to 439 nm, $\varepsilon = 2.62 \times 10^4$, H_2L^{4-}, λ_{max}, 578 nm, $\varepsilon = 6.09 \times 10^4$).[5]

Methyl Thymol Blue (MTB)

Commercially available samples are mono- or disodium salt, dark violet to black crystalline powder. It is hygroscopic and readily soluble in water, but insoluble in ethanol and common organic solvents. The aqueous solution is blue (reddish in passing light), and very dilute solution is yellow. As in the case of XO, the commercial samples are a mixture of MTB and semi-MTB (up to 50% of semi-MTB). Acid dissociation constants of purified MTB are summarized in Table 2.

The color of aqueous solution turns from yellow to blue when it becomes alkaline. The change of absorption spectra in aqueous solution with pH is illustrated in Figure 2 ($H_6L \sim H_3L^{3-}$, λ_{max}, 435 nm, $\varepsilon = 1.89 \times 10^4$, H_2L^{4-}, λ_{max}, 607 nm, $\varepsilon = 2.15 \times 10^4$).[8] Such spectral change is due to the protonations or deprotonations of the reagent itself, and the dependencies of the maxima of absorbance of MTB (and XO) on pH or H_o values are shown in Figure 3. As the spectral change caused by the complexation with metal ions is quite similar to that due to the deprotonation of the reagent, the pH range for the photometric use is restricted to 0 to 6 for XO and MTB, with an additional range of 11.5 to 12.5 for MTB.

In the presence of cationic surfactant, however, the intensity of the red color of XO in alkaline range is suppressed to give an almost colorless solution (or pale greyish pink at higher XO concentration). Upon an addition of metal ion to this solution, the decolorized XO again forms deep-colored complexes. From the analytical viewpoint, the color reactions with Ca, Ba, Mg, Sr, Cd, Co(II), La, Mn(II), Ni, and Zn are most

Table 1
XYLENOL ORANGE AND METHYL THYMOL BLUE

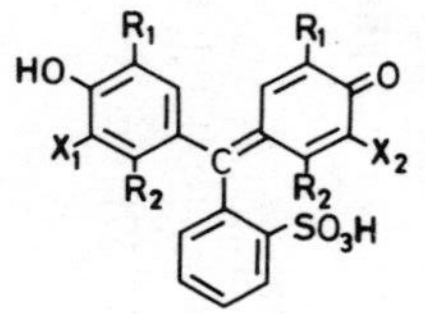

Reagent	Structure	Synonym	Molecular formula	Mol wt	Ref.
Xylenol Orange	$R_1 = CH_3$, $R_2 = H$ $X_1 = X_2 = CH_2 N \begin{smallmatrix} CH_2COOH \\ CH_2COOH \end{smallmatrix}$	3,3′-*Bis*[*N,N*-di(carboxymethyl) aminomethyl)]-*o*-cresolsulfon- phthalein, *o*-Cresolphthalexon S, XO	$C_{31}H_{32}N_2O_{13}S \cdot 2H_2O$	706.70	4
Semi-Xylenol Orange	$R_1 = CH_3$, $R_2 = H$ $X_1 = H, X_2 = CH_2 N \begin{smallmatrix} CH_2COOH \\ CH_2COOH \end{smallmatrix}$	3-[*N,N*-Di(carboxymethyl)- aminomethyl]-*o*-Cresolsulfon- phthalein, Semi-XO	$C_{26}H_{25}NO_9S \cdot H_2O$	545.56	4
Methyl Thymol Blue	$R_1 = CH(CH_3)_2$, $R_2 = CH_3$ $X_1 = X_2 = CH_2 N \begin{smallmatrix} CH_2COOH \\ CH_2COOH \end{smallmatrix}$	3,3′-*Bis*[*N,N*-di(carboxy- methyl) aminomethyl]- thymolsulfonphthalein, Thymolphthalexon S, MTB	$C_{37}H_{49}N_2O_{13}S$	761.85	
Semi-Methyl Thymol Blue	$R_1 = CH(CH_3)_2$, $R_2 = H$ $X_1 = H, X_2 = CH_2 N \begin{smallmatrix} CH_2COOH \\ CH_2COOH \end{smallmatrix}$	3-[*N,N*-Di(carboxymethyl)- aminomethyl]-thymol- sulfonphthalein, Semi-MTB	$C_{32}H_{37}NO_9S$	611.71	

Table 2
ACID DISSOCIATION CONSTANTS OF XO AND MTB

Reagent	1	2	3	4	5	6	7	8	9	Condition	Ref.
					pKa						
XO(H$_6$L^{3+})	-1.74	-1.09	0.76	1.15	2.58	3.23	6.40	10.5	12.6	$\mu = 0.2$ (NO$_3^-$)	M1, R1
MTB (H$_9$L^{3+})	-1.76	-1.11	0.78	1.13	2.60	3.24	7.2	11.2	13.4	$\mu = 0.2$ (NO$_3^-$)	M1, 2
Semi-XO(H$_6$L^{2+})	-0.65	0.53	<1.5	—	2.60	—	7.47	10.9	—	25°, $\mu = 1$ (KNO$_3$)	5, 6
Semi-MTB (H$_6$L^{2+})	—	—	2.0	—	2.81	—	7.6	12.1	—	25°, $\mu = 0.1$ (KNO$_3$)	7
MXB (H$_9$L^{3+})	<-2.8	-1.90	0.08	2.0	3.4	4.3	7.00	10.6	12.5	25°, $\mu = 0.2$	122
Functional group	SO$_3$H			-COOH			OH	-N$^+$H			

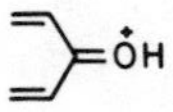

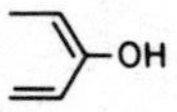

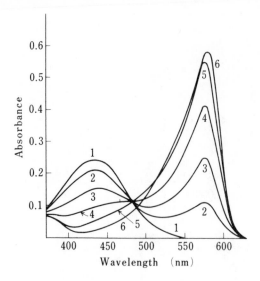

FIGURE 1.   Absorption spectra of XO in aqueous solution at various pH; dye concentration 9.20×10^{-6} M. (1) pH = 1.68 to 4.42; (2) pH = 6.09; (3) pH = 6.93; (4) pH = 7.18; (5) pH = 9.38; (6) pH = 14.2. (From Murakami, M., Yoshino, T., and Harasawa, S., *Talanta,* 14, 1293, 1967. With permission.)

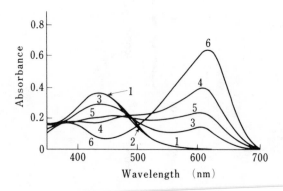

FIGURE 2.   Absorption spectra of MTB in aqueous solution at various pH; dye concentration 1.99×10^{-5} M. (1) pH = 1 to 1.82; (2) pH = 5.08; (3) pH = 6.66; (4) pH = 8.70; (5) pH = 11.51; (6) pH = 13.52. (From Yoshino, T., Imada, H., Murakami, S., and Kagawa, *Talanta,* 21, 211, 1974. With permission.)

interesting. The color ranges from bright-red to bluish-violet, with an exception of La which forms pure blue chelate. The absorption spectra of this system are shown in Figure 4. Thus, with an aid of cationic surfactant, the pH range of XO can be extended to the alkaline side. Such color reactions can be utilized for use as a metal indicator[10] or a photometric reagent.[11] MTB and other structurally related reagents may also be expected to behave similarly.

Complexation Reaction and Properties of Complexes

The yellow color of the reagent solution at pH 0 to 6 turns to red-violet (XO) or to

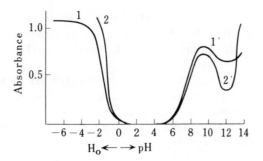

FIGURE 3. Dependence of the maxima of absorbance on pH (and H_o) values. (1,1') XO (1, 580 nm; 1' 518 nm); (2,2') MTB (2, 595 nm; 2', 550 nm). (From Budesinsky, B., *Chelates in Analytical Chemistry*, Vol. 1, Flaschka, H. A. and Barnard, A. J., Jr., Eds., Marcel Dekker, New York, 1967. With permission.)

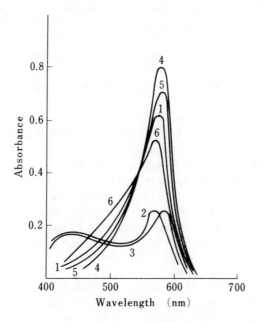

FIGURE 4. Absorption spectra of XO in presence of cationic sufactant. (1) XO 2.0×10^{-5} M at pH 10.5; (2) XO 2.0×10^{-5} M at pH 6.4; (3) (1) plus 5.0×10^{-4} M cethylpyridinium bromide; (4) (3) plus excess Ca; (5) (3) plus excess Zn; (6) (3) plus excess Mn. (From Svoboda, V. and Chromy, V., *Talanta*, 12, 431, 1965. With permission.)

blue (MTB) upon complexation with metal ions which may be differentiated into three groups according to their reaction with the reagents.[M1]

1.   Metal ions that are hydrolyzed within the pH range 0.0 to 6.0 where the indicators are yellow (see Figure 4) (The reactions of these metal ions with XO or MTB are usually slow, and warming to 60 to 80° increases their rate.): Ag, Al, Au(III), Be, Bi, Cr(III), Fe(III), Hf, Ga, In, Mo(VI), Nb(V), Pd(II), Re(IV), Rh(III) (IV), Ru(IV), Sc, Sn(II) (IV), Ta(V), Th, Ti(IV), T1(III), U(IV)(VI), V(IV)(V), W(VI), and Zr.

2.   Metal ions that react with XO or MTB within the pH range 0.0 to 6.0, but are hydrolyzed only at pH > 6.0 (For such metals, an optimum pH value exists for their photometric determinations.): Cd, Co(II), Cu(II), Fe(II), Hg(II), Mg, Mn(II), Ni, Pb, Y, Zn, and rare earths

3.   Metal ions that do not react with XO or MTB in the pH range of 0.0 to 6.0, but do react at a higher pH range (The reagents are of no use with this group of metal ions which include Ca, Ba, Sr, and Ra. In the case of MTB, however, the color reaction due to complexation may be of analytical use in a narrow pH range of pH 11.5 to 12.5 (see Figure 3).

As XO and MTB behave as polybasic ligands with two coordination sites, the compositions of the resulting chelates are rather complicated. The results by the earlier workers with the impure reagents containing substantial amount of semi-XO or semi-MTB may be sometime contradictory to the recent results with the purified reagents.[5]

In most cases, the metal chelates have ML-type composition (or M_2L_2), but ML_2 and M_2L-types are not uncommon. In general, the metal chelates can be expressed as $M_iH_jL_k$, where L, H, and M stand for the fully deprotonated ligand, proton, and the metal ion, respectively. The structures of some XO-metal chelates are illustrated in Figure 5.

Due to the complex equilibria in the solution, only a few data are available on the chelate stability constants with XO[16,17] and MTB,[8] but the conditional stability constants may be of practical value on the analytical applications of XO and MTB. Those values are summarized in Table 3.

The absorption spectra of metal chelates of XO and MTB are quite similar to that of the deprotonated free ligands (L^{6-}). The spectral characteristics of the metal chelates at their optimum pH range are summarized in Table 4, and the absorption spectra of Zr-XO and Zr-MTB chelates are shown in Figures 6 and 7, respectively. The shape of the spectra of metal chelates change very slightly from metal to metal.

The anionic chelates of XO and MTB may be extracted into immiscible solvents in the presence of a long chain alkylammonium ion[21] or diphenylguanidinium ion.[22] Thus, it becomes possible to determine metal ions by the extraction photometry using XO or MTB. Examples are summarized in Tables 5 and 6.

Purification and Purity of Reagents

The commercial grade of XO (or MTB) is generally contaminated with a starting material (Cresol Red or Thymol Blue) and semi-XO (or semi-MTB). Although such impure samples are quite satisfactory for use as a metal indicator or photometric reagent in the routine analytical purposes, the contaminants may cause serious erroneous results in the precision works or the physicochemical study.

The purification to obtain pure XO or MTB may be done most conveniently by the chromatography of sodium salt on a cellulose powder column (for XO, elution with butanol saturated with 10% acetic acid[5], and for MTB, elution with butanol containing 0.1% acetic acid).[20] The order of elution for the crude XO is Cresol Red, semi-XO, and XO. MTB behaves similarly. A column of DEAE cellulose was also used for the same purpose.[16] The free acid can be obtained by treating the aqueous solution of sodium salt with cation exchange resin (H form).

The presence of impurities can be checked by the paper chromatography using *n*-butanol saturated with 10% acetic acid as an eluent. R_f values for XO are 0.05 to 0.1 (XO), 0.6 to 0.8 (semi-XO) and, 0.8 to 1.0 (Cresol Red). MTB is expected to behave similarly.

The molar absorptivity is another criteria of purity. In aqueous solution, the following values are reported: XO, pH 3.11, λ_{max}, 435 nm, $\varepsilon = 2.62 \times 10^4$,[5] MTB, pH 5.45, λ_{max}, 435 nm, $\varepsilon = 1.89 \times 10^4$.[20]

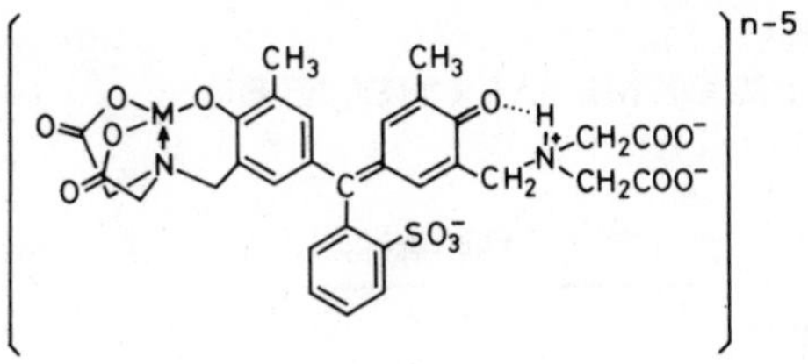

(A)

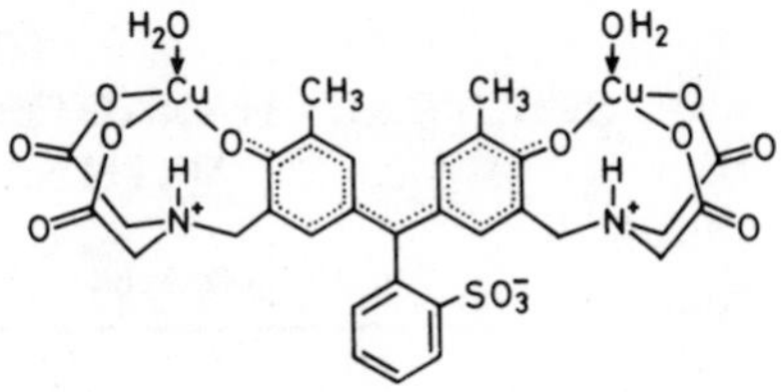

(B)

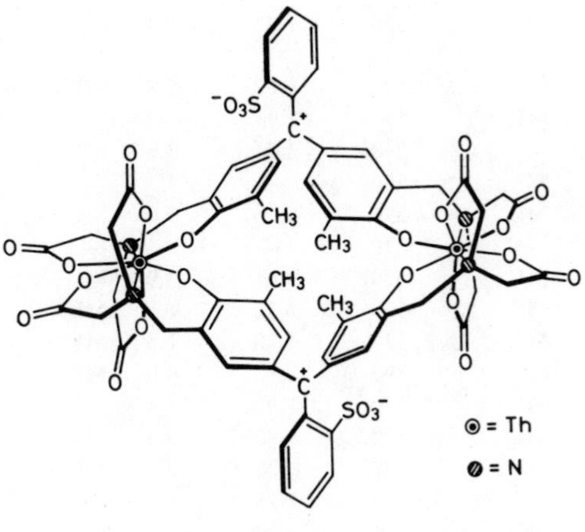

⊚ = Th

⦸ = N

(C)

FIGURE 5. (A) MHL^{n-5}- type; (B) Cu$_2$H$_2$L;[13] (C) Th$_2$L$_2^{4-}$.[14]

Table 3
CONDITIONAL CHELATE STABILITY CONSTANTS[15]

Metal ion	XO		MTB		Ref.
	$\log K'_{ML}$	pH	$\log K'_{ML}$	pH	
Al	14.3	2.3	—	—	
Ca	—	—	~5.5	10	
Cd	3.78	5.5	~3.3	5.5	
Co(II)	6.78	5.9	—	—	
Cr(III)	3.57	1.7	—	—	
Fe(III)	3.91	0.15 NHClO$_4$	—	—	
Hg(II)	—	—	~13.4	5.5	
Mg	—	—	~5.2	10	
Mn(II)	3.23	5.9	~8.0	10	
Mo(VI)	17.0	3.6	—	—	
Pb	—	—	~6.4	5.5	
Ru(III)	9.6	4.0	—	—	49
Sc(III)	5.95	2.6	—	—	
Th(IV)	11.6	4.0	—	—	
Zn	—	—	~5.5	5.5	
Rare earths	6.3 ~ 6.6	6	~4.9	5.5	

Table 4

SPECTRAL CHARACTERISTICS OF METAL CHELATE
WITH XO AND MTB[M1,R1]

Metal ion	XO chelate				MTB chelate			
	pH	Ratio	λ_{max} (nm)	$\varepsilon\,(\times 10^4)$	pH	Ratio	λ_{max} (nm)	$\varepsilon(\times 10^4)$
Al	3.4	ML	536	2.17	3.4	ML	584	1.42
Ba	—	—	—	—	11.6	ML	610	0.38
Be	5.8	ML$_2$	495	1.37	5.0	ML$_2$	500	0.32
Bi	1.9	ML	530	1.60	1.9	ML	550	—
Ca	—	—	—	—	11.4	ML	610	2.03
Cd	5.8	ML	570	1.22	5.8	ML	—	—
Ce(III)	5.8	ML	570	1.95	—	—	—	—
Co(II)	5.8	ML	570	2.44	5.8	ML	600	0.91
Cr(III)	4.0	ML$_{1\sim2}$	530	—	3	—	560	1.15
Cu(II)	5.8	ML	570	5.26	5.8	ML	600	3.20
Fe(II)	5.8	—	530	—	—	—	—	—
Fe(III)	3.0	ML	550	2.66	3.0	ML	580	1.5
Ga	1.5	ML	530	2.32	1.5	ML	570	1.28
Hf	1.5	ML	535	4.78	1.5	ML	570	1.87
Hg(II)	7.0	ML	590	3.1	6.0	ML	630	1.01
In	3.4	ML	530	1.90	3.4	ML	600	1.80
La	5.8	ML	570	1.90	5.8	ML	610	1.78
Mg	5.8	ML	570	0.20	10.8	ML	610	1.52
Mn(II)	5.8	ML	570	1.78	5.8	—	—	—
Mo(VI)	3.6	ML	490	0.52	—	—	—	—
Nb(V)	2.6	ML$_{1\sim2}$	535	2.30	1	ML	560	1.3
Ni	5.8	ML	570	2.65	5.8	—	600	1.10
Pd(II)	1.1	ML$_2$	518	2.61	—	—	—	—
Pb	5.5	MH$_2$L	570	1.67	5.8	600	1.14	—
Ru(IV)	4.0	ML$_2$	530	—	—	—	—	—
Sc	2.8	ML	560	2.16	2.9	ML	570	0.88
Sn(II)	3.0	—	536	0.22	—	—	—	—
Sn(IV)	3.0	M$_2$L	530	—	—	—	—	—
Sr	—	—	—	—	11.5	ML	610	1.15
Ta(V)	2.7	ML$_{1\sim2}$	535	—	—	—	—	—
Th	4.0	ML	560	3.80	4.0	ML$_2$	568	3.94
Ti(IV)	4.0	ML	502	1.44	4.0	ML	—	—
Tl(III)	3.0	ML	520	2.13	—	—	—	—
U(IV)	1.3	ML$_{1\sim2}$	550	2.60	1.3	—	—	—
U(IV)	3.6	ML$_2$	568	5.10	—	—	—	—
U(VI)	5.8	ML	510	1.05	—	—	—	—
V(IV)	2.8	ML	560	2.40	3.5	ML	5.90	1.6
V(V)	2.7	ML	560	1.85	2.7	ML	—	—
W(VI)	4.0	—	530	—	—	—	—	—
Y	5.8	ML	570	1.95	5.8	ML	610	1.85
Zn	6.0	ML	574	1.70	6.0	ML	—	—
Zr	1.5	ML	535	2.70	1.5	ML	580	2.17

Analytical Applications

Use as a Photometric Reagent

XO and MTB form strongly colored chelates with many metal ions in the pH range of 0 to 6.0, but lack selectivity. Certain metals, such as Bi, Fe(III), Hf, Nb, Pd(II), and Zr react in solutions of fairly high acidity, so that they can be determined selectivity with the combined use of masking agents.

In the presence of cationic surfactant, metal chelates of higher ligand to metal ratio are formed to give higher photometric sensitivity.

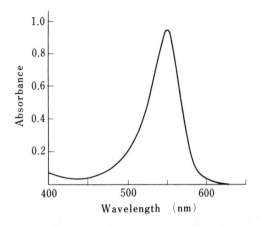

FIGURE 6. Absorption spectra of Zr - XO Chelate in 1 M HClO$_4$ (XO, 5×10^{-6} M; Zr, $5 \times 10^-$ M). (From Sonoda, K., Otomo, M., and Kodama, K., *Bunseki Kagaku*, 27, 429, 1978. With permission.)

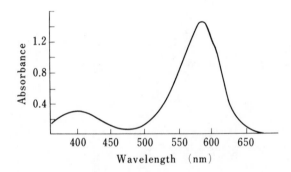

FIGURE 7. Absorption spectra of Zr - MTB chelate in 0.2 M HCl. (MTB, 4×10^{-5} M; ZrOCl$_2$, 10^{-4} M. (From Hems, R. V., Kirkbright, G. F., and West, T. S., *Talanta*, 16, 789, 1969. With permission.)

In most cases, metal chelates of XO and MTB are negatively charged, and they can be extracted into organic solvent as an ion pair as stated above, to attain higher selectivity and sensitivity. Tables 5 and 6 summarize the examples of photometric determination with XO and MTB, respectively.

The indirect photometric determination of nonmetallic species using metal-XO or MTB are reported: F$^-$ (Zr-XO),[101] H$_2$O$_2$ (Ti-XO),[102] SO$_4^{2-}$ (Th-XO[103] Ba-MTB,[104] Zr-MTB),[19] CO$_3^{2-}$ (Cr-XO),[105] halides,[106] CN$^-$,[107] SCN$^-$ (Hg-MTB-Zephiramine®),[108] and NH$_3$ (Hg-MTB).[109]

Use as a Metal Indicator in Chelatometry

The addition of EDTA to the colored solution of metal-XO (red-violet) or metal-MTB(blue) chelates brings about a decoloration to the original yellow color. However, on using the mixed indicator (XO-MTB, 3:1), the color change at the end point is from violet to orange-yellow at pH 4.[110] The color changes of these single or mixed indicators are instantaneous and sharp with a few exceptions of metal ions, so that they are widely used as metal indicators in chelatometry. Metal ions that can titrated with XO or MTB as an indicator are[M2,M3] Pu(IV) (0.1 to 0.2 N HNO$_3$), Bis, Hf, Fe(III),[116] Th, Zr (pH 1

Table 5
APPLICATIONS OF XO AS A PHOTOMETRIC REAGENT

Metal ion[a]	Condition (pH, buffers)	λ^b (nm)	Range of determination (ppm)	Remarks	Ref.
Al	3.4, acetate, 100°	536	0.2 ~ 1.0	Fe(III) and Th interfere seriously	23—25
Be	5.6 ~ 6.2, hexamine, EDTA	495	0 ~ 0.24	Al, Bi, Co, Th, Zr interfere	26
Bi	0.08 ~ 0.12 N HNO₃ F⁻, ascorbic acid	530	0.4 ~ 3.2	Common elements do not interfere	27
Cu(II)	3.5 ~ 6.4, hexamine or MES[c]	575 ~ 80	0 ~ 1.8	Many metals interfere	32, 33
Fe(III)	0.10 ~ 0.06 N HClO₄	545	0.12 ~ 1.8	Zr interfere	34
Hf	0.2 ~ 0.4 N HClO₄, H₂O₂, Na₂SO₄	530	0.7 ~ 4.0	Hf(100 µg) in presence of Zr(5 ~ 50 µg)	36
Pd(II)	1.1 ~ 1.7 N HClO₄, 100°	518	0.2 ~ 4.0	Cl⁻, EDTA, Zr interfere	46
Pu(IV)	0.1 N HNO₃	560	0.5 ~ 4.0	U(VI)(6×10^3-fold) does not interfere	47
Th	3.9 ~ 4.3, acetate	560	0.2 ~ 2.8	Th in U	52
Th	2.5 ~ 4.0, cetyltrimethylammonium bromide	560 ~ 570	0.04 ~ 4.0	Al, Bi, Ce, Cr, Cu, Fe, Ti, Tl, U, V interfere	53
Th	3, diphenylguanidine, extraction with butanol	578	0.14 ~ 0.83	U, excess oxalate tartarate, F⁻ interfere	54—56
Ti	0.04 ~ 0.06 N HClO₄, H₂O₂	535	0.3 ~ 2.8	Bi, Fe, Ti, Th, Tl, Zr interfere	57
U(VI)	3.5 ~ 3.7, acetate, ascorbic acid, 100°	568	0.8 ~ 4.0	Cu, Fe, Pd interfere	59, 60
Zr	0.3 ~ 0.8 N HClO₄, 0.15 ~ 0.25 M H₂SO₄, ascorbic acid	535	0.14 ~ 2.5	Specific for Zr; only Co, Cu, Cr interfere	36, 64

Zr	Acetate buffer, extraction with tri-*n*-octylamine	550	0.03 ~ 2	Mo, Pd, Ti give positive and Cd, Cu, Hg give negative errors	65,66
Rare earth	5.5 ~ 6.0, hexamine-HNO_3	570	0.2 ~ 1.0	After ion-exchange separation	67
Rare earth	7.5 ~ 9, cetylpyridinium bromide	610 or 625	0.08 ~ 0.8		68
Rare earth	5.7 ~ 6.9, trioctylethyl-ammonium bromide, extraction with xylene	590 ~ 605	0 ~ 0.9		69, 70

[a] Other metals determined by the photometry with XO are Cd,[28] Ce(III)[29] Co(II),[30] Cr(III),[31] Fe(III),[30] Ga,[35] Hg(II),[37] In,[38-40] La,[41] Mn(II),[30] Mo(VI),[30] Nb,[23] Ni,[30,43] Np,[44] Pb,[45] Rh,[48] Ru,[48] Sc,[49] Sn(IV),[51] Tl(III),[58] V(IV) (V),[61,62] and Zn.[63]

[b] Observing wavelength.

[c] MES: 2-(*N*-Morphorino) ethansulfonic acid.

Table 6
APPLICATION OF MTB AS A PHOTOMETRIC REAGENT

Metal ion[a]	Condition (pH, buffer)	λ^b (nm)	Range of determination	Remarks	Ref.
Be	4.5, diphenylguanidine, extraction with butanol	500	0.01 ~ 0.7	Cd, Co, Cu, Fe (III), Hg,Ni, Sn, Zn, Zr interfere	72, 73
Bi	0.5 ~ 2.1	550	0 ~ 1.7	—	74
Bi	0.6 ~ 2.4, diphenyl-guanidine, extraction with butanol	560	0 ~ 1.1	—	75
Ca	11 ~ 12, KOH	610	7 ~ 29	Mg does not interfere	76
Fe(III)	0.07 ~ 0.1 NHClO₄	580	0.2 ~ 2.0	Common metals do not interfere	80
Hf	3.0, formic acid	570	0.7 ~ 3.6	Bi, Fe(III), Mo(VI), Ni, Sn(IV), Th, Ti, V, Zr interfere	36
Th	4.0 ~ 4.5, acetate	568	0.4 ~ 4.0	Al, Bi, Fe, Ga, Ta, U, Zr, interfere	93
Th	9.2 ~ 10, EDTA, triethanolamine	535	0.5 ~ 2.8	Ag, Be, Sb, Ti, U interfere	94
Th	2.5 ~ 3, diphenyl-guanidine, extraction with butanol	590	0.3 ~ 3.0	—	95
U(VI)	6.2 ~ 6.9	510		—	96
Zr	0.3 ~ 1.2 NHClO₄	580	0.4 ~ 1.8	Fe(III) is masked with ascorbic acid	36, 98
Zr	1 ~ 2.2, BPA extraction with butanol	600	0.4 ~ 8	Al(200-fold), Cr, Ga, In, Nb, Pb, Ti(50-fold) do not interfere	99
Rare earths	5 ~ 6.5 for Ce groups 4.35 ~ 5.0 for Y group, diphenylguanidine, extraction with iso-Am OH	610	0.04 ~ 0.07	—	100

[a]   Other metals determined by the photometry with MTB are Al,[71] Cr(III),[77] Cu(II),[78] Er,[79] Ga,[81,82] Hg(II),[83,84] In,[82,85] La,[86] Mg,[87,88] Nb ,[89] Ni,[90] Pb,[91] Pd(II),[92,93] V(IV),[97] and Y[86]

[b]   Observing wavelength.

to 2), Ga(pH 1.2 to 1.4, 100°), In,[113] Nb, Sc(pH 2 to 3), Am,[111] Ce, Tl(III),[114] Y, rare earths, actinides (pH 4 to 5),[115,117] Cu(II) (pH 4 to 5, phenanthroline),[112] Cd, Co(II), Hg(II), Pb, Zn (pH 5 to 6), and Ca, Cd, Mg, Mn(II), Zn(pH 10, Zephiramine®).[10] Additional metals can be titrated with MTB; Ca, Cd, Mn(II), Pb, Zn (pH ≃ 12), and Mg (pH 10 to 11). The color change is from blue to grey in this case.

A 0.1% aqueous solution of XO and a dry powder preparation of MTB diluted with KNO₃ or K₂SO₄ (1:100) are recommended for the indicator use.

Other Reagents with Related Structure
Semi-XO and Semi-MTB

The structures and acid dissociation constants of both reagents are shown in Tables 1 and 2, respectively. They can be separated from the commercial products of XO[3-6] and MTB[118-120] as major impurities. These reagents behave similarly to their parent compounds, but with different metal to ligand ratio. According to the recent work, a higher sensitivity for the photometric determination of Zr (pH 1.6 to 1.9, ML₂, λ_{max}, 550 nm, $\varepsilon = 1.48 \times 10^5$)[121] than that by the conventional XO method (pH 1.5, λ_{max}, 535 nm, $\varepsilon = 3.38 \times 10^4$)[64] is reported.

Methyl Xylenol Blue

3,3′-*Bis* [*N,N*-di (carboxmethyl) aminomethyl]-*p*-xylenolsulfonphthalein, sodium salt ($R_1=R_2=R_3=R_4=CH_3$), MXB,[1] Grey-brown powder. Soluble in water and methanol, but insoluble in higher alcohols, ether, and acetone. The values of pKa are shown in Table 2. It behaves similarly to XO or MTB and can be used as the photometric reagent for Al (pH 2.4, 580 nm),[123] Be (pH 6.0, 515 nm),[118] Bi (pH 1.2, 575 nm),[124] Ca (pH 11.8, 610 nm),[125] Ce(III) (pH 8, cetylpyridinium chloride, 645 nm),[126] Fe(III) (pH 1.4, 610 nm),[127] and Hg(II) (pH 6.8, 615 nm).[128] Semi-MXB has also been investigated in some detail.[129-131]

Glycinecresol Red and Glycinethymol Blue:

Glycinecresol Red (GCR)[132] and Glycinethymol Blue (GTB) [133]are analogues of XO and MTB, respectively, in which iminodiacetic acid groups are replaced by glycine. Both reagents are claimed to be highly specific chromogens for Cu(II).

MONOGRAPHS

M1. **Budesinsky, B.**, Xylenol Orange and Methylthymol Blue as Chromogenic Reagents, in *Chelates in Analytical Chemistry*, Vol. 1, Flaschka, H. A. and Barnard, A. J., Jr., Eds., Marcel Deker, New York, 1967, 15.

M2. **Ueno, K.**, *Chelatemetric Titration*, 2nd ed., Nankodo, Tokyo, 1972.

M3. **Pribil, R.**, *Analytical Applications of EDTA and Related Compounds*, Pergamon Press, New York, 1972.

M4. **Jablonski, W. Z. and Johnson, E. A.**, Xylenol Orange, and **Ainsworth, L. R., Jablonski, W. Z., Johnson, E. A.**, and **Yardley, J. T.**, Methylthymol Blue, *in Organic Reagents for Metals*, Vol. 2, Johnson, W. C., Ed., Hopkin & Williams, Chadwell Heath, England, 1964, 239 and 249.

REVIEW

R1. **Otomo, M.**, Xylenol Orange and its analogs, *Bunseki Kagaku,* 21, 436, 1972.

REFERENCES

1. **Körbl, J. and Pribil, R.**, *Chem. Ind.,* 233, 1957.
2. **Körbl, J.**, *Chem. Listy,* 51, 1680, 1957; *Collect. Czech. Chem. Commun.,* 23, 889, 1958.
3. **Nakada, S., Yamada, M., Ito, T., and Fujimoto, M.**, *Bull. Chem. Soc. Jpn.,* 50, 1887, 1977.
4. **Yamada, M., and Fujimoto, M.**, *Bull. Chem. Soc. Jpn.,* 49, 693, 1976.
5. **Murakami, M., Yoshino, T., and Harasawa, S.**, *Talanta,* 14, 1293, 1967.
6. **Yamada, M. and Fujimoto, M.**, *Bull. Chem. Soc. Jpn.,* 44, 294, 1971.
7. **Yoshino, T., Murakami, S., and Kagawa, M.**, *Talanta,* 21, 199, 1974.
8. **Yoshino, R., Imada, H., Murakami, S., and Kagawa, M.**, *Talanta,* 21, 211, 1974.
9. **Svoboda, V. and Chromy, V.**, *Talanta,* 12, 431, 1965.
10. **Chromy, V. and Svoboda, V.**, *Talanta,* 12, 437, 1965.
11. **Svoboda, V. and Chromy, V.**, *Talanta,* 13, 237, 1966.

12. Otomo, M., *Res. Rep. Hachinohe Tech. Coll.*, 1, 26, 1966.
13. Yamada, M., *Bunseki Kagaku*, 850, 1976.
14. Pribil, R., *Talanta*, 13, 1711, 1966.
15. Ringbom, A., *Complexation in Analytical Chemistry*, John Wiley & Sons, New York, 1963.
16. Sato, H., Yokoyama, Y., and Momoki, K., *Anal. Chim. Acta*, 94, 217, 1977.
17. Sato, H., Yokoyama, Y., and Momoki, K., *Anal. Chim. Acta*, 99, 167, 1979.
18. Sonoda, K., Otomo, M., and Kodama, K., *Bunseki Kagaku*, 27, 429, 1978.
19. Hems, R. V. Kirkbright, G. F., and West, T. S., *Talanta*, 16, 789, 1969.
20. Yoshino, T., Imada, H., Kuwano, T., and Iwasa, K., *Talanta*, 16, 151, 1969.
21. Pribil, R. and Vesely, V., *Talanta*, 17, 801, 1970.
22. Shestidesyatnaya, N. L., Kotelyanskaya, L. I., and Milyaeva, N. M., *Org. Reagenty Anal. Khim. Tezisy Dokl. Vses. Konf.*, 4th, 2, 5, 1976; *Chem. Abstr.*, 87, 177045r, 1977.
23. Budesinsky, B., Fresenius *Zh. Anal. Khim.*, 18, 1071, 1963.
24. Tikhonov, V. N. and Andreeva, T. V., *Izv. Vyssh. Uchebn. Zaved. Khim. Khim. Tekhnol.*, 19, 1615, 1976; *Chem. Abstr.*, 86, 182426z, 1977.
25. Otomo, M., *Bull. Chem. Soc. Jpn.*, 36, 809, 1963.
26. Otomo, M., *Bull. Chem. Soc. Jpn.*, 38, 730, 1965.
27. Onishi, H. and Ishiwatari, N., *Bull. Chem. Soc. Jpn.*, 33, 1581, 1960.
28. Otomo, M., *Bull. Chem. Soc. Jpn.*, 37, 504, 1964.
29. Tonosaki, K. and Otomo, M., *Bull. Chem. Soc. Jpn.*, 35, 1683, 1962.
30. Otomo, M., *Bunseki Kagaku*, 14, 45, 1965.
31. Tonosaki, K., Otomo, M., and Tanaka, K., *Bunseki Kagaku*, 15, 683, 1966.
32. Hung, S. C. and H. S., Chiang, *Hua Hsueh Tung Pao*, 46, 1964; *Chem. Abstr.*, 61, 2467f, 1964.
33. Wada, H., Ishizuki, T., and Nakagawa, G., *Talanta*, 23, 669, 1976.
34. Otomo, M., *Bunseki Kagaku*, 14, 677, 1965.
35. Otomo, M., *Bull. Chem. Soc. Jpn.*, 38, 624, 1965.
36. Cheng, K. L., *Anal. Chim. Acta*, 28, 41, 1963.
37. Cabrera Martin, A., Peral Fernandez, J. L., Vicente Perz, S., and Burriel Marti, F., *Talanta*, 16, 1023, 1969.
38. Ishiwatari, N., Nagai, H., and Toita, Y., *Bunseki Kagaku*, 12, 603, 1963.
39. Pyatnitskii, I. V. and Pinaeva, S. G., *Zh. Anal. Khim.*, 28, 671, 1973.
40. Pyatnitskii, I. V., Kolomiets, L. L., and Sadovskaya, I. L., *Zh. Anal. Khim.*, 30, 2131, 1975, *Ukr. Khim. Zh.*, 43, 283, 1977; *Chem. Abstr.*, 87, 123479s, 1977; *Ukr. Khim. Zh.*, 43, 865, 1977; *Anal. Abstr.*, 34, 3B68, 1978.
41. Matsushita, T., Kaneda, M., and Shono, T., *Anal. Chim. Acta*, 104, 145, 1979.
42. Ishiwatari, N. and Onishi, H., *Bunseki Kagaku*, 11, 576, 1962.
43. de Wet, W. J. and Bohrens, G. B., *Anal. Chem.*, 40, 200, 1968.
44. Ermolaev, N. P., Kovalenko, G. S., Krot, N. N., and Blokhina, V. I., *Zh. Anal. Khim.*, 20, 1336, 1965.
45. Cabrera Martin, A., Fernandez M. H., and Peral Fernandez, J. L., Burriel Marti, F., *Quim. Anal.*, 28, 38, 1974.
46. Otomo, M., *Bull. Chem. Soc. Jpn.*, 36, 889, 1963.
47. Milyukova, M. S. and Nemodruk, A. A., *Radiokhimiya*, 8, 246, 1966; *Zh. Anal. Khim.*, 21, 296, 1966.
48. Otomo, M., *Bunseki Kagaku*, 17, 125, 1968.
49. Srivastawa, S. C. and Dey, A. K., *Chim. Anal.* 51, 131, 1969.
50. Konkova, O. V., *Zh. Anal. Khim.*, 19, 73, 1964.
51. Danilova, V. N., *Zavod. Lab.*, 29, 407, 1963.
52. Budesinsky, B., *Collect. Czech. Chem. Commun.*, 28, 1858, 1963.
53. Ramakrishna, T. V. and Murthy, R. S. S., *Talanta*, 26, 499, 1979.
54. Contarini, M., Pasquinelli, P., and Rigali, L., *Anal. Chim. Acta*, 89, 397, 1977.
55. Golentovskaya, I. P., Kirillov, A. I., and Vlasov, N. A., Deposited Doc., 1974, VINITI 489; *Chem. Abstr.*, 86, 182529k, 1977.
56. Otomo, M. and Wakamatsu, Y., *Nippon Kagaku Zasshi*, 89, 1087, 1968.
57. Otomo, M., *Bull. Chem. Soc. Jpn.*, 36, 1341, 1963.
58. Otomo, M., *Bull. Chem. Soc. Jpn.*, 38, 1044, 1965.
59. Budeskinsky, B., *Collect. Czech. Chem. Commun.*, 27, 226, 1962.
60. Otomo, M., *Bull. Chem. Soc. Jpn.*, 36, 140, 1963.
61. Otomo, M., *Bull. Chem. Soc. Jpn.*, 36, 137, 1963.
62. Budesinsky, B., *Collect. Czech. Chem. Commun.*, 28, 1858, 1963.
63. Ishihara, Y., Naniwa, T., Yokokura, S., and Uchida, S., *Bunseki Kagaku*, 17, 991, 1968.
64. Cheng, K. K., *Talanta*, 2, 61, 186, 266, 1959.

65. Kabrt, K., Radii, V., Suchanek, M., Uhrova, M., and Urner, Z., *Sb. Vys. Sk. Chem. Technol. Praze Anal. Chem.* H12, 251, 1977; *Chem. Abstr.*, 90, 15790q, 1979.
66. Cerrai, E. and Testa, C., *Anal. Chim. Acta*, 26, 204, 1962.
67. Budesinsky, B. and Bezdekova, A., *Fresenius Z. Anal. Chem.*, 196, 172, 1963.
68. Otomo, M. and Wakamatsu, Y., *Bunseki Kagaku*, 17, 764, 1968.
69. Shijo, Y., *Bunseki Kagaku*, 25, 680, 1976.
70. Shijo, Y., *Bunseki Kagaku*, 23, 884, 1974.
71. Tikhonov, V. N., *Zh. Anal. Khim.*, 21, 275, 1966.
72. Srivastava, K. C. and Banerji, S. K., Analusis, 1, 132, 1972; *Anal. Abstr.*, 24, 46, 1973.
73. Beshetnova, E. T., Anisimova, L. G., and Mataeva, S. S., *Fiz. Khim. Metody Anal. Kontrolya Proizvad.*, 12, 1975; *Chem. Abstr.*, 87, 126626y, 1977.
74. Enoki, T., Mori, I., and Izumi, Y., *Bunseki Kagaku*, 18, 963, 1969.
75. Shestidesyatnaya, N. L., Milyaeva, N. M., and Kotelyanskaya, L. I., *Zh. Anal. Khim.*, 31, 1176, 1976.
76. F. Garcia Montelongo, Herrera, C., and Arias, J. J., *Inf. Quim. Anal.*, 25, 26, 1971.
77. Cheng, K. L., *Talanta*, 14, 875, 1967.
78. Karadakov, B., Alekieva, A., and Venkova, D., *God. Vissh. Khim. Tekhnol. Inst. Sofia*, 23, 265, 1977; *Chem. Abstr.*, 90, 114408c, 1979.
79. Poluektov, N. S., Ovchar, L. A., and Lauer, R. S., *Zh. Anal. Khim.*, 28, 1958, 1973.
80. Tonosaki, K., *Bull. Chem. Soc. Jpn.*, 39, 425, 1966.
81. Tonosaki, K. and Sakai, K., *Bunseki Kagaku*, 14, 495, 1965.
82. Pyatnitskii, I. V., Kolomiets, L. L., and Lemesh, L. V., *J. Anal. Chem. USSR*, 29, 1761, 1974; *Ukr. Khim. Zb.*, 43, 277, 1977; *Chem. Abstr.*, 87, 145230f, 1977.
83. Iritani, N. and Miyahara, T., *Bunseki Kagaku*, 12, 1183, 1963.
84. Nomura, T. and Kimura, G., *Bunseki Kagaku*, 22, 576, 1973; *Bunseki Kagaku*, 23, 1501, 1974.
85. Tonosaki, K. and Ishido, H., *Bunseki Kagaku*, 18, 1096, 1969.
86. Okada, H., Kaneko, K., and Atozeki, S., *Bunseki Kagaku*, 12, 822, 1963.
87. Metcalfe, J., *Analyst*, 90, 409, 1965.
88. Maekawa, S. and Kato, K., *Bunseki Kagaku*, 16, 422, 1967.
89. Elinson, S. V. and Mirzoyan, N. A., *Zh. Anal. Khim.*, 21, 1436, 1966.
90. Ozawa, T., *Bunseki Kagaku*, 21, 1359, 1972.
91. Tataev, O. A. and Magaramov, M. N., *Zh. Neorg. Khim.*, 22, 120, 1977; *Chem. Abstr.*, 86, 96824w, 1977.
92. Srivastava, K. C. and Banerji, S. K., *Microchem. J.*, 18, 288, 1973.
93. Otomo, M., *Bunseki Kagaku*, 14, 229, 1965.
94. Adam, J. and Pribil, R., *Talanta*, 16, 1596, 1969.
95. Otomo, M., *Nippon Kagaku Zasshi*, 89, 503, 1968.
96. Srivastava, K. C. and Banerji, S. K., *J. Prakt. Chem.*, 311, 769, 1969.
97. Tikhonov, V. N., Grankina, M. Ya., and Vernigora, V. P., Zh. Anal. Khim., 21, 1172, 1966; *Zh. Anal. Khim.*, 22, 359, 1967.
98. Shtokalo, M. I. and Ostrovskaya, M. I., Zavod. Lab., 44, 925, 1978; *Chem. Abstr.*, 89, 190443e, 1978.
99. Shtokalo, M. I., Ostrovskaya, M. S., Pyzhenko, V. L., and Tolok, V. N., *Zh. Anal. Khim.*, 33, 2383, 1978.
100. Akhmedli, M. K., Granovskaya, P. B., and Neimotova, R. A., Izv. Vyssh. Uchebn. Zaved. Khim. Khim. Tekhnol., 13, 1093, 1970; Uchen. Zap. Azerb. Gos. Univ. Ser. Khim. Nauk. 25, 1972; *Anal. Abstr.*, 26, 773, 1974.
101. Cabello Tomes, M. L. and West, T. S., *Talanta*, 16, 781, 1969.
102. Nordschow, C. D. and Tammes, A. R., *Anal. Chem.*, 40, 465, 1968.
103. Palaty, V., *Talanta*, 10, 307, 1963.
104. Colovos, G., Panesar, M. R., and Parry, E. P., *Anal. Chem.*, 48, 1693, 1976.
105. Pantalar, R. P. and Pulyaeva, I. V., *Zh. Anal. Khim.*, 32, 394, 1977.
106. Nomura, T. and Kimura, M., *Bunseki Kagaku*, 23, 1501, 1974.
107. Nomura, T., *Bull. Chem. Soc. Jpn.*, 41, 1619, 1968.
108. Nomura, T., Takemura, Z., Nakamura, T., and Komatsu, S., *Bunseki Kagaku*, 22, 576, 1973.
109. Komatsu, S., Nomura, T., Nakamura, T., and Suzuki, H., *Nippon Kagaku Zasshi*, 91, 865, 1970.
110. Geyer, R. and Bormann, R., Z. Chem., 7, 30, 1967; *Chem. Abstr.*, 66, 72069q, 1967.
111. Buijs, and Bartsher, *Anal. Chim. Acta*, 88, 403, 1977.
112. Wada, H., Ishizuka, T., and Nakagawa, G., *Bull. Chem. Soc. Jpn.*, 50, 2104, 1977.
113. Tschetter, M. J., Bachman, R. Z., and Banks, C. V., *Talanta*, 18, 1005, 1971.
114. Kulba, F. Ya. and Platunova, N. B., *Zh. Anal. Khim.*, 28, 1009, 1973.

115. Timofeev, G. A., Simakin, G. A., Baklanova, P. F., Kuznetsov, G. F., and Ivanov, V. I., Zh. Anal. Khim., 31, 2337, 1976; *J. Anal. Chem. USSR,* 31, 1717, 1976.
116. Yang, H. H., Chang, C. C., Yang, K. C., and Chen, C. T., Semento Gijutsu Nempo (Japan) 30, 74, 1976; *Chem. Abstr.,* 87, 140198s, 1977.
117. Zielinski, S. and Lomozik, L., *Chem. Anal. (Warsaw),* 23, 815, 1978; *Anal. Abstr.,* 37, 1B103, 1979.
118. Yoshino, T., Imada, H., Kuwano, T., and Iwasa, K., *Talanta,* 16, 151, 1969.
119. Yoshino, T., Okazaki, H., Murakami, S., and Kagawa, M., *Talanta,* 21, 673, 1974.
120. Kosenko, N. F., Malkova, T. V., and Yatsimirskii, K. B., *Zh. Anal. Khim.,* 30, 2245, 1975; *J. Anal. Chem. USSR,* 30, 1883, 1975.
121. Olson, D. C. and Margerum, D. W., *Anal. Chem.,* 34, 1299, 1962.
122. Vytras, K. and Vytrasova, J., Chem. Zvesti, 28, 252, 779, 1974; *Chem. Abstr.,* 81, 72160x, 1974; *Chem. Abstr.,* 83, 21423u, 71003c, 1975.
123. Ueda, J., *Nippon Kagaku Kaisshi,* 273, 1974.
124. Deguchi, M., Okumura, I., and Sakai, K., *Bunseki Kagaku,* 19, 836, 1970.
125. Kitano, M. and Ueda, J., *Nippon Kagaku Zasshi,* 92, 168, 1971.
126. Mori, I. and Enoki, T., *Yakugaku Zasshi,* 90, 494, 1970.
127. Deguchi, M., Yamabuki, S., and Yashiki, M., *Bunseki Kagaku,* 20, 891, 1971.
128. Deguchi, M. and Sakai, K., *Bunseki Kagaku,* 19, 241, 1970.
129. Ueda, J., *Nippon Kagaku Kaishi,* 350, 1977.
130. Ueda, J., *Bull. Chem. Soc. Jpn.,* 51, 773, 1978.
131. Ueda, J., *Nippon Kagaku Kaishi,* 1115, 1979.
132. Budesinsky, B. and Gurovic, J., *Collect. Czech. Chem. Commun.,* 28, 1154, 1963.
133. Vytras, K. and Langmyhr, F. J., *Anal. Chim. Acta,* 92, 155, 1977.

ALIZARIN COMPLEXONE

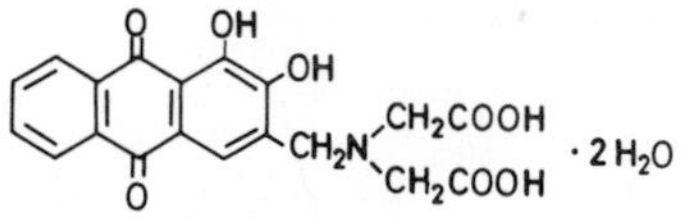

$C_{19}H_{15}NO_8 \cdot 2H_2O$
mol wt = 421.36

H₄L

Synonyms

3-[Di(carboxymethyl)aminomethyl]-1,2-dihydroxyanthraquinone, 1,2-dihydroxy-anthraquinon-3-ylmethylamine-N,N-diacetic acid, Alizarin Fluorine Blue, Alizarin Complexane, ALC

Source and Method of Synthesis

Commercially available. Prepared by the Mannich condensation of alizarin with iminodiacetic acid.[1,2]

Analytical Uses

Metal indicator in chelatometric titration and photometric reagent for Al, Co, Cu, In, Mn, Ni, Zn, and rare earths. Lanthanum complex is well known as a highly sensitive and specific photometric reagent for F⁻.

Properties of Reagent

Usually obtained as a free acid, dihydrate; yellow-brown crystalline powder, mp 190° (with decomposition); almost insoluble in water, ether, alcohol and other non-polar organic solvents, but easily soluble in aqueous alkali. The color of aqueous solution changes with pH which corresponds to the proton dissociation steps of the reagent as shown below ($\mu = 0.1$).[4]

$$H_4L \xrightarrow[\text{(—COOH)}]{pKa_1 = 2.40} H_3L^- \xrightarrow[\text{(2—OH)}]{pKa_2 = 5.54} H_2L^{2-} \xrightarrow[\text{(≡N⁺H)}]{pKa_3 = 10.07} HL^{3-} \xrightarrow[\text{(1—OH)}]{pKa_4 = 11.98} L^{4-}$$

Yellow　　　　　Red　　　　　Red　　　　　Blue

The absorption spectra of each deprotonated species in aqueous solution are illustrated in Figure 1.

Complexation Reaction and Properties of Complexes

The metal ions which form colored chelates with Alizarin Complexone may be divided into two groups as listed below.[2] The first group of metal ions forms red chelate (MHL) at pH 4.3 to 4.6, where the free reagent is yellow. The second group of metal ions forms red-violet chelate (ML) at pH 10, where the free reagent is red.

Group 1　— Al, Cd, Ce(III), Co(II), Cu(II), Fe(III), Ca, Hg(II), In, La, Mn(II), Ni, Pb, Th, Ti(III), (IV), Zn, Zr, and rare earths

Group 2　— Ba, Ca, Cd, Mg, Mn(II), Ni, and Sr

The structure of MHL-type chelate is shown below (1).

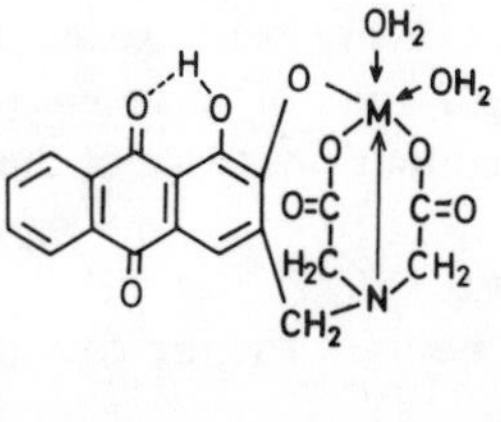

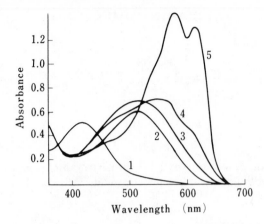

FIGURE 1.   Absorption spectra of Alizalin Complexone at various pH. (1) pH 4.6; (2) pH 7.0; (3) pH 10.0; (4) pH 11.5; (5) pH 13.0; dye, 9.63 ppm in water. (From Ingman, F., *Talanta*, 20, 135, 1973. With permission.)

(1)

The red rare earth chelate at pH $\sim$ 4 turns to blue in the presence of F⁻ due to the formation of a ternary complex (MLF).[2,5] Of the rare earth elements, the La(III) or Ce(III) chelate gives the most sensitive color change, and this reaction is utilized as a highly sensitive and specific photometric method for the determination of trace of fluoride ion.[3,6,7]

The similarity of the absorption spectrum of blue (MLF)-type chelate to that of free reagent (L^{4-}) suggests the mechanism of color reaction with F⁻. The coordination of F⁻ to the complexed rare earth ion may result in the weakening of O-H bond as shown in (2), thus making it possible to dissociate even at pH 4, where the free reagent does not dissociate.

(2)

The color reaction of rare earth chelate with F⁻ has been investigated by various workers, and different compositions ($Ce_5L_4F_4$,[8] $La_2L_2F_2$,[9] or $(La_5L_5F_2)_n$,[10]) have been proposed for the ternary chelates. Absorption spectra of LaHL and LaLF chelates are illustrated in Figure 2. The blue ternary chelate can be extracted into organic solvents if the coordinatated water on the rare earth ion is replaced with the hydrophobic ligand such as alkylamine or arylamine.[11] The extractability is dependent upon the nature of extracting solvents and amines,[12] and iso-butanol or iso-amylalcohol containing 5% *N,N*-diethylaniline gave the best result.[13,14] Figure 3 illustrates the absorption spectra of extracted species. Because most of the excess reagent remains in aqueous phase, the lower reagent blank (one tenth of that of aqueous photometry) contributes to the higher sensitivity of fluoride determination (twice of that of aqueous photometry).

Purification and Purity of Reagent

Most commercial samples can be used for the routine analytical works without pu-

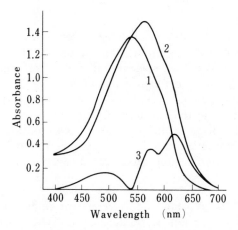

FIGURE 2. Absorption spectra of La-ALC complex and La-ALC-F ternary complex. (1) LaHL; (2) LaLF; (3) difference between LaHL and LaLF; dye concentration, 0.002 M; F⁻, 20 μg. (From Greenhalgh, R. and Riley, J P., *Anal. Chim. Acta,* 25, 179, 1961. With permission.)

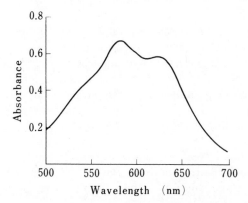

FIGURE 3. Absorption spectrum of La-ALC-F ternary complex in isoamylalcohol solution containing 5% *N,N*-diethylaniline; F, 0.5 ppm. As to the condition of color reaction, refer to the analytical procedure in the text. (From Hirano, S., Fujinuma, H., and Kasai, T., *Bunseki Kagaku,* 15, 1339, 1966. With permission.)

rification. However, it is necessary to purify them for the physicochemical measurements. The recommended procedure is as follows.[4]

> Suspend finely powdered 1-g sample in 50 mℓ of 0.1 M NaOH, filter the solution, and then extract alizarin with five successive portions of methylenedichloride. Precipitate the free reagent by dropwise addition of hydrochloric acid while stirring the solution in an ice bath. Filter the orange precipitate on a glass filter, wash with ice-water, and dry in a vacuum dessicator over solid potassium hydroxide.

Purity of the free acid can be checked either by alkalimetry or by photometric titration with a standard Al solution at pH 4.

Analytical Applications

Use of Ce or La Chelate as a Photometric Reagent for Fluoride

A mixture of aqueous solution containing 1:1 or 2:1 of La to reagent is recommended for this purpose. Premixed reagent preparations containing LaHL chelate and buffering reagents are also commercially available.[15]

The optimum pH for the color reaction is 4.5 to 4.7, and the nature of buffering reagents also affects the sensitivity. Succinate buffer is recommended for the general purpose, as the mixed reagent solution is stable for 9 months,[16] but acetate buffer is recommended for the extraction photometry. The presence of miscible organic solvent improves the sensitivity, and acetone (20 to 25%) gave the best result.[6,17,18] The color is stable for 1 hr.[2,18] The sensitivity may also be improved by the solvent extraction as discussed previously.

Cations, such as Al, Be, Th, and Zr, interfere the color reaction as they compete with the reagent for fluoride ion. Cations, such as Al, Cd, Co(II), Cr(III), Cu(II), Hg(I)(II), Fe(II)(III), Mn(II), Ni, Pb, V(IV), and Zn, also interfere as they form more stable chelate with Alizarin Complexone than the rare earth ion does. Anions, such as BO_3^- and PO_4^{3-}, also interfere as they form stable complexes with rare earth ions.[5]

Photometric Procedure for Fluoride in Aqueous Solution[R1]

Reagent solution — Instead of employing separate reagent solutions, a single mixed reagent solution of the following composition is used.

Transfer a few milliliters aqueous solution containing 47.9 mg Alizarin Complexone, 0.1 mℓ concentrated aqueous ammonia, and 1 mℓ of 20% ammonium acetate into a 200-mℓ volumetric flask containing a solution of 8.2 g sodium acetate and 6 mℓ glacial acetic acid in a minimum amount of water. To this solution, add 100 mℓ acetone with stirring, then add a solution of 40.8 mg La_2O_3 in 2.5 mℓ of 2 N HCl, and finally make to the volume with water. This solution is stable for 2 weeks.

Procedure — Place a weakly alkaline or neutral sample solution containing 3 to 30 µg F⁻ in a 25-mℓ volumetric flask. To this, add 8.00 mℓ of mixed reagent solution and dilute it to the volume. After 30 min, measure the absorbance at 620 nm against a reagent blank using 1-cm cell.

Extraction Photometric Procedure for Fluoride[13]

Place a 20- to 30-mℓ sample solution containing 0.5 to 10 µg F⁻ in a 100-mℓ separatory funnel, add successively with stirring 3 mℓ of 0.001 M Alizarin Complexone solution, 3 mℓ of acetate buffer solution (pH 4.4), 8 mℓ of acetone, and 3 mℓ of 0.01 M lanthanum chloride solution, and dilute the mixture to 50 mℓ. Extract the blue complex by shaking the aqueous solution with 10 mℓ of iso-amyl alcohol containing 5 vol% of N,N-diethylaniline for 3 min with a mechanical shaker. After phase separation, measure the absorbance of organic phase at 570 nm against a reagent blank. Apparent molar absorptivity is 2.35×10^4, and Beer's law is observed in a range of 0 to 10 ppm F⁻.

Other Uses

Alizarin Complexone has been recommended for the photometric determination of metal ions, such as Al (pH 4.1 to 4.3, 455 nm),[19] Cu(II) (pH 3.5),[20] Mn(II) (in alkali, 570 nm),[21] and Co(II), Ni(II) (pH 4.5, 500 nm),[22] but nothing is more advantageous with the use of this reagent.

It is also used as a metal indicator in the chelatometric titration of Co(II), Cu(II), In, Pb, and Zn, but Xylenol Orange and Methyl Thymol Blue are the better indicators in such cases .

Other Reagents with Related Structure

Sulfonated Alizarin Complexone (ALC-5S, AFBS)

Sulfonate group is introduced on 5 position of alizarin. As expected, this reagent is more soluble in water than the parent compound.[23,24] The behavior of La chelate with F^- has been investigated in some detail, and this reagent is claimed to be slightly more sensitive than the parent compound.[25,26]

Quinalizarin Complexone

$C_{19}H_{15}NO_{10}$
mol wt = 417.33

Rare earth (Ce, Nd, Pr, and Sm) chelates of the reagent behave similarly to those of Alizarin Complexone. Sensitivity for the fluoride determination is reported to be 1.5-fold higher than that of the latter.[27,28]

MONOGRAPH

M1 **Leonard, M. A.**, Alizarin Fluorine Blue, in *Organic Reagents for Metals,* Vol. 2, Johnson, W. C., Ed., Hopkin & Williams, Chadwell Heath, England, 1964.

REVIEW

R1 **Hashitani, H.**, Direct-photometric determination of fluoride in river and sea water using Alizarin Complexone, *Kogyo Yosui* 85, 23, 1965.

REFERENCES

1. Belcher, R., Leonard, M. A., and West, T. S., *J. Chem. Soc.,* 2390, 1958.
2. Leonard, M. A. and West, T. S., *J. Chem. Soc.,* 4477, 1960.
3. Analytical Methods Committee, *Analyst,* 96, 384, 1971.
4. Ingman, F., *Talanta,* 20, 135, 1973.
5. Belcher, R. and West, T. S., *Talanta,* 8, 853, 1961.
6. Belcher, R. and West, T. S., *Talanta,* 8, 863, 1961.
7. Greenhalgh, R. and Riley, J. O., *Anal. Chim. Acta,* 25, 179, 1960.
8. Jeffery, P. G. and Williams, D., *Analyst,* 86, 590, 1961.
9. Langmyhr, F. J., Klausen, K. S., and Nouri-Nekoui, M. H., *Anal. Chim. Acta,* 57, 341, 1971.

10. Anfält, R. and Jagner, D., *Anal. Chim. Acta,* 70, 365, 1974.
11. Belcher, R., Leonard, M. A., and West, T. S., *J. Chem. Soc.,* 3577, 1959.
12. Hall, R. J., *Analyst,* 88, 76, 1963.
13. Hirano, S., Fujinuma, H., and Kasai, T., *Bunseki Kagaku,* 15, 1339, 1966.
14. Hirano, S., Fujinuma, H., Yoshida, Y., and Makita, Y., *Bunseki Kagaku,* 18, 516, 1969.
15. For example, Amadac F® (Burdick & Jackson Laboratories Inc., Michigan) and Dotite Alfusone® (Dojindo Laboratories, Ltd., Kumamoto, Japan).
16. Okutani, T. and Utsumi, S., *Bunseki Kagaku,* 24, 196, 1975.
17. Hashitani, H., Yoshida, H., and Muto, H., *Bunseki Kagaku,* 16, 44, 1967.
18. Yamamura, S. S., Wade, M. A., and Sikes, J. H., *Anal. Chem.,* 34, 1308, 1962.
19. Ingman, F., *Talanta,* 20, 999, 1973.
20. Negoiu, D., Panait, Cl., and Pantaji, D., *An. Univ. Bucuresti Chim.,* 18, 27, 1969; *Chem. Abstr.,* 74, 91819r, 1971.
21. Capitan, F., Roman, M., and Guiraum, A., *Quim. Ind. (Madrid),* 17, 15, 1971; *Chem. Abstr.,* 75, 58318z, 1971.
22. Leonard, M. A. and Nagi, F. I., *Talanta,* 16, 1104, 1969.
23. Leonard M. A. and Murray, G. T., *Analyst,* 99, 645, 1974.
24. Leonard, M. A., *Analyst,* 100, 275, 1975.
25. Deane, S. F. and Leonard, M. A., *Analyst,* 102, 340, 1977.
26. Deane, S. F., Leonard, M. A., McKee, V., and Svehla, G., *Analyst,* 103, 1134, 1978.
27. Barmina, G. M., Zhivopitsev, V. P., and Minin, A. A., *Tr. Estestvennonauchn. Inst. Permsk. Gos. Univ.,* 13, 240, 1975; *Anal, Abstr.,* 32, 5A6, 1976.
28. Minin, A. A., Barmina, G. A., and Khahkalkina, I. G., *Zh. Anal. Khim.,* 30, 2196, 1975.

8-HYDROXYQUINOLINE

C_9H_7NO
mol wt = 145.16

HL

Synonyms
8-Quinolinol, Oxine

Source and Method of Synthesis
Commercially available. Obtained by the sulfonation of quinoline, followed by alkali fusion.[1]

Analytical Uses
As a solvent extraction reagent, extraction photometric reagent, and precipitation reagent for a wide variety of metal ions other than univalent cations. The selectivity can be improved by choosing a proper pH range and masking agents.

Properties of Reagent
Colorless cyrstals or crystalline powder, mp 74 to 76°[M4], bp about 267°, sublimes above 31° (2×10^{-2} Torr);[2] almost insoluble in water and ether; easily soluble in common organic solvents and acids such as acetic acid or mineral acids. A solution in anhydrous solvents is colorless, but becomes yellow in the presence of moisture. 8-Hydroxyquinoline is stable in solid as well as in solution, but should be kept in a dark place; pKa (N^+H) = 4.85 and pKa (OH) = 9.95 (μ = 0.1 $NaClO_4$, 25°).[3]

Absorption spectral data of the reagent in UV region are summarized in Table 1. As expected, marked changes of spectra are observed on each species at various stage of protonation.[R3] Absorption spectra of neutral oxine in organic solvents are very much dependent upon the nature of solvents, since the following equilibrium is affected by the solvent porality.

(neutral)   (Zwitter ion)

Oxine has been widely used as an extraction reagent for metals, and the distribution coefficients of oxine in various organic solvent-aqueous systems are summarized in Table 2.

Complexation Reactions and Properties of Complexes
Oxine behaves as a bidentate (N,O^-) univalent anionic ligand to form uncharged chelates with wide range of metal ions. Cations of +n charge and coordination number of 2n form so-called "coordination saturated uncharged chelates" which are insoluble in water, but easily soluble in organic solvents (structure (1) for iron (III)-oxinate as an example). When the coordination number of the metal is greater than 2n, the uncoordinated sites of metal ion are often occupied by water as illustrated by the structure (2) for nickel-oxinate as an example. Although this type of chelate is fairly insoluble

Table 1
SPECTRAL CHARACTERISTICS OF OXINE[4]

Species	λ_{max} (nm)	ε ($\times 10^3$)
Neutral (pH 7.6)	239	32.4
HL	305	2.63
Zwitter ion	270	2.84
HL±	431	0.064
Cation (pH = 1)	251	3.16
H_2L^+	308	1.48
	319	1.55
Anion (pH = 12)	252	30.2
L^-	334	2.88
	352	2.82

Reprinted with permission from Anderson, P. D. and Hercules, D. M., *Anal. Chem.*, 38, 1702, 1966. Copyright 1966 American Chemical Society.

Table 2
DISTRIBUTION COEFFICIENT OF OXINE AND 2-METHYLOXINE[3]
($25 \pm 0.2°$, $\mu = 0.10$)

Organic solvent	K_D(org solv/water)	
	Oxine	2-Methyloxine
1-Butanol	45.4	82.6
Carbon tetrachloride	116	435
MIBK	136	314
Toluene	162	557
o-Dichlorobenzene	303	1003
Dichloromethane	377	1248
Chloroform	433	1670

in water, it is also not easily soluble in organic solvents due to the presence of coordinated water (solubility of $ZnL_2 \cdot 2H_2O$ in $CHCl_3$, 2×10^{-6} *M*).[M3]

(1)   (2)

For example, when a solution of Cd, Zn, Ni, or Mg is shaken with a chloroform solution of oxine at the proper pH, these metals may be extracted as ML_2-type chelate, but in a few minutes, they become hydrated and precipitate out of the chloroform phase. However, such chelates can be successfully extracted into organic phases, either by adding auxiliary uncharged ligand, such as pyridine, 1,10-phenanthroline, alkylamine, or excess oxine, or by using coordinating solvent, such as MIBK or amylalcohol, as an extracting solvent.

As the phenolic group of oxine is an weak acid, the extent of complexation reaction with metal ion is very much dependent upon the pH of reaction medium. The optimum pH for the complete precipitation and for the quantitative extraction with chloroform are summarized in Table 3. The extent of extraction is also dependent upon the concentration of oxine in the organic phase. The pH values shown in Table 3 are those using 0.001 to 0.01 M oxine solutions. The extraction constants with chloroform, and pH values for 50% extraction are also included in the same table. Extraction is usually more rapid at higher oxine concentration and higher pH values.

Some of the uncharged, coordination saturated oxinates tend to sublime under the reduced pressure, and this property is utilized for the separation of metal oxinates by the fractional sublimation.[2] Sublimation temperatures of several metal oxinates are shown in Table 3.

Stability constants of metal oxinates are usually observed in aqueous organic solvents due to their poor solubility in water. Representative values are compiled in Table 4.

The chloroform solution of metal oxinates are usually yellow and show an absorption band in the visible range, where free oxine absorbs very little. Exceptions are oxinates of Fe(III) (green-black) and V(V) (magenta black) which show multiple bands.[R3] The visible absorption spectra of some metal oxinates in chloroform are illustrated in Figure 1, and the spectral characteristics of metal oxinates which are of analytical importance are summarized in Table 5. Although their molar absorptivities are not so high, oxine forms chelates with wide variety of metal ions, so that oxine has been accepted as one of the commonest organic reagents for metal analysis. The selectivity may be improved by the proper selection of pH and masking agents in the solvent extraction step.

In chloroform, oxine itself shows very weak fluorescence under UV light, however, oxinates of diamagnetic metal ions show strong fluorescence in the same condition. Based on this principle, the fluorimetric determination of traces of metals have been proposed. The metal oxinates which fluoresce strongly enough for trace determination are Al (λ_{em},510 nm), Ga(526 nm), In(528 nm),[58] Nb(580 nm),[67] Sc(533 nm),[68] Zn(530 nm), and Zr(520 nm).[76]

The chloroform solutions of metal oxinates show some photodecomposition resulting in the slow decrease in absorbance. The reason for this may be attributed to the formation of phosgene which is a photodecomposition product of chloroform.[80]

Purification and Purity of Reagent

Oxine is a well-defined compound with a sharp melting point. Crude oxine may be purified through copper oxinate followed by the liberation of free ligand with H_2S or by the steam distillation after acidification with H_2SO_4. The melting point (73 to 75°) is a good criterion for the purity.

Analytical Applications

Use as an Extraction and Photometric Reagent

The extraction behaviors of metal oxinates in a water-chroloform system are summarized in Table 3. The conditions for the extraction of metal ions may also be found

Table 3

PRECIPITATION AND EXTRACTION BEHAVIOR OF METAL OXINATES[M3]

Metal ion[a]	Precipitation[M4,R1]		Extraction[M6,R2,R3,6,7] into CHCl₃			Remarks	Ref.
	pH for complete precipitation	Composition	$\log K'_{ex}$[b]	X	$pH_{1/2}$		
Ag	$6.1 \sim 11.6$	$ML \cdot HL$	-4.5	—	6.5	$\sim$90% extraction at pH 8 $\sim$ 9(0.1 MHL)	
Al	$4.2 \sim 9.8$	$ML_3 \cdot H_2O$	-5.2	0	2.9	$\log K_{sp} \simeq -32.3(18°)$; quantitative extraction at pH $4.5 \sim 10$ (0.01 MHL)	2,56
Ba	—[c]	—	-20.9	2	—	Incomplete extraction at pH $\simeq$11 (1 MHL)	
Be	9^d	ML_2	-9.6	—	5.8	Maximum 87% extraction at pH $6 \sim 10$ (0.5 MHL)	
Bi	$5.0 \sim 8.3^d$	$ML_3 \cdot H_2O$	-1.2	0	2.1	Quantitative extraction at pH $4 \sim 5$(0.1 MHL)	
Ca	$9 \sim 13$	$\{ML_2 \cdot 2H_2O, M_2 \cdot HL$	-17.9	1	10.4	Quantitative extraction at pH $\simeq 11$ as $ML_2 \cdot HL$(0.5 MHL), or in presence of $BuNH_2$ as $ML_2 \cdot BuNH_3L$	
Cd	$5.7 \sim 14.6$	$\{ML_2 \cdot 2H_2O, ML_2 \cdot 2HL$	-5.3	2	4.7	$\log K_{sp} = -26.2(25°)$. quantitative extraction at pH $5 \sim 9$(0.1 MHL); Zephiramine® prevents extraction	9,25
Co(II)	$4.3 \sim 14.5$	$\{ML_2 \cdot 2H_2O, ML_2 \cdot HL, ML_2 \cdot 2HL$	-2.2	2	3.2	$\log K_{sp} = -24.1(25°)$, sublimes at 2,10-12,53 $288°/2 \times 10^{-2}$ Torr. quantitative extraction, at pH $4.5 \sim 10.5$; (0.1 MHL)	

Cu(II)	5.3～14.6	{ ML$_2$·2H$_2$O ML$_2$	1.8	0	1.5	log K$_{sp}$ = −28.8(20°), sublimes at 196°/2×10^{-2} Torr. Quantitative extraction at pH 2～12 (0.1 MHL)	3,121
Fe(III)	2.8～11.2	ML$_3$	4.1	0	1.0	log K$_{sp}$ ≃ −47(18°), Sublimes 252°/2×10^{-2} Torr. Quantitative extraction at pH 2～11 (0.01 MHL MHL)	2,15,16,56,81
Ga	3.6～11.0	ML$_3$	3.7	0	1.1	Quantitative extraction at pH 3～12 (0.01 MHL)	6,82
Hf	4.5～11.3^d	ML$_4$		0	1.3	Quantitative extraction at pH 4.5～11.3	
In	2.5～3.0	ML$_3$·H$_2$O	0.9～1.4	0	1.5	Sublimes 242°/2×10^{-2} Torr Quantitative extraction at pH 4～12 (0.01 MHL)	
La	6.5～11.3^d	ML$_3$	−16.4	0	6.5	Quantitative extraction at pH 7～10 (0.1 MHL)	3,54
Mg	9.4～12.7	{ ML$_2$·2H$_2$O ML$_2$·4H$_2$O	−15.3	0	9.6	log K$_{sp}$ = −15.2(18°). Hydrated oxinate is not extd. in CHCl$_3$, but extd. quantitatively in presence of BuNH$_2$, Pyridine or Zephiramine as ML$_2$X. 18～20	18—20
Mn(II)	5.9～9.5	{ ML$_2$·2H$_2$O ML$_2$·HL	−9.3	0	5.7	Quantitative extraction at pH 9～12 in presence of NH$_2$NH$_2$ or NH$_2$OH	
Mo(VI)	3.3～7.6	MO$_2$L$_2$	9.9	—	0.5	Quantitative extraction at pH 2～5 (0.01 MHL), but slow.	
Ni	3.8～9.2	{ ML$_2$·2H$_2$O ML$_2$·HL	−2.2	2	2.4	Sublimes 256° (2×10^{-2} Torr); Quantitative extraction at pH 5～8.5 (0.1 M HL), but very slow	23—26,68
Pb	8.4～12.3	{ ML$_2$1～2H$_2$O ML$_2$	−8.0	0	5.0	Quantitative extraction at pH 8.2～11.0(0.1 MHL)	

Table 3 (continued)
PRECIPITATION AND EXTRACTION BEHAVIOR OF METAL OXINATES[M3]

Metal ion[a]	Precipitation[M4.1]		Extraction[M6,R2,R3,6,7] into CHCl$_3$			Remarks	Ref.
	pH for complete precipitation	Composition	log K'_{ex}[b]	X	pH$_{1/2}$		
Pd(II)	3.0 ~ 11.6	{ ML$_2$ M(HL)$_2$Cl$_2$	~15	0	<0	Quantitative extraction at pH 3 ~ 11 (1 ~ 2% HL), but very slow	27,59
Sc	6.5 ~ 8.5	{ ML$_3$·HL (ML$_3$)$_2$HL	−6.6	1 ~ 0	3.6	Quantitative extraction at pH 4.5 ~ 10 (0.1 MHL)	30,31
Sm	—	ML$_3$	−13.4	—	5.7	Quantitative extraction at pH 6 ~ 8.5 (0.5 MHL)	
Sr	—[c]		−19.7	2	12.1	Quantitative extraction at 11.3 in presence of 4% BuNH$_2$	60
Th	4.4 ~ 4.8	{ ML$_4$ ML$_4$·HL	−7.2	0	2.9	Quantitative extraction, at pH 4 ~ 10 (0.1 MHL)	36
Tl(I)	>4[d]	ML·HL			11		
Tl(III)	4 ~ 8	ML$_3$	~5	0	2.1	Reduced to Tl(I)L at >130° or by light; quantitative extraction at pH 5.5 (0.01 MHL)	38
U(VI)	4.1 ~ 4.8	{ MO$_2$L$_2$·HL MO$_2$L$_2$·2HL	−1.6	1	2.6	Quantitative extraction at pH 7 ~ 9 (0.1 MHL)	39—41
V(V)	2.9 ~ 5.5	{ MO$_2$L MO(OH)L$_2$ MO(OH)L$_2$·HL	1.7	—	0.9	Quantitative extraction as MO(OH)L$_2$ at pH 2 ~ 4. No extraction above pH 9	42—45
Y	4.8 ~ 8.5	{ ML$_3$ ML$_3$ HL	—	—	6	Nearly quantitative extraction at pH 7 ~ 10 (0.2 M HL)	48

Zn	4.6 ~ 13.4	$\begin{cases} ML_2 \cdot 2H_2O \\ ML_2 \cdot HL \end{cases}$	−2.4	2	3.3	$\log K_{sp} = -27.7(25°)$; sublimes at 246° (2×10^{-2} Torr); hydrate precipitates out from $CHCl_3$ phase on continued shaking at higher pH, but is prevented in presence of pyridine or Zephiramine®.	2,25,49,50,52
Zr	5.2 ~ 10.0	MOL_2	2.7	—	1.0	Quantitative extraction at pH 4.5 ~ 11.3	

[a] Other elements investigated with oxine include As,[8] Au,[8] Ce, Cr(III),[13] Eu,[14] Ge,[14] Hg,[17] Mo(V),[21] Nb[22,35] Nd,[14] Pt,[27] Pu,[28] Rh,[59] Ru,[29] Sb(III), Sn,[33,34] Tb, Ti[37] and W(VI).[46,47]

[b] As some metal oxinates are extracted as oxine adducts($ML_n \cdot XHL$), the effective constant as defined below is more convenient to evaluate the extractability of such system.

$$K'_{ex} = \frac{[ML_n X \, HL]_{org} \cdot [H]^n}{\Sigma \, [M] \, [HL]_{org}^{n+x}}$$

The values for x are indicated in Table 3. $\Sigma\{M\}$ represents the sum of concentrations of aqueous species containing metal ion.

[c] No precipitation.

[d] Precipitation is not stoichiometric.

Table 4
CHELATE STABILITY CONSTANT OF METAL OXINATES[5]

Metal ion	log K_{ML}	log K_{ML2}	log K_{ML3}	Condition		Ref.
				Temp (°)	μ, medium	
Ag	5.20	4.36	—	18∿25	0.1 NaClO$_4$	
Ba	2.07	—	—	20	→ 0	
Be	3.36	—	—	—	—	
Ca	7.3	5.9	—	30	75% dioxane	
Cd	9.43	7.68	—	25	50% dioxane	
Ce	9.15	7.98	—	25	50% dioxane	
Co(II)	10.55	9.11	β_3 24.35	25	50% dioxane	10
Cu(II)	13.49	12.73	—	25	50% dioxane	
Fe(II)	8.71	β_2 16.83	β_3 22.13	25	0.3 M NaClO$_4$	
Fe(III)	—	—	β_3 38.00	25	0.3 M NaClO$_4$	
Ga	14.51	13.50	12.49	20	0.1 NaClO$_4$	
In	12.0	β_3 24.0	β_3 35.5	25	0.1 NaClO$_4$	54
La	8.66	7.74	—	25	50% dioxane	
Mg	6.38	5.43	—	25	50% dioxane	
Mn(II)	8.28	7.17	β_3 26.80	25	50% dioxane	24
Ni	11.44	9.94	—	25	50% dioxane	
Pb	10.61	8.09	—	25	50% dioxane	
Sc	—	—	≃9	—	Absolute ethanol	31
Sm	6.84	—	β_3 19.50	25	0.1 NaClO$_4$	
Sr	2.89	—	β_3 3.19	25	0.1 NaClO$_4$	
Th	—	β_2 21.3	9.42	25	0.1 NaClO$_4$	
UO$_2^{2+}$	11.25	9.64	—	20	50% dioxane	
VO^{2+}	10.97	β_2 20.19	—	25	0.085	
Zn	9.96	8.90	—	25	50% dioxane	

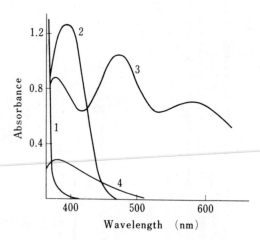

FIGURE 1. Absorption spectra of Al, Fe(III), and UO$_2$ oxinates in chloroform. (1) Oxine; (2) Al 50 μg; (3) Fe(III) 100 μg; (4) UO$_2^{2+}$ (100 μg as U). (From Motojima, K., *Bunseki Kagaku*, 8, 66, 1959. With permission.)

in Table 5 which summarizes the applications of oxine as a photometric reagent. The proper choice of pH value for the extraction and of the masking agents are necessary for the higher selectivity. Chloroform is the most preferred solvent for extraction. The

Table 5
APPLICATION OF OXINE AS A PHOTOMETRIC REAGENT

Metal ion[a]	Condition[b]	Chelate			Range of determination (ppm)	Remarks	Ref.
		Ratio	λ max (nm)	ε(×10³)			
Al	pH 4.5~11.3, tartarate, EDTA, H₂O₂	ML₃	386~390	6.97	0.5~5	—	56,57
Co(II)	pH 3.3~9.9, H₂O₂	ML₃	420	9.48	0.3~3	No extraction at pH <4	55
Cr(III)	pH 6.0~8.0	ML₃	420	7.28	0.5~8		61
Ga	pH 3.0~11.7	ML₃	393	6.47	3.5~10	Extn. with iso-Am-OH + CHCl₃	63
Ga	pH 2.9~3.3, 70° MTB(X)	ML₂X	600	19.5	—		64
Hf	pH 4.5~11.3	ML₄	385	22.9	0.6~12	Al(by F⁻) and U(by acetate) can be masked.	65
In	pH >4.0	ML₃	395	6.67	0.8~18	No extraction at pH <6	63
Mn(II)	pH 9.0~10.5	ML₂	395	10.3	0.3~7	Ag,As,Bi,Cd,	33
Mo(VI)	pH 0.7~1.6	MO₂L₂	368	7.93	0.3~10	Co,Mn,Pb,Sb,V interfere	
Nb(V)	pH 2.8~10.5, EDTA, Oxalate	MO₂L₃	385	9.75	2.5~6		22
Pd(II)	pH 3.0~11.2	ML₂	435	7.15	0.5~15	Cl⁻,NH₄⁺, reductant interfere seriously	59
Rh	pH 4.2~8.2	ML₃	426	10.8	0.3~10	Cl⁻,NH₄⁺ interfere	59
Ru(IV)	pH 4.0~6.5	—	430	11.9	0.2~8	Co,Cr,Mo,Rh give positive error.	29
Th	pH 7.5~9.5	ML₄	380	11.6	1~20		61
Tl(III)	pH >4.0	ML₃	402	7.8	1.5~30		71.72
U(VI)	pH 2~2.5	MO₂L₂ HL	380	6.8~7.1	2~40		73
V(III)	pH 8.5~10	ML₃	420~5	12.7	~7	V(V) is reduced to V(III) with Na₂S₂O₄. Extraction with CCl₄	

a Many other elements have been determined by the oxine photometry; Au(III),[8] Be,[60] Bi,[60] Ca,[60] Cu(II),[62] Fe(III),[15,16,56,61] Mg,[18,66] Ni,[61,68] Pb, Sn(IV),[33] Ta(IV),[69] V(V),[70] Ti,[69] Zr,[77] and rare earths,[79] but the molar absorptivity of these oxinates is not higher than 6 × 10³.

b Otherwise mentioned, metal-oxinates were extracted into chloroform.

sensitivity of photometric determination of metals is not so high, as the molar absorptivity is in a order of 10^3 to 10^4, but oxine is still a useful photometric reagent because of its wide applicability.

Use as a Precipitating Reagent

Since the introduction of oxine as an analytical reagent by Berg in 1927,[32] the main use in early years has been as a precipitating reagent for the separations and gravimetric determination of metal ions. The precipitate of metal oxinates can also be determined volumetrically. Oxine reacts readily and quantitatively with bromine to form 5,7-dibromoxine. The metal oxinates are dissolved in warm hydrochloric acid and treated with KBr and an excess of $KBrO_3$ solution. After adding KI, the excess bromate is determined by titrating the freed iodine with a standard $Na_2S_2O_3$ solution.[83]

PFHS (precipitation from homogeneous solution) technique has been recommended for the preparation of metal oxinates which are more dense, more easily filtered, and less contaminated with excess ligand and accompanying elements than those prepared by the conventional procedures.[M2,R4]

In PFSH method, oxine is generated by the hydrolysis of 8-acetoxyquinoline (mp 56.2 to 6.5°)[84] to precipitate metal ions (AlL_3[85,86] BiL_3,[87] CuL_2,[88] GaL_3,[89] InL_3[90], MgL_2,[91] $SbOL \cdot 2HL$,[92] ThL_4 or $ThL_4 \cdot HL$,[93] $UO_2L_2 \cdot HL$ or $(UO_2L_2)_2 \cdot HL$,[94,95] and ZnL_2).[96] Alternatively, pH of the reaction medium is raised by heating the aqueous solution containing oxine, metal ion, and urea (AlL_3[99] BeL_2,[98] CrL_3[99] MgL_2[100] and $NbOL_3$[35]). If urease is added to this mixture, the reaction can be carried out at room temperature.[112]

Chromatographic separations of metal oxinates have been investigated in some details. Metal oxinates can be separated on a column[101,146] or on a thin layer.[102] Metal ions may also be separated on a filter paper impregnated with oxine,[103,104] silica beads with immobilized oxine,[105] or on a polymer with oxine functional group.[106,107]

Other Reagents with Related Structure
2-Methyloxine (8-hydroxyquinaldine)

$$C_{10}H_9NO$$
$$\text{mol wt} = 159.19$$

HL

Colorless flakes, mp 73 to 74,[108] bp 266 to 267°[M4]; $pKa(N^+H) = 4.58$ and $pKa(OH) = 11.71$ (50% dioxane, 25°).[147] The values for K_D are summarized in Table 2. It shows quite similar physical properties and chelating behavior to oxine, with the exceptions for Al and Be. Aluminum can not be precipitated with 2-methyloxine.[109] This is explained to be due to the steric effect of methyl group on 2 position which does not allow to form ML_3-type chelate with Al^{3+} which is rather small in ionic radius. Thus, 2-methyloxine is recommended as a substitute for oxine when a large excess of Al interferes the metal determination.[110]

According to the recent work, however, Al is partially extracted with 2-methyloxine in acetate buffer. This may be explained to be due to the presence of acetate ion which forms an ion pair with the bis-chelate (AlL_2^+) in the chlorofo extraction.[111]

Beryllium can be precipitated quantitatively with 2-methyloxine as BeL_2, whereas precipitation is incomplete with oxine.[109,112] The reagent has been also used as an extraction photometric reagent for various metal ions, including Mn(II),[113] Pd(II),[114] and V(V).[6]

8-Acetoxyquinaldine (mp 63 to 64°)[115] is used as a precursor of 2-methyloxine in PFHS method to precipitate ThL_4,[116] ZnL_2,[117] CuL_2,[118] and InL_3.[119]

Oxine-5-sulfonic Acid

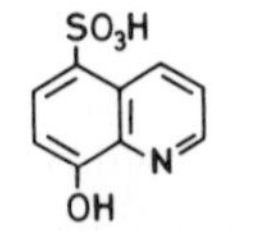

$$SO_3H$$

$C_9H_7NO_4S$
mol wt = 225.22

OH

H_2L

Pale yellow crystalline powder. Easily soluble in water; $pKa(N^+H) = 4.10$ and $pKa(OH) = 8.76$ ($\mu \rightarrow 0$, 25°). [120] The chelating behavior is similar to that of oxine, but the chelates are soluble in water. Hence, the reagent has been used as a photometric reagent (Fe(III),[122] Ta(V),[121] and V[122]) or as in fluorimetric reagent (Al,[123] Cd,[124] Ce(IV),[125] Ga, In,[123,126] Mg[127] and Zn[128]) for metal ions in aqueous solution. As the extraction step is unnecessary, the procedure is simpler than with oxine. On the other hand, sensitivity and selectivity may be sacrificed by the elimination of the extraction step. The resulting metal chelates in aqueous phase may be extracted into aromatic solvents or chloroform as an ion-pair with long-chain alkylamines[129] or Zephiramine® [130,131] for the subsequent photometry.

Ferron (8-Hydroxy-7-iodoquinoline-5-sulfonic acid, 7-Iodooxine-5-sulfonic acid)

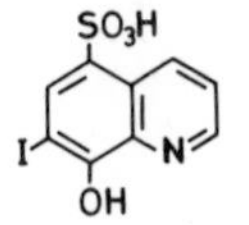

$$SO_3H$$

$C_9H_6NO_4SI$
mol wt = 351.12

OH

H_2L

Pale yellow powder, mp 225° (with decomposition). Sparingly soluble in water (~0.2%), slightly soluble in alcohol, and insoluble in other organic solvents; $pKa(N^+H) = 2.50$ and $pKa(OH) = 7.10$ ($\mu = 0.1$ KCl).[132] Chelating behavior is quite similar to that of oxine-5-sulfonic acid. Although Ferron has been originally developed as a medicine, it reacts with Fe(III) and V(V) to form colored soluble chelates, so that the reagent can be used as a photometric reagent for these and some other elements: Al,[133] Ca,[134] Fe(III),[135] Mo(VI),[136,138] Nb,[137] U(VI),[138,139] and V(V).[138]

*Kelex 100® ***

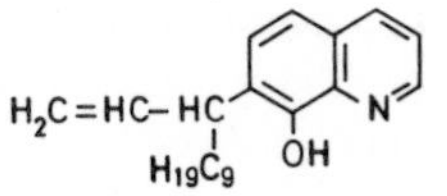

$$H_2C=HC-HC$$
$$H_{19}C_9 \quad OH$$

$C_{21}H_{29}NO$
mol wt = 311.47

HL

The main component of Kelex 100® is 7-dodecenyl-8-hydroxyquinoline as shown above. Viscous oil which is soluble in nonpolar solvents such as hexane or kerosene. The purification and characterization of the commercial product has been reported.[140] Due to the long side-chain attached to quinoline ring, the partition of the reagent as well as the metal chelates are highly favorable to the organic phase, and Kelex 100® is expected to be a successful extraction reagent for the analytical separation,[141-143]

* Trade name of Ashland Chemical Co., Columbus, Ohio.

although it has been developed as a solvent extraction reagent for the industrial separation of metal ions.[144,145]

Azodye Derivatives of Oxine

A large number of azodye derived from oxine has been synthesized and evaluated as an analytical reagent. Details may be found in the review.[R5]

8-Mercaptoquinoline (Thio-oxine) (See page 253.)

MONOGRAPHS

M1. Hollingshead, R. G. W., *Oxine and Its Derivatives,* Vols. 1 to 4, H. K. Lewis, London, 1954 to 1956.

M2. Gordon, L., Salutsky, M. L., and Willard, H. H., *Precipitation from Homogeneous Solution,* John Wiley & Sons, New York, 1959.

M3. Sandell, E. B. and Onishi, H., *Photometric Determination of Traces of Metals,* 4th ed., John Wiley & Sons, New York, 1978, 415.

M4. Welcher, F. J., 8-Hydroxyquinoline, in *Organic Analytical Reagents,* Vol. 1, Von Nostrand, Princeton, N.J., 1947, 264.

M5. Motojima, K., Oxine and Its Derivatives, in *Shin Bunseki Kagaku Koza,* Vol. 6, Nippon Bunseki Kagakukai, Ed., Kyoritsu, Tokyo, 1959.

M6. Stary, J., *The Solvent Extraction of Metal Chelates,* Pergamon Press, New York, 1964.

REVIEWS

R1. Bock, R. and Umland, F., Über die pH-Bereiche der Fällung einiger Metal-Oxychinolin-Verbindungen, *Angew. Chem.,* 67, 420, 1955.

R2. Umland, F., Distribution of metal 8-quinolinolates between water and organic solvents. VII. Review of the pH range of partition of metal 8-quinolinolates between water and chloroform, *Fresenius Z. Anal. Chem.,* 190, 186, 1962.

R3. Motojima, K., The analytical use of oxine, *Bunseki Kagaku,* 8, 66, 1959.

R4. Firsching, F. H., Precipitation of metal chelates from homogeneous solution, *Talanta,* 10, 1169, 1963.

R5. Ivanov, V. M. and Rudometkina, T. F., Application of 8-hydroxyquinoline azodye derivatives in analytical chemistry, *Zh. Anal. Khim.,* 33, 2426, 1978.

REFERENCES

1. Bedall, K. and Fisher, O., *Ber.,* 14, 442, 1366, 1881; *Ber* 15, 684, 1882.

2. Honjo, T., Imura, H., Shima, S., and Kiba, T., *Anal. Chem.,* 50, 1545, 1978.

3. Mottola, H. A. and Freiser, H., *Talanta,* 13, 55 1966.

4. Anderson, P. D. and Hercules, D. M., *Anal. Chem.,* 38, 1702, 1966.

5. Sillen, L. G. and Martell, A. E., *Stability Constants of Metal-Complexes,* Suppl. 1, The Chemical Society, London, England, 1964, 1971.

6. Motojima, K. and Hashitani, H., *Bunseki Kagaku,* 9, 151, 1960.

7. Stary, J., *Anal. Chim. Acta*, 28, 132, 1962.
8. Stevens, H. M., *Anal. Chim. Acta*, 18, 359, 1958.
9. Hellwege, H. and Schweitzer, G. K., *Anal. Chim. Acta*, 28, 236, 1963.
10. Oki, S., *Anal. Chim. Acta*, 50, 465, 1970.
11. Munakata, M., Niina, S., and Syozawa, S., *Bunseki Kagaku*, 26, 30, 1977.
12. Sekido, E. and Fujita, K., *Bull. Chem. Soc. Jpn.*, 49, 3073, 1976.
13. Beyermann, K., Rose, H. J., and Christian, R. P., *Anal. Chim. Acta*, 45, 51, 1969.
14. Shakhova, N. V., Kriseleva, O. A., and Zolotov, Yu. A., *Zh. Anal. Khim.*, 29, 1229, 1974.
15. Horiuchi, Y. and Nishida, H., *Bunseki Kagaku*, 17, 1325, 1968.
16. Horiuchi, Y. and Nishida, H., *Buneki Kagaku*, 19, 830, 1970.
17. Rudenko, N. P. and Kovtum, L. V., *Vestn. Mosk. Univ. Khim.* 24, 103, 1969; *Chem. Abstr.* 71, 95542f, 1969; *Zh. Anal. Khim.*, 24, 1390, 1969.
18. Umland, F. and Hoffman, W., *Anal. Chim. Acta*, 17, 234, 1957.
19. Nakaya, S. and Nishimura, M., *Bunseki Kagaku*, 22, 733, 1973.
20. Noriki, S. and Nishimura, M., *Anal. Chim. Acta*, 72, 339, 1974.
21. Basev, A. I. and Fan, C., *Zh. Anal. Khim .*, 15, 455, 1960.
22. Motojima, K. and Hashitani, H., *Anal. Chem.*, 33, 48, 1961.
23. Sekido, E., *Nippon Kagaku Zasshi*, 80, 1011, 1959.
24. Oki, S., *Anal. Chim. Acta*, 49, 455, 1970; *Anal. Chim. Acta*, 66, 201 (1973); *Anal. Chim. Acta*, 69, 220, 1974.
25. Takiyama, K. and Kozen, T., *Bunseki Kagaku*, 22, 291, 1973.
26. Sekido, E. and Kunikida, K., *Anal. Chim. Acta*, 92, 183, 1977.
27. Kukushkin, Yu. N., Sedova, G. N., Vlasova, R. A., and Belyaev, A. N., *Zh. Prikl. Khim. (Leningrad)*, 51, 2451, 1978; *Chem. Abstr.*, 90, 80171c, 1979.
28. Harrvey, G. B., Heal, H. G., Meddoch, A. G., and Rowley, E. L., *J. Chem. Soc.*, 1010, 1947.
29. Hashitani, H., Katsuyama, K., and Motojima, K., *Talanta*, 16, 1553, 1969.
30. Cardwell, T. J., and Magee, R. J., *Anal. Chim. Acta*, 36, 180, 1966.
31. Petronio, J. N. and Ohnesorge, W. E., *Anal. Chem.*, 39, 460, 1967.
32. Berg, B., *J. Prakt Chem.*, 115, 178, 1927; *Fresenius Z. Anal. Chem.*, 76, 191, 1929.
33. Eberle, A. R. and Lerner, M. W., *Anal. Chem.*, 34, 627, 1962.
34. Rakovskii, E. E. and Krylova, T. D., *Zh. Anal. Khim.*, 29, 910, 1974.
35. Kosta, L. and Dular, M., *Talanta*, 8, 265, 1961.
36. Corsini, A. and Abraham, J., *Talanta*, 17, 439, 1970.
37. Yamamoto, K., Sakai, F., and Ohashi, K., *Bunseki Kagaku*, 22, 918, 1973.
38. Goto, H., Kakita, Y., and Ichinose, N., *Nippon Kagaku Zasshi*, 88, 640, 1967.
39. Bordner, J. and Gordon, L., *Talanta*, 9, 1003, 1962.
40. Oki, S., *Anal. Chim. Acta*, 44, 315, 1969.
41. Motojima, K., Yamamoto, T., and Kato, Y., *Bunseki Kagaku*, 18, 208, 1969.
42. Shigematsu, T., Matsui, M., Munakata, M., and Naemura, T., *Bull. Inst. Chem. Res. Kyoto Univ. (Japan)*, 46, 262, 1968.
43. Yuchi, A., Yamada, S., and Tanaka, M., *Bull. Chem. Soc. Jpn.*, 52, 1643, 1979.
44. Tanaka, M. and Kojima, I., *J. Inorg. Nucl. Chem.*, 29, 1769, 1967.
45. Rao, V., Pandu, R., and Anjaneyulu, Y., *Mikrochim. Acta*, 481, 1973.
46. Vinogradov, A. V. and Dranova, M. I., *Zh. Anal. Khim.*, 23, 696, 1968.
47. Awad, K., Rudenko, N. P., Kuznetsov, V. I., and Gudym, L. S., *Talanta*, 18, 279, 1971.
48. Cardwell, T. J. and Magee, R. J., *Anal. Chim. Acta*, 43, 321, 1968.
49. Chou, F. C. and Freiser, H., *Anal. Chem.*, 40, 34, 1968.
50. Oki, S. and Terada, I., *Anal. Chim. Acta*, 61, 49, 1972.
51. Yamamoto, D., Tsukada, M., and Lynn, S., *Meiji Daigaku Nogakubu Kenkyu Hokoku*, 44, 25, 1978; *Chem. Abstr.*, 90, 114443k, 1979.
52. Sekido, E., Yoshimura, Y., and Masuda, Y., *J. Inorg. Nucl. Chem.*, 38, 1183, 1187, 1976.
53. Akaiwa, H., Kawamoto, H., and Saito, T., *Nippon Kagaku Zasshi*, 92, 1156, 1971.
54. Zolotov, Yu. A., and Lambrev, V. G., *Zh. Anal. Khim.*, 20, 1153, 1965, *J. Anal. Chem. USSR*, 20, 1204, 1965.
55. Sanbhla, D. S., Arora, O. P., and Misra, S. N., *Curr. Sci.*, 48, 253, 1978; *Chem. Abstr.*, 90, 214443d, 1979.
56. Motojima, K., *Nippon Kagaku Zasshi*, 76, 903, 1955.
57. Burke, K. E., *Anal. Chem.*, 38, 1608, 1966.
58. Nishikawa, Y., Hiraki, K., Morishige, K., Takahashi, K., and Shigematsu, T., *Bunseki Kagaku*, 25, 459, 1976.
59. Hashitani, H., Yoshida, H., and Motojima, K., *Bunseki Kagaku*, 18, 136, 1969.

60. Holzbecher, Z., Divis, L., Kral, M., Sucha, L., and Ulacil, F., *Handbook of Organic Reagent in Inorganic Analysis,* Ellis Horwood, Chichester, England, 1976.
61. Motojima, K. and Hashitani, H., *Bunseki Kagaku,* 7, 28 1958.
62. Motojima, K. and Hashitani, H., *Bunseki Kagaku,* 8, 526, 1959.
63. Moeller, T. and Cohen, A. J., *J. Am. Chem. Soc.,* 72, 3546, 1950.
64. Akhmedli, M. K., Gluschenko, E. L., and Gasanova, Z. L., *Uch. Zap. Azerb. Gos. Univ. Ser. Khim. Nauk,* 14, 1972; *Anal. Abstr.,* 25, 2183, 1973.
65. Motojima, K., Hashitani, H., and Yoshida, H., *Bunseki Kagaku,* 11, 659, 1962.
66. Haar, R. and Umland, F., *Fresenius Z. Anal. Chem.,* 191, 81, 1962.
67. Kirkbight, G. F., Thompson, J. V., and West, T. S., *Anal. Chem.,* 42, 782, 1970.
68. Oki, S. and Terada, I., *Anal. Chim. Acta,* 66, 201, 1973.
69. Issa, I. M., Issa, R. M., and Awadallah, R. M., *Egypt. J. Chem.,* 18, 215, 1975; *Anal. Abstr.,* 35, 4B125, 1978.
70. Hashitani, H. and Motojima, K., *Bunseki Kagaku,* 7, 478, 1958.
71. Motojima K., Yoshida, H., and Izawa, K., *Anal. Chem.,* 32, 1083, 1960.
72. Keil, R., *Fresenius Z. Anal. Chem.,* 244, 165, 1969.
73. Yatirajam, V. and Arya, S. P., *Talanta,* 23, 596, 1976.
74. Motojima, K., *Nippon Kagaku Zasshi,* 78, 533, 1957.
75. Kodama, M. and Tominaga, T., *Chem. Lett.,* 789, 1976.
76. Schneider, H. O. and Roselli, M. E., *Analyst,* 96, 330, 1971.
77. Van Santen, R. T., Schlewitz, J. H., and Toy, C. H., *Anal. Chim. Acta,* 33, 593, 1965. 78. Pyatnitsky, I. V. and Gavrilova, E. F., *Zh. Anal. Khim.,* 25, 445, 1970.
79. Keil, R., *Fresenius Z. Anal. Chem.,* 245, 362, 1969.
80. Linnel, R. H. and Raab, F. F., *Anal. Chem.,* 33, 154, 1961.
81. Rakovskii, E. E., Kalinichenko, V. I., and Petrukhin, O. M., *Tr. Samark. Gos. Univ.,* 206, 276, 1973; *Chem. Abstr.,* 79, 111472f, 1973.
82. Sudo, E. and Goto, H., *Trans. Natl. Res. Inst. Metals (Tokyo),* 5, 120, 1963; *Chem. Abstr.,* 60, 3481b, 1964.
83. Taketa, G. and Motojima, K., *Bunseki Kagaku,* 3, 493, 1954.
84. Salesin, E. D. and Gordon, L., *Talanta,* 4, 75, 1960.
85. Howick, L. C. and Trigg, W. W., *Anal. Chem.,* 33, 302, 1961.
86. Marec, D. J., Solesin, E. D., and Gordon, L., *Talanta,* 8, 293, 1961.
87. Takiyama, K. and Kozuki, E., *Bunseki Kagaku,* 17, 1412, 1968.
88. Tanigawa, Y., Minami, S., and Takiyama K., *Bunseki Kagaku,* 14, 1055, 1965.
89. Jones, J. P., Hilman, O. E., Jr., Townshend, A., and Gordon, L., *Talanta,* 11, 855, 1964.
90. Jones, J. P., Hilman, O. E., Jr., and Gordon, L., *Talanta,* 11, 861, 1964.
91. Corkin, J. T., Pietrzak, R. F., and Gordon, L., *Talanta,* 9, 49, 1962.
92. Cardwell, T. J., Nasouri, F. G., McConnell, H. H., and Magee, R. J., *Aust. J. Chem.,* 21, 359, 1968.
93. Takiyama, K., Salesin, E. D., and Gordon, L., *Talanta,* 5, 231, 1960.
94. Bordner, J., Salesin, E. D., and Gordon, L., *Talanta,* 8, 579, 1961.
95. Bordner, J. and Gordon, L., *Talanta,* 9, 1003, 1962.
96. Jones, J. P., Hileman, O. E., Jr., and Gordon, L., *Talanta,* 10, 111, 1963.
97. Stumpf, K. E., *Freseniush Z. Anal. Chem.,* 138, 30, 1953.
98. Sastri, M. N. and Prasad, T. P., *Talanta,* 14, 481, 1967.
99. Kitono, M., Yamagata, K., and Yamasaki, U., *Nippon Kagaku Zasshi,* 90, 190, 1969.
100. Heyn, A. H. A. and Finston, H. L., *Anal. Chem.,* 32, 328, 1960.
101. Hillard, L. B. and Freiser, H., *Anal. Chem.,* 24, 752, 1952.
102. Takitani, S., Hata, Y., and Suzuki, M., *Bunseki Kagaku,* 18, 626, 1969.
103. Nagai, H., *Bunseki Kagaku,* 11, 694, 1962; *Bunseki Kagaku,* 15, 50, 1966.
104. Fernand, Q. and Phillips, F. P., *Anal. Chem.,* 25, 819, 1953.
105. Jezorek, J. R. and Freiser, H., *Anal. Chem.,* 51, 366, 1979.
106. Sugii, A., Ogawa, N., and Hashizume, H., *Talanta,* 26, 189, 1979.
107. Buono, J. A., Buono, J. C., and Fasching, J. L., *Anal. Chem.,* 47, 1926, 1975.
108. Fresco, J. and Feiser, H., *Anal. Chem.,* 36, 372, 1964.
109. Motojima, K., *Bull. Chem. Soc. Jpn.,* 29, 28, 71, 75, 1956.
110. Merritt, L. L. and Wolker, J. K., *Ind. Eng. Chem. Anal. Ed.,* 16, 387, 1944.
111. Zolotov, Yu. A., Demina, L. A., and Petrukhim, O. M., *Zh. Anal. Khim.,* 25, 1487, 1970; *J. Anal. Chem. USSR,* 25, 1283, 1970.
112. Bacon, J. R. and Ferguson, R. B., *Anal. Chem.,* 44, 2149, 1972.
113. Motojima, K., Hashitani, H., and Imahashi, T., *Anal. Chem.,* 34, 571, 1962.
114. Majumdar, S. K. and Paria, P. K., *J. Indian Chem. Soc.,* 55, 998, 1978; *Chem. Abstr.,* 90, 132245b, 1979.

115. Buchi, J., Aebi, A., Deforin, A., and Hurni, H., *Helv. Chim. Acta,* 201, 1676, 1956.
116. Billo, E. J., Robertson, B. E., and Graham, R. P., *Talanta,* 10, 757, 1963.
117. Hikime, S. and Gordon, L., *Talanta,* 11, 851, 1964.
118. Graham, R. P., Billo, E. J., and Thomson, J. A., *Talanta,* 11, 1641, 1964.
119. Jones, J. P., Hileman, O. E., Jr., and Gordon, L., *Talanta,* 11, 861, 1964.
120. Nasanen, R. and Unsitalo, E., *Acta. Chem. Scand.,* 8, 112, 835, 1954.
121. Sharma, Y. and Shivahare, G. C., *Monatsh. Chem.,* 106, 695, 1975; *Anal. Abstr.,* 30, 6B156, 1976.
122. Molland, J., *Tidskr. Kjemi Bergves.,* 19, 119, 1939.
123. Hiraki, K., Morishige, K., and Nishikawa, Y., *Anal. Chim. Acta,* 97, 121, 1978.
124. Ryan, D. E., Pitts, A. E., and Cassidy, R M., *Anal. Chim. Acta,* 34, 491, 1966.
125. Pal, B. K., Toneguzzo, F., and Corsini, A., *Anal. Chim. Acta,* 88, 353, 1977.
126. Nishikawa, Y., Hiraki, K., Morishige, K., and Katagi, T., *Bunseki Kagaku,* 26, 365, 1977.
127. Schachter, D., *J. Lab. Clin. Med.,* 58, 495, 1961.
128. Bishop, J. A., *Anal. Chim. Acta,* 29, 178, 1963.
129. Sugawara, M., Hanagata, G., and Kambara T., *Bunseki Kagaku,* 27, 683, 1978.
130. Sugawara, M. and Kambara, T., *Bull. Chem. Soc. Jpn.,* 46, 3789, 1973.
131. Kambara, T. and Sugawara, M., *Bull. Chem. Soc. Jpn.,* 46, 500, 1973.
132. Langmyhr, F. J. and Storm, A. R., *Acta. Chem. Scand.,* 15, 1461, 1961.
133. Goto, K., Tamura, H., Onodera, M., and Nagayama, M., *Talanta,* 21, 183, 1974.
134. Ekman, A. and Nässänen, R., *Acta Chem. Scand.,* 7, 1261, 1953.
135. Swank, H. W. and Mellon, M. G., *Ind. Eng. Chem. Anal. Ed.,* 9, 406, 1937.
136. Rao, V. P. R. and Rao, K. V., *J. Inorg. Nucl. Chem.,* 30, 2445, 1968.
137. Shivahare, G. C., Mathur, M. K., and Mathur, S., *Indian J. Chem.,* 11, 1327, 1973; *Anal. Abstr.,* 27, 3227, 1974.
138. Bertoglio, C., Fulle Saldi, T., and Spini, G., *Anal. Chim. (Rome),* 58, 3, 1968; *Chem. Abstr.,* 68, 92680b, 1968.
139. Rusheed, A., Ahmed, S., and Eiaz, M., *J. Radioanal. Chem.,* 49, 205, 1979.
140. Ashbrook, A. W., *J. Chromatgr.,* 105, 151, 1975; *Coord. Chem. Rev.,* 16, 285, 1975.
141. Vernon, F., *Sep. Sci. Technol.,* 13, 587, 1978.
142. Flett, D. S., Hartlage, J. A., Spink, D. R., and Okuhara, D. N., *J. Inorg. Nucl. Chem.,* 37, 1967, 1975.
143. Flett, D. S., Cox, M., and Heels, J. D., *J. Inorg. Nucl. Chem.,* 37, 2197, 1975.
144. Ritcey, G. M., *CIM Bull.,* 66, 75, 1973; *Chem. Abstr.,* 79, 95159b, 1973.
145. Flett, D. S., *Inst. Min., Metall. Trans. Sect. C.,* 30, 83, 1974; *Chem. Abstr.,* 81, 94322d, 1974.
146. Vernon, F. and Nyo, K. M., *J. Inorg. Nucl. Chem.,* 40, 887, 1978.
147. Johnston, W. D. and Freiser, H., *Anal. Chim. Acta,* 11, 201, 1954.

GLYOXAL *BIS*(2-HYDROXYANIL) AND SCHIFF BASE REAGENTS

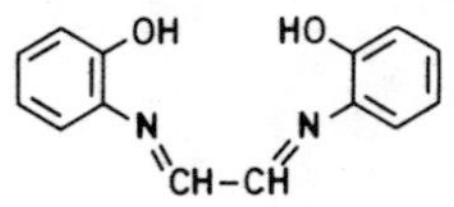

(1) 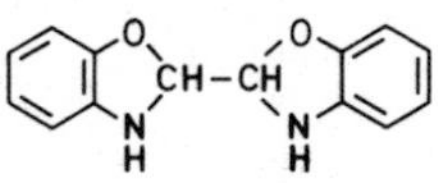   (2)

$C_{14}H_{12}N_2O_2$
mol wt = 240.26

H₂L

Synonyms

2,2'-(Ethanediylidenedinitrilo)diphenol, *N,N-Bis* (*o*-hydroxyphenyl) ethylenedi-imine, Di-(*o*-hydroxyphenylimino) ethane, GHA

The correct name for the solid reagent supplied under these name should be 2,2'-benzoxazoline (2) which rearranges to glyoxal *bis* (2-hydroxyanil) (1) in the reaction medium.

Source

Commercially available as a compound (2); synthesized by the condensation of *o*-aminophenol with glyoxal.[1,2]

Analytical Uses

Detecting reagent for Al, Ca, Cd, Ga, Sn, and rare earths; photometric reagent for Ca, Cd, Sn, UO₂²⁺, and rare earths; metal indicator for the chelatometric titration of Ca.

Properties of Reagent

Colorless or pale yellow needles, mp 210 to 214° (decomposition); nearly insoluble in water, but soluble in aqueous alkaline solution, fairly soluble in alcohol, benzene, and dioxane, and slightly soluble in chloroform. It is more soluble in methanol than in ethanol ($\simeq 8.5$ mg/mℓ)[3], but can be recrystallized from methanol. A very weak absorption band ($\simeq 450$ nm, $\varepsilon \simeq 1000$) is observed in visible region on a 1×10^{-3} M (1) in 50% methanol solution (0.04 N NaOH).[4]

Complexation Reactions and Properties of Complexes

Glyoxal *bis* (2-hydroxyanil) forms deeply violet to red chelates with alkaline earths, Cd, Co(II), Cu(II), Mn(II), Ni, UO₂²⁺, and Zn. The chelates of Cu(II), Ni, and UO₂²⁺ are smoothly and rapidly formed in weakly acidic medium at room temperature, and those of Cd, Co(II), Zn, and the alkaline earth ions are formed in weakly alkaline solution. The reagent behaves as a quadridentate ligand, forming a 1:1 chelate as shown below.

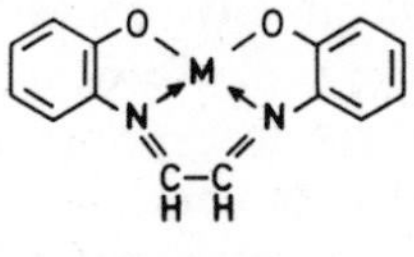

(3)

When an alcoholic solution of (1) is added to an aqueous solution of Ba, Ca, and Sr, followed by the addition of sodium hydroxide, a color change to red or a red precipitate at higher concentration is observed. Only the calcium compound, however,

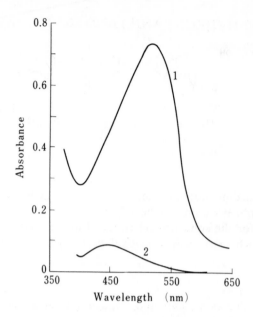

FIGURE 1. Absorption spectra of GHA and Ca chelate; pH 12.6 (borate buffer); medium, water-ethanol-*n*-butanol (2:1:1 by volume). The spectra were taken 30 min after mixing at room temperature. (1) CaL (30 μg Ca in 21.5 mℓ; (2) GHA (2.5 mg in 21.5 mℓ). (From Kerr, J. R. W., *Analyst*, 85, 867, 1960. With permission.)

is stable to alkali carbonate, hence the reaction is highly selective for Ca. However, the color of Ca chelate (CaL) in aqueous alkaline solution is not stable enough for the photometry.

Thus, improvements were made either by stabilizing it in aqueous organic solvent mixture (water-alcohol) or by extracting it into a mixture of chloroform with higher alcohol (hexanol or iso-pentanol)[5] or pyridine.[3]

The color reactions in homogeneous phase has been investigated by several workers. Although the reaction is accelerated at higher temperature (30°), the result is not reproducible due to hydrolyses of the reagent and the chelate. The recommended condition is to carry out the reaction at 0° for 30 to 50 min in a mixture of water-ethanol-*n*-butanol (2:1:1) which is 0.02 to 0.35 N in NaOH.[6,7] The absorption curve is illustrated in Figure 1 ($\varepsilon = 1.6 \times 10^4$ at λ_{max}, 520 nm). The strontium chelate (SrL) has been investigated in some details (log $K_{SrL} \simeq 5.9$, $\varepsilon = 7.5 \times 10^3$ at λ_{max}, 520 nm).[24] The colored chelate is also stable in organic phase. However, extraction with chloroform alone or inactive solvent is not satisfactory. This may be attributed to the presence of coordinated water on calcium, and it was recently found that the chelate could be extracted with dichloroethane in the presence of tetradecyldimethylbenzylammonium chloride (Zephiramine®).[8] The extracted species does not contain Zephiramine®, having a composition of CaL(H$_2$L). In contrast to the chelate in aqueous phase or extracted with chloroform(CaL), the color of CaL(H$_2$L) in dichloroethane is much more stable ($\varepsilon = 1.3 \times 10^4$ at λ_{max}, 530 nm). Cadmium behaves similarly to form a chelate CdL(H$_2$L) ($\varepsilon = 5.2 \times 10^3$ at λ_{max}, 560 nm).

Purification and Purity of Reagent

The reagent is a well-defined compound and can easily be purified by recrystallization from methanol or ethanol.

The purity may be checked by observing the melting point (210 to 3° with decomposition) or the absorptivity of ethanol solution in UV region ($\varepsilon = 9880$ at 294 nm.)[2]

Analytical Application
Use as a Photometric Reagent

Glyoxal *bis* (2-hydroxyanil) is a highly selective photometric reagent for Ca, but it has been also used for the photometric determination of some other metals, as summarized in Table 1.

Photometric Determination of Ca[8,11]

> **Reagent solution** — Glyoxal *bis* (2-hydroxyanil) 5×10^{-3} *M* solution (0.12 g in 100 m*l* methanol). Store in dark place. It can be kept for 1 week; Zephiramine® 5×10^{-3} *M* solution (0.18 g in 100 m*l* water); Buffer solution (0.1 *M* NaOH − 0.1 *M* Na$_3$BO$_3$, pH 12.8).
>
> **Procedure** — Place sample solution (0 to 1.25 μ mol of Ca) in a 100 m*l* separatory funnel, add 2 m*l* of buffer solution, dilute to 20 m*l* with water, and then add 3 m*l* of reagent solution. After 1 min, add 1 m*l* of Zephiramine® solution and 20 m*l* of 1,2-dichloroethane. Shake for 1 min. After phase separation, dry the organic layer with sodium sulfate, and measure the absorbance at 530 nm against a reagent blank.

Other Uses

GHA is used as a detecting reagent for Ca (detection limit, 0.05 μg; dilution limit, 1:10^6),[16] Al (0.5 μg; 1:10^5),[17] rare earths (1:10^3 to 10^5),[18] Sn(II) (1:10^4),[18] and Cd (0.05 μg; 1:10^6).[19]

GHA has also been used as a metal indicator in the chelatometric titration of Ca[20,21] and back titration with Ca.[22] The color change at the end point on the direct titration is from red to pale yellow (pH 11.7, NaOH).

Other Schiff Base Reagents
o-Salicylidene-aminophenol (SAPH, Manganon)

Orange red crystalline powder, mp 183 to 185°. Used as extraction photometric or fluorimetric reagent for Cu(ML, λ_{max}, 423 nm, $\varepsilon = 1.62 \times 10^4$ in MIBK),[25] Al(ML, λ_{exit}, 410nm, λ_{emit} 520 nm in ethylacetate),[26,27] and Ga.

$$C_{13}H_{11}O_2N$$
mol wt = 213.24

H$_2$L

N,N-Bis (salicylidene)-2,3-diaminobenzofurane,(SABF)

Yellow orange crystalline powder, mp 179 to 181°. Used as extraction photometric reagent for Cu (pH >3.7, ML, λ_{max}, 510 nm, $\varepsilon = 1.52 \times 10^4$ in benzene),[28] Ni(NH$_3$, ML, λ_{max}, 535 nm, $\varepsilon = 1.22 \times 10^4$ in benzene or CHCl$_3$),[29] and Zn(pH 8.8 to 9.8, pyridine, ML, λ_{max}, 535 nm, $\varepsilon = 1.92 \times 10^4$)[30] and as fluorimetric reagent for Mg(pH 10.5, ML, λ_{exit}, 475 nm, λ_{emit} 545 nm in 50% water methanol).[31]

Table 1

PHOTOMETRIC DETERMINATION OF METAL IONS USING GLYOXAL-*BIS* (2-HYDROXYANIL)

Metal ion	Condition	Metal Chelate			Range of determination (ppm)	Ref.
		Ratio	λ_{max} (nm)	$\varepsilon(\times10^4)$		
Ca	pH 12.8 (borate), H_2O-ethanol-*n*-butanol (2:1:1)	ML	520	1.6	$0 \sim 4$	4,6,7
Ca	1.6 NNaOH + 0.08 NNa$_2$CO$_3$, extraction with CHCl$_3$ ethanol	ML	535	~ 2	$0 \sim 10$	9,10
Ca	pH 12.8, Zephiramine®, extraction with 1,2-dichloroethane	ML·H$_2$L	530	1.3	$0 \sim 2.5$	8, 11
Cd	pH 10.6 $\sim$ 13,6 extraction with CHCl$_3$-pyridine (5:1)	ML	610	2.6	$0.1 \sim 4$	3
Cd	pH 13.0, Zephiramine® extraction with 1,2-dichloroethane	ML·H$_2$L	560	0.52	—	8
In(III)	pH 2.2 $\sim$ 3.8	ML or ML$_2$	580	—	$0.08 \sim 10$	23
Sc(III)	pH 4.5 $\sim$ 5.5, H_2O-ethanol (5:20) (80°)	ML$_2$	560	1.4	$0.6 \sim 7$	12
UO$_2^{2+}$	0.04 NNaOH, H_2O-methanol (1:1) EDTA, triethanolamine	ML$_2$	575	1.46	$0.4 \sim 20$	13
UO$_2^{2+}$	Zephiramine®, extraction with 1,2-dichloroethane	ML$_2$X	620	1.41	$0 \sim 10$	14
PbEt$_2$(II)	pH 9.5 $\sim$ 9.8 (NH$_3$-NH$_4$-citrate)	ML	680	1.9	$0.03 \sim 8$	15
PbMe$_2$(II)	pH 8.5 $\sim$ 8.8 (NH$_3$-NH$_4$-citrate)	ML	715	1.4	$0.04 \sim 8$	15

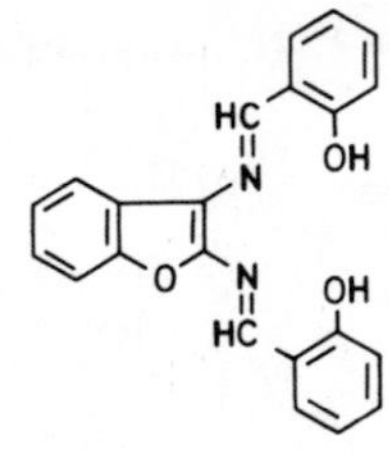

$C_{22}H_{16}O_3N_2$
mol wt = 356.38

H₂L

3-Hydroxypicolinaldehyde Azine (3-OH-PAA)

Yellow crystals, mp 162°. Fluoresces under UV light. Investigated as a photometric reagent for Ag, Cu(II,I), Hg(II), Pt(IV), Co(II), Ni, Zn, Pd(II), Cd, Mn(II), and Fe(II,III).[32]

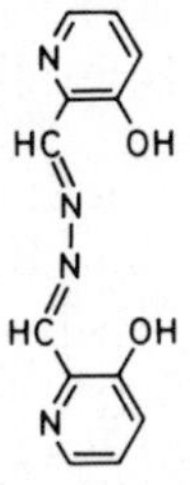

$C_{12}H_{10}O_2N_4$
mol wt = 242.24

H₄L

Azomethine H (5-Hydroxy-4-salicylidene aminonaphthalene-2,7-disulfonic Acid)

Bright orange crystalline powder, mp 300° (decomposition). Highly selective photometric reagent for B (pH ≃ 5, λ_{max}, 415 nm.)[33,34]

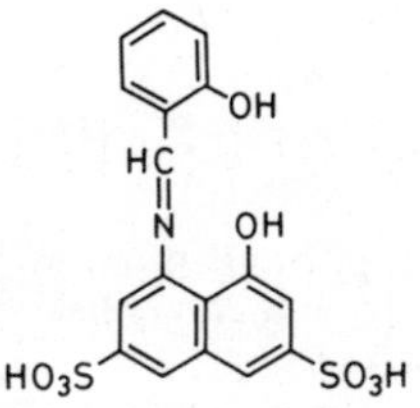

$C_{17}H_{15}NO_8S_5$
mol wt = 425.43

H₄L

Cuprizone (Biscyclohexanone-oxaldihydrazone)

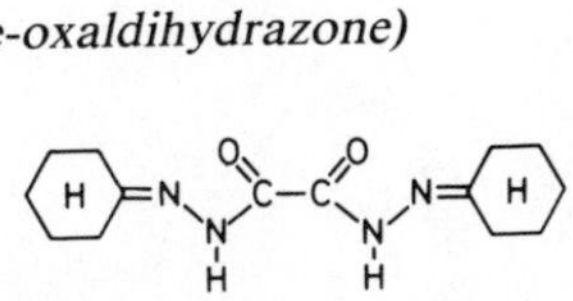

$C_{14}H_{22}O_2N_4$
mol wt = 278.35

A colorless crystalline powder, mp 213°.[35] Almost insoluble in water, but soluble in hot 50% ethanol.[36] Cuprizone has been introduced as a highly sensitive photometric reagent for Cu(II) (pH 8.5, 606 nm, $\varepsilon = 1.71 \times 10^4$, 0 to 4 ppm),[37] but it is less selective for Cu than the Cuproin reagents,[38] as it also reacts with Hg and Sn to form colored precipitates and with Co, Cr, Fe, and Ni to form colored soluble chelates.

Schiff Base Reagents Derived From Pyridinealdehyde:

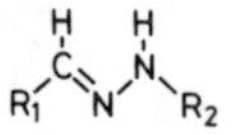

PAPH: $R_1 = R_2 =$

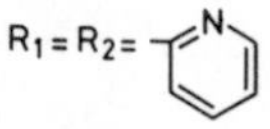

PAQH: $R_1 =$ $R_2 =$

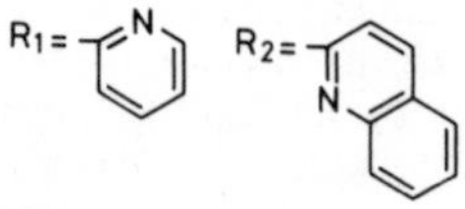

QAQH: $R_1 = R_2 =$

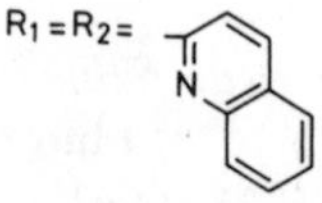

PAPH (pyridine-2-aldehyde-2'-pyridylhydrazone, $C_{11}H_{10}N_4$, mol wt. = 198.23), is a pale yellow crystalline powder, mp 178 to 180°.[39,40] PAQH (pyridine-2-aldehyde-2'-quinolylhydrazone, $C_{15}H_{12}N_4$, mol. wt. = 248.29) is an almost colorless crystalline powder, mp 204 to 204.5°.[41]

Both reagents have been proposed as an extraction photometric reagent for Pd(II): PAPH (pH <3.1, Cl^- or SCN^-, extraction with dichlorobenzene, λ_{max}, 562 nm, $\varepsilon = 1.65 \times 10^4$),[39,42] and PAQH (extraction with $CHCl_3$, λ_{max}, 592 nm, $\varepsilon = 1.58 \times 10^4$).[43]

QAQH (quinoline-2-aldehyde-2'-quinolyhyldrazone, $C_{19}H_{14}N_4$, mol wt = 298.35) is an yellow-brown crystalline powder, mp 268 to 270°,[44] and is known as a highly sensitive extraction photometric reagent for Cu(II) (pH >6.9, extraction with benzene, λ_{max}, 540 nm, $\varepsilon = 5.8 \times 10^4$).[45,46] The only disadvantage of this reagent in comparison with the Cuproin reagents is the lack of selectivity.

A number of analogous reagents have been investigated as an analytical reagent.[47-49]

MONOGRAPHS

M1 Jungries, E., Analytical applications of Schiff bases, in *Chelates in Analytical Chemistry*, Vol. 2, Flaschka, H. A. and Barnard, A. J., Jr., Eds., Marcell Dekker, New York, 1969, 149.
M2 Galliford, D. J. B. and Newman, E. J., Di-(o-hydroxyphenylimino)-ethane, in *Organic Reagents for Metals*, Vol. 2, Johnson, W. C., Ed., Hopkin & Williams, Chadwell Heath, England, 1964.

REVIEW

R1 Murase, I., Benzoxazoline derivatives and their metal chelates, *Kagaku No Ryoiki (Tokyo)*, 15, 796, 1961.

REFERENCES

1. Bayer, E., *Chem. Ber.*, 90, 2325, 1957.
2. Murase, I., *Bull. Chem. Soc. Jpn.*, 32, 827, 1959.
3. Oi, N., *Bunseki Kagaku*, 9, 770, 1960.
4. Umland, F. and Meckenstock, K. U., *Fresenius Z. Anal. Chem.*, 176, 96, 1960.
5. King, H. G. C. and Prunden, G., *Analyst*, 94, 39, 1969.
6. Kerr, J. R. W., *Analyst*, 85, 867, 1960.
7. Shigematsu, T., Matui, M., Ota, A., and Fujino, O., *Nippon Kagaku Kaishi*, 2226, 1974.
8. Nishimura, M., Noriki, S., and Muramoto, S., *Anal. Chim. Acta*, 70, 121, 1974.
9. Williams, K. T. and Wilson, J. R., *Anal. Chem.*, 33, 244, 1961.
10. Kuczerpa, A. V., *Anal. Chem.* 40, 581, 1968.
11. Nishimura, M. and Noriki, S., *Bunseki Kagaku*, 21, 640, 1972.
12. Okac, A. and Vrchlabsky, M., *Fresenius Z. Anal. Chem.*, 195, 338, 1963.
13. Wilson, A. D., *Analyst*, 87, 703, 1962.
14. Fujinaga, T., Kuwamoto, S., and Ozaki, T., *Nippon Kagaku Kaishi*, 1852, 1976.
15. Imura, S., Fukutaka, K., and Kawamura, T., *Bunseki Kagaku*, 18, 1008, 1969.
16. Goldstein, D. and Mayer, C. S., *Anal. Chim. Acta*, 19, 437, 1958.
17. Jungreis, E. and Lerner, A., *Anal. Chim. Acta*, 25, 199, 1961.
18. Okac, A. and Vrchlabsky, M., *Fresenius Z. Anal. Chem.*, 182, 425, 1961.
19. West, P. W. and Diffee, J., *Anal. Chim. Acta*, 25, 399, 1961.
20. Goldstein, D., *Anal. Chim. Acta*, 21, 339, 1959.
21. Tsunogai, S., Nishimura, M., and Nakaya, S., *Talanta*, 15, 385, 1968.
22. Osman, M. M., *Fresenius Z. Anal. Chem.*, 284, 43, 1977.
23. Bocek, P. and Vrchlabsky, M., *Chem. prum.*, 10, 625, 1966; *Chem. Abstr.*, 66, 16310z, 1967. 16310Z, 1967.
24. Tsvetanov, K. and Simeonova, T., *Monatsh. Chem.*, 104, 80, 1973.
25. Ishii, H. and Einaga, H., *Bunseki Kagaku*, 18, 230, 1969.
26. Dagnall, R. M., Smith, R., and West, T. S., *Talanta*, 13, 609, 1966.
27. Stolyarov, K. P., Grigorev, N. N., and Khomenok, G. A., *Zh. Anal. Khim.* 26, 1890, 1971.
28. Ishii, H., *Analyst*, 44, 1038, 1969.
29. Ishii, H. and Einaga, H., *Bunseki Kagaku*, 19, 1351, 1970.
30. Ishii, H. and Sawaya, T., *Bunseki Kagaku*, 20, 1298, 1971.
31. Dagnall, R. M., Smith, R., and West T. S., *Analyst*, 92, 20, 1967; *Analyst*, 93, 638, 1968.
32. de Torres, A. G., Valcarcel, M., and Pino, F., *Talanta*, 20, 919, 1973.
33. Capelle, R., *Anal. Chim. Acta*, 24, 555, 1961.
34. Petrovsky, V., *Talanta*, 10, 125, 1963.
35. Nilsson, G., *Acta Chim. Scand.*, 4, 205, 1950.
36. Wetleson, C. U., *Anal. Chim. Acta*, 16, 268, 1957.
37. Smith, G. F., *The Trace Elements Determination of Copper and Mercury in Pulp and Paper*, G. F. Smith Chemical Co., Columbus, Ohio, 1954.
38. Peterson, R. E. and Bollier, M. E., *Anal. Chem.*, 27, 1195, 1955.
39. Bell, C. F. and Rose, D. R., *Talanta*, 12, 697, 1965.
40. Lions, F. and Martin, K. V., *J. Am. Chem. Soc.*, 80, 3858, 1958.
41. Jensen, R. E. and Pflaum, R., *Anal. Chem.*, 38, 1268, 1966.
42. Cameron, A. J. and Gibson, N. A., *Anal. Chim. Acta*, 40, 413, 1968.
43. Jensen, R. E. and Pflaum, R. T., *Anal. Chim. Acta*, 37, 397, 1967.
44. Jensen, R. E., Bergham, N. C., and Helvig, R. J., *Anal. Chem.*, 40, 624, 1968.
45. Sims, G. G. and Ryan, D. E., *Anal. Chim. Acta*, 44, 139, 1969.
46. Zatka, V., Abraham, J., Holzbecher, J., and Ryan, D. E., *Anal. Chim. Acta*, 54, 65, 1971.
47. Kodama, K. and Fukui, K., *Bunseki Kagaku*, 21, 432, 1972.
48. Katyal, M. and Dutt, Y., *Talanta*, 22, 151, 1975.
49. Schilt, A. A., *Talanta* 26, 85, 373, 1979.

DIPHENYLCARBAZIDE AND DIPHENYLCARBAZONE

(1) H₄L
Diphenylcarbazide
$C_{13}H_{14}N_4O$ mol wt = 242.28

(2) H₂L
Diphenylcarbazone
$C_{13}H_{12}N_4O$ mol wt = 240.26

Synonyms
(1): *sym*-Diphenylcarbazide, 1,5-Diphenylcarbohydrazide
(2): Phenylazoformic acid-2-phenylhydrazide

Source and Method of Synthesis
Both are commercially available, however, the reagents in the market are often impure. Some of the products sold as diphenylcarbazide were found to contain 30 to 40% phenylsemicarbazide as impurity.[1] Also, some of the products sold as diphenylcarbazone were found to be the mixture of (1), (2), phenylsemicarbazide, and diphenylcarbodiazone in various ratios. As to the assay and the method of purification, refer to the following text.

Diphenylcarbazide (1) is synthesized by the reaction of diphenylcarbonate (not urea) with phenylhydrazine under heat.[4] Diphenylcarbazone (2) is obtained by the oxidation of (1) with hydrogen peroxide or air under alkaline condition.[4] Improved procedures for the synthesis of (1) and (2) were discussed in detail.[5]

Analytical Uses
(1) is chiefly used as a photometric reagent for Cr(VI) and also for Au(III), Cu(II), Hg(II), Re, Se(IV), Tc(III) etc. It is also used as a detecting reagent for various heavy metals and oxidizing materials. (2) is used as a photometric reagent for Cr(III), Hg(II), and many other heavy metals.

Properties of Reagents[M1-M3]
Diphenylcarbazide (1)
Colorless cyrstalline powder, mp 172 to 175°;[6] gradually becomes pinkish in air and should be kept in a dark place; only slightly soluble in water (∿0.024 g/100 mℓ), but soluble in alcohol, acetone, and other organic solvents except ether. Solutions are more easily oxidized than solid, gradually becoming yellow and brown. The acidic properties of enol form of (1) are so weak that the acid dissociation constants have not been reported.

Diphenylcarbazone (2)
Orange-red needles. Although analytically pure sample melts at 126 to 127°, most of commercial samples melt at 153 to 158°.[5] Very slightly soluble in water (0.09 g/100 mℓ),[7] more soluble in ethanol (∿0.24 g/100 mℓ), chloroform (∿0.24 g/100 mℓ), carbon tetrachloride (0.67 g/100 mℓ), benzene, toluene (2.4 g/100 mℓ), and other organic solvents, giving a red solution. It sublimes in vacuo.[8] Enol form of the reagent behaves as a weak acid (pKa_1 = 8.54).[7] Distribution coefficient of uncharged species (HL) is reported to be K_D(CCl₄/H₂O)= 7.5 (μ = 0.1, ClO₄⁻),[7] K_D (benzene/H₂O) = 12.6,[9] and K_D (toluene/H₂O) = 39.[7]

Absorption spectra of (2) in various solvents are illustrated in Figure 1. The spectrum of (2) in MIBK in equilibrium with the aqueous solutions of various pH values indicate

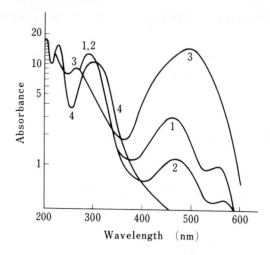

FIGURE 1.   Absorption spectra of diphenylcarbazone in various organic solvents and in water; reagent concentration, 10^{-3} *M*. (1) Carbon tetrachloride; (2) toluene; (3) water (pH 10.5, HL$^-$); (4) water (pH 1.0, H$_2$L). (From Balt, S. and van Dalen, Z., *Anal. Chim. Acta*, 27, 188, 1962. With permission.)

that (2) exists in keto and enol forms in the MIBK phase.[10] At higher pH, the enol form predominates.

keto form   enol form
(λ_{max}, 460 nm)   (λ_{max}, 505nm)

Complexation Reactions and Properties of Complexes

Due to the impure nature of the reagents used by earlier workers, some uncertainties remain in the papers concerning the reactions of (1) and (2) with various metal ions.

Diphenylcarbazide (1)

The color reactions of (1) with various metal ions have been reported to be Ag(violet), Cd(red), Co(II) (blue), Cu(II) (red), Cr (VI) (violet), Fe(III) (red), Hg(II) (blue), Mg(violet), Mn(II) (violet), Ni(blue-violet), Pb(orange-blue), and Zn(mauve).[M2] However, it is doubtful that these color reactions are true with the pure (1).

Although the highly sensitive color reaction of (1) with Cr(VI) has been known since 1900, the nature of colored species remained unknown until recently. The reaction may be written as

$$2CrO_4^{2-} + 3H_4L + 8H^+ \rightarrow Cr(III)(HL')_2^+ + Cr^{3+} + H_2L' + 8H_2O,$$

where H$_4$L and H$_2$L′ represent (1) and (2), respectively. Thus, the reactions occurring seem to be the simultaneous oxidation of reagent (from (1) to (2)), reduction of Cr(VI) to Cr(III), and the chelate formation of (2) with Cr(III).[11,12] The structure of the resulting chelate, however, still remains to be elucidated. This intensely colored soluble chelate cation can also be extracted as an ion-pair with chloride into amylalcohol.

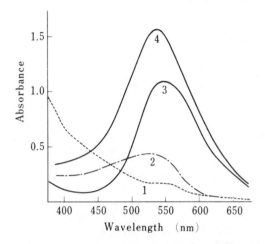

FIGURE 2. Absorption spectra of diphenylcarbazide-Cr(VI) system. (1) 1.0×10^{-3} M diphenylcarbazide in water, pH 1.5; (2) 1.0×10^{-3} M diphenylcarbazide in DMF; (3) 8.0×10^{-5} M Cr(III) and dephenylcarbazide in DMF; (4) 4.0×10^{-6} M Cr(III) and dephenylcarbazide in water. (From Pflaum, R. T. and Howick, L. C., *J. Am. Chem. Soc.*, 78, 4862, 1956. With permission.)

Absorption spectra of Cr(III) chelate are shown in Figure 2(λ_{max}, 550 nm, $\varepsilon = 2.6 \times 10^4$ in DMF).[13]

The color reactions of (1) with other metal ions, such as Cu(II),[14] Fe(III), Hg(I) (II),[15,16] Se(IV),[17] and Tc(VII),[18] are also attributed to the similar combination of redox reactions and complex formation, to give the colored chelate with (2). Accordingly, the spectral characteristics of the resulting chelates may be similar to those of the corresponding chelates with (2).[15,16]

Diphenylcarbazone (2)

Enol form of (2) reacts with divalent metal ions to form 1:1 and 2:1 chelates, depending upon the solution condition and the kind of metal ions as in the case of dithizone. The coordination structure of the $M(HL)_2$-type chelate is illustrated below (3).

(3)

The metal chelates are highly colored and can be extracted into immiscible solvents, so that (2) is widely used as an extraction photometric reagent for metals. The reactions with metal ions, the spectral characteristics of the metal chelates, and the extraction behaviors are summarized in Table 1. The chelate stability constants are reported on a few metal ions: log K_{CuL}, 9.8; log β_2, 19.5; log β_3, 29 (28°, $\mu \simeq 0$), log K_{NiL}, 6.02; and log K_{ZnL}, 5.76 (25°, $\mu = 0.1$ in 50% dioxane).[23]

Table 1
REACTION OF DIPHENYLCARBAZONE (2) WITH METAL IONS[19]

Metal ion	Condition	Ratio	Extraction with	λ_{max} (nm)	ε ($\times 10^4$)	log Kex	Ref.
Cd(II)	pH 9.5 ~ 10 TBP	M(HL)$_2$·TBP	Benzene	530	4	100% extraction	20
Co(II)	pH 4.3 ~ 4.7	M(HL)$_2$	Toluene	555	4.5	−7.00	
Cu(I)	pH 1.9 ~ 2.5	M(HL)	Toluene	550	3.75	4.13	9,21
Cu(II)	pH 1.9 ~ 3.5	M(HL)$_2$	MIBK	530	7.6	−1.11	10
Cu(II)	pH 1.9 ~ 3.5	M(HL)$_2$	Toluene	550	6.8	1.27	14
Fe(II)	pH 3.8 ~ 4.3	M(HL)$_2$	Benzene	508	3.57	−4.70	
Fe(III)	pH 1.6 ~ 2.7	M(HL)$_3$	Benzene	507	5.03	1.05	
Hg(I)	pH 3.0 ~ 3.5	M$_2$(HL)$_2$	Benzene	560	11.4	3.11	15,16
Hg(II)	pH 3.0 ~ 3.5	M(HL)$_2$	Benzene	565	8.6	5.26	15,16
Mn(II)	pH 6.7 ~ 7.7	M(HL)$_2$	Benzene	495	2.4	−11.5	
Mn(II)	pH 9.5 ~ 10, TBP	M(HL)$_2$·TBP	Benzene	525		100% extraction	20
Ni(II)	pH 4.3 ~ 5.1	M(HL)$_2$	Toluene	480	4.9	−6.08	15
Ni(II)	pH 8.7 ~ 9.2, TBP	M(HL)$_2$·TBP	Benzene	530	7	100% extraction	20
Pb(II)	pH 4.1 ~ 4.7	M(HL)$_2$	Benzene	535	6.13	−5.60	
Re(VII)	1 N H$_2$SO$_4$ extraction with diantripi-ryl propylmethane, then reaction with H$_2$L	M(HL)	Chloroform	540	2.14	—	22
Sn(II)	pH 1.5 ~ 2.1	M(HL)$_2$	Benzene	545	4.3	2.06	
Zn(II)	pH 4.6 ~ 5.0	M(HL)$_2$	Toluene	550	3.3	−6.76	
Zn(II)	pH 7 ~ 9 (NH$_3$)	M(HL)$_2$	MIBK	520	5.8	−7.10	10

Metal chelate

It should be noted that the direct complexation reaction between Cr(III) and (2) occurs only in nonaqueous media where Cr(III) ion is not aquated, otherwise the robust aquo Cr(III) ion does not react with (2) in aqueous solution. However, if a redox reaction is accompanied with complexation, as in the case of the reaction of Cr(VI) with (1), Cr(III) ion is formed in *statu nascendi*, and the complexation occurs even in the aqueous medium.[12]

Purification and Purity of Reagents
Diphenylcarbazide(1)

The commercially available reagent grade "diphenylcarbazide" is often contaminated with substantial amount of phenylsemicarbazide. The contaminant can be separated chromatographically on a column packed with polyamide powder (Nylon 11 or 66). A mixed solvent (water-methanol-acetic acid, 1:3:0.04) is used as an eluent. When eluted with a fivefold diluted solution of the mixed solvent, the first component (phenylsemicarbazide) flows down the column. Then, undiluted mixed solvent is used to elute (1).[1]

The samples prepared from diphenylcarbonate, instead of urea, do not contain phenylsemicarbazide.

Diphenylcarbazone(2)

The commercial samples are mixture of (1), (2), phenylsemicarbazide, and diphenylcarbodiazone. (2) can be separated from other contaminants by dissolving the sample in alcoholic sodium hydroxide, followed by the extraction with ether. After acidification of the aqueous phase, the precipitate was recrystallized from alcohol-water or acetone-water.[24]

The presence of various contaminants in (1) and (2) can be most conveniently detected and estimated by a thin layer chromatography.[2,25] The samples are developed on a thin layer of polyamide 11 powder with a mixed solvent (water-methanol-acetic acid, 1:3:0.04). The spots can be detected by irradiating with UV light at 366 nm for several hours. Rf values are 0.50 for (1) and 0.32 for (2). When a commercial TLC plate (DC Alufolien Polyamide 11, F 254, E. Merk) is used, the instant spot detection is possible by mean of light at 254 nm.

Analytical Applications
Use as a Photometric Reagent

Diphenylcarbazide (1) is most exclusively used as a highly sensitive and selective chromogenic reagent for Cr(VI), although it is also suggested as a photometric reagent for some oxidizing ions, such as Cu(II), Fe(III), Hg(I)(II), Se(IV), and Tc(VII).

Diphenylcarbazone (2) is widely used as a chromogenic reagent for various heavy metal ions, as it forms blue, violet, or red chelates. The metal chelates can be extracted with immiscible solvents for the photometry or solubilized in the aqueous organic solvents. The spectral characteristics of such metal chelates are summarized in Table 1. The chelate of Hg(II)-(2) or the ternary complex Hg-(2)-Br is used for the indirect photometric determination of anions, such as Cl⁻,[26,27] Br⁻,[28] I⁻,[29] and SO_4^{2-} in various samples.

Photometric Determination of Cr(III) with Diphenylcarbazide[M3,35]

Reagent — Diphenylcarbazide solution, 0.2 to 1% acetone, aqueous acetone (1:1), or ethanol solution, of which the acetone solution is the most stable. (Prepare daily and store in a dark cool place.) 5 to 6 N sulfuric acid, treat with dropwise addition of KMnO₄ solution until it remains faint pink under heating.

Procedure — To 10- to 20-m*l* sample solution containing 1 to 50 µg of Cr, add sufficient amount of sulfuric acid to make its concentration approximately 0.2 *N* when diluted to 25 m*l*, mix, and dilute it to 30 m*l*. To this add 0.5 m*l* of 0.1 *N* KMnO₄ solution and boil the solution gently for 10 to 20 min. After cooling, add 5 m*l* of 20% urea solution, mix, and add 10% NaNO₃ solution drop by drop until the mixture becomes colorless and transparent. Transfer the solution into a 50-m*l* volumetric flask, add 1 m*l* of diphenylcarbazide solution at 15°, and dilute to the volume. After 5 min (if sample contains vanadium, after 15 min), observe the absorbance at 540 nm against a reagent blank.

Elements, such as Fe(III), Mo(VI), and V(V) interfere, but Mo(VI) and Fe(III) can be masked with oxalate and phosphate, respectively. The color of vanadium chelate fades after 15 min, hence it does not interfere up to tenfold amount of Cr.[32]

Other Uses

(1) and (2) are recommended as indicators in the complexometric titration with a standard marcuric or lead solution. Anions, such as Cl⁻, Br⁻, I⁻, CN⁻, and SO₄²⁻, are titrated with a Hg(II) solution,[M1] and molybdate, metavanadate, and phosphate are titrated with a Pb solution.[M2] These reagents are also used as an indicator in the chelatometric titration of Hg and VO₂⁺.[32] (1) is used as a detection reagent for oxidizing species which include Cu(II), Cr(VI), Fe(III), Hg(II), Mo(VI), V(V), and H₂O₂.[M2,M3] A filter paper treated with Hg-(2) chelate can be used for the detection of chloride.[33,34]

MONOGRAPHS

M1. Casapieri, P., Diphenylcarbazide, and Diphenylcarbazone, in *Organic Reagent for Metals*, Vol. 1, Johnson, W. C., Ed., 61,66. Hopkins & Williams, Chadwell Heath, England, 1955.

M2. Welcher, F. J., *Organic Analytical Reagents*, Vol. 3 Van Nostrand, Princeton, N.J., 1947, 456.

M3. Sandell, E. B., *Colorimetric Determination of Traces of Metals*, 3rd ed., John Wiley & Sons, New York, 1959.

REFERENCES

1. Willems, G. J., Lantie, R., and Seth-Paul, W. A, *Anal. Chim. Acta*, 51, 544, 1970.
2. Allen, T. L., *Anal. Chem.*, 30, 447, 1958.
3. Willems, G. J. and De Ranter, C. J., *Anal. Chim. Acta*, 68, 111, 1974.
4. Slotta, K. H. and Jacobi, K. R., *Fresenius Z. Anal. Chem.*, 77, 344, 1929.
5. Friese, B. and Umland, F., *Fresenius Z. Anal. Chem.*, 286, 107, 1977.
6. Analar Standards for Laboratory Chemicals, 7th ed., Analar Standards, England, 1977.
7. Balt, S. and van Dalen, E., *Anal. Chim. Acta*, 27, 188, 1962.
8. Honjo, T., Imura, H., Shima, and Kiba., T., *Anal. Chem.*, 50, 1547, 1978.
9. Geering, H. R. and Hodyson, J. H., *Anal. Chim. Acta*, 36, 537, 1966.
10. Einaga, H. and Ishii, H., *Analyst*, 98, 802, 1973.
11. Sano, H., *Anal. Chim. Acta*, 27, 398, 1962.
12. Willems, J., Blaton, N. M., Peeters, O. M., and Ranter, C. J., *Anal. Chim. Acta*, 88, 345, 1977.
13. Pflaum, R. T. and Howick, L. C., *J. Am. Chem. Soc.*, 78, 4862, 1956.
14. Stoner, R. E. and Dasler, W., *Anal. Chem.*, 32, 1207, 1960.

15. Van Dalen, E. and Balt, S., *Anal. Chim. Acta,* 25, 507, 1961.
16. Balt, S. and Van Dalen, E., *Anal. Chim. Acta,* 27, 416, 1962.
17. Sushkova, S. G. and Murashova, V . I., *Tr. Ural. Politekh. Inst.,* 35, 1967; *Anal. Abstr.,* 16, 1233, 1969.
18. Miller, F. J. and Zittel, H. E., *Anal. Chem.,* 35, 299, 1963.
19. Balt, S. and van Dalen, E., *Anal. Chim. Acta,* 30, 434, 1964.
20. Gavrilova, L. G. and Zolotov, Yu. A., *Zh. Anal. Khim.,* 25, 1054, 1970; *J. Anal. Chem. USSR,* 25, 914, 1970.
21. Lapin, L. N. and Reis, N. V., *Zh. Anal. Khim.,* 13, 426, 1958.
22. Akimov, V. K., Busev, A. I., and Kliot, L. Ya., *Zh. Anal. Khim.,* 28, 1014, 1973; *J. Anal. Chem. USSR,* 28, 903, 1973.
23. Sillen, L. G. and Martell, A. E., *Stability Constants of Metal Ion Complexes, Suppl. 1,* The Chemical Society, London, 1964, 1971.
24. Krumholz, P. and Krumholz, E., *Monatsh. Chem.,* 70, 431, 1937.
25. Willems, G. J., Lontie, R. A., and Seth-Paul, W. A., *Anal. Chim. Acta,* 51, 544, 1970.
26. Gerlach, J. L. and Frazier, R. G., *Anal. Chem.,* 30, 1142, 1958.
27. Izawa, K., Aoyagi, H., Yoshida, Z., and Takahashi, M., *Bunseki Kagaku,* 22, 1046, 1973.
28. Tomonari, A., *Nippon Kagaku Zasshi,* 83, 459, 1962.
29. Okutani, T., *Nippon Kagaku Zasshi,* 88, 737, 1967; *Bull. Chem. Soc. Jpn.,* 41, 1728, 1968.
30. Machida, W., Okutani, T., and Utsumi, S., *Bunseki Kagaku,* 17, 1128, 1968.
31. Onishi, H. and Koshima, H., *Bunseki Kagaku,* 27, 726, 1978.
32. Sajo, L., *Magy. Kem. Foly.,* 62, 176, 1967; *Chem. Abstr.,* 52, 9850i, 1958.
33. Tanaka, Y. and Yamamoto, S., *Bunseki Kagaku,* 9, 8, 1960; *Bunseki Kagaku,,* 10, 182, 1961.
34. Boesch, H. and Weingerl, H., *Mikrochim. Acta,* 45, 1977.
35. Saltman, B. E., *Anal. Chem.,* 24, 1016, 1952.

ZINCON

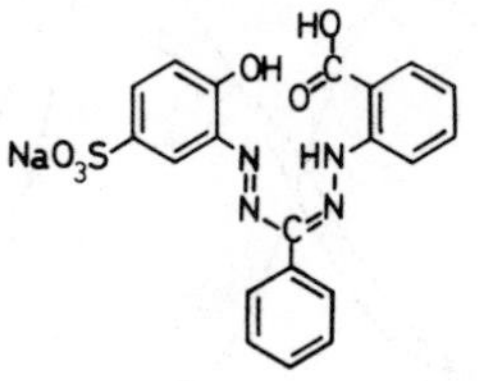

$C_2H_{15}N_4O_6SNa$
mol wt = 462.41

H₄L (Na H₃ L)

Synonyms

o- { 2-[α-(2-Hydroxy-5-sulfophenylazo)-benzylidene]-hydrazino } -benzoic acid, 2-carboxy-2′-hydroxy-5′-sulfoformazylbenzene, 5-o-carboxyphenyl)-1-(2-hydroxy-5-sul-fophenyl)-3-phenylformazan

Source and Method of Synthesis

Commercially available. It is prepared by the coupling of diazotized 2-amino-1-phenol-4-sulfonic acid with benzaldehyde-o-carboxyphenylhydrazone.[1]

Analytical Uses

It was first introduced as a chromogenic reagent for Cu(II) and Zn,[M1,2] but now it is also used as a metal indicator in the chelatometry.

Properties of Reagent

Usually supplied as monosodium salt (NaH₃L), dark violet powder. It is slightly soluble in water and ethanol, is easily soluble in aqueous alkali to give a deep red solution, and is insoluble in common organic solvents. It rapidly decomposes in acidic solution.

Zincon is structurally related with diphenylcarbazone, and the proton dissociation equilibria in the aqueous solution are[3,4]

$$H_4L \rightleftharpoons H_3L^- \xrightarrow{pKa_1 = 4\ to\ 4.5} H_2L^{2-} \xrightarrow{pKa_2 = 7.9\ (or\ 8.3)} HL^{3-} \xrightarrow{pKa_3 = 13\ (or\ >14)} L^{4-}$$

| Red | Pink or | Yellow | Orange-Yellow | Violet |
| | Red-Violet | | or Red-Orange | |

The slight discrepancies in pKa values and colors by different workers may be due to the impure nature of the samples. Absorption spectrum of Zincon at pH 9 (HL³⁻) is illustrated in Figure 1.

Complexation Reactions and Properties of Complexes

Yellow (H₂L²⁻) or orange (HL³⁻) colored species of Zincon at pH 5 to 9 forms blue ML-type soluble chelates with Co(II), Cu(II), Hg(II), Ni, and Zn. The molar absorptivities of Cu and Zn chelates are so high (CuL, pH 5.0 to 9.5, λ_{max}, 600 nm, $\varepsilon = 1.9 \times 10^4$; ZnL,pH 8.5 to 9.5, λ_{max}, 620 nm, $\varepsilon = 2.3 \times 10^4$)[M1,5] that Zincon has been recommended as a photometric reagent for these metals. Absorption spectra of Cu(II) and Zn chelates are illustrated in Figure 1.

The structure of metal chelate has not been elucidated, however, it is likely that Zincon behaves as a quadridentate ligand to coordinate metal ion with two oxygen(OH, COOH) and two nitrogen (N=N, NH). No data on complexation equilibrium is available except for the following:

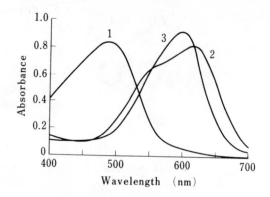

FIGURE 1. Absorption spectra of Zincon and Cu, Zn-Zincon chelates; dye concentration 4.0×10^{-3} M, at pH 9. (1) dye; (2) Zn chelate; (3) Cu chelate. (Reprinted with permission from Rush, R. M. and Yoe, J. H., *Anal. Chem.*, 26, 1345, 1954. Copyright 1954 American Chemical Society.)

$$HL^{3-} + Zn^{2+} \rightleftharpoons ZnL^{2-} + H^+ \qquad \log K = -1 \text{ (or } -1.8)^{2,3}$$

Purification and Purity of Reagent

Zincon from the commercial supply sources may be used without further purification for the general analytical purposes. However, it is necessary to check the purity for the physicochemical studies.[8] The main contaminants are inorganic salts which can be removed by treating with dilute acetic acid. Organic contaminants can be removed by refluxing with ether.

Analytical Applications

Use as a Photometric Reagent

Zincon is a fairly sensitive photometric reagent for Cu(II) and especially for Zn. Main drawbacks of Zincon are the lack of selectivity and the poor stabiity of the reagent. Thus, the photometry with Zincon is applied to the sample after separation of Zn by solvent extraction or anion exchange process.

Traces of nitrilotriacetic acid (NTA) in environmental aquatic samples can be determined by measuring the decrease of absorbance of Zn-Zincon chelate at 620 nm according to the following stoichiometry (pH 8.5 to 9.5).[6]

$$NTA + Zn\text{-Zincon} \rightleftharpoons Zn\text{-NTA} + Zincon$$

As low as 0.2 ppm of NTA can be determined by this method.

A simultaneous determination of Cu (Bathocuproine-sulfonic acid, λ_{max}, 484 nm), Fe (TPTZ, λ_{max}, 600 nm), and Zn (Zincon, λ_{max},620 nm) in one sample is possible because the absorption peaks do not overlap each other.[7] This method is applied to a flow photometric procedure.

Determination of NTA in Waste Water.[6]

Reagents —

1. Buffer — Dissolve 31 g boric acid and 37 g KCl in 800 mℓ of distilled water, adjust pH of the solution to 9.2 with 6 N NaOH, and dilute it to 1000 mℓ.
2. Zinc solution — Dissolve 0.440 g ZnSO$_4$·7H$_2$O in 100 mℓ of 2 N HCl, transfer to a 1000-mℓ volumetric flask, and dilute to the mark.

3. Zinc-Zincon reagent — Dissolve 0.130 g of Zincon in 2 mℓ of 1 N NaOH, transfer to a 1000-mℓ volumetric flask, and add 300 mℓ buffer and 15 mℓ zinc solution. Dilute to the volume. The reagent should be prepared fresh weekly and stored in the dark.

4. Standard solutions — Prepare the working standards in the range of 0.5 to 10 mg/ℓ by diluting a 1000 ppm NTA stock solution. Standards may be stored in either polyethylene or glass bottles. New working standards should be prepared weekly.

Procedure — Pipette 25 mℓ of sample into a 50-mℓ beaker. With each batch of samples, run a blank of distilled water and a standard of 5 ppm NTA. Add approxiately 2.5 g cation exchange-resin and a stirring bar to each beaker. Cover with watch glass and stir on a magnetic stirrer for 15 min. Filter sample through a filter paper into a second beaker. Do not wet the paper before the filtration and do not wash the precipitate. Pipette 15 mℓ of the filtrate into a third 50-mℓ beaker and add 35 mℓ of zinc-Zincon reagent by pipette. Read the absorbance of the solution in a 2-cm cell against a water blank at 620 nm.

Zincon is also used as a photometric reagent for Cu(II) (pH 7.5 to 9.5, Zephiramine®, extraction with CHCl$_3$, 625 nm, 0.2 to 1.6 ppm),[8] Ga (pH 6, extraction with TBP or TOA in amyl alcohol, 630 nm, $\varepsilon = 2.2 \times 10^4$, 0.25 to 3.5 ppm),[9] Hg(II) (pH 7.2, 600 nm, 2 to 8 ppm),[10] Ni (pH 4 to 11, 665 nm, $\varepsilon = 9.76 \times 10^4$),[11] and Pd(II) (pH 8.0, 595 nm, $\varepsilon = 2.6 \times 10^4$, 0.2 to 6 ppm).[22]

Use as a Metal Indicator
Zincon is used as a metal indicator in the chelatometric titrations of Zn (borate, pH 9.5 to 10),[M1,13] Cd, Hg(II), and Pb (pH 6).[12] A mixture of Zn-EDTA and Zincon is used as a Ca-sensitive indicator in the selective GEDTA titration of Ca in the presence of excess Mg.[4]

Other Reagents with Related Structures
Formazan Derivatives
A large number of formazan dyes has been synthesized and evaluated as analytical reagents. They are reviewed in various articles,[R3,R6,14] but very few were found to be of practical importance.

Tetrazolium Salts
Tetrazolium salts of the general formular of (1) and (2) can be considered to be a precursor of formazan dyes.

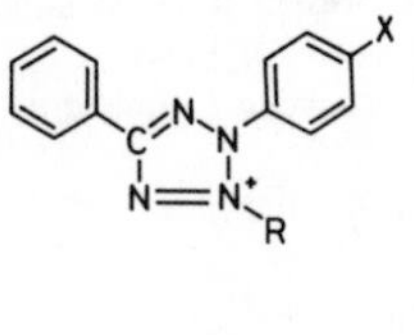

(1)

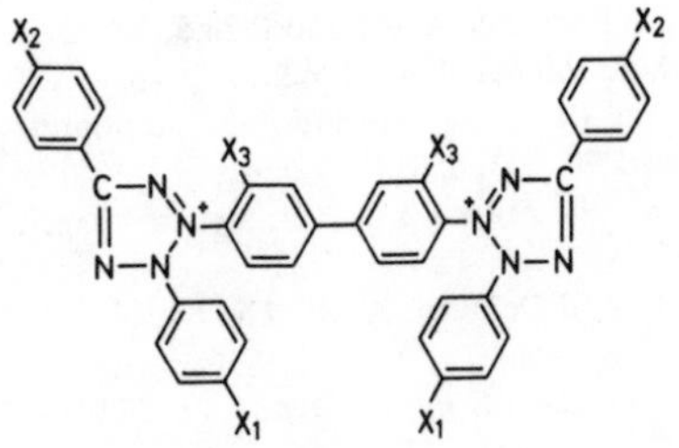

(2)

The tetrazolium salts are usually colorless, but are easily reduced to formazans which are highly colored. The redox potential of tetrazolium salt — formazan couple is fairly small as listed below, and they can be enzymatically reduced by various dehydrogenase.[M2] Accordingly, these tetrazolium salts are widely accepted as the specific reagent for the detection and determination of various enzymes in the biological samples.[45] Some of the important tetrazolium salts are listed below:

1. INT (R = p-iodophenyl in (1) 3-(p-iodophenyl)-2-(p-nitrophenyl)-5-phenyl-2H-tetrazolium chloride; E = −0.09 v. Formazan (λ_{max}, 490 nm, ε = 1.5 × 10⁴)[15,18]

2. MTT (R = 4,5-dimethyl-2-thiazolyl in (1), 3-(4,5-dimethyl-2-thiazolyl)-2,5-diphenyl-2H-tetrazolium bromide. Formazan (λ_{max}, 565 nm, ε = 2 × 10⁴)[16]

3. Neo-TB ($X_1 = X_2 = X_3 = H$ in (2), 3,3′-(4,4′-biphenylene)-[bis(2,5-diphenyl)-2H-tetrazolium chloride]. E = −0.17 v; Formazan (λ_{max}, 505 nm, ε = 1.4 × 10⁴)[17]

4. TB ($X_1 = X_2 = H$, $X_3 = OCH_3$ in (2), 3,3′-(3,3′-dimethoxy-4,4′-biphenylene)-[bis (2,5-diphenyl)-2H-tetrazolium chloride]; Tetrazolium Blue, E = −0.16 v; Formazan (λ_{max}, 525 to 530 nm)[18]

5. Nitro TB ($X_1 = NO_2$, $X_2 = H$, $X_3 = OCH_3$ in (2), 3,3′-(3,3′-dimethoxy-4,4′-biphenylene)-bis [2-(p-nitrophenyl)-5-phenyl-2H-tetrazolium chloride]; E = −0.05 v; Formazan (λ_{max}, 530 nm, ε = 3.6 × 10⁴)[19]

6. TNTB ($X_1 = X_2 = NO_2$, $X_3 = OCH_3$ in (2)), 3,3′-(3,3′-dimethoxy-4,4′-biphenylene)-bis[2,5-bis(p-nitrophenyl)-2H-tetrazolium chloride], E ≈ −0.05 v[20]

7. Tc-Nitro-TB ($X_1 = NO_2$, $X_2 = OCH_3$, $X_3 = CSNH_2$ in (2)), 3,3′-(3.3′-dimethoxy-4,4′-biphenylene)-bis[2-(p-nitrophenyl)-5-(p-thiocarbamylphenyl)-2H-tetrazolium chloride]; E ≈ −0.05 v[21]

MONOGRAPHS

M1. Jablonski, W. Z. and Johnson, E. A., Zincon, in *Organic Reagents for Metals*, Vol. 2, Johnson, W. C., Ed., Hopkin & Williams, Chadwell Heath, England, 1964, 264.

M2. Brustons, M. S., *Enzyme Histochemistry and Its Application in the Study of Neoplasms*, Academic Press, New York, 1962.

REVIEWS

R1. Iida, H., Formazan metal complex dyes, *Yuki Gosei Kagaku Kyokaishi*, 18, 269, 1960.

R2. Johnston, M. B., Barnard, A. J., and Broad, W. C., Zincon, present status as an analytical reagent, *Rev. Univ. Ind. Santander*, 4, 43, 1962.

R3. Suenaga, E., Studies on the coordination compounds of heterocyclic formazyl derivatives, *Nippon Yakugaku Zasshi*, 79, 803, 850, 1959.

R4. Podchainova, V. N., Formazans as analytical reagents, *Zh. Anal. Khim.*, 32, 822, 1977; *J. Anal. Chem.*, 32, 650, 1977.

R5. Anderson, G. L. and Deinard, A. S., The Nitroblue tetrazolium (NBT) test, *Am. J. Med. Technol.*, 345, 1974.

R6. Kawase, A., Formazans, *Bunseki Kagaku*, 21, 578, 1972.

R7. Cheronis, N. D. and Stein, H., Tetrazolium salts as chemical reagents, *J. Chem. Educ.*, 33, 120, 1956.

REFERENCES

1. Brocks, R. A. and Salem, N. J., U.S. Patent 2,662,074 and 2,662,075, 1953.
2. Yoe, J. H. and Rush, R. M., *Anal. Chim. Acta,* 6, 526, 1952.
3. Ringbom, A., Pensar, G., and Wanninen, E., *Anal. Chim. Acta,* 19, 525, 1958.
4. Sadek, F. S., Schmid, R. W., and Reilley, C. N., *Talanta,* 2, 38, 1959.
5. Rush, R. M. and Yoe, J. H., *Anal. Chem.,* 26, 1345, 1954.
6. Thomson, J. E. and Duthie, J. R., *J. Water Pollut. Control Fed.,* 40, 306, 1968.
7. Zuk, B., Cohen, J. S., and Williams, L. A., *Microchem. J.,* 6, 67, 1962.
8. Sugawara, M., Niiyama, K., and Kambara, T., *Bunseki Kagaku,* 22, 1219, 1973.
9. Dosal Gomez, M. A., Perez Bustamante, J. A. and Burriel Marti, F., *Inf. Quim. Anal.,* 27, 78, 1973; *Anal. Abstr.,* 25, 2993, 1973.
10. Morris, A. G., *Analyst,* 82, 34, 1957.
11. Dosal Gomez, M. A. and Perez Bustamante, J. A., *An. Quim.,* 73, 712, 1977; *Chem. Abstr.,* 87, 173552f, 1977.
12. Singhal, S. K. and Reilley, K. N., *Talanta,* 14, 1351, 1967.
13. Reilley, C. N. and Sheldon, M. V., *Chemist-Analyst,* 46, 59, 1957.
14. Kawase, A., *Bunseki Kagaku,* 16, 1364, 1967; Matsushima, T., and Kawase, A., *Bunseki Kagaku,* 20, 156, 1310, 1971; Kiyokawa, M. and Kawase, A., *Bunseki Kagaku,* 21, 244, 1972.
15. Nachlas, M. M., Margulies, S. I., and Seligman, A. M., *J. Biol. Chem.,* 235, 2739, 1960.
16. Pearse, A. G. E. and Hess, R., *Experientia,* 17, 136, 1961; *Chem. Abstr.,* 56, 698b, 1962.
17. Oda, T., Seki, S., Shibata, T., Sakai, A., and Okazaki, H., *Okayama Igakkai Zasshi,* 70, 128, 1958; *Chem. Abstr.,* 53, 495a, 1959.
18. Rutenburg, A. M., Gofstein, R., and Seligman, A. M., *Cancer Res.,* 10, 113, 133, 1951.
19. Nachlas, M. M., Tsou, K. C., De Souza, E., Cheng, C. S., Seligman, A. M., *J. Histochem. Cytochem.,* 5, 420, 1957; *J. Am. Chem. Soc.,* 78, 6139, 1956.
20. Sedar, A. W., Rosa, C. G., and Tsou, K. C., *J. Histochem. Cytochem.,* 10, 506, 1962.
21. Seligman, A. M., Ueno, H., Morizono, Y., Wasserkrug, H. L., Katzoff, L., and Hanker, J. S., *J. Histochem. Cytochem.,* 15, 1, 1967.
22. Dosal, M. A., Perez Bustamante, J. A., and Burriel Marti, F. *An. Quim.,* 68, 1241, 1972; *Chem. Abstr.,* 78, 131719z, 1973.

MUREXIDE

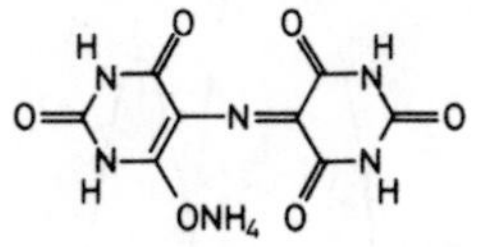

$$C_8H_4O_6N_5 \cdot NH_4$$
mol wt = 284.19

H$_5$L (NH$_4$·H$_4$L)

Synonyms

5-[(Hexahydroxy-2,4,6,-trioxo-5-pyrimidinyl) imino]-2,4,6, (1H, 3H, 5H)-pyrimi-dinetrione, monoammonium salt; 5,5-nitrilodibarbituric acid monoammonium salt; acid ammonium purpurate; ammonium purpurate; C. I. 56085

Source and Method of Synthesis

Commercially available. Prepared by the reaction of alloxantine with ammonium acetate in boiling glacial acetic acid.[1]

Analytical Uses

Metal indicator in chelatometric titration of Ca, Cu(II), Co(II), and Ni; photometric reagent for Ca.[13]

Properties of Reagent

Purple red crystalline powder, lustrous green by reflected light; does not melt, nor decompose below 300°. The aqueous solution is deep purple, but decolorizes fairly rapidly, is sparingly soluble in cold water and ethyleneglycol, more in hot water, and is insoluble in alcohol and ether.

The proton dissociation constants of hydroxy and imino groups of purpuric acid were determined as indicated below.[2]

$$H_5L \xrightleftharpoons[\quad]{pKa_1 \simeq 0} H_4L^- \xrightleftharpoons[\quad]{pKa_2 = 9.2} H_3L^{2-} \xrightleftharpoons[\quad]{pKa_3 = 10.5} H_2L^{3-}$$

Yellow   Red Violet   Violet   Blue Violet

Absorption spectra of H$_4$L$^-$ and H$_2$L^{3-} in aqueous solution are illustrated in Figure 1.

Complexation Reactions and Properties of Complexes

In neutral or alkaline solution, purpurate ion forms highly colored chelates (1) with metal ions, such as Ca, Cu(II), Co(II), Ni, and Zn. The color of the chelates depends on the kind of metal ion and on the pH of solution.

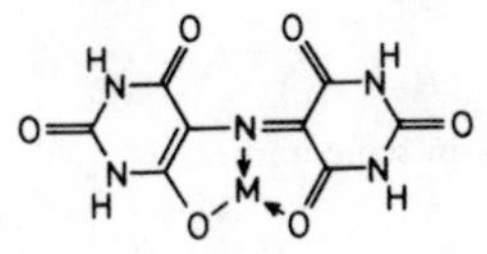

(1)

In the case of calcium chelate, the following equibrium is reported.[2]

$$CaH_4L^+ \xrightleftharpoons[\quad]{pKa_2 = 8.2} CaH_3L \xrightleftharpoons[\quad]{pKa_3 = 9.5} CaH_2L^-$$

Orange Yellow   Orange Red   Red

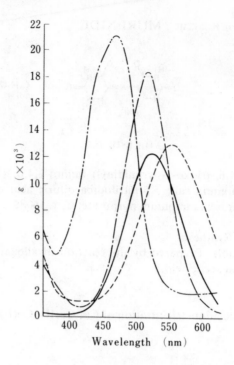

FIGURE 1. Absorption spectra of murexide and metal-murexide chelates; ———— , Murexide (neutral solution, H_4L^-); -------- , Murexide (strong alkaline solution, H_2L^{3-}); —·—·— , Ca-murexide (CaH_2L^-) at pH 13; —————, Cu(II)-murexide (CuH_2L^-) at pH 8. (From Schwarzenbach, G. and Gysling, H., *Helv. Chim. Acta*, 32, 1314, 1949. With permission.)

Table 1
CONDITIONAL STABILITY CONSTANTS OF MUREXIDE CHELATES (LOG K'_{ML})[a]

Metal ion	4	5	6	7	8	9	10	11	12	Ref.
Ca	—	—	—	2.6	2.8	3.4	4.0	4.6	5.0	3
Cu(II)	—	—	6.4	8.2	10.2	12.2	13.6	15.8	17.9	3
Ni	—	—	4.6	5.2	6.2	7.8	9.3	10.3	11.3	3
Eu	5.34	5.42	—	—	—	—	—	—	—	4
Gd	4.90	5.00	—	—	—	—	—	—	—	4
La	4.49	4.55	—	—	—	—	—	—	—	4
Tb	4.89	4.98	—	—	—	—	—	—	—	4

[a] Values for Reference 3, at room temperature, $\mu = \sim 0.1$; values for Reference 4, at 25° $\pm 0.1°$, $\mu = 0.1$.

The conditional stability constants for some metal ions were calculated by Ringbom[3] and Balaji[4] as summarized in Table 1.

Copper(II), nickel(II), and cobalt(II) chelates are yellow, and zinc and cadmium chelates are orange yellow. Absorption spectra of calcium chelate (CaH_2L^- at pH 13) and copper chelate (CuH_2L^- at pH 8) are illustrated in Figure 1.

Purification and Purity of Reagent

Solid reagent sometimes contains a small percentage (20%) of murexide (acid ammonium purpurate) and large amount of uramil, alloxantine, etc.[5] However, it is rather difficult to purify impure sample. Instead, the sample of high purity may be prepared from pure alloxantine according to the standard procedure.[1]

The purity of the reagent is roughly determined by titration with $TiCl_3$[5] or can be evaluated by measuring the absorbance of aqueous solution (0.5 mg/100 ml) (λ_{max}, 520 nm; $\varepsilon = 1.20 \times 10^4$).[6]

Analytical Applications:

Murexide has been used as a metal indicator in the EDTA titration of Ca (pH 12, NaOH, color change at the end point is from red to violet), Co(II) (pH 8, NH_3, from yellow to violet), Cu(II) (pH 8, NH_3, from yellow to red-violet)[9] and Ni(pH 10, NH_3, from yellow to blue-violet),[10] but better indicators are now available. As the aqueous solution of murexide is unstable, a 0.1% anhydrous ethyleneglycol solution is recommended for a long shelf life.[7,8] Dry powder preparation diluted with NaCl or K_2SO_4 (1:100 to 500) is also widely used.

Murexide is also used as a photometric reagent for Ca (pH 11.3, at 506 nm) in various samples.[4,8,11,12]

REFERENCES

1. Davidson, D., *J. Am. Chem. Soc.,* 58, 1821, 1936.
2. Schwarzenbach, G. and Gysling, H., *Helv. Chim. Acta,* 32, 1314, 1949.
3. Ringbom, A., *Complexation in Analytical Chemistry,* Interscience, New York, 1963.
4. Balaji, K. S., Kumar, S. D., and Gupta-Bhaya, P., *Anal. Chem.,* 50, 1972, 1978.
5. Karsten, P., Kies, H. L., van Engelen, H. Th. J., and de Hoog, P., *Anal. Chim. Acta,* 12, 64, 1955.
6. Purity Specification of Organic Reagents, Dojindo Laboratories, Kumamoto, Japan.
7. Brunischolz, G., Genton, M., and Plattner, E., *Helv. Chim. Acta,* 36, 782, 1953.
8. Rathje, W., German Offen. 2,337,811; *Chem. Abstr.,* 82, 118590p, 1975.
9. Schwarzenbach, G. and Flaschka, H., *Die komplexometrishe Titration,* 2nd ed, Ferdinand Enke, Stuttgart, 1965.
10. Flashka, H., *Mikrochem. Ver. Mikrochim. Acta,* 39, 38, 1952.
11. Williams, M. B. and Moser, J. H., *Anal. Chem.,* 25, 1414, 1953.
12. Pollard, F. H. and Mortin, J. V., *Analyst,* 81, 348, 1956.
13. Gysling, H. and Schwarzenbach, G., *Helv. Chim. Acta,* 32, 1484, 1949.

NITROSONAPHTHOLS AND NITROSOPHENOLS

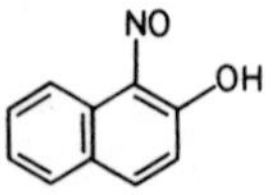

(1) HL

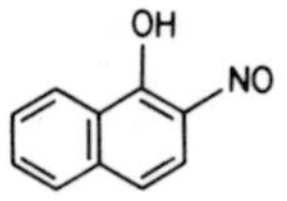

(2) HL

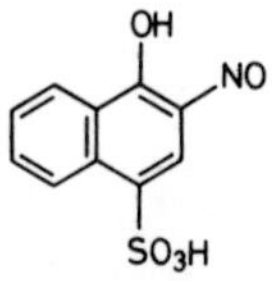

(3) H$_2$L

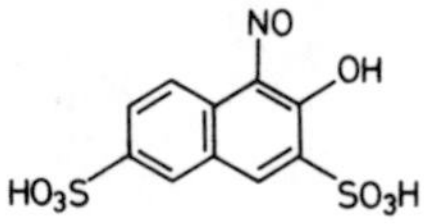

(4) H$_3$L

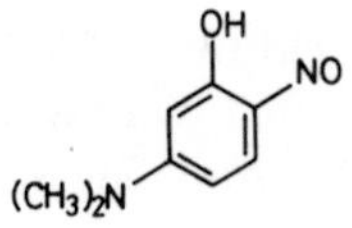

(5) HL

Synonyms

Reagents covered in this section are listed in Table 1, together with their synonyms.

Source and Methods of Synthesis

All are commercially available. These reagents can be prepared by the nitrosation of corresponding naphthol or phenol with nitrous acid.[1-4]

Analytical Uses

Reagents of this type are well known as highly selective detection and photometric reagents for Co(II). The reagents react with Co(II) to form very stable Co(III) chelate which does not dissociate, even in mineral acid or alkali. Reagents (1) and (2) are used for the extraction photometry, while (3) and (4) are for the photometry in aqueous solution; (5) is used in both ways. Applications as photometric reagent for Cu(II) and Fe(III) and as precipitating reagent for various elements are also reported.

Properties of Reagents

1-Nitroso-2-naphthol (1) — Orange-brown or brown crystalline powder,[M1] mp 110.4 to 110.8°;[5] easily sublimes,[6] slightly soluble in water (0.02 g/100 g at 20°), but soluble in organic solvents such as ethanol (2.4 g/ 100 g at 13°), benzene, ether, glacial acetic acid, chloroform (23.4 g/ 100 mℓ), carbon tetrachloride, MIBK (7.3 g/100 mℓ) and carbon disulfide; also freely soluble in caustic alkali solution. Solid reagent is not so unstable as reagent solution, but should be kept tightly closed in a cool dark place; pKa = 9.47 (in 50% dioxane, 30°) or 11.60 (in 75% dioxane, 30°).[7] log K$_D$(CHCl$_3$/water) = 2.97,[9] log K$_D$(chlorobenzene/water) = 2.73,[5] and log K$_D$ (MIBK/water) = 2.55.[9]

2-Nitroso-1-naphthol (2) — Yellow or greenish-yellow crystalline powder,[M1] mp 147

Table 1
ANALYTICALLY IMPORTANT NITROSOPHENOLS

Reagent	Name of reagent	Synonym	Molecular formula	Mol wt
(1)	1-Nitroso-2-naphthol	α-Nitroso-β-naphthol	$C_{10}H_7NO_2$ (HL)	173.17
(2)	2-Nitroso-1-naphthol	β-Nitroso-α-naphthol	$C_{10}H_7NO_2$ (HL)	173.17
(3)	2-Nitroso-1-naphthol-4-sulfonic acid	—	$C_{10}H_7NO_5S \cdot 3.5\ H_2O$ (H_2L)	316.28
(4)	1-Nitroso-2-naphthol-3,6-disulfonic acid disodium salt	Nitroso R acid, disodium salt	$C_{10}H_5NO_8S_2Na_2$ (Na_2HL)	377.25
(5)	2-Nitroso-4-dimethylaminophenol, hydrochloride	Nitroso-DMAP, hydrochloride	$C_8H_{10}O_2N_2 \cdot$ HCl (HL $\cdot \cdot$ HCl)	202.64

to 148°; slightly soluble in cold water, but fairly soluble in hot water; fairly soluble in alcohol, but less soluble in ether, chloroform (1.7 g/100 mℓ), and MIBK (2.2 g/100 mℓ). Solid reagent should be kept tightly closed in a cool dark place; pKa = 8.90 (in 50% dioxane, 30°) or 11.20 (in 75% dioxane, 30°);[7] log K_D(CHCl₃/water) = 2.11 and log K_D(MIBK/water) = 2.33.[9]

2-Nitroso-1-naphthol-4-sulfonic acid **(3)** — Yellow-brown crystals with a composition of $H_2L \cdot 3.5\ H_2O$; can be dehydrated by heating at 115° to give red powder (H_2L); easily soluble in water and alcohol. Aqueous solution is stable for many months; pKa(OH) = 6.51 ($\mu = 0$, 25°).[R3]

Nitroso R acid **(4)** — Disodium salt is golden-yellow, with fan-shaped crystals;[28] easily soluble in water (acidic or alkaline), but slightly soluble in alcohols. Aqueous solution is stable for 2 weeks; pKa(OH) = 7.51 ($\mu = 0$, 25°).[R3]

Nitroso DMAP **(5)** — Hydrochloride is yellow needles,[R1,R2] Soluble in water to give a yellow solution which turns to red above pH 2.5. The solid reagent is stable for at least 2 years, but an aqueous solution (in 0.01 M HCl) can be kept for a month;[R2,8] pKa_1(N⁺H) = 2.69, pKa_2(OH) = 8.40 ($\mu = 0.1$, KCl, 25°);[R3] log K_D(dichlroethane/water) = 1.42.[8]

While **(3)** and **(4)** are used for the photometry in aqueous solutions, **(1)**, **(2)**, and **(5)** are often used for the extraction photometry. Due to the amphoteric nature of **(5)**, excess reagent in organic phase can be extracted into aqueous phase by washing with aqueous acid (pH <−0.8) or alkali (pH >12),[R1] whereas excess **(1)** and **(2)** are removed by washing with aqueous alkali.

Although these reagents do not show any strong absorption band in the visible region, the end of UV absorption peak appears in the visible region, giving yellow or orange-brown coloration. Absorption spectra of **(1)**, **(3)** and **(5)** are illustrated in Figures 1, 2, and 3, respectively.

Complexation Reactions and Properties of Complexes

The reagents having a chelating site of 1-nitroso-2-hydroxyaryl group behave as a N,O-donating ligand toward metal ions such as Co(II), (III), Fe(II), Ni, and Pd, whereas it behaves as a O,O-donating ligand toward the other metals (e.g., Zr). The structures of each chelate are schematically shown below.

Although **(1)** to **(5)** react with various metal ions to give colored precipitates or colored soluble chelates, their reactions with Co(II), Fe(II), and Pd are of particular in-

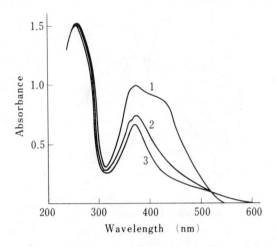

FIGURE 1.   Absorption spectra of 1-nitroso-2-naph-
thol (1); solvent, water (25°). (1) pH 11.34; (2) pH 6.91;
(3) pH 0.83. (From Suzuki, N. and Yoshida, H., *Nippon Kagaku Zasshi*, 80, 1005, 1959. With permission.)

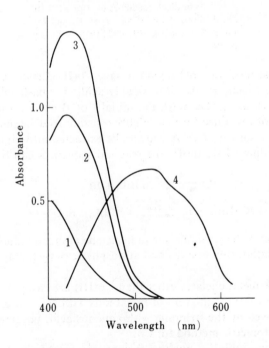

FIGURE 2.   Absorption spectra of 2-nitroso-1-naph-
thol-4-sulfonic acid (3) and its Co(III) chelate in water;
reagent, 2.00×10^{-4} M; Co, 5.00×10^{-5} M. (1) Reagent
at pH 5.0; (2) reagent at pH 7.0; (3) reagent at pH 8.0;
(4) Co(III) chelate at pH 7.0 to 10.0, against reagent
blank. (Reprinted with permission from Wise, W. M.
and Brandt, W. W., *Anal. Chem.*, 26, 693, 1954. Copyright 1954 American Chemical Society.)

terest. **(1)** and **(2)** form colored precipitates with various metal ions, however, the reactions are not always quantitative, and most of the precipitates are not suitable for
gravimetry. The reaction of these reagents with Co(II) is quite unique. The reagent

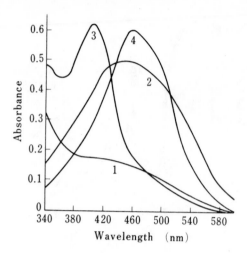

FIGURE 3. Absorption spectra of nitroso-
DMAP and its Co(III) chelate. (1) Reagent $3 \times$
10^{-5} M in water at pH 6.3; (2) reagent 3×10^{-5} M
and Co 1×10^{-5} M in water at pH 6.3; (3) reagent
5×10^{-5} M in dichloroethane; (4) CoL_3 1×10^{-5}
M in dichloroethane. (From Toei, K. and Mo-
tomizu, S., *Analyst*, 101, 497, 1976. With permission.)

reacts with Co(II) at moderate pH region to give Co(II) chelate (CoL_2) which is soon
oxidized to Co(III) chelate (CoL_3) in a weakly acidic (or basic citrate) solution. The
chelating agents such as EDTA mask the initial Co(II) chelate formation. However,
the resultant red CoL_3 is, once formed, highly resistant to dissociation in mineral acid
or base medium or even with EDTA and can be extracted into organic solvent such as
chloroform. According to Kolthoff, the reaction sequences can be written as below.

$$Co^{2+} + 2HL \rightleftharpoons Co(II)L_2 + 2H^+$$

$$2Co(II)L_2 + 2H^+ \rightleftharpoons Co(III)L_3 + H_2L \cdot + Co^{2+}$$

where $H_2L \cdot$ represents the semiquinone radical, one electron reduction product of the
reagent.[12] The oxidation step is slow, and atmospheric oxygen may also oxidize Co(II)
chelate.

Since most of the metal chelates other than Co(III) are easily decomposed in acidic
conditions, the extraction of mixture of metal ions with (1), (2), or (5) into an immiscible solvent, followed by the stripping with aqueous acid, becomes a highly selective
separation and photometric method for Co.

(3) and (4) behave similarly, but the chelates are soluble in water. Of the metals
other than Co, (4) forms colored soluble chelates with Ag(lemon yellow), Ba(orange),
Ca(green), Cu(II)(green), Fe(II)(green in basic solution), Fe(III)(brown), Ni(brownish
red), and Pb(red).

Chelate stability constants of (1) and (2) in 75% aqueous dioxane and of (4) in water
are summarized in Table 2. The overall stability constant of (5) with Co(III) is reported
to be log $\beta_3 = 26.77$ ($\mu = 0.1$, KCl).[8] Extraction constant of CoL_3 for (5) in 1,2-dichlo-
roethane/water is $10^{0.42}$ ($\mu = 0.1$, KCl).[8]

Spectral characteristics of Co(III) and Fe(II) chelates are summarized in Table 3.
Charged chelate anions can be extracted into organic phase as an ion-pair with Zephir-
amine® (tetradecyldimethylbenzylammonium chloride). Absorption spectra of

Table 2
CHELATE STABILITY CONSTANTS OF NITROSO REAGENTS[13]

Metal ion	(1) $\log K_{ML}$	(1) $\log K_{ML_2}$	(1) $\log K_{ML_3}$	(1) Temp. (°)	(1) Medium (% dioxane)	(2) $\log K_{ML}$	(2) $\log K_{ML_2}$	(2) $\log K_{ML_3}$	(2) Temp. (°)	(2) Medium (% dioxane)	(4) $\log K_{ML}$	(4) $\log K_{ML_2}$	(4) $\log K_{ML_3}$	(4) Temp. (°)	(4) Medium μ(water)	Ref.
Ag(I)	7.74	—	—	30	75	7.74	—	—	30	75	—	—	—	—	—	
Cd(II)	6.18	5.20	—	30	50	8.64	7.31	—	—	75	3.4	β_2 6.0	—	25	0.1(KCl)	
Ce(III)	—	—	—	—	—	—	—	—	—	—	4.42	—	—	25	0.1(KCl)	
Ce(II)	10.67	12.14	—	30	75	11.70	10.01	—	30	75	—	—	—	—	—	
Cu(II)	12.52	10.85	—	30	75	—	—	—	—	—	7.7	β_2 15.0	β_3 17.99	25	0.1(NaClO₄)	14
Fe(II)	—	—	—	—	—	5.62	4.35	—	30	75	—	—	—	—	—	
Mg(II)	6.05	4.72	—	30	75	6.78	5.42	—	30	75	2.7	—	—	25	0.1(KCl)	
Mn(II)	—	—	—	—	—	8.51	7.6	—	30	75	—	—	—	—	—	
Nd(III)	9.5	8.2	7.86	30	75	10.07	9.33	7.05	30	75	—	—	—	—	—	
Ni(II)	10.75	10.54	6.80	30	75	8.93	7.14	—	30	75	6.9	β_2 12.5	β_3 17.3	25	0.1(KCl)	
Pb(II)	9.73	7.58	—	30	75	8.48	7.3	—	30	75	4.64	β_2 7.37	—	25	0.1(KCl)	
Pr(III)	9.04	8.02	6.79	30	75	10.0	—	6.36	30	75	—	—	—	—	—	
Ru(III)	10.2	—	β_3 24.2	—	30% ethanol	—	—	β_3 24.0	30	30% ethanol	—	—	—	—	—	
Th(IV)	—	9.02	7.89 (log K_{ML_4} = 6.26)	25	0.1 (NaClO₄)	—	—	7.50 (log K_{ML_4} = 6.22)	25	0.1 (NaClO₄)	—	—	—	—	—	
V(IV)	—	—	—	—	—	8.3	7.6	7.4	30	75	6.71	—	—	25	—	49
Y(III)	9.02	8.72	7.30	30	75	—	—	—	—	—	4.48	β_2 7.83	β_3 11.25	25	0.1(KCl)	
Yb(III)	—	—	—	—	—	—	—	—	—	—	4.74	—	—	25	0.1(KCl)	
Zn(II)	9.32	7.70	—	30	75	8.40	7.02	—	30	75	4.5	β_2 27.1	—	25	0.1(KCl)	
Zr(IV)	3.6	—	—	32	50% ethanol	3.7	—	—	25	50% ethanol	—	—	—	—	—	

Table 3
SPECTRAL CHARACTERISTICS OF CO(III) AND FE(II) CHELATES

Reagent	Co(III) chelate[R1] Condition	Ratio	λ_{max} (nm)	$\varepsilon (\times 10^4)$	Fe(II) chelate[R2] Condition	Ratio	λ_{max} (nm)	$\varepsilon (\times 10^4)$
(1)	Aqueous dioxane, pH 6 ~ 8	ML_3	424	3.1	Water, pH 4.5 ~ 8.5	ML_3	732	1.60
	$CHCl_3$, pH 7 ~ 8	ML_3	414	3.4	$CHCl_3$, Zephiramine®, pH 4.5 ~ 8.5	ML_3X	730	1.80
(2)	Aqueous dioxane, pH 5 ~ 8	ML_3	368	2.9	Water, pH 5.0 ~ 8.5	ML_3	724	1.49
	$CHCl_3$, pH 5 ~ 8	ML_3	362	3.5	$CHCl_3$, Zephiramine®, pH 5.0 ~ 8.5	ML_3X	724	1.65
(3)	Water, pH 5 ~ 8	ML_3	368	3.3	Water, pH 4.0 ~ 8.5	ML_3	720	2.12
	$CHCl_3$, Zephiramine®, pH 7 ~ 8	ML_3X_n	368	4.0	$CHCl_3$, Zephiramine®, pH 5.5 ~ 8.5	ML_3X_4	716	2.12
(4)	Water, pH 6 ~ 7	ML_3	410	3.5	Water, pH 4.0 ~ 8.0	ML_3	724	2.02
	$CHCl_3$, Zephiramine®, pH 5 ~ 8	ML_3X_n	420	3.5	$CHCl_3$, Zephiramine®, pH 5.0 ~ 8.0	ML_3X_7	720	2.02
(5)	Water, pH 5 ~ 7	ML_3	445	4.9	Water, pH 6.5 ~ 9.5	ML_4	750	3.95
	$CHCl_3$, dichloroethane, pH 5 ~ 7	ML_3	456	6.0	$CHCl_3$, Zephiramine®, pH 6.5 ~ 9.5	$ML_3(HL)X$	730	2.87

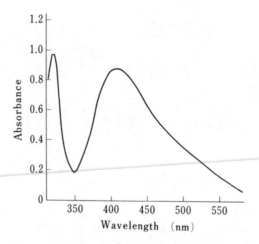

FIGURE 4.   Absorption spectrum of Co(III) chelate of 1-nitroso-2-naphthol (1) in carbon tetrachloride; Co chelate, 21.1 ppm. (From Shimura, H., *Nippon Kagaku Zasshi*, 83, 1, 1962. With permission.)

Co(III) chelates with (1), (3), (4) and (5) are illustrated in Figures 4, 2, 5, and 3, respectively.

Purification and Purity of Reagents

Commercial samples are usually pure enough for general analytical purposes. (1)

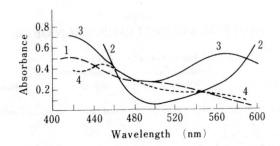

FIGURE 5. Absorption spectra of Co(III) chelate of nitroso-R salt in water. (1), Reagent 0.02% plus Co 1 ppm; (2), (1) plus Ni 1000 ppm; (3), (1) plus Cr(III) 1000 ppm; (4), (1) plus Fe(III) 1000 ppm; (From Wünsh, G., *Talanta*, 26, 179, 1979. With permission.)

and (2) may be purified by recrystallization from ligroin or by dissolving in hot ethanol, followed by the successive addition of small portions of water, (3) and (4) are purified by dissolving in aqueous alkali, followed by the addition of hydrochloric acid. (5) is purified by recrystallization from aqueous hydrochloric acid to obtain hydrochloride.

Purity of (1) and (2) can be checked by measuring their melting points. Assay of these reagents can be done by iodometry of nitroso group in a standard manner.[17]

Analytical Applications

Use as a Photometric Reagent

As described before, this class of reagents are highly selective for Co, and a number of papers have appeared on the use of these reagents for the separation and photometric determination of Co in various samples.

With the use of (1) or (2), Co is separated from the matrix by solvent extraction, then determined by photometry of the Co(III) chelate. Both reagents behave almost similarly, but K_{ex} for (2) is reported to be higher than that for (1) by 10^2.[10]

The formation of Co(III) chelate is slow and can be accelerated at higher temperature. The excess reagent in organic phase can be removed by washing with aqueous alkali, and most elements other than Co can be stripped by washing with aqueous mineral acid.

When (3) or (4) is used as a chromogen, the complexing reaction with Co is carried out at neutral pH range, and after completing the formation of Co(III) chelate, the solution is acidified to destroy the chelates with interfering elements. Co(III) chelate is not extractable into inert solvents, but can be extracted as an ion-pair in the presence of Zephiramine®.

(5) can be used with or without extraction. When used in aqueous system, a large amount of Fe interferes because the chelate with Fe(II) causes a substantial absorption at the measuring wavelength. The procedure with solvent extraction (1,2-dichloroethane) is similar to that for (1) or (2), but molar absorptivitiy is the highest of all, attaining the highest sensitivity for Co. The typical examples for Co are summarized in Table 4. The examples for the determination of elements other than Co and Fe are summarized in Table 5.

Determination of Co in steel and iron with nitroso-DMAP (5)[R1] —

Dissolve 0.5 to 1-g sample (0.05 to 0.1 mg Co) in 3 to 5 mℓ of aqua regia with heating, dilute it to 100 mℓ, transfer 5.0 mℓ aliquot into a stoppered test-tube, add 1 mℓ of citrate buffer (2 M, pH 5.3), and mix. Then add 1

Table 4
PHOTOMETRIC DETERMINATION OF CO

Reagent	Condition	Metal chelate Ratio	λ^a (nm)	Range of determination (ppm)	Remarks	Ref.
(1)	pH 8∿9, citrate + H_2O_2, extraction with $CHCl_3$, strip. of excess reagent with base, strip. of Cu and Ni with 2 N HCl	ML_3	410	0.004 ∿ 0.02	Detection of Co in sea water	18
(2)	pH 2.5∿5, citrate + H_2O_2, extraction with $CHCl_3$, strip. of Cu with HCl	ML_3	530	0.2 ∿ 4	Large amount of Ni interferes	19
(2)	pH 3.4, citrate + acetate + H_2O_2, extraction with $CHCl_3$	ML_3	367	0.05 ∿ 0.6	Detection of Co in plants	20
(3)	pH 7.0	ML_3	525	0.2 ∿ 7	Small amount of Cu and Fe are masked with citrate	11
(3)	pH 7∿8, Zephiramine® extraction with $CHCl_3$	ML_3X_n	312	—	—	R3
(4)	Reaction in sodium acetate, then add HNO_3	ML_3	546	0.001 ∿ 5%	Analysis of steel and ores; oxidation with Br_2 eliminates interference due to Cr(III), Fe(II), Sn(II), and V(IV)	21
(4)	Reaction in sodium acetate, then boil with NHO_3	ML_3	500	0.2 ∿ 0.8	Detection of Co in organometallic compounds	22
(4)	pH 6.5 ∿ 7.5, citrate	ML_3	500	0.4 ∿ 4	About 10 mg of Cr(III) is tolerable if the boiling time is extended from 1 to 5 min	16
(4)	pH 5 ∿ 8, Zephiramine®, extraction with $CHCl_3$	—	323	—	—	R3
(4)	pH 7 ∿ 8, diphenylguanidine, extraction with $CHCl_3$	—	520	0 ∿ 10	Cu, Fe, Ni, Sn, and Ti (2000-fold) do not interfere	50
(5)	Reaction at pH 5 ∿ 7 (citrate), then to pH 1 with HCl	ML_3	445	0 ∿ 2	Ni does not interfere	R1,R3
(5)	pH 5.3 citrate, extraction with dichloroethane, strip. with 4 NHCl	ML_3	456	0 ∿ 0.8	Detection of Co in inorganic salts, steel, sea water, and uranium compounds	R1,45

^{a} λ indicates the observing wavelength which does not always coincide with λ_{max}.

mℓ of nitroso-DMAP solution (5×10^{-3} M in 0.01 M HCl), mix thoroughly, and allow the mixture to stand for 5 min. Shake the mixture with 5 mℓ of 1,2-dichloroethane for 30 sec and back extract the excess reagent and the chelate of interfering elements with 5 mℓ of (1 + 2) HCl. Filter the organic phase through a dry filter-paper and measure its absorbance at 456 nm. Down to 0.0001% level of Co can be determined.

Other Uses

These reagents are used as highly sensitive detecting reagent for Co and other metals, such as Pd[4,25] and Zr.[4] (1) has been recommended as coprecipitating agent for collecting traces of Ce, Co, Fe, U, Zn, and Zr in natural water.[18,35] Cross-linked polystyrene

Table 5
PHOTOMETRIC DETERMINATION OF ELEMENTS OTHER THAN Co AND Fe

Reagent	Metal ion	Condition	Metal chelate			Range of determination (ppm)	Remarks	Ref.
			Ratio	λ^a (nm)	ε ($\times 10^4$)			
(1)	Ru(III)	0.2 NH_2SO_4, extraction with CCl_4	—	645	1.83	0.16 ~ 2.2	In Zn-Mg alloy	23
(1)	ZrO²⁺	pH 3.5	ML	490	0.06	~60	Fe(III), Mn(II), Th(IV), and U(VI) interfere	24
(2)	Pd(II)	pH 2.5(HCl), EDTA then add (2), extraction with toluene from NH_3 alkaline soln	ML_2	370	2.1	1 ~ 5	Co, Cr, Cu, Fe, and Ni are masked with EDTA	25,44
(3)	Cu	pH 4 ~ 9.5, Zephiramine®, extraction with $CHCl_3$	ML_3X_4	307.5	4.5	0 ~ 0.6	Ag gives negative, but Co and Ni give positive errors	46
(3)	Ni(II)	pH 8.6, Zephiramine®, extraction with $CHCl_3$	ML_3X_4	307.5	5.3	0 ~ 1	Large amounts of Fe are extracted with MIBK from 5 ~ 8 NHCl	47
(3)	Pd(II)	pH 1 ~ 11	—	420		0.35 ~ 10.5	—	26
(3)	VO²⁺	pH 4.4	ML_3	462	0.55	—	—	49
(4)	Cu(II)	Zephiramine®, extraction with $CHCl_3$		406	2.0	—	—	R3
(4)	Ni(II)	pH 6 ~ 7, Zephiramine®, extraction with $CHCl_3$	ML_3X_4	310	3.7	—	—	R3
(4)	Os(VIII)	1 NHCl	—	520—580	—	0.8 ~ 15	Pd, Rh and Ru do not interfere, but Ag interferes	27,29
(4)	Pd(II)	pH 2 ~ 2.6	ML_2	510	1.2	1.3 ~ 4.3	Rh and Ru interfere; Ti does not interfere in 2 ~ 3 N H_2SO_4	31,48
(4)	Rh(III)	pH 4	ML_3	510	2.4	1.5 ~ 8.6	Cu, Pd, and Ru interfere	32
(4)	Ru(III)	pH 4		520—580	—	0.4 ~ 6.8	Pd and Rh interfere	30
(4)	VOII	pH 4.4	ML_2	462	0.56	—	—	49
(5)	Pd(II)ᵇ	4 ~ 5 NH_2SO_4	ML_2	540	2.33	0 ~ 3.2	Sn(II) and W(VI) interfere	33
(5)	Pd(II)	2.5 MH_2SO_4, extraction with $CHCl_3$	ML_2	486	4.38	0 ~ 2.1	Au(III), Ir(III), and W(VI) give negative error	34

ᵃ λ indicates the observing wavelength which does not always coincide with λ_{max}.

ᵇ 2-Nitroso-5-diethylaminophenol is used.

beads soaked with chlorobenzene solution of (1) was used as a adsorbent for trace of Co in sea water in a column process.[5] In the presence of EDTA, nitrosonaphthols are highly selective for Pd.[25]

Reagents with Related Structure

o-Nitrosophenol,[36,37] o-nitrosocresol,[38] 3-nitrososalicylic acid,[39] o-nitrosoresorcinol monomethylether (3-methoxy-6-nitrosophenol),[40,41] 2-nitroso-4-chlorophenol,[42] 3-nitroso-2,6-pyridinediol,[43] and p-nitrosodimethylaniline[28] have been proposed as the reagents for Co, Fe, Ni, and Rh.

MONOGRAPHS

M1. Welcher, F. J., *Organic Analytical Reagent*, Vol. 3, Von Nostrand, Princeton, N.J., 1947, 331.

M2. Casapieri, P., Reynolds, R. J., and Jablonski, Z., Nitroso R Salt, in *Organic Reagent for Metals*, Johnson W. C., Ed., Hopkin & Williams, Chadwell Heath, England, 1955, 139.

M3. Feigl, F., Angar, V., and Oesper, R. E., *Spot Tests in Inorganic Analysis*, 6th ed., Elsevier, New York, 1972.

REVIEWS

R1. Toei, K. and Motomizu, S., New colorimetric reagent of cobalt, *Bunseki Kagaku*, 22, 1079, 1973; Properties and uses of the colorimetric reagents, 2-nitroso-5-dimethylaminophenol and 2-nitroso-5-diethylaminophenol for cobalt, *Analyst*, 101, 497, 1976.

R2. Toei, K., Motomizu, S., and Korenaga, T., Nitrosophenol and nitrosonaphthol derivatives as reagents for the spectrophotometric determination of iron and determination of micro-amounts in waters with 2-nitroso-5-dimethyl-aminophenol, *Analyst*, 100, 629, 1975.

R3. Motomizu, S. and Toei, K., Synthesis and analytical application of nitroso-related compounds, *Bunseki Kiki (Tokyo)*, 15, 1, 1977.

REFERENCES

1. Henriques, R. and Ilinsky, M., *Ber.*, 18, 704, 1885.

2. Witt, O. N. and Kaufmann, H., *Ber.*, 24, 3157, 1891.

3. *Organic Synthesis, Coll.*, Vol. 1, 2nd ed., John Wiley & Sons, New York, 1941, 411.

4. Toei, T., Motomizu, S., and Korenaga, T., *Nippon Kagaku Zasshi*, 92, 92, 1971; *Nippon Kagaku Kaishi*, 2445, 1972.

5. Kubo, M., Yano, T., Kobayashi, H., and Ueno, K., *Talanta*, 24, 519, 1977.

6. Honjo, T., Imura, H., Shima, S., and Kiba, T., *Anal. Chem.*, 50, 1547, 1978.

7. Van Uitert, L. G. and Fernelius, W. C., *J. Am. Chem. Soc.*, 76, 375, 1954.

8. Motomizu, S., *Anal. Chim. Acta*, 56, 415, 1971.

9. Zolotov, Yu. A., *Extraction of Chelate Compounds*, Humphrey Science, Ann Arbor, Mich., 1970.

10. Suzuki, N. and Yoshida, H., *Nippon Kagaku Zasshi*, 80, 1005, 1959.

11. Wise, W. M. and Brandt, W. W., *Anal. Chem.*, 26, 693, 1954.

12. Kolthoff, I. M. and Jocobsen, E., *J. Am. Chem. Soc.*, 79, 3677, 1957.

13. Sillén, L. G. and Martell, A. E., *Stability Constants of Metal-Complexes*, and Suppl. No. 1, The Chemical Society, London, England, 1964, 1971.

14. Jedrzejewski, W., Jozwiak, A., and Klaja, W., *Acta Univ. Lodz.*, Ser., 2, 13, 95, 1976; *Chem. Abstr.*, 87, 91579f, 1977.
15. Shimura, H., *Nippon Kagaku Zasshi*, 83, 1, 1962.
16. Wünsh, G., *Talanta*, 26, 179, 1979.
17. AnalaR Standards for Laboratory Chemicals, 7th ed., Analar Standards, England, 1977.
18. Kentner, E. and Zeitlin, H., *Anal. Chim. Acta*, 49, 587, 1970.
19. Funke, A. and Laukner, H. J., *Fresenius Z. Anal. Chem.*, 249, 26, 1970.
20. Ssekaalo, H., *Anal. Chim. Acta*, 51, 503, 1970.
21. Koch, K. H., Ohls, K., Sebastiani, E., and Reimer, G., *Fresenius Z. Anal. Chem.*, 249, 307, 1970.
22. Crossland, B. and Fennel, T. R. F. W., *Talanta*, 17, 112, 1970.
23. Kesser, G., Meyer, R. J., and Larsen, R. P., *Anal. Chem.*, 38, 221, 1966.
24. Patil, S. V. and Raju, J. R., *J. Indian Chem. Soc.*, 53, 635, 1976.
25. Cheng, K. L., *Anal. Chem.*, 26, 1894, 1954.
26. Sangall, S. P. and Dey, A. K., *Fresenius Z. Anal. Chem.*, 202, 348, 1964.
27. Miller, D. J., Srivastava, S. C., and Good, M. L., *Anal. Chem.*, 37, 739, 1965.
28. Wilson, R. and Jacob, W. D., *Anal. Chem.*, 33, 1652, 1961.
29. Nath, S. K. and Agarwal, R. P., *Chim. Anal. (Paris)*, 47, 257, 1965.
30. Nath, S. K. and Agarwal, R. P., *Chim. Anal. (Paris)*, 48, 439, 1966.
31. Shamir, J. and Schwartz, A., *Talanta*, 8, 330, 1961.
32. Saxena, K. K. and Agarwala, B. V., *Indian J. Chem.*, 14A (8), 634, 1976; *Chem. Abstr.*, 86, 132958d, 1977.
33. Toei, K., Motomizu, S., and Hamada, S., *Bunseki Kagaku*, 27, 668, 1978.
34. Toei, K., Motomizu, S., and Hamada, S., *Anal. Chim. Acta*, 101, 169, 1978.
35. Weiss, H. V., Lai, and Gillespie, A., *Anal. Chim. Acta*, 25, 550, 1961.
36. Gronheim, G., *Ind. Eng. Chem. Anal. Eds.*, 14, 445, 447, 1942.
37. Shimura, H., *Nippon Kagaku Zasshi*, 76, 867, 1955.
38. Ellis, G. H. and Thompson, J. F., *Ind. Eng. Chem. Anal. Ed.*, 17, 254, 1945.
39. Perry, M. H. and Serfass, E. J., *Anal. Chem.*, 22, 565, 1950.
40. Reach, S. M., *Analyst*, 81, 371, 1956.
41. Iimura, F. and Torii, T., *Bunseki Kagaku*, 4, 177, 1955; *Nippon Kagaku Zasshi*, 76, 328, 333, 675, 680, 707, 825, 1955.
42. Korenaga, T., Motomizu, S., and Toei, K., *Anal. Chim. Acta*, 65, 335, 1973; *Talanta*, 21, 645, 1974.
43. McDonald, C. W. and Carter, R., *Anal. Chem.*, 41, 1478, 1969.
44. Posta, S. and Kukula, F., *Radioisotopy*, 17, 559, 1976; *Anal. Abstr.*, 33, 4B185, 1977.
45. Motomizu, S., *Analyst*, 100, 39, 1975.
46. Motomizu, S. and Toei, K., *Bunseki Kagaku*, 27, 213, 1978.
47. Motomizu, S. and Toei, K., *Anal. Chim. Acta*, 97, 335, 1978.
48. Banerjee, S. and Dutta, P. K., *Fresenius Z. Anal. Chem.*, 277, 379, 1975.
49. Makitie, O. A. and Lajunen, K. V. O., *Talanta*, 22, 1053, 1975.
50. Shustpva, M. B. and Nazarenko, V. A., *Zavod. Lab.*, 39, 18, 1973; *Chem. Abstr.*, 78, 118890d, 1973.

N,N-DONATING CHELATING AGENTS

BIPYRIDINE AND OTHER FERROIN REAGENTS

(1)

(3)

(2)

(4a) R = H   (4b) R = SC$_3$Na

Synonyms

Reagents covered in this section are listed in Table 1, together with their synonyms.

Source and Method of Synthesis

All are commercially available.(1) and (2) are prepared by the dehydrogenation of pyridine with anhydrous FeCl$_3$[1] or with Ziegler catalyst at 100 to 200°.[2] (3) is prepared by heating o-phenylenediamine with glycerol, nitrobenzene, and concentrated H$_2$SO$_4$[3] or by Skraup reaction from 8-aminoquinoline.[M2,4] (4a) is prepared from 8-amino-4-phenylquinoline by Skraup reaction with β-chloropropiophenone.[M1,5] (4b) is a sulfonation product of (4a) with chlorosulfonic acid, followed by hydrolysis.[6,7]

Analytical Uses

(1) to (4) are mainly used as photometric reagents for Fe(II) and as masking agents and detecting reagents for various metal ions. Their iron chelates are also used as redox indicators.

Properties of Reagents

2,2′-Bipyridine (1) is a white crystalline solid, mp 70 to 2° and bp 272 to 275°.[M9] It has modest vapor pressure at room temperature and a characteristic, rather pleasant odor. It can readily be sublimed or steam distilled, is soluble in water ($\sim$5 g/ℓ at room temperature), and is very soluble in alcohol, ether, benzene, petroleum ether, chloroform, and dilute acids.[M5] pKa_1 = −0.2[8] and pKa_2 = 4.4 (μ = 0.1 KNO$_3$, 25°).[9]

2,2′2″-Terpyridine (2) is a white crystalline solid, mp 88 to 89°. It is not volatile with steam and is only sparingly soluble in water, but dissolves readily in most organic solvents and dilute acids. pKa_1 = −1.6,[10] pKa_2 = 3.99, and pKa_3 = 4.69 (μ = 0.1 K$_2$SO$_4$, 25°).[11]

1,10-Phenanthroline (3) is a white crystalline powder. It is usually obtained as monohydrate which melts over the range of 98 to 100° with loss of some water. The anhydrous form melts at 117°[3] and is soluble in water ($\sim$3.3 g/ℓ at room temperature) and in benzene ($\sim$14 g/ℓ at room temperature). It is very soluble in alcohol (540 g/ℓ),[10] acetone, and dilute acids.[M2] pKa_1 = 0.70[12] and pKa_2 = 4.98 (μ = 0.1 KCl, 25°).[13]

Bathophenanthroline (4a) is a white or pale yellow crystalline powder, mp 215 to 216° and is slightly soluble in dilute acids, but practically insoluble in water in neutral or alkaline region. It is soluble in organic solvents such as alcohol, acetone, and benzene;[M4] pKa_2 = 4.30 (μ = 0.3 KNO$_3$, 50% dioxane, 25°).[11,13]

Sulfonated Bathophenanthroline (4b) is usually supplied as disodium salt which is a pale yellow or light pink crystalline powder. Color deepens depending upon the degree of contamination with trace of Fe. It is hygroscopic, but after drying at 110° for 2 hr,

Table 1
BIPYRIDINE AND RELATED REAGENTS

Number	Name of reagent	Synonym	Molecular formula	Mol wt
(1)	2,2′-Bipyridine	α,α'-Bipyridyl, Dipyridyl, 2,2′-Bipyridyl	$C_{10}H_8N_2$	156.19
(2)	2,2′2″-Terpyridine	2,2′,2″-Terpyridyl	$C_{15}H_{11}N_3$	233.27
(3)	1,10-Phenanthroline	*o*-Phenanthroline, 4,5-Phenanthroline	$C_{12}H_8N_2 \cdot H_2O$	198.23
(4a)	4,7-Diphenyl-1,10-phenanthroline	Bathophenanthroline	$C_{24}H_{16}N_2$	332.41
(4b)	4,7-Diphenyl-1,10-phenanthroline disulfonic acid, disodium salt	Bathophenanthroline disulfonic acid, Na$_2$ salt	$C_{24}H_{14}N_2O_6S_2Na_2$	536.48

it shows no change in weight up to 275°.[7] It shows a light blue fluorescence under UV light and is very easily soluble in water, but not soluble in many organic solvents. The free acid is a syrupy liquid, highly hygroscopic, and is difficult to obtain in solid form; $pKa(SO_3H) = 2.83$ and $pKa_2 = 5.20$.[7]

Reagents of this group are colorless and do not show any absorption band in the visible region.

Complexation Reactions and Properties of Complexes

The reagents of this class form stable colored chelates with transition metal ions.[14,15] However, many form no more intensely colored chealtes than those in aquo ion state. Some give nearly colorless chelates (Cd, Zn, Mn(II), and Ag). Copper(I) and Fe(II) are the exceptions which form highly colored chelates. In these chelates, (1), (3), and (4) behave as a bidentate uncharged ligand to form ML_3^{n+} and ML_2^{n+}-type chelates with a metal ion (M^{n+}) of coordination number 6 and 4, respectively, as illustrated in Figure 1.

Terpyridine (2) behave as a terdentate ligand, forming ML_2^{n+}-type chelate with a metal ion of coordination number 6.

A relatively large number of stability constants have been determined for the metal chelates of bipyridine, terpyridine, and phenanthroline. These values are summarized in Table 2 to 4. The values for Bathophenanthroline (4a) are Cu(II), $\log K_{ML2} \sim 5.7$ and $\log K_{ML3} \sim 3.75$ (25°, $\mu = 0.3$ KNO$_3$, 50% dioxane),[11] and Fe(II), $\log \beta_3$ 21.8 (18°, $\mu = 0.1$, NaCl, 10% ethanol),[27] and for sulfonated Bathophenanthroline (4b) are Fe(II), $\log \beta_3$ 22.3.[28]

Reagents of this class are widely used for the determination of traces of Fe and Cu because they form intensely colored chelates with these metals. The spectral characteristics of the Co,Cu,Fe, and Ru chelates are listed in Table 5, and absorption curves of Fe(II) chelates of (1), (3) and (4a) are illustrated in Figure 2. Few other metal ions give colored chelates, but their spectral characteristics are so different from those of Fe(II) and Cu(I) chelates that they do not interfere the photometric determination of Fe and Cu.

In contrast to the intense orange-red color of Fe(II) chelates (ferroin), the corresponding Fe(III) chelates (ferriin) are colorless or pale blue. The ferroin-ferriin redox couple,

$$FeL_3^{3+} + e \rightleftharpoons FeL_3^{2+}$$

Colorless   Orange-Red

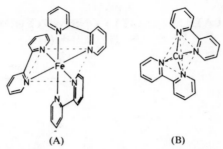

(A)   (B)

FIGURE 1.   (A) Structure of tris(1,10-phenan-
throline)-iron(II);   (B)   bis(1,10-phenanthroline)
copper(I).

Table 2
CHELATE STABILITY CONSTANTS OF BIPYRIDINE (1)

| | | | | | Condition | |
Metal ion	log K_{ML}	log K_{ML_2}	log K_{ML_3}	Temp (°)	μ	Ref.
Ag(I)	3.7	β_2 7.22	—	25	0.1 (KNO$_3$)	16
Ag(II)	—	β_2 6.8	—	25	—	17
Cd	4.25	3.6	2.7	20	0.1(NaNO$_3$)	18
Co(II)	6.06	5.36	4.60	20	0.1(NaNO$_3$)	20
Cr(II)	∿4	∿6.4	3.5	—	0.1	19
Cu(I)	—	β_2 14.2	—	25	0.1(KNO$_3$)	20
Cu(II)	8.0	5.60	3.48	20	0.1(NaNO$_3$)	18
Fe(II)	4.20	3.70	9.55	25	0.1	13
Fe(III)	4.2	>5	β_3 17.06	25	—	21
Hg(II)	9.64	7.10	2.8	20	0.1(NaNO$_3$)	18
Mg	0.5	—	—	27	0.5(LiClO$_4$)	14
Mn(II)	2.62	2.00	∿1.1	25	0.1(KCl)	13
Ni	7.13	6.88	6.53	20	0.1(NaNO$_3$)	18
Pb	2.9	—	—	20	0.1(NaNO$_3$)	18
Tl(I)	—	β_2 3	—	25	1.0(NaNO$_3$)	22
Tl(III)	9.40	6.70	—	25	1.0(NaNO$_3$)	22
V(II)	4.91	4.67	3.85	—	0.1	19
Zn	5.04	4.35	3.57	25	0.1(KCl)	13

Table 3
CHELATE STABILITY CONSTANTS OF TERPYRIDINE (2)

| | | | | Condition | |
Metal ion	log K_{ML}	log K_{ML_2}	Temp (°)	μ	Ref.
Cd	5.1	—	25	—	23
Co(II)	8.4	β_2 18.3	25	—	23
Cu(I)	∿9.3	—	25	0.1(K$_2$SO$_4$)	11
Cu(II)	∿13.0	—	25	0.1(K$_2$SO$_4$)	11
Fe(II)	7.1	β_2 20.9	25	—	23
Mn(II)	4.4	—	25	—	23
Ni	10.7	β_2 21.8	25	—	23
Zn	6.0	—	25	—	23

Table 4
CHELATE STABILITY CONSTANTS OF 1,10-PHENANTHROLINE (3)

				Condition		
Metal ion	$\log K_{ML}$	$\log K_{ML_2}$	$\log K_{ML_3}$	Temp (°)	μ	Ref.
Ag	5.02	7.05	—	25	0.1	24
Ca	0.7	—	—	20	0.1(NaNO₃)	18
Cd	5.78	5.04	4.10	20	0.1(NaNO₃)	18
Co(II)	7.25	6.70	5.95	20	0.1(NaNO₃)	18
Cu(I)	—	β₂ 15.82	—	25	0.1(K₂SO₄)	11
Cu(II)	9.25	6.75	5.35	20	0.1(NaNO₃)	18
Fe(II)	5.86	5.25	10.03	20	0.1(NaNO₃)	13
Fe(III)	6.5	β₂ 11.4	β₃ 23.5	20	0.1(NaNO₃)	25
Hg(II)	—	β₂ 19.65	3.7	20	0.1(NaNO₃)	18
Mg	1.2	—	—	20	0.1(NaNO₃)	18
Mn(II)	4.13	3.48	2.7	20	0.1(NaNO₃)	18
Ni	8.8	8.3	7.7	20	0.1(NaNO₃)	18
Pb	4.65	—	—	20	0.1(NaNO₃)	18
Tl(I)	—	β₂ 4	—	25	1.0(NaNO₃)	22
Tl(III)	11.57	6.73	—	25	1.0(NaNO₃)	22
VO(II)	5.47	4.22	—	25	0.082	26
Zn	6.55	5.80	5.20	20	0.1(NaNO₃)	18

Table 5
SPECTRAL CHARACTERISTICS OF BIPYRIDINE AND RELATED COMPOUNDS

	CuL_2^+			FeL_3^{2+}			RuL_3^{2+}			CoL_3^{2+}		
Reagent	λ_{max} (nm)	ε(× 10³)	Ref.	λ_{max} (nm)	ε(× 10³)	Ref.	λ_{max} (nm)	ε(× 10³)	Ref.	λ_{max} (nm)	ε(× 10³)	Ref.
(1)	435	4.5	29	522	8.65	30	418	14.3	31	∼435	0.09ᵃ	
(2)	430	3.25	29	552	11.5	30	475	12.68	32	505	1.36	33
										510	2.90ᶜ	34
(3)	435	7.0	29	510	11.1	35	448	18.5	36	∼425	0.10	
(4a)	457	12.14ᵈ	37	533	22.35ᵈ	37	460	27.0ᶜ	38	—	—	
(4b)	483	12.25	7	535	22.14	7	—	—		—	—	

ᵃ   Shoulder on tail of an UV band extending into visible region.
ᵇ   Forms FeL₂²⁺ or CoL₂²⁺-type chelate as terdentate ligand.
ᶜ   In *n*-hexylalcohol.
ᵈ   In iso-amylalcohol.

is highly reversible and meets all the requirements of the ideal redox indicator. The redox potential of the system can be expressed by the Nernst equation:

$$E = E_o + 0.0591 \log \frac{[FeL_3^{3+}]}{[FeL_3^{2+}]}$$

The value of E_o which is known as the formal potential varies with the composition of the solution and depends on the extent of influences of solution composition on activity coefficients and the stabilities of the Fe(II) and Fe(III) chelates. Table 6 summarizes the formal potential of the ferroin-ferriin systems at various acid concentration, along with their color change. Some derivatives of bipyridine (1) and phenanthroline (3) which are important as the redox indicator are also included in the same table.

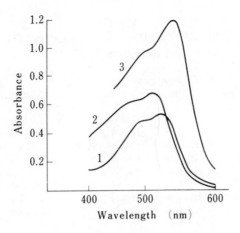

FIGURE 2. Absorption spectra of Fe(II)-2,2'-Bipyridine, 1,10-phenanthroline, and Bathophenanthroline chelates. (1) 2,2'-Bipyridine(water); (2) 1,10-Phenanthroline(water); (3) Bathophenanthroline(nitrobenzene); Fe, 6×10^{-5} M; reagent, 4×10^{-4} M. (From Yamamoto, Y., *Bunseki Kagaku*, 21, 418, 1972. With permission.)

Table 6
FORMAL POTENTIAL OF FERROIN - FERIIN (FeL$_3^{2+}$ - FeL$_3^{3+}$) COUPLE

Reagent	Formal potential (E$_o$) in H$_2$SO$_4$(V)					Color change		Ref.
	0.1M	0.5M	1.0M	2.0 M	4.0M	From	To	
(1)	1.062^a	—	1.026	1.00	0.95	Red	Faint blue	39
(2)	1.076	1.054	0.927	—	—	Violet	Green	40
(3)	1.10^b	—	1.06	1.03	0.96	Red	Pale blue	41
4-Methyl (3)	—	—	1.02	1.00	0.93			42
5-Chloro (3)	—	—	1.11	1.10	1.04			42
5-Nitro (3)	—	1.26	1.25	1.22	1.12	Red	Pale greenish blue	42,43
Dicyano-5-nitro (3)	—	—	0.90	—	0.93^c			44
5-Sulfo (3)	—	—	1.20 $\sim$ 1.26	—	—			45
5,6-Dimethyl (3)	1.00	0.97	—	—	—	Red	Yellow-green	41
3,5,7-Trimethyl (3)	0.93	0.89	—	—	—			41
3,4,6,8-Tetramethyl (3)	0.89	—	—	—	—			41
(4a)	—	—	1.24	—	1.13^d			43
(4b)	—	—	1.09	—	—	Red	Green	7

^{a} In 0.11 MH$_2$SO$_4$.
^{b} In 0.05 MH$_2$SO$_4$.
^{c} In 6.0 MH$_2$SO$_4$.
^{d} In 4.6 MH$_2$SO$_4$.

As discussed above, (1) to (4a) behave as an uncharged ligand, resulting in the formation of cationic chelates which retain the positive charge of central metal ion. These cationic chelates tend to form highly insoluble precipitates with mono- or divalent bulky anions, such as ClO$_4^-$, SCN$^-$, or CdI$_4^{2-}$. The solubilities are very much dependent upon the combination of anions and the cationic chelates, and the reaction can be utilized in the detection or separation of anions by precipitation. Table 7 summarizes the dilution limit for the detection of various anions with Fe(bipy)$_3$SO$_4$.[46] In contrast

Table 7
PRECIPITATION OF ANIONS WITH
0.1 M [Fe(bipy)$_3$]SO$_4$[46]

Anion	Dilution limit	Anion	Dilution limit
Cl$^-$	1:10	Fe(CN)$_6^{4-}$	1:5,000
Br$^-$	1:50	Fe(CN)$_6^{3-}$	1:30,000
I$^-$	1:1,000	Fe(CN)$_5$(NO)$^{2-}$	1:20,000
SCN$^-$	1:5,000	PtCl$_6^{2-}$	1:100,000
VO$_3^-$	1:40,000	BiCl$_4^-$	1:10,000
Cr$_2$O$_7^{2-}$	1:6,000	SnCl$_6^{2-}$	1:2,000
MnO$_4^-$	1:12,500	HgCl$_4^{2-}$	1:50,000
ReO$_4^-$	1:1,000	CdI$_4^{2-}$	1:1,000,000[47]

Table 8
SOLUBILITIES OF Fe(PHEN)$_3$X$_2$ IN VARIOUS SOLVENTS[48]

(Moles Per Liter at 25°)

Solvent	X ClO$_4^-$	I$^-$	Br$^-$	Cl$^-$	SCN$^-$
Water	8.8 × 10^{-4a} 7.90 × 10^{-4}	6.15 × 10^{-3}	5.08 × 10^{-2}	2.55 × 10^{-1}	3.3 × 10^{-3}
Nitrobenzene	2.95 × 1^{-2}	6.26 × 10^{-3}	6.7 × 10^{-4}	6.2 × 10^{-4}	6.04 × 10^{-3}
Water saturated with Nitrobenzene	4.50 × 10^{-4}	3.29 × 10^{-3}	5.31 × 10^{-2}	2.71 × 10^{-1}	1.59 × 10^{-3}
Nitrobenzene saturated with water	3.55 × 10^{-2}	6.0 × 10^{-3}	1.27 × 10^{-3}	2.32 × 10^{-4}	6.66 × 10^{-3}

[a] The value obtained by using a membrane electrode. The solubility for Fe(bipy)$_3$ (ClO$_4$)$_2$, 1.9 × 10^{-2} M.[49]

to the water insolubility of these precipitates, they are soluble in the organic solvents of high dielectric constant.

The solubilities of Fe(phen)$_3$X$_2$ (X = ClO$_4^-$, I$^-$, Br$^-$, Cl$^-$) in various solvents are summarized in Table 8. Thus, certain anions can be extracted into organic phase as an ion pair. The extraction coefficient is again very much dependent upon the combination of cationic chelate, anions to be extracted, and the solvent. Table 9 summarizes the distribution ratios for the Fe(bipy)$_3$X$_2$ and Fe(phen)$_3$X$_2$ systems.

The cationic chelates formed by biquinoline and related reagents (p.315) can also be used for this purpose. The examples of the ion-pair extraction of various anions with ferroin-type chelates are summarized in Table 10.

Purification and Purity of Reagents

(1) to (4a) are all well-characterized crystals and are easily purified by recrystallization from appropriate solvents, (1) from dilute ethanol and (2) and (4a) from benzene. (3) is obtained as monohydrate when recrystallized from ethanol-water or moist benzene. Anhydrous material can be obtained by recrystallization from benzene or petroleum ether after distilling off the hydrated water.

Table 9
DISTRIBUTION RATIO (D) OF FERROIN ION-PAIRS,
$(FeL_3)^{2+}(X^-)_2$ $(25°)$, [50]

X⁻	[Fe(bipy)₃]²⁺(X⁻)₂		[Fe(phen)₃]²⁺(X⁻)₂
	Nitrobenzene/water	Dichloroethane/water	Nitrobenzene/water
ClO₄⁻	$(6 \times 10^{-5}\ M)^b$ 4.22	$(5 \times 10^{-4}\ M)^b$ 0.077	$(10^{-4}\ M)^b$ 13.5
SCN⁻	$(2.5 \times 10^{-4}\ M)$ 0.39	$(5 \times 10^{-3}\ M)$ 0.012	$(10^{-4}\ M)$ 1.71
I⁻	$(3 \times 10^{-4}\ M)$ 0.19	$(3 \times 10^{-3}\ M)$ 0.004	$(10^{-4}\ M)$ 0.83
Br⁻	$(1 \times 10^{-3}\ M)$ 0.06	—	$(10^{-3}\ M)$ 0.03
Cl⁻	—	—	$(10^{-2}\ M)$ 0.003

[a] Equilibrated in the presence of $1/15\ M$ phosphate buffer (pH 5.3).

[b] Number in the parentheses is the original concentration of ferroin salts in the aqueous phase.

Table 10
ION-PAIR EXTRACTION OF VARIOUS ANIONS WITH
FERROIN-TYPE CATIONIC CHELATES[R3]

Anion to be extracted	Pairing cationic chelate[a]	Solvent	Ref.
I⁻	Fe(phen)₃²⁺	Nitrobenzene	
ClO₄⁻	Fe(phen)₃²⁺	Nitrobenzene	
	Fe(bipy)₃²⁺	Nitrobenzene	
	Cu(cup)₂⁺	Chloroform	
	Cu(neocup)₂⁺	Chloroform	
	Fe(bathophen)₃²⁺	n-Butyl acetate	
NO₃⁻	Cu(cup)₂⁺	Chloroform, MIBK, Chlorobenzene	
	Cu(neocup)₂⁺	MIBK	
ReO₄⁻	Fe(bipy)₃²⁺	Nitrobenzene	
RuO₄	Fe(bipy)₃²⁺	Nitrobenzene	
HCrO₄⁻	Fe(phen)₃²⁺	Nitrobenzene	
SCN⁻	Fe(bipy)₃²⁺	Nitrobenzene	
	Fe(phen)₃²⁺	Nitrobenzene	
AuCl₄⁻ PtCl₆²⁻	Fe(phen)₃²⁺	Nitrobenzene	
HgI₄²⁻ CdI₄²⁻	Fe(bipy)₃²⁺	Dichloroethane	
HgBr₄²⁻ TlBr₄⁻	Fe(bipy)₃²⁺	Dichloroethane	
Ni(CN)₄²⁻ Ag(CN)₂⁻	Fe(phen)₃²⁺	Nitrobenzene	
Sn(C₂O₄)₃²⁻	Fe(phen)₃²⁺	Nitrobenzene	
Mo-C₂O₄	Fe(phen)₃²⁺	Nitrobenzene	
BF₄⁻ TaF₆⁻ AsF₆	Fe(phen)₃²⁺	Nitrobenzene	
Phosphomolybdate	Fe(phen)₃²⁺	Nitrobenzene	
Arsenomolybdate	Fe(phen)₃²⁺	Nitrobenzene	
Trichloroacetate	Fe(phen)₃²⁺	Nitrobenzene	

Table 10 (continued)
ION-PAIR EXTRACTION OF VARIOUS ANIONS WITH
FERROIN-TYPE CATIONIC CHELATES[R3]

Anion to be extracted	Pairing cationic chelate[a]	Solvent	Ref.
Picrate	Fe(bipy)$_3^{2+}$	Nitrobenzene	
Dehydroacetate	Fe(phen)$_3^{2+}$	Nitrobenzene	
Cyclamate	Fe(phen)$_3^{2+}$	Nitrobenzene	
Saccharin	Fe(phen)$_3^{2+}$	Nitrobenzene	
Phthalate (monohydrogen)	Cu(cup)$_2^+$	Chloroform	
	Cu(neocup)$_2^+$	MIBK	
Salicylate	(Fe(phen)$_3^{2+}$	Nitrobenzene	
	Fe(bathophen)$_3^{2+}$	n-Butyl acetate	
Alkylbenzenesulfonate	Fe(phen$_3^{2+}$	Nitrobenzene	
Lauryl sulfonate	Fe(phen)$_3^{2+}$	Chloroform	51
Dodecylbenzene-sulphonate	Fe(phen)$_3^{2+}$	Chloroform	51
Maleate	Fe(phen)$_3^{2+}$	Chloroform	
Pentachlorophenolate	Fe(bipy)$_3^{2+}$	Chloroform	
Anionic dyes (Orange II, III, Acid Orange 8, 19, Acid red 88)	Fe(phen)$_3^{2+}$	1,2-Dichloroethane	52

[a] Key for the ligands: bipy, bipyridine (1); bathophen, Bathophenanthroline (4a); cup, Cuproine (or biquinoline); neocup, Neocuproine (or dimethylphenanthroline); phen, phenanthroline (2).

(4b) of low purity is brown or tan powder, and it is not so easy to remove the colored contaminants. Aqueous solution of (4b) is so sensitive to iron, that utmost care has to be taken to avoid the iron contamination from water, reagents, and glass wares. Inorganic salt and some of the colored contaminants may be removed by dissolving the crude sample into minimum amount of water, followed by an addition of ethanol to precipitate the contaminants. Pure (4b) is obtained by careful evaporation of the filtrate.

The purity of (1) to (4a) can be checked most conveniently by observing their melting points or by titrating with perchloric acid in glacial acetic acid using naphtholbenzein as an indicator.[53] The purity of (4b) is determined by absorption photometric method. The sample of satisfactory purity should pass the following requirement.

$$A \ (at \ 278.5 \ nm) > 0.70$$

$$A \ (at \ 535 \ nm) < 0.01$$

where A is an absorbance of 10 mg/ℓ aqueous solution of (4b) in a 10-mm standard silica cell against water as a reference.[54]

Analytical Applications

Use as Photometric Reagents

Many chromogenic reagents have been reported for the photometric determination of a trace of Fe; none is superior to the ferroin-type reagents.

Bipyridine (1), terpyridine (2), and phenanthroline (3) are recommended for general purposes, where extreme sensitivity is not required and unusual conditions are not involved.[R1R4,55,56] Bathophenanthroline (4a) and sulfonated derivative (4b) are recommended for the determination of ultra trace iron.[M4,57] The determination can be conducted in the aqueous phase for (1), (2), (3), and (4b). However, the extraction of colored complex with organic solvent is necessary for (4a).

With the use of (1) to (3), the orange-red ferrous chelate forms quantitatively over the pH range of 2 to 9 (optimum range 4 to 6).[56] of the various reducing agents for converting Fe(III) to Fe(II), hydroxylamine hydrochloride or ascorbic acid is most widely used in this pH range. The rate of color development is very rapid as far as Fe(III) exists as labile species. The order of addition among reagents is rather important, and the following order is recommended; reductant, chromogen, and buffer.

Absorbance follows Beer's law over the range of 0 to 8 ppm of Fe, and 5 to 50 μg of Fe in a final volume of 10 mℓ can be measured. The color is stable for many months.

Although (1) to (3) are highly selective for Fe, some elements, if present in excess, may interfere.[56] The amount of interfering elements that can be tolerated depends on the pH, amount of chromogen, and the type and concentration of masking agent employed. When citrate and EDTA are used at pH 5.0 to 6.5, the following interferences and maximum tolerance ratio (interfering ion to Fe molar ratio) are found: Ag (50), Co (<12), Cu (25), Cr (<200), Mo (<40), Ni (12.5), Ru (<30) and W (<7).[M5,58]

(4a) is recommended when unusually high sensitivity and selectivity are required. With the use of this reagent, 0.001 to 0.1 ppm of Fe can be determined. It is also highly selective for Fe because only Fe(II) gives a colored chelate at pH 4 in aqueous solution, and the chelate can be extracted into iso-amylalcohol[59] or a mixture of iso-amylalcohol and iso-propylether (1:1).[60] Copper(I) forms a colorless extractable monochelate at or below pH 4, but does not interfere in the determination of Fe, provided that sufficient amount of chromogen is added to complex both Fe and Cu.

(4b) forms the intense red tris-chelate which is stable over the pH range 2 to 9. (4b) is less selective than (4a) for Fe. Relatively small amounts Cu, Co, and Cr interfere by forming colored chelates, but (4b) may be more convenient for the routine Fe determination because the extraction step can be avoided with the use of (4b).[M4,M7]

Use as a Fluorimetric Reagent

(3) is useful as a fluorimetric reagent for the determination of trace amounts of Ag, Cd, Cu, Pd, Re, Sc, Zn, and rare earths. The principle of the method is based on the appearance of highly intense fluorescence in the mixed ligand complexes formed in the presence of the second ligand, such as salicylic acid,[61] 2-phenyl-4-quinolinecarboxylic acid,[62] Eosine,[64,65] dibromofluoresceine,[63] or TTA.[66]

Use as a Masking Agent

The reagents of this class form colorless chelates with various metal ions and can be used as masking agents in chelatometry and photometry.[67-69]

(4a) is also very useful as a chelating agent for the removal of trace of Fe, Cu, and certain other metal ions from reagent solutions that are to be used in trace metal analysis. After treating with (4a), the resulting chelates and unreacted (4a) are simply removed by extracting with an immiscible solvent.

Use as an Ion-Pair Extraction Agent

As described previously, the ferroin-type cationic chelates are highly stable, and various anions can be extracted into organic phase as an ion-pair (see Table 10).[R3] In the favorable condition, the extraction is quantitative, and the concentration of respective anion in the aqueous phase can be determined by observing the color intensity of ferroin in the organic phase, as explained by the following equation:

$$\text{Fe(phen)}_3{}^{2+}{}_{aq} + 2X^-{}_{aq} \rightleftharpoons \text{Fe(phen)}_3{}^{2+} \cdot X_2{}^{2-}{}_{org}$$

Use as a Redox Indicator

As described earlier, the ferroin $\rightleftharpoons$ ferriin reversible reaction can be utilized as an indicator in the redox titration.[M3] Although the transition potential, at which indicator color change is observed, is closely related to the formal potential, its estimated value varies somewhat dependent upon the individuals. In general, since the orange-red color of ferroin is much more intense than the pale blue color of ferriin, approximately nine tenths of the ferroin must be converted to ferriin before one can detect the color change. Accordingly, the transition potential is approximately $0.05V(0.0591$ log[Ferriin]/ [Ferroin]) greater than the formal potential in Table 7. The indicator derived from 5-nitro-1,10-phenanthroline is ideally suited for the titration with Ce(IV) (perchlorate or nitrate) in perchloric acid or nitric acid solution.[70]

Other Reagents with Related Structures
4,7-Dihydroxy-1,10-phenanthlorine (Snyder Reagent)

H_2L

Usually supplied as hydrochloride ($H_2L \cdot HCl$, mol wt = 248.67); yellow crystalline powder (decomposes at 475°). It is very slightly soluble in water (pH 1 to 8) and in common organic solvents, but is easily soluble in aqueous alkaki (pH 8). The aqueous solution is stable for many months, even under heat.[71] pKa_1 (N^+H) = 2.55, pKa_2 (OH) = 7.28, and pKa_3 (OH) = 11.5.[75]

Snyder reagent is of special interest as it forms a red chelate with Fe(II) in 6 to 10 N NaOH λ_{max}, 520 nm; ε = 1.48 × 10⁴).[73,74] Thus, it is recommended as a photometric reagent for Fe(II) in strongly alkaline solution. Similarly, 2,9-dimethyl derivatives have been recommended as a copper reagent (λ_{max}, 400 nm; ε = 1.15 × 10⁴ in 0.5 to 7 N NaOH).[75]

Phenyl-2-pyridylketoxime[M4] (2-Benzoylpyridine oxime, PPKO)

$C_{12}H_{10}N_2O$,
mol wt = 198.22

Syn form mp 151 to 152°   Anti form mp 165 to 167°

Of the two stereoisomers, syn form reacts with various metal ions to form colored chelates which are extractable into chloroform. It is almost insoluble in water and cold ethanol, but easily soluble in hot ethanol, benzene, chloroform, dioxane, iso-amylalcohol and dilute mineral acids. The reagent, both solid and solution, is sensitive to light and hence has to be protected from direct sunlight; pKa_1 (H_2L^+) = 2.84 and pKa_2 (HL) = 12.19 (μ = 0.1 NaClO₄, 40% acetone, 20°).[76]

It forms colored soluble chelate in a neutral or alkaline solution with Fe(II), (III) (red), Cu(I) (orange), Cu(II) (green), Au(III), Co(II), Pd(II), Pt(II) and UO_2^{2+} (yellow), and Cr(III) (colorless). In a strongly alkaline solution (pH 10 to 5 N NaOH), Fe(II) forms a deep red chelate (ML_3; λ_{max}, 545 nm; ε = 1.52 × 10⁴, that can be extracted into iso-amylalchohol. Thus, PPKO is also recommended for the determination of trace of Fe in alkaline sample.[M5] It is also used for the determination of Au (III), Cu (II), and Pd (II).[77-79]

MONOGRAPHS

M1. Case, F. H., *A Review of Synthesis of Organic Compounds Containing the Ferroin Group*, G. F. Smith Chemical Co., Columbus, Ohio, 1960.

M2. Smith, G. M. and Richter, F. P., *Phenanthroline and Substituted Phenanthroline Indicators*, G. F. Smith Chemical Co., Columbus, Ohio, 1944.

M3. Smith, G. F., *Cerate Oxidimetry*, G. F. Smith Chemical Co., Columbus, Ohio, 1964.

M4. Diehl, H. and Smith, G. F., *The Iron Reagent: Bathophenanthroline, 2,4,6-Tripyridyl-s-triazine, Phenyl-2-pyridylketoxime*, G. F. Smith Chemical Co., Columbus, Ohio, 1960.

M5. Schilt, A. A., *Analytical Applications of 1,10-Phenanthroline and Related Compound*, Pergamon Press, London, 1969.

M6. Hans, J. J. H. and Johnson, W. C., 1,10-Phenanthroline, Newman, E. J. and Peters, G., 4,7-Diphenyl-1,10-phenanthroline, Newman, E. J. and Tishall, M., Sym-phenyl-2-pyridyl ketoxime, in *Organic Reagents for Metals*, Johnson, W. C., Ed., Hopkin & Williams, Essex, England, 1955 and 1964.

M7. Sandell, E. B. and Onishi, H., *Photometric Determination of Traces of Metals*, 4th ed., John Wiley & Sons, New York, 1978.

M8. Welcher, F. J., *Organic Analytical Reagents*, Vol. 3, Van Nostrand, Princeton, N.J., 1947.

REVIEWS

R1. Smith, G. F., The Ferroine, Cuproine and Terroine reacting organic analytical reagents, *Anal. Chem.*, 26, 1534, 1954.

R2. Vydra, F. and Kopanica, M., 1,10-Phenanthroline as an analytical reagent, *Chemist-Analyst*, 52, 88, 1963.

R3. Yamamoto, Y., Extraction-spectrophotometric study of anions by the ternary systems between anions, metal ions (Fe or Cu) and Ferroine and its related compounds, *Bunseki Kagaku*, 21, 418, 1972.

REFERENCES

1. Hein, F. and Retter, N., *Ber.*, 61, 1790, 1928.

2. Waddan, D. Y., and Williams, D., (Imperial Chem. Ind. Ltd.), Ger. Offen, 1950074; *Chem. Abstr.*, 73, 3799k, 1970.

3. Blau, F., *Monatsh. Chem.*, 19, 647, 1898.

4. Smith, G. F., *J. Am. Chem. Soc.*, 52, 397, 1930.

5. Case, F. H. and Sasin, R., *J. Org. Chem.*, 20, 1330, 1955.

6. Trinder, P., *J. Clin. Pathol.*, 9, 170, 1956.

7. Blair, D. and Diehl, H., *Talanta*, 7, 163, 1961.

8. Perkampus, H. H. and Kohler, H., *Z. Elektrochem.*, 64, 365, 1960.

9. Yamasaki, K. and Yasuda, M., *J. Am. Chem. Soc.*, 78, 1324, 1956.

10. Linnel, R. H. and Kaczmarczyk, *J. Phys. Chem.*, 65, 1196, 1961.

11. James, B. R. and Williams, R. J. P., *J. Chem. Soc.*, 2007, 1961.

12. Matijevic, W., Kolak, N., and Catone, D. J., *J. Phys. Chem.*, 73, 3556, 1969.

13. Irving H. and Mellor, D. H., *J. Chem. Soc.*, 5222, 1962.

14. Sone, K., Krumholz, P., and Stammrich, H., *J. Am. Chem. Soc.*, 77, 777, 1955.

15. Cabrera, A. and Rubio Barroso, S., *Quim. Anal.*, 31, 295, 1977; *Chem. Abstr.*, 89, 84114p, 1978.

16. Cabani, S. and Scrocco, E., *Ann. Chim. (Rome)*, 48, 85, 99, 1958; *Chem. Abstr.*, 52, 10693h, 1958; *J. Inorg. Nucl. Chem.* 8, 332, 1958.

17. Scrocco, E. and Salvetti, O., *Boll. Sci. Fac. Chim. Ind. Bologna*, 12, 98, 1954; *Chem. Abstr.*, 49, 8026h, 1955.

18. Anderegg, G., *Helv. Chim. Acta*, 46, 2397, 1963.

19. Crabtree, J. M., Marsh, D. W., Tomkinson, J. C., Williams, R. J. P., and Fernelius, W. C., *Proc. Chem. Soc.*, 336, 1961.
20. Onstott, E. I. and Laitinen, H. A., *J. Am. Chem. Soc.*, 72, 4724, 1950.
21. Baxendale, J. H., and George, P., *Trans. Faraday Soc.*, 46, 55, 1950.
22. Kulba, E. Ya., Makashev, Ya. A., and Mironov, V. E., *Zh. Neorgan. Khim.*, 6, 630, 1961; *Chem. Abstr.*, 57, 430lb, 1962.
23. Holyer, R. H., Hubbard, C. D., Kettle, S. F. A., and Wilkins, R. G., *Inorg. Chem.*, 5, 622, 1966.
24. Dale, J. M. and Banks, C. V., *Inorg. Chem.*, 21, 591, 1963.
25. Anderegg, G., *Helv. Chim. Acta*, 45, 1643, 1962.
26. Trujillo, R. and Brito, F., *An. R. Soc. Esp. Fis. Quim.* 53B, 249, 1957; *Chem. Abstr.*, 53, 21343b, 1959.
27. Nakashima, F. and Sakai, K., *Bunseki Kagaku*, 10, 94, 1961.
28. Dantani, F. and Ciantelli, G., *Ric. Sci.*, 38, 953, 1968; *Chem. Abstr.*, 71, 7040n, 1969.
29. Pflaum, and Brandt, W. W., *J. Am. Chem. Soc.*, 77, 2019, 1955.
30. Moss, M. L. and Mellon, M. G., *Ind. Eng. Chem. Anal. Ed.*, 14, 862, 1942.
31. Miller, R. R., Brandt, W. W., and Puke, M. Sr., *J. Am. Chem. Soc.*, 77, 3178, 1955.
32. Ciantelli, G., Legittimo, P., and Pantani, F., *Anal. Chim. Acta*, 53, 303, 1971.
33. Moss, M. L. and Mellon, M. G., *Ind. Eng. Chem. Anal. Ed.*, 15, 74, 1943.
34. Miller, R. R. and Brandt, W. W., *Anal. Chem.*, 26, 1968, 1954.
35. Foutune, W. B. and Mellon, M. G., *Ind. Eng. Chem. Anal. Ed.*, 10, 60, 1938.
36. Banks, C. V. and O'Laughlin, F. W., *Anal. Chem.*, 29, 1412, 1957.
37. Smith, G. F., McCurdy, W. H., Jr., and Diehl, H., *Analyst*, 77, 418, 1952.
38. Vita, O. A. and Trivisonno, C. F., *Nucl. Appl.*, 1, 375, 1965; *Chem. Abstr.*, 64, 13378f, 1966.
39. Schilt, A. A., *Anal. Chem.*, 35, 1599, 1963.
40. Dwyer, F. P., *J. Proc. R. Soc. N.S.W.*, 83, 134, 1949.
41. Brandt, W. W. and Smith, G. F., *Anal. Chem.*, 21, 1313, 1949.
42. Smith, G. F. and Richter, F. P., *Ind. Eng. Chem. Anal. Ed.*, 16, 580, 1944.
43. Smith, G. F. and Banick, W. M., Jr., *Talanta*, 2, 348, 1959.
44. Schilt, A. A. and Bacon, J., *Anal. Chem.*, 41, 1669, 1969.
45. Blair, D. and Diehl, H., *Anal. Chem.*, 33, 867, 1961.
46. Poluektov, N. S. and Nazarenko, V. A., *J. Appl. Chem. USSR*, 10, 2105, 1937.
47. Feigl, F. and Miranda, L. I., *Ind. Eng. Chem. Anal. Ed.*, 16, 41, 1944.
48. Yamamoto, Y., *Anal. Chim. Acta*, 84, 217, 1976.
49. Cakrt, M., Bercik, J. and Hladky, Z., *Fresenius Z. Anal. Chem.*, 28, 295, 1976.
50. Yammamoto, Y., Tarumoto, T., and Iwamoto, E., *Chem. Lett.*, 255, 1972.
51. Taylor, C. G. and Fryer, B., *Analyst*, 94, 1106, 1969.
52. Knizek, M. and Musilova, M., *Talanta*, 15, 479, 1968.
53. *AnalaR Standards for Laboratory Chemicals*, 7th ed., Analar Standards, England, 1977.
54. Purity Specification of Organic Reagents, Dojindo Laboratories, Kumamoto, Japan.
55. Zemeikova, M. and Sommer, L., *Spisy Prirodoved. Fak. Univ. J. E. Purkyne Brne*, 381, 1970; *Anal. Abstr.*, 22, 137, 1972.
56. The Analytical Methods Committee, Standardized general method for the determination of iron with 1,10-phenanthroline, *Analyst*, 103, 391, 1978.
57. The Analytical Methods Committee, General method for determination with bathophenanthroline, *Analyst*, 103, 521, 1978.
58. Vydra, F. and Pribil, R., *Talanta*, 10, 339, 1963.
59. Takagi, S. and Kumai, K., *Bunseki Kagaku*, 16, 958, 1967.
60. Miyamoto, M., *Bunseki Kagaku*, 9, 753, 1960.
61. Kononenko, L. I., Lauer, R. S., and Poluektov, N. S., *Zh. Anal. Khim.*, 18, 1468, 1963.
62. Poluektov, N. S., Vitkun, R. A., and Kononenko, L. I., *Ukr. Khim. Zh.*, 30, 629, 1964; *Chem. Abstr.*, 61, 10029a, 1969.
63. Stolyarov, K. P. and Firyulina, V. V., *Zh. Anal. Khim.*, 33, 2102, 1978.
64. El-Ghamry, M. T., Frei, R. W., and Higgs, G. W., *Anal. Chim. Acta*, 47, 4, 1969.
65. Schcherbov, D. P., Ivankova, A. I., Lisitsyna, D. N., and Vvedenskaya, I. D., *Zh. Anal. Khim.*, 32, 1932, 1977.
66. Kononenko, L. I., Poluektov, N. S., and Nikonova, N. P., *Zavod. Lab.*, 30, 779, 1964; *Chem. Abstr.*, 61, 8891f, 1964.
67. Pribil, R. and Vydra, F., *Collect. Czech. Chem. Commun.*, 24, 3103, 1959.
68. Pribil, R., *Talanta*, 3, 91, 1959.
69. Nakahara, K., Hayashi, K., Danzuka, T., Nagamura, S., and Imamura, T., *Benseki Kagaku*, 12, 854, 1963.
70. Smith, G. F., and Getz, C. A., *Ind. Eng. Chem. Anal. Ed.*, 10, 304, 1938.

71. Snyder, H. R. and Freier, H. E., *J. Am. Chem. Soc.,* 68, 1320, 1946.
72. James, B. R., and Williams, R. J. P., *J. Chem. Soc.,* 2007, 1961.
73. Schilt, A. A., Smith, G. F., and Heimbuch, A., *Anal. Chem.,* 28, 809, 1956.
74. Poe, D. P. and Diehl, *Talanta,* 23, 141, 1976.
75. Dunbar, W. E., and Schilt, A. A., *Talanta,* 19, 1025, 1972.
76. Shuman, D. C. and Sen, B., *Anal. Chim. Acta,* 33, 487, 1965.
77. Bhaskare, C. K., and Kawakar, S. G., *Anal. Chim. Acta,* 73, 405, 1974.
78. Bhaskare, C. K. and Kawakar, S. G., *Talanta,* 22, 189, 1975.
79. Bhashare, C.K. and Kawakar, S.G., *J. Indian Chem. Soc.,* 52, 520, 523, 1975.

TRIPYRIDYLTRIAZINE (TPTZ) AND PYRIDYLDIPHENYLTRIAZINE (PDT)

(1) L

(2)L, X = H
(3)H₂L, X = SO₃H

Synonyms

Reagents covered in this section are listed in Table 1, together with their synonyms.

Source and Methods of Synthesis

All commercially available. **(3)** is usually supplied as mono- or disodium salt containing 1 to 2 mol of water of crystallization.

(1) is prepared by the self-condensation of 2-cyanopyridine in the presence of sodium hydride.[M3,1] **(2)** is prepared by the stepwise reactions of 2-cyanopyridine with hydrazine, then with benzyl.[2] **(3)** is a sulfonated product of **(2)** with fuming sulfuric acid.[3,4]

Analytical Applications

As photometric reagents for Fe(II) and Cu(I).[M2,M4] They are also used for the photometric determination of other metals, such as Co, Cr, Ni, and Ru. **(1)** and **(3)** are suitable for use in an aqueous system.

Properties of Reagents

TPTZ **(1)** is pale yellow needles, mp 245 to 248°, and is slightly soluble in water ($\sim$0.03 g/100 mℓ),[5] but more soluble in ethanol. It is easily soluble in dilute hydrochloric acid and insoluble in nonpolar solvents. The pKa value of HL^+ is 3.10(25°, $\mu = 0.1$ KCl),[5] or in other work, the values of two-step dissociations of H_2L^{2+} are 2.82 and 2.75 (25°, $\mu = 0.23$ NaCl).[6] Distributions of **(1)** between various organic solvents and acid solutions were investigated, and the $pH_{1/2}$ values are 1.93(nitrobenzene), 0.38(phenylcarbinol), 1.20(chloroform), 2.16(dichloroethane), 2.50(trichloroethylene), 1.72(n-butanol), and 2.28(n-pentanol).[35]

PDT **(2)** is pale yellow needles, mp 191 to 192° and is slightly soluble in water, but more soluble in alcohol and other organic solvents; pKa (LH^+) = 2.95 (25°).[7] Distribution ratios of **(2)** between chloroform and various acid solutions are summarized in Table 2.

PDTS **(3)** is a pale yellow or greenish yellow powder, decomposing above 350°,[4] is usually supplied as mono- or disodium salt containing 1 to 2 mol of water of crystallization, and is easily soluble in water to give a colorless or pale pink solution (0.7 g/100 mℓ for NaHL at 20°).[4]

No characteristic absorption band is observed on **(1)**, **(2)**, and **(3)** in the visible region.

Complexation Reactions and Properties of Complexes

TPTZ **(1)** behaves as a terdentate ligand like terpyridine, and PDT **(2)** and PDTS **(3)** behave as a bidentate ligand like bipyridine, forming ML_2- and ML_3-type chelates with hexacoordinate metal ions, respectively. The chelate with **(1)** and **(3)** are characteristic in their water solubility. Iron(II) chelate wih **(2)** ($[FeL_3]^{2+}$) can be extracted into

Table 1
TPTZ AND RELATED COMPOUNDS

Reagent	Name	Synonym	Molecular formula	Mol wt
(1)	Tripyridyltriazine	2,4,6-Tris(2-pyridyl)-s-triazine, TPTZ	$C_{18}H_{12}N_6$	312.33
(2)	Pyridyldiphenyl triazine	3-(2-Pyridyl)-5,6-diphenyl-1,2,4-triazine,PDT	$C_{20}H_{14}N_4$	310.36
(3)	Pyridyldiphenyl triazine sulfonic acid, sodium salt	3-(2-Pyridyl)-5,6-diphenyl-1,2,4-triazine, disulfonic acid, disodium salt, 3-2(-Pyridyl)-5,6-di(4-sulfophenyl)-1,2,4-triazine, disodium salt, PDTS. Ferrozine® [a]	$C_{20}H_{12}N_4O_6Na_2 \cdot 2H_2O$	486.35

[a]   Trade name of Hach Chemical Co., Inc., Ames, Iowa.

Table 2
DISTRIBUTION RATIO OF PDT(2) BETWEEN CHLOROFORM AND VARIOUS ACID SOLUTIONS

Acid conc (M)	$D(CHCl_3/acid)$				
	$HClO_4$	H_2SO_4	HCl	HNO_3	$HClO_4$ + NaSCN (0.5 M)
0.5	110	150	190	120	120
1.0	40	150	110	60	70
2.0	13	28	16	13	26
5.0	1.6	0.20	4.2	3.2	10
8.0	0.04	0.05	0.66	1.5	1.4

Reprinted with permission from Chriswell, C. D. and Schilt, A. A., *Anal. Chem.*, 46, 992, 1974. Copyright 1974 American Chemical Society.)

iso-amylalcohol,[9] or, if in the presence of SCN⁻, into chloroform from fairly acidic aqueous phase, being the basis for the determination of trace of Fe in acids.[7] The conditional equilibrium constant for the following reaction.

$$Fe^{2+} + 3(LH^+X^-)_{org} \xrightarrow{K_{ex}} ([FeL_3] X_2)_{org} + 3H^+ + X^-$$

$K' = K_{ex}/[H^+]^3 \cdot [X^-]$, was evaluated as $(9.0 \pm 0.4) \times 10^{11}$ (in 1 M H_2SO_4 and 0.5 M NaSCN).[7] The chelate stability constants are observed on a limited number of metals; TPTZ, log β_2(Co(II)),~9.11,[11] log β_2(Fe(II)),11.4;[20,21] PDTS, log β_3(Fe(II)),15.56 (conditional constant at pH 3.5 to 4.5).[4]

The spectral characteristics of the metal chelates are summarized in Table 3. Absorption spectrum of Fe-TPTZ chelate (FeL_2^{2+}) in aqueous solution is illustrated in Figure 1.

Purifications and Purity of Reagents:

TPTZ (1) and PTD (2) can be purified by repeated recrystallization from aqueous ethanol and ethanol-DMF, respectively, until indicated melting points are observed.

Table 3

SPECTRAL CHARACTERISTICS OF METAL CHELATES WITH TPTZ (1) PDT (2) AND PDTS (3)

Reagent	Metal ion	Condition	Metal Chelate			Remarks	Ref.
			Ratio	λ_{max} (nm)	$\varepsilon (\times 10^4)$		
(1)	Co(II)	pH 8.5, 66% ethanol	ML_2	485	0.28	Ir(III),Os(III) and Ru(III) interfere	11
	Fe(II)	pH 3.4 ~ 5.8	ML_2	593	2.26	See Table 4 for interfering ions	M1, 12
	Fe(II)	pH 4.5 ~ 5	ML_2	595	2.41	Extraction with nitrobenzene or propylene carbonate	M1, 12,13
	Ru(III)	pH 2.2 ~ 4.0 20 ~ 55% ethanol, (87%)	ML_2	510	1.81	Fe(II)(III),Pt(IV),Rh(III) Os(III) and EDTA interfere, 1 ~ 4 ppm	14
	Ru(III)	0.5 m $HClO_4$ (100°, 35 min), glycerol-water (2:1)	ML_2	535	2.7	Extraction with nitrobenzene Co(II),Cu(II),Fe(II)(III) and Ni(II) do not interfere	15
(2)	Cu(I)	pH 3.5 ~ 6	ML_2	488	0.8	In iso-amyl alcohol	9
	Fe(II)	pH 3.5 ~ 6	ML_3	555	2.35	In iso-amyl alcohol	9
	Fe(II)	$HClO_4$ or H_2SO_4, <4M SCN^-	ML_3X_2	555	2.45	Extraction with chloroform	7
	Ru(II)	pH 5, ethanol	ML_2	485	2.1	0.5 ~ 3.4 ppm, NO_3^-, ClO_4^-, SO_4^{2-}, PO_4^{3-} and F^- do not interfere	16
(3)	Co(II)	pH 5.5 ~ 7.0	ML_3	500	0.46	0 ~ 6 ppm	17
	Cu(I)	pH 6.5 ~ 8.8	ML_2	470	0.43[a]	0.5 ~ 20 ppm	13,17,18
	Fe(II)	pH 3.5 ~ 4.5, thioglycolic acid or hydroxylamine	ML_3	562	2.86	See Table 5 for interfering ions	4,8
	Os(VIII)	pH 2.0 ~ 3.4 (boiling, 4 hr)	ML_2	510	1.44	Co,Cu,Cr,Fe,Ir,Pd,Pt and Rh interfere, 0 ~ 6 ppm	19
	Ru(III)	pH 2.0 ~ 3.0 (boiling, 4 hr)	ML_2	480	3.15	Co,Cu,Cr and Fe interfere, 0 ~ 2.8 ppm	19

[a] The color of Cu-PDTS chelate fades by the oxidation, but can be protected by the addition of sodium sulfite.

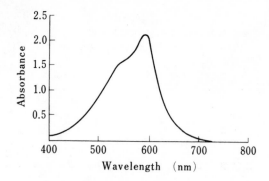

FIGURE 1. Absorption spectrum of Fe(II)-TPTZ chelate in acetate buffer solution; Fe, 2.0×10^{-5} *M*; 5-cm cell. (From Diehl, H. and Buchanan, E. B., Jr., *Anal. Chem.*, 32, 1117, 1960. With permission.)

PDTS (**3**) may be purified by recrystallization from water or by dissolving into a minimum amount of water, followed by the addition of ethanol to precipitate the pure material.

The assay of (**3**) is carried out by observing the absorbance of Fe(II) chelate at 562 nm, using a tenfold excess of Fe(II) at pH 4.5 (hydroxylamine-acetate buffer) ($\varepsilon = 2.86 \times 10^4$).[10] The iron blank present in the reagent is determined by scanning a 2.5% aqueous solution from 500 to 700 nm using a 1-cm cell. The difference in absorbance between 562 and 700 nm is used to calculate the iron content.[4]

Analytical Uses

Of the many triazine derivatives, (**1**), (**2**), and (**3**) were found to be the iron reagents of the most practical usefulness.[22-28]

While application of (**2**) is usually combined with solvent extraction, (**1**) and (**3**) can be used in the aqueous system. The optimum conditions for the color developments are summarized in Table 3.

The advantages of TPTZ are that the absorption peak of the chelate appears in a relatively longer wavelength region and the reagent is less expensive, However, the reagent is not so selective for Fe because Co, Cu, Cr, Ni, and Ru also form colored chelates. Anions such as MnO_4^-, NO_2^-, CN^-, and CrO_4^{2-} also give serious interference.[M1]

PDT and PDTS also form colored chelate with Cu,[9,13,17,18] but the absorption peak for Cu(I) chelate is less intense and appears at 488 nm, where it does not interfere the iron determination.[26,28] The interfering ions and approximate tolerance levels for the determination of Fe(II) with TPTZ and PDTS are summarized in Tables 4 and 5, respectively.

TPTZ and PDTS have been used for the determination of Fe in serum with or without preliminary deproteination.[30-32] A small amount of Cu can be masked with Neocuproine,[33] thiourea, or thioglycolic acid.[34]

Reagents with Related Structure

Various derivatives of *asym*-triazine have been investigated for the chromogenic reagent of Fe, Cu, or Co. Some of the successful reagents are summarized in Table 6.

Table 4
INTERFERING IONS FOR THE DETERMINATION OF Fe(1.1 PPM) WITH TPTZ (1)

Metal ion	Concentration (ppm)	Relative error (%)	Anion	Concentration (ppm)	Relative error (%)
Ag	102	Ppt	F^-	502	+0.2
Ba	101	+0.2	Br^-	556	+0.2
Bi	100	Ppt	CN^-	500	Very large
Cd	100	−0.7	NO_2^-	500	Large
Co(II)	4.8	+3.6	NO_3^-	504	+0.2
Cr(III)	20.8	+2.4	PO_4^{3-}	528	+0.2
Cu(II)	6.3	+4.8	ClO_3^-	548	+0.2
Hg(II)	100	Ppt	SCN^-	507	+0.2
K	1067	+0.4	$S_2O_3^{2-}$	528	+0.4
Mn(II)	110	+0.2	MoO_4^{3-}	34	Very large
Ni	10.6	+2.7	$C_2H_3O_2^-$	14400	−0.2
Pb	101	+0.2			
Sn(II)	100	−0.2	No interference —		
Sr	99	+0.2	Al ,Be ,Ca ,Li ,Mg,		
Th	120	−0.2	NH_4 ,I^-,SO_4^{2-},$S_2O_3^{2-}$, ClO_4^-,BO_3^{3-},		
UO_2^{2+}	115	+0.4	BrO_3^-.		

(From Collins, P. F., Diehl, H., and Smith, G. F., *Anal. Chem.*, 31, 1862, 1959. With permission.)

Table 5
INTERFERING IONS FOR THE DETERMINATION OF Fe(1.0 PPM) WITH PDTS (3)[8]

Metal ion	Concentration (ppm)	Relative error (%)
Co(II)	5.0	+3.0
	8.0	+5.0
Cu(I)	1.0	+3.0
	5.0	+15.0
CN^-	1.0	−1.0
	10.0	−5.0

Note: The only anionic interferences were oxalate in concentration over 500 ppm, cyanide and nitrite. The latter two were destroyed by heating with the acid reagent solution and did not interfere with the final photometry.

Table 6
SPECTRAL CHARACTERISTICS OF METAL CHELATES OF SOME ASYM.-
TRIAZINES

Reagent	Metal ion	Metal chelate		Remarks	Ref.
		λ_{max} (nm)	$\varepsilon(\times 10^4)$		
3-(4-Phenyl-2-pyridyl)-5,6- diphenyl-1,2,4-triazine (PPDT or PAT)	Co(II)	516	0.22	—	
	Cu(I)	480	0.79	—	
	Fe(II)	561	2.87	FeL₃ at pH 3 ∿ 7, in ethanol-water (1:1) 0.27 ∿ 2.44 ppm Fe EDTA, CN⁻ interfere	22, 23, 29
2,4-Bis(5,6-diphenyl-1,2,4- triazeno-3-yl)pyridine	Co(II)	475	0.01	—	26
	Cu(I)	510	1.52	—	
	Fe(II)	563	3.2	Extraction with CHCl₃ at pH 7 (violet)	
3-(4-Phenyl-2-pyridyl)-3 phenyl-1,2,4-triazine	Cu(I)	495	1.05		28
	Fe(II)	560	3.48	FeL₃ at pH 3 ∿ 12, (violet); highest sensitivity; Co chelate is colorless	

MONOGRAPHS

M1. **Diehl, H. and Smith, G. F.,** *The Iron Reagents, Bathophenanthroline, 2,4,6-Tripyridyl-s-triazine, Phenyl-2-Pyridyl ketoxime,* G. F. Smith Chemical Co., Columbus, Ohio, 1960.

M2. **Schilt, A. A.,** *Analytical Applications of 1,10-Phenanthroline and Related Compounds,* Pergamon Press, London, 1969.

M3. **Case, F. H.,** *A Review of Synthesis of Organic Compounds Containing the Ferroin Group,* G. F. Smith Chemical Co., Columbus, Ohio, 1960.

M4. **Sandell, E. B. and Onishi, H.,** *Photometric Determination of Traces of Metals,* 4th ed., John Wiley & Sons, New York, 1978.

REFERENCES

1. **Case, F. H. and Koft, E.,** *J. Am. Chem. Soc.,* 81, 905, 1959.
2. **Case, F. H.,** *J. Org. Chem.,* 30, 931, 1965.
3. **Stookey, L. L.,** U.S. Patent 3,838,331 *Chem. Abstr.,* 81, 163076r, 1974.
4. **Gibbs, C. R.,** *Anal. Chem.,* 48, 1197, 1976.
5. **Buchanan, E. B., Jr., Crichton, D., and Bacon, J. R.,** *Talanta,* 13, 903, 1966.
6. **Prasad, J. and Peterson, N. C.,** *Inorg. Chem.,* 10, 88, 1971.
7. **Chriswell, C. D. and Schilt, A. A.,** *Anal. Chem.,* 46, 992, 1974.
8. **Stookey. L. L.,** *Anal. Chem.,* 42, 779, 1970.
9. **Schilt, A. A., and Taylor, P. J.,** *Anal. Chem.,* 42, 220, 1970.
10. **Diehl, H. and Buchanan, E. B., Jr.,** *Anal. Chem.,* 32, 1117, 1960.
11. **Janmohamed, M. J. and Ayres, G. H.,** *Anal. Chem.,* 44, 2263, 1972.
12. **Collins, P. F., Diehl, H., and Smith, G. F.,** *Anal. Chem.,* 31, 1862, 1959.

13. Stephens, B. G., Fekel, H. L., Jr., and Spinelli, W. M., *Anal. Chem.*, 46, 692, 1974.
14. Embry, W. A. and Ayres, G. H., *Anal. Chem.*, 40, 1499, 1968.
15. Sasaki, Y., *Anal. Chim. Acta*, 98, 335, 1978.
16. Kamra, L. C. and Ayres, G. H., *Anal. Chim. Acta*, 78, 423, 1975.
17. Kundra, S. K., Katyal, M., and Singh, R. P., *Anal. Chem.*, 46, 1605, 1974.
18. Anusiem, A. C. I. and Ojo, G., *Anal. Chem.*, 50, 531, 1978.
19. Kundra, S. K., Katyal, M., and Singh, R. P., *Curr. Sci.*, 44, 548, 1975; *Chem. Abstr.*, 83, 201537t, 1975.
20. Fraser, F. H., Epstein, P. M., and Macero, D. J., *Inorg. Chem.*, 11, 2031, 1972.
21. Legittimo, P., Pantani, F., and Ciantelli, G., *Gazz. Chim. Ital.*, 101, 465, 1971; *Chem. Abstr.*, 76, 18640n, 1972.
22. Schilt, A. A., *Talanta*, 13, 895, 1966.
23. Schilt, A. A. and Hoyle, W. C., *Anal. Chem.*, 39, 114, 1967.
24. Schilt, A. A. and Kluge, K. R., *Talanta*, 15, 475, 1968.
25. Schilt, A. A., Dunbar, W. E., and Warren, S. E., *Talanta*, 17, 649, 1970.
26. Schilt, A. A., Chriswell, C. D., and Fang, T. A., *Talanta*, 21, 831, 1974.
27. Traister, G. L., and Schilt, A. A., *Anal. Chem.*, 48, 1216, 1976.
28. Schilt, A. A., Yang, T. A., Wu, J. F., and Nitzki, D. M., *Talanta*, 24, 685, 1977.
29. Kiss, E., *Anal. Chim. Acta*, 72, 127, 1974.
30. Ruutu, R., *Clin. Chim. Acta*, 61, 229, 1975.
31. Rosner, E. and Molnar, A., *Kiserl. Orvostud.*, 23, 220, 1971; *Anal. Abstr.*, 22, 2455, 1972.
32. Manasterski, A., Watkins, R., Baginski, E. S., and Zak, B., *Z. Klin. Chem. Klin. Biochem.*, 11, 335, 1973; *Anal. Abstr.*, 26, 1633, 1974.
33. Carter, P., *Anal. Biochem.*, 40, 450, 1971.
34. Yee, H. Y. and Goodwin, J. F., *Clin. Chem.*, 20, 188, 1974.
35. Sasaki, Y., *Bunseki Kagaku*, 27, 729, 1978.

2,2′-BIQUINOLINE AND OTHER COPPER REAGENTS

(1)   (2)

(3a) R=H
(3b) R=SO₃Na

(3a) R=H
(3b) R=SO$_3$Na

Synonyms

Reagents covered in this section are listed in Table 1, together with their synonyms.

Source and Method of Synthesis

All reagents are commercially available. (1) is prepared by the Freidlander condensation of biacetyl and o-aminobenzaldehyde[1] or the catalytic dehydrogenation of quinoline.[2] (2) is prepared by the double Skraup reaction of o-phenylenediamine with crotonaldehydediacetate.[3,4] Synthesis of (3a) needs lengthy steps, starting from o-nitroaniline and phenylpropylketone to obtain 2-methyl-8-nitro 4-phenylquinoline. This intermediate is reduced to the corresponding amino compound which is finally condensed with phenylpropylketone by a modified Skraup reaction to give (3a).[5] (3b) is a sulfonated product of (3a).[6,7]

Analytical Uses

(1) to (3) are highly specific reagents for Cu(I) and are exclusively used as chromogenic reagents for the photometric determination and the detection of trace of Cu in various samples.

Properties of Reagents

Cuproine (1) is colorless flakes, mp 196°, and is almost insoluble in water (6.4 × 10^{-5} g/ℓ), but soluble in dilute acid, methanol, iso-amylalcohol (2 g/ℓ), acetic acid, isoamyl acetate, dioxane, acetonitrile, and carbon tetrachloride. It is very soluble in DMF;[M2,M4] pKa = 3.10 (μ = 0.1 K$_2$SO$_4$, 25°).[9]

Neocuproine (2) is obtained as anhydrous material by recrystallization from nonpolar solvent such as benzene and is a white powder, mp 167 to 168°. The hemihydrate is obtained when recrystallized from water, melting at about 140 to 141°.[M2,8] Most of the commercial samples contain various amount of water of crystallization, and the melting point of the hydrated material (mp 158 to 162°) varies with the difference in the degree of hydration, as illustrated in Figure 1.[8] Free base is slightly soluble in cold water, but easily soluble in various organic solvents[4] and dilute mineral acid; pKa = 5.85 (μ = 0.1 KCl, 25°).[10]

Dihydrochloride is also commercially available and pale yellow hygroscopic crystalline powder, mp 218 to 220°. It is finding increasing use in the automatic analyzer, as it is soluble in water.

Bathocuproine (3a) is a colorless or pale yellow crystalline powder, mp 278 to 280°, and is almost insoluble in water, but soluble in common organic solvents.[M2]

Bathocuproine disulfonic acid(disodium salt) (3b) is a colorless or pale yellow crystalline powder which is slightly hygroscopic. It shows a light blue fluorescence under

Table 1
BIQUINOLINE AND RELATED REAGENTS

Number	Name of Reagents	Synonym	Molecular formula	Mol wt
(1)	2,2'-Biquinoline	2,2'-biquinolyl, Cuproine	$C_{18}H_{12}N_2$	256.31
(2)	2,9-Dimethyl-1,10-phenanthroline	Neocuproine	$C_{14}H_{12}N_2$	208.26
(3a)	2,9-Dimethyl-4,7-diphenyl-1,10-phenanthroline	Bathocuproine	$C_{26}H_{20}N_2$	360.46
(3b)	2,9-Dimethyl-4,7-diphenyl-1,10-phenanthroline disulfonic acid disodium salt	Bathocuproine disulfonic acid, disodium salt	$C_{26}H_{18}N_2(SO_3Na)_2$	564.54

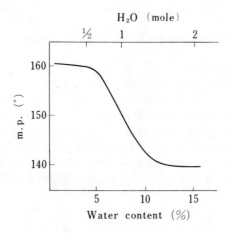

FIGURE 1. Variation of mp with water content of Neocuproine. (From Saito, M. and Iwano, H., *Talanta*, 18, 989, 1971. With permission.)

UV light and is very easily soluble in water, but not soluble in nonpolar organic solvents;[6] pKa $(SO_3H) = 2.65$ and pKa $(N^+H) = 5.80$.[6]

(1) to (3) are colorless and do not show any absorption band in the visible region.

Complexation Reactions and Properties of Complexes

In contrast to 2,2'-bipyridine and related reagents (p.309), the compounds of this class fail to give the color reaction with ferrous ion, but produce intense color with cuprous ion. This selectivity for Cu(I) is understood to be due to the steric hindrance of the substituent groups on carbon atoms adjacent to the ring nitrogen atoms. Copper(I) ion is chelated with two molecules of the reagent, both nitrogen atoms of each molecules being bound to the copper, so that two five-membered rings are formed, and the planes of the two rings lie at right angles to each other, as illustrated below.[11]

Table 2
CHELATE STABILITY CONSTANTS OF
BIQUINOLINE (1)[9] AND
DIMETHYLPHENANTHROLINE (2)[10]

Metal ion	log K_{ML}	log K_{ML2}	Condition		Ref.
			Temp	μ	
		2′,2′-Biquinoline			
Ag	—	log β_2 10.29	20	In ethanol	12
Cu(I)	—	log $\beta_2 \curvearrowright$16.5	25	0.1 (K$_2$SO$_4$)	
Cu(II)	4.27	3.46	25	0.1 (K$_2$SO$_4$)	
		2,9-Dimethyl-1,10-phenanthroline			
Cd	4.1	3.3	25	0.3 (K$_2$SO$_4$)	
Co(II)	4.2	2.8	25	0.3 (K$_2$SO$_4$)	
Cu(I)	—	log β_2 19.1	25	0.3 (K$_2$SO$_4$)	
Cu(II)	5.2	5.8	25	0.1 (KCl)	
Fe(II)	≪4	—	25	0.1 (KCl)	
Mn(II)	≪3	—	25	0.1 (KCl)	
Ni	5.0	3.5	25	0.1 (KCl)	
Zn	4.1	3.6	25	0.1 (KCl)	

However, in the case of the Fe(11) ion which is hexacoordinate, three molecules of the reagent can not coordinate to the central metal ion because of the steric hinderance of the substituents.

These reagents are not only highly specific chromogens for Cu because they react with no other metals to give colored chelates, but also these chelates can be extracted into certain immiscible solvents. (3b) gives a water-soluble Cu(I) chelate and is used only in aqueous solution.

Not many data on chelate stability constants are available for (1) and (2). Nothing is reported for (3a) and (3b). Reported values are summarized in Table 2.

(1) — (3a) are not sufficiently soluble in water to react with Cu(I) in aqueous solution. When a certain water-miscible solvent, such as alcohol, dioxane, or dimethylformamide, is present, the reagents form deep-colored bis-chelates with Cu(I). These chelates can also be simultaneously formed and extracted from water into immiscible solvents, such as iso-amylalcohol or n-hexyl-alcohol. Absorption spectral characteristics of Cu(I) chelates are listed in Table 3, and the representative absorption curves are shown in Figure 2.

When Cu(I) ion is extracted with the solution of (1) to (3a) in immiscible solvent, the resulting cationic chelate (CuL$_2$$^+$) should be accompanied with an appropriate anion to neutralize the charge. This principle can be applied to the ion-pair solvent extraction of various inorganic and organic anions. Thus, a colorless anion is extracted with the colored copper chelate for the extraction photometric determination of the anion. The extraction constant is very much dependent upon the combination of ion-pair and the nature of solvent. Table 4 summarizes the dependency of K_{ex} of Cu(biquinoline)$_2$$^+$ upon the nature of dye anion.[14] The extraction is more effective for the dyes which contain bulky substituents in the ortho positions to the phenolic oxygens.

Examples of the ion-pair extraction of colorless anions with Cu(I) chelates are summarized in Table 10 (p.315).

Many of the common anions also cause precipitation with the copper(I) chelates (CuL$_2$$^+$), but they are soluble in organic solvents such as nitrobenzene.

Table 3
SPECTRAL CHARACTERISTICS OF
COPPER(I) CHELATES $(CuL_2{}^+)^{M3,13}$

Reagent	Media	λ_{max} (nm)	$\varepsilon(\times 10^3)$
(1)	Iso-amyl alcohol	546	6.22
	n-Hexyl alcohol	545	—
	Aqueous DMF	545	6.45
	Aqueous THF	550	—
	Acrylonitrile	540	—
	Nitrobenzene	553	—
(2)	Iso-amyl alcohol	454	7.95
	Chloroform-ethanol	457	—
(3a)	Aqueous ethanol	479	13.9
	n-Hexyl alcohol	479	14.1
	Iso-amyl alcohol	479	14.2
(3b)	Water	483	12.2

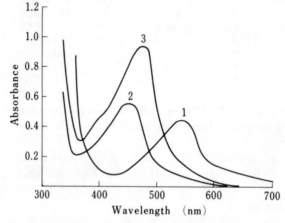

FIGURE 2.   Absorption spectra of copper(I) chelates in iso-amylalcohol. (1) Cuproine — λ_{max}, 546 nm; (2) Neocuproine — λ_{max}, 454 nm; (3) Bathocuproine — λ_{max}, 479 nm; Cu, 4.0 ppm. (From Schilt, A. A. and McBride, L., *The Copper Reagents, Cuproine, Neocuproine, Bathocuproine*, 2nd ed., G. F. Smith Chemical Co, Columbus, Ohio, 1972. With permission.)

Purification and Purity of Reagents

(1) to (3a) are well-characterized materials and are easily purified by recrystallization from appropriate solvent: (1) from ethanol, (2) (hemihydrate) from water, and (2) (anhydrous) and (3a) from benzene.

(3b) of low purity is brown or tan powder, and it is not so easy to remove the colored contaminants. Inorganic salts and some of the colored components may be removed by dissolving the crude sample into minimum amount of water, followed by the addition of ethanol to precipitate the contaminants. Pure (3b) is obtained by careful evaporation of the filtrate. As the aqueous solution of (3b) is extremely sensitive to copper, the care has to be taken to avoid the copper contaminant during the purification.

The purity of (1) to (3a) can be checked most conveniently by observing their melting points or by titrating with perchloric acid in glacial acetic acid using 1-naphtholbenzein as indicator.[17]

Table 4
EXTRACTION CONSTANTS OF Cu(BIQUINOLINE)$_2^+$ ION-PAIRS IN CHLOROFORM-ISOPENTANOL (20:1)[14]

Dye	R$_1$	R$_2$	R$_3$	log K$_{ex}$
Phenol red	H	H	H	4.68
Cresol red	H	CH$_3$	H	5.33
Metacresol purple	CH$_3$	H	H	4.58
Chlorphenol red	H	Cl	H	4.86
Bromophenol red	H	Br	H	5.20
Xylenol blue	CH$_3$	H	CH$_3$	4.92
Bromophenol blue	H	Br	Br	6.76
Bromocresol purple	H	Br	CH$_3$	8.14
Bromocresol green	CH$_3$	Br	Br	5.91

Analytical Applications

Use as a Photometric Reagent

Many chromogenic reagents have been reported for the photometric determination of trace of Cu; none is superior to this class of reagent.[M1,M2,R1] These reagents are highly specific chromogens for Cu because they react with no other metals to give colored products. The Cu chelates, with an exception of (3b), are also soluble in certain solvents immiscible in water, so that they can been concentrated by extraction into such solvents.

Cuproine (1) and Neocuproine (2) are recommended for the general purposes, where extreme sensitivity is not required and unusual conditions are not involved. Bathocuproine (3a) and Bathocuproine disulfonic acid (3b) are recommended for the determination of ultratrace amount of Cu. (3b) can be employed in an aqueous system and conveniently be used in the routine work by elimination of the extraction step. On the other hand the ability to separate the Cu chelate from other colored substances and the concentrating effect of the extraction are both lost. In all cases, hydroxylamine hydrochloride or ascorbic acid is used as a reductant for Cu(II) ion.

The purple Cu(I) chelate of Cuproine can be extracted into iso-amylalcohol, benzyl alcohol, hexyl alcohol, benzene, carbon tetrachloride, chloroform, ethyl acetate, and amyl acetate. The distribution coefficient is greatest in iso-amylalcohol which is recommended in the practical applications. For the extraction from large volume of the aqueous solution, n-hexyl alcohol is preferable.

The bright orange Cu(I) chelate of Neocuproine (2) can be extracted into a chloroform-ethanol mixture or iso-amylalcohol, although the former offers some advantage over the latter. A small amount of alcohol must be present on the extraction with chloroform for the maximum development of the color. The composition of the solvent does not affect the spectrum, if a minimum of 2 mℓ of ethanol is contained in 25 mℓ of chloroform. Thus, it is practical to use ethanol solution of Neocuproine for the extraction photometry with chloroform.

Procedures for Cu Determinations

Procedure with Neocuproine (2)[M2] — Transfer an aliquot of the sample solution containing 20 to 200 μg of Cu to a separatory funnel. Add 5 mℓ of 10% hydroxylamine hydrochloride and 10 mℓ of 30% sodium citrate. Adjust the pH to 4 to 6 with dilute ammonia. Add 10 mℓ of 0.1% (2) (in ethanol) and 10 mℓ of chloroform. Shake for 30 sec, allow the layers to separate, and draw off the chloroform layer into a 25-mℓ volumetric flask. Repeat the extraction of aqueous phase with another 5 mℓ of chloroform. Dilute the combined extracts to the volume with ethanol. Measure the absorbance at 457 nm against a reagent blank.

Procedure with Bathocuproine (3a)[M2] — Transfer an aliquot of the sample solution containing 0.06 to 2 μg of Cu to a separatory funnel. Add 2 mℓ of 10% hydroxylamine hydrochloride solution and add 10% ammonium acetate solution (pH 7.1) to bring the volume to about 25 mℓ. Add 1 mℓ of 0.01% (3a) (in hexyl alcohol) and 5 mℓ of hexyl alcohol. Shake for 2 min, and after 5 min, transfer the organic layer to a 10-mℓ volumetric flask. Dilute to the volume with the same solvent and measure the absorbance at 479 nm against a reagent blank.

Other Uses

As discussed in Complexation Reactions and Properties of Complexes, various anions can be extracted into immiscible solvent as an ion-pair with cationic chelates of Cuproine type reagents. Based on this principle, the extraction photometric determination of colorless anions and anionic complexes can be conducted. Examples are summarized in Table 10 (p.315).

The color reaction of (3a) with Cu(I) is so sensitive, that it has been used as a spot test reagent.[M4] The reaction becomes more sensitive when test solution is dropped on cation exchange resin beads impregnated with (3b).[15] A polyvinyl chloride film impregnated with (3a) is recommended for the semiquantitative determination of trace of Cu.[16]

MONOGRAPHS

M1. **Case, F. H.,** *A Review of Synthesis of Organic Compounds Containing The Ferroin Group,* G. F. Smith Chemical Co, Columbus, Ohio, 1960.

M2. **Diehl, H. and Smith, G. F.,** The Copper Reagents: Cuproine, Neocuproine, Bathocuprine, 1st ed, 1958, Schilt, A. A. and McBride, L., 2nd Ed., (1972). G. F. Smith Chemical Co., Columbus, Ohio.

M3. **Schilt, A.A.** *Analytical Application of 1,10-Phenanthroline and Related Compounds,* Pergamon Press, London, 1969.

M4. **Sandell, E. B. and Onishi, H.,** *Photometric Determination of Traces of Metals,* 4th ed., John Wiley & Sons, New York, 1978, 366.

M5. **Ainsworth, L. R. and Harrap, K. R.,** 2,2′-Diquinolyl, Jones, D. D., Newman, E. J., and Peters, G., 2,9-Dimethyl-1,10-Phenanthroline, Jones, D. D. and Newman, E. J., 2,9-Dimethyl-4,7-diphenyl-1,10-Phenanthroline, in *Organic Reagents for Metals,* Johnson, W. C., Ed., Vol. 2, Hopkin & Williams, Essex, England, 1964, 55.

REVIEW

R1. Smith, G. F., The Ferroine, Cuproine, and Terroine reacting organic analytical reagents, *Anal. Chem.*, 26, 1534, 1954.

REFERENCES

1. Smirnoff, A. P., *Helv. Chim. Acta,* 4, 806, 1921.
2. Wibaut, J., Willink, J., and Nieuwenhuis, W., *Rec. Trav. Chim.,* 54, 804, 1935.
3. Case, F. H., *J. Am. Chem. Soc.,* 70, 3994, 1948.
4. Smith, G. F. and McCurdy, W. H., Jr., *Anal. Chem.,* 24, 371, 1952.
5. Case, F. H. and Brennan, J. A., *J. Org. Chem.,* 19, 919, 1954.
6. Blair, D. and Diehl, H., *Talanta,* 7, 163, 1961.
7. Cryberg, R. L. and Diehl, H., Proc. Iowa Acad. Sci., 70, 184, 1963; *Chem. Abstr.,* 61, 8285f, 1964.
8. Saito, M. and Iwano, H., *Talanta,* 18, 989, 1971.
9. James, B. J. and Williams, R. J. P., *J. Chem. Soc.,* 2007, 1961,
10. Irving, H. and Mellor, D. H., *J. Chem. Soc.,* 5237, 1962.
11. Ueno, K., *Introduction to Chelate Chemistry,* Nankodo, Tokyo, 1969, 149.
12. Peard, W. J. and Pflaum, R. T., *J. Am. Chem. Soc.,* 80, 1593, 1958.
13. Pflaum, R. and Brandt, W. W., *J. Am. Chem. Soc.,* 77, 2019, 1955.
14. Skorobogatov, V. M., Gershuns, A. L., and Adamovich, L. P., *Zh. Anal. Khim.,* 32, 1450, 1977.
15. Kakihana, H. and Kato, K., *Benseki,* 231, 1977.
16. Tanaka, T., Hiiro, K., and Kawahara, A., *Bunseki Kagaku,* 27, 247, 1978.
17. *Anala R Standards for Laboratory Chemicals,* 7th ed., Analar Standards Ltd., England, 1977.

α-DIOXIMES

Synonyms

α-Dioximes covered in this section are listed in Table 1, together with their synonyms and structural formulas.

Source and Methods of Synthesis

Commercially available. Prepared by the reactions of corresponding α-diketone and hydroxylamine,[1,2] with an exception for (1) which is obtained from methyl ethyl ketone via biacetylmonoxime.[3]

Analytical Uses

Dimethylglyoxime (1) is one of the first selective organic reagents applied in analytical chemistry. Dioximes (1) to (3) are widely used as a selective precipitating reagent, detecting reagent, and photometric reagent for Ni, Pd(II), Pt(II), and some other metal ions. Of the three possible geometric isomers, only the anti- form is capable of forming chelates with metal ions.

anti- (α)   syn-(β)   amphi-(γ)

Properties of Reagents

Dimethylglyoxime (1) is a colorless needle-like crystalline powder, mp 238 to 240° (with decomposition), is very slightly soluble in water (0.632 g/ℓ, 25°)[11] and chloroform (0.052 g/ℓ, 25°),[11] and is more soluble in 96% ethanol (16.3 g/ℓ, 25°).[R2] It is easily soluble in alkaline solution. In acid solution (pH < 1), it is slowly hydrolyzed at room temperature. It sublimes at 26 to 68° (2×10^{-2} Torr).[6] D(CHCl$_3$/H$_2$O) = 0.083 (25°); pKa (H$_3$L$^+$) = 8.7, pKa (H$_2$L) = 10.6, pKa (HL$^-$) = 1 1.9 (μ = 0.01, KClO$_4$, 25°).[R2]

In Furildioxime (2), most commercial samples are a 1:1 mixture of anti- and syn-isomers, satisfactory for most analytical uses. It is colorless or pale yellow needles, mp 166 to 168° (with decomposition).[1] The pure anti-isomer melts at 192°,[4,5] and it is almost insoluble in water, but easily soluble in acetone, alcohol, and ether and slightly soluble in chloroform (10.6 g/ℓ), benzene, and petroleum ether. D(CHCl$_3$/H$_2$O) = 0.35, pK$_a$(H$_2$L) = 11.6 (50% dioxane, 25°).[R2]

Cyclohexanedionedioxime (3) is colorless crystals, mp 187 to 188° (with decomposition). (3) is more soluble in water than (1) (8.2 g/ℓ, 21.5°) and can be recrystallized from hot water; pKa(H$_2$L) = 12.0 (50% dioxane, 25°).[R2]

Complexation Reactions and Properties of Complexes

α-Dioximes (H$_2$L), in anti- configuration, form the M(HL)$_2$-type chelates with many of the divalent transition metal ions. As the analytical reagent, the most important property of (1), (2), and (3) is their ability to form water-insoluble, colored square-planer chelates (4) with Ni, Pd(II), and Pt(II).[7,8]

Table 1
α-DIOXIMES

Number	Structural formula	Synonyms	Molecular formula and mol wt
(1)	H₃C—C—C—CH₃ / HO—N N—OH	2,3-Butandionedioxime, Dimethylglyoxime, Diethyldioxime, DMG	$C_4H_8N_2O_2$, 116.12
(2)	(furildioxime structure) HO—N N—OH	Furildioxime	$C_{10}H_8N_2O_4$, 220.18
(3)	(cyclohexanedione structure) N—OH / N—OH	1,2-Cyclohexanedione-dioxime, Nioxime	$C_6H_{10}N_2O_2$, 142.16

(4)

The intramolecular hydrogen bonds stabilize the $M(HL)_2$-type chelate, and, hence, the chelate is not readily solvated. In the case of Ni chelate, the formation of weak metal to metal bonds in the solid state also contribute to the water insolubility. On the other hand, Cu(II) and Co(II) dimethylglyoximates are slightly soluble in water because the fifth and sixth coordination position of the central metal ion are occupied by water molecules.[9] In the case of Pd(II) chelate, it is insoluble in aqueous acidic or neutral medium, but is soluble in alkaline medium because the central atom is capable of coordinating a hydroxide ion. Because the Ni chelate is unable to coordinate a hydroxide ion even in a strongly alkaline medium, this difference in behavior can be utilized for the separation of Ni and Pd.[M1]

The complexation behaviors of reagents (1), (2), and (3) are almost similar, but (2) has advantages over (1) in that the metal chelates are more soluble in organic solvents and the color reaction is more sensitive. An advantage of (3) over (1) is higher water solubility of reagent and higher molecular weight which are desirable for gravimetry.

The chelate stability constants in water or aqueous dioxane are summarized in Table 2. It is interesting to note that, in most cases, $K_{ML}/K_{ML_2} \leqslant 1$ which is due to the intramolecular hydrogen bonds that stabilize the $M(HL)_2$-type chelate in solution. The favored stability of the $M(HL)_2$-type chelate assures the almost exclusive formation of $M(HL)_2$ chelate, offering a considerable advantage for use as an analytical reagent. The solubilities and solubility products of metal chelates in various solvents are summarized in Table 3.

Since the chelates with Ni, Pd(II), and Pt(II) are extractable into immiscible organic solvents, they can be separated from other elements by solvent extraction and can be determined by the photometric method. The most common solvent for the extraction is chloroform, and the distribution coefficients of dimethylglyoxime chelates between $CHCl_3$-H_2O system are reported to be log $K_D = -9.93$ [$Cu(HL)_2$], 1.85 [$Pd(HL)_2$], and 2.60[$Ni(HL)_2$].[9,14]

The Ni and Pd(II) chelates are yellow in chloroform and can be determined by absorption photometry. However, the molar absorptivity of Ni-dimethylglyoximate in

Table 2
CHELATE STABILITY CONSTANTS OF α-DIOXIME CHELATES[M1,M2,R2,10]

Reagent	Metal ion	$\log K_{ML}$	$\log K_{ML_2}$	Condition
(1)	Cd	5.7	5.0	50% dioxane, 25°
	Co(II)	10.8	10.2	50% dioxane
	Co(II)	8.7	9.0	Water, 25°
	Co(III)	15.0	10.0	—
			$(\log K_{ML_3}\ 7.2)$	
	Cu(II)	11.0	12.5	50% dioxane, 25°
	Fe(II)	10.8	9.2	50% dioxane, 25°
	La(III)	6.6	5.9	50% dioxane, 25°
	Mn(II)	8.6	8.6	50% dioxane, 25°
	Ni	11.16	10.54	50% dioxane, 25°
	Pd(II)	$K_1 \ll K_2$	$\beta_2\ 34.1$	Water, 25°
	Pb	7.3	—	50% dioxane, 25°
	Zn	8.1	9.3	50% dioxane, 25°
(2)	Co(II)	8.2	7.2	75% dioxane, 25°
	Cu(II)	10.3	10.1	50% dioxane, 25°
	Fe(II)	$K_1 \ll K_2$	$\beta_2\ 21.8$	75% dioxane, 25°
	Mn(II)	6.4	5.4	75% dioxane, 25°
	Ni	6.9	7.2	75% dioxane, 25°
	Zn	$K_1 \ll K_2$	$\beta_2\ 15.7$	75% dioxane, 25°
(3)	Co(II)	13.0	12.5	50% dioxane, 25°
	Cu(II)	13.2	12.5	50% dioxane, 25°
	Mn(II)	8.2	7.2	50% dioxane, 25°
	Ni	11.08	11.38	75% dioxane, 25°
	Zn	$K_1 \ll K_2$	$\beta_2\ 14.9$	75% dioxane, 25°

Table 3
SOLUBILITY(S) AND SOLUBILITY PRODUCTS(K_{sp}) OF α-DIOXIME CHELATES

Reagent	Metal chelate	Solvents	S (mol/l)	Condition		Ref.
(1)	Cu(HL)$_2$	Water	7.9×10^{-3}	$\mu = 0.1$ NaClO$_4$	25°	9
		CHCl$_3$	1.2×10^{-3}		25°	12
		Benzene	4.1×10^{-4}		25°	12
	Ni(HL)$_2$	Water	9.8×10^{-7}	$\mu = 0.05$ NaCl, $\log K_{sp}\ -23.37$	25°	11
		CHCl$_3$	4.8×10^{-4}		25°	11
		Benzene	7.6×10^{-5}		25°	12
		n-Heptane	3.6×10^{-8}		25°	12
	CCl$_4$		6.0×10^{-6}		25°	13
	Pd(HL)$_2$	CHCl$_3$	5.0×10^{-4}		25°	73
(2)	Ni(HL)$_2$	Water	6.3×10^{-7}		25°	73
		CHCl$_3$	1.6×10^{-3}		25°	73
	Pd(HL)$_2$	CHCl$_3$	8.1×10^{-4}		25°	73
(3)	Ni(HL)$_2$	Water	1.6×10^{-6}		25°	73
				$\log K_{sp}\ -28.39$		
		CHCl$_3$	6.8×10^{-5}		25°	73
	Pd(HL)$_2$	CHCl$_3$	1.4×10^{-4}		25°	73

chloroform is not so high that it is of little practical importance for the trace analysis. As summarized in Table 4, the Ni chelate with (2) is more sensitive. Another sensitive method for Ni with (1) is based on the formation of brown or red soluble chelate in alkaline medium in the presence of oxidizing agent such as bromine or persulfate.[M2]

Table 4
APPLICATION OF α-DIOXIMES AS A PHOTOMETRIC REAGENT

Reagent	Metal ion	Condition	Metal Chelate				Range of determination (ppm)	Ref.
			Ratio	λ_{max} (nm)	$\varepsilon(\times 10^4)$	Solvent		
(1)	Co(II)	KI at pH 4, then (1) at pH 6	M(HL)$_2$I$_2$	435	1.06	Aqueous	2 ~ 20	16
	Co(II)	pH 5.5 ~ 10, benzidine (amines)	M(HL)X	420	1.20	Benzyl alcohol	~12	26
	Fe(II)	pH 8(NH$_3$), pyridine, tartarate	M(HL)$_2$	500	0.56	CHCl$_3$	1 ~ 4	27
	Ni	pH 6 ~ 12	M(HL)$_2$	326	0.49	CHCl$_3$	1 ~ 15	28
	Ni			360~370	(0.32)^a	CHCl$_2$	0.2 ~ 4	28, 29
	Ni	pH > 11(NH$_3$ or NaOH), Br$_2$ or persulfate	[ML$_3$]$^{2-}$	445	1.4	Aqueous		M1
	Pd(II)	pH 1 ~ 6	—	380	0.17	CHCl$_3$	~12	30
	Re(VII)	3 N HCl, after reduction to Re(IV) with SnCl$_2$	—	445	0.38	Aqueous	~4	31
	Re(IV)	1 N HCl, SnCl$_2$, tartaric acid	—	445	1.45	CHCl$_3$	1 ~ 20	32
(2)	Au(III)	pH 1.8, pyridine	—	330	3.16	CHCl$_3$	1 ~ 4.5	33
	Co(II)	pH 7 ~ 8	M(HL)$_2$	350	1.84	Aqueous	~4	34
	Cu(II)	pH 7.2 ~ 9.5, tartarate	—	445	—	CHCl$_3$	0.4 ~ 2.5	35
	Fe(II)	pH 2.4, pyridine	—	570	0.89	CHCl$_3$	5 ~ 25	36
	Ni	pH >7.2	M(HL)$_2$	438	—	CHCl$_3$	0.05 ~ 0.5	37
	Os(VIII)	pH 1.8 ~ 2.8	M(HL)	450	1.06	Aqueous	4 ~ 12	38
	Pd(II)	0.1 ~ 1.4 N HCl	H(HL)$_2$	420	1.4	10 % ethanol	0.5 ~ 3	39
	Pd(II)	0.1 ~ 1.4 N HCl	M(HL)$_2$	380	2.3	CHCl$_3$	0.5 ~ 3	39
	Pd(II)	pH 2.4 ~ 4.9	M(HL)$_2$	297 or 436	0.2	Benzene	0.1 ~ 3	40
	Re(VII)	0.65 ~ 0.9 N HCl, SnCl$_2$	M(HL)$_2$	532	4.13	Benzene	0.8 ~ 6	41
	Re(VII)	~1 N HCl, after reduction to Re(IV) with SnCl$_2$	—	530	3.39	CHCl$_3$	1 ~ 5	31

(3)	U(VI)	pH 6.0 ~ 8.6, pyridine	$MO_2(HL)_2$	428	0.32	Aqueous	2.5 ~ 40	42
	Ni	pH 4 ~ 6, citrate, gum arabic	$M(HL)_2$	550	0.61	Aqueous	0.5 ~ 8	43
	Ni	pH 3.5, quinoline	—	430	0.18	Benzene	0.4 ~ 7, in presence of 3 × 10⁴ fold Co.	44
	Pd(II)	pH > 0.5	—	280	1.35	$CHCl_3$		45
	Re(VII)	≈0.6 N HCl, $SnCl_2$	—	440	0.4	$CHCl_3$	20 ~ 50	31, 46

Table 5

DETECTION OF VARIOUS METALS WITH DIMETHYLGLYOXIME (1)[75]

Metal ion	Condition	Color change	Formula of complex	Detection limit	Dilution limit	Ref.
Bi	NH$_3$ alkali	Yellow ppt	—	14 µg/1 ml	1:7.0 × 10^4	17
Fe(II)	NH$_3$ alkali	Red, soluble	M(HL)$_2$	0.4 µg/0.05 ml	1:1.25 × 10^5	18
Fe(III)	0.2 N NaOH	Red-brown, soluble	M(HL)$_2$$^+$	—	—	
Ni	NH$_3$ alkali	Red ppt	M(HL)$_2$	0.16 µg/0.05 ml	1:3 × 10^5	19
Pd(II)	pH ≈2(HCl or H$_2$SO$_4$)	Yellow ppt	M(HL)$_2$	—	—	20
Pt(II)	Acidic, after reduction of Pt(IV) with formate	Ppt	H(HL)$_2$	Pd and Pt chelates are soluble in NaOH, but excess Pt causes purple ppt	—	21
Tc	SnCl$_2$-HCl	Green	—	0.04 µg/100 ml	—	22
V(IV)	NH$_3$ alkali in presence of Fe(III)	Brown soln	—	1 µg/0.05 ml	1:5 × 10^4	23

The resulting chelate is thought to be a Ni(IV) chelate [(NiL$_3$)$^{2-}$] and can also be extracted as an ion-pair with diphenylguanidine.[15] According to Zolotov, the composition of the chelate is independent of the nature of the oxidizing agent, but changes according to the medium. An unstable neutral chelate is formed in ammoniacal solutions, and a stable anionic chelate is formed in NaOH or KOH solutions.[15]

α-Dioximes can form a chelate of square-planer N$_4$ configuration with metal ions, such as Co(II), (III), Cu(II), and Fe(II), in which the coordination positions on z axis are accessible by various ligands, such as coordinating anions (OH$^-$, Cl$^-$, Br$^-$, I$^-$, and CN$^-$) or uncharged donor molecules (NH$_3$, pyridine etc.), to form mixed ligand chelates [M(HL)$_2$X or M(HL)$_2$X$_2$]. Dimethylglyoximates of Ni, Pd, and Pt do not form such adducts or do very weakly. For example, when Co(II) is treated with KI at pH 4, then with (1) at pH 6, a red-brown soluble mixed ligand chelate [Co(HL)$_2$I$_2$.]$^{2-}$ (λ_{max}, 435 nm; $\varepsilon = 1.06 \times 10^4$) is formed. The reaction is known to be highly selective for Co.[16]

Analytical Applications

Use as a Detection Reagent

Dimethylglyoxime (1) forms colored precipitates or colored soluble chelates with a limited number of metal ions, and these reactions are so senstive that they have been utilized for the detection of such metals. Examples are summarized in Table 5.

Use as a Reagent for Separation of Metals

As described previously, α-dioximes form water-insoluble chelates with Ni, Pd(II), Pt(II), and some other elements, and the chelate can be extracted into immiscible organic solvent. As the conditions for the precipitation are different for each metal, they can be separated from each other or from other matrix elements by precipitation or solvent extraction. Nickel in the organic phase can be back extracted with dilute hydrochloric acid. Dimethylglyoxime(1) or cyclohexanedionedioxime (3) is the specific precipitating agent for Ni in ammoniacal solution in the presence of tartarate or citrate as a masking agent. In the presence of Co(II), Mn(II), or Zn, however, the precipitation is carried out at pH ≈ 5 (acetate buffer).

It is a common practice to add alcoholic solution of dimethylglyoxime to a hot aqueous solution to precipitate Ni chelate. However, when the hot solution is more than 50% in alcohol, an appreciable amount of Ni chelate dissolves. Its solubility becomes negligible at lower alcohol concentration (<33%). Consequently, in the use of an ethanolic solution of dimethylglyoxime, the volume added must not exceed half that of the aqueous nickel solution. Boiling of the alcohol containing solution must be avoided because the eventual loss in alcohol content results in the precipitation of excess reagent (1), giving an erroneous result in the gravimetric finish. In the case of cyclohexanedionedioxime (3), the reagent is soluble in water, so that the undesirable effect of alcohol is avoided. This reagent is also an excellent gravimetric reagent for Bi which can be precipitated at pH 12 in the presence of EDTA. Bismuth could be quantitatively precipitated from aqueous solution containing more than 30 elements, including Ni and Pd.[24,25]

The Pd(II) chelate of (1) or (3) also precipitates quantitatively from slightly mineral acid medium (HCL or H_2SO_4), where other metal ions do not precipitate.

Use as a Photometric Reagent

As discussed before, the extraction photometry for Ni and Pd (II) is most sensitive and simple in procedure with furildioxime (2). The absorption photometry of Ni in aqueous alkaline solution in the presence of oxidizing agent has also been accepted for the trace determination. In this method, the absorption intensity is very much dependent upon the reaction condition, such as the nature of medium (alkali hydroxide or ammonia), oxidizing agent (bromine or persulfate), and elapse of time. Consequently, the color reaction should be carried out under strictly controlled condition to get the reproducible result. The detailed procedures have been discussed in various literature.[M2,74]

The applications of the reagents for the photometric determination of metals are summarized in Table 4.

Other Uses

These reagents have been used as the titrant in the precipitation titration of Ni with an amperometric end point detection.[47,48] The Co(II)-dimethylglyoximate catalyzes the evolution of hydrogen on the dropping mercury electrode and the catalytic hydrogen wave is useful for the determination of traces of Co(II) (5 to 20 ppm).[49]

The Ni chelates of (1) and (3) were investigated as a stationary phase in the gas-liquid partition chromatography of various hydrocarbons.[50]

Other Oximes of Analytical Interest

Dioxime and Trioxime

Following is a list of α-dioximes and trioximes which have been investigated as an analytical reagent for metals.

> 4-Methyl-1,2-cyclohexanedionedioxime(Ni, Pd, and Re)[51]
> 4-Isopropyl-1,2-cyclohexanedionedioxime(Ni and Pd)[52]
> 4-tert-Butyl-1,2-cyclohexanedionedioxime(Ni)[53]
> 1,2-Cycloheptanedionedioxime(Heptoxime) (Co, Cu, Fe, Mo, Ni, and Pd)[53,54]
> 1,2-Cyclodecanedionedioxime (Ni and Pd)[55]
> Benzildioxime (diphenylglyoxime) (Ni, Pd, and Re)[R4,56]
> Pentane-2,3,4-trionetrioxime (Co, Cu, Fe, Ni, and Pd)[57]
> 1,2,3-Cyclohexanetrionetrioxime (Niconoxime) (Co, Cu, Fe, and Ni)[58,59]
> 2,2′ Dipyridyl-α-glyoxime (Fe)[60]
> Dimedonedioxime (Co, Cu, and Ni)[61]

Monoximes

Bidentate ligands, in which the oxime group is introduced on the 2 position of pyridine, behaves as the ferroin reagent toward Fe(II), and such reagents have been discussed under ferroin reagents (p.309).

Other bidentate ligands having an oxime group include salicylaldoxime (5) (H_2L, mp 57 to 59°, pKa = 9.18, 25°)[62-65] and benzoin monoxime (6) (H_2L, mp 153 to 155°, pKa = 12.0, 50% dioxane, 25°)[66-69] which is also known as Cupron. Both reagents form $M(HL)_2$-type uncharged chelates with Cu(II), and some other divalent metal ions and they can be extracted into immiscible organic solvents. Thus, they are useful in the separation of metals.

(5)   (6)

Recently, various α-monoximes derived from acetophenone,[70,71] acenaphthenequinone,[72] etc.[R1,76] were investigated as analytical reagents.

MONOGRAPHS

M1 **Burgar, K.,** Selectivity and analytical application of dimethylglyoxime and related diomes, in *Chelates in Analytical Chemistry*, Flashka, H. and Barnard, A. J., Jr., Eds., Vol. 2, Marcel Dekker, New York, 1969, 179.
M2. **Sandell, E. B. and Onishi, H.,** *Photometric Determination of Traces of Metals*, 4th ed., John Wiley & Sons, New York, 1978.
M3. **Fletcher, A. W.,** Dimethylglyoxime, Nioxime, Furildioxime, in *Organic Reagents for Metals*, Johnson, W. C., Ed., Vol. 1, Hopkin & Williams, Essex, England, 1955.

REVIEWS

R1. **Singh, R. B., Gang, B. S., and Singh, R. P.,** Oximes as spectrophotometric reagent, *Talanta*, 26, 425, 1979.
R2. **Dyrssen, D.,** Stability constants and solubilities of the metal dioximes, in *Kungl. Teknika Högskolan Handlinger*, Stockholm, Sweden, No. 220, 1964.
R3. **Diehl, H.,** The Application of the Dioximes to Analytical Chemistry, G. F. Smith Chemical Co., Columbus, Ohio, 1940.

REFERENCES

1. Reed, S. A., Banks, C. V., and Diehl, H., *J. Org. Chem.*, 12, 792, 1947.
2. Hach, C. C., Banks, C. V., and Diehl, H., *Organic Synthesis*, 4, 229, 1963.
3. Semon, W. L. and Damerell, V. R., *Organic Synthesis*, 2, 204, 1950.
4. Fryer, F. A., Galliford, D. J. B., and Yardley, J. T., *Analyst*, 88, 188, 191, 1963.
5. Shinra, K., Ishikawa, K., and Shimura, H., *Nippon Kagaku Zasshi*, 74, 353, 1953.

6. Honjo, T., Imura, H., Shima, S., and Kiba, T., *Anal. Chem.*, 50, 1547, 1978.
7. Godycki, L. E. and Rundle, R. E., *Acta Crystallogr.*, 6, 487, 1953.
8. Rundle, R. E. and Parasol, M., *J. Chem. Phys.*, 20, 1489, 1952.
9. Dyrssen, D. and Hemnichs, M., *Acta Chem. Scand.*, 15, 47, 1961.
10. Sillen, L. G. and Martell, A. E., *Stability Constants of Metal Compounds,* Suppl. 1, The Chemical Society, London, 1964; 1971.
11. Christopherson, H. and Sandell, E. B., *Anal. Chim. Acta,* 10, 1, 1954.
12. Fleischer, D. and Freiser, H., *J. Phys. Chem.*, 66, 389, 1962; Thesis, University of Pittsburgh, 1959, University Microfilms, 59-2396.
13. Banks, C. V. and Anderson, S., *J. Am. Chem. Soc.*, 84, 1486, 1962.
14. Burger, K. and Dyrssen, D., *Acta Chem. Scand.*, 17, 1489, 1963.
15. Zolotov, Yu. A. and Vlasova, G. E., *Zh. Anal. Khim.*, 28, 1540, 1973; *J. Anal. Chem. USSR,* 28, 1373, 1973.
16. Burger, R. and Ruff, I., *Acta Chim. Acad. Sci. Hung.*, 45, 77, 1965; *Chem. Abstr.*, 64, 15b, 1966.
17. Kubina H. and Plichta, J., *Fresenius Z. Anal. Chem.*, 72, 12, 1927.
18. Tschugaef, L. and Orelkin, B., *Z. Anorg. Allgem. Chem.*, 89, 401, 1914.
19. Feigl, F. and Kapulitzas, H. T., *Mikrochemie,* 8, 244, 1930.
20. Holzer, H., *Fresenius Z. Anal. Chem.*, 95, 392, 1933.
21. Copper, R. A., *J. Chem. Metall. Min. Soc. S. Afr.*, 25, 296, 1925; *Chem. Abstr.*, 20, 1042, 1926.
22. Jasmin, F., Magee, R. J., and Wilson, C L., *Talanta,* 2, 93, 1959.
23. Ephraim, F., *Helv. Chim. Acta,* 14, 1266, 1931.
24. Wiersma, L. D. and Lott, P. F., *Anal. Chim. Acta,* 40, 291, 1968.
25. Bassett, J., Letow, G. B., and Vogel, A. I., *Analyst,* 92, 279, 1967.
26. Pyatnitskii, I. V., Mikhelson, P. B., and Moskovskaya, L. T., *Ukr. Kim. Zh.*, 41, 739, 1975; *Chem. Abstr.*, 83, 212156w, 1975.
27. Oi, N., *Nippon Kagaku Zasshi,* 75, 1067, 1069, 1969.
28. Taylor, C. G., *Analyst,* 81, 369, 1956.
29. Nielsch, W. and Giefer, L., *Microchim. Acta,* 522, 1956.
30. Davis, D. W., *Talanta,* 16, 1330, 1969.
31. Pollock, E. N. and Zopatti, L. P., *Anal. Chim. Acta,* 32, 418, 1965.
32. Kozlicka, M., Wojtourcz, M., and Adamiec, I., *Chem. Anal. (Warsaw),* 15, 701, 1970; *Chem. Abstr.*, 74, 19066a, 1971.
33. Patil, P. S. and Shinde, V. M., *Microchim. Acta,* 331, 1977.
34. Jones, J. L. and Gastfield, J., *Anal. Chim. Acta,* 51, 130, 1970.
35. Benediktova-Lodochnikova, N. V., *Zh. Anal. Khim.*, 18, 1322, 1963.
36. Patil, P. S. and Shinde, V. M., *Microchim. Acta,* 151, 1971.
37. Kirillova, Z. P., Merisov, Yu. I., and Petrova, E. I., *Zavod. Lab.*, 43, 1308, 1977; *Chem. Abstr.*, 89, 36003e, 1978.
38. Bhowal, S. K., *Indian J. Chem.*, 13, 1225, 1975; *Anal. Abstr.*, 31, 2B242, 1976.
39. Menis, O. and Rains, T. C., *Anal. Chem.*, 27, 1932, 1955.
40. Peshkova, V. M., Shlenskaya, V. I., and Sokolov, S. S., *Tr. Kom. Anal. Khim. Akad. Nauk SSR,* 11, 328, 1960; *Chem. Abstr. ,* 55, 8171c, 1961.
41. Meloche, V. W., Martin, R. L., and Webb, W. H., *Anal. Chem.*, 29, 527, 1957.
42. Spacu, P. and Popea, E., *Acad. Repub. Pop. Rom. Stud. Cercet. Chim.*, 9, 139, 1962; *Chem. Abstr.*, 56, 13544f, 1962.
43. Ferguson R. C. and Banks, C. V., *Anal. Chem.*, 23, 448, 1951.
44. Monnier, D. and Haerdi, W., *Anal. Chim. Acta,* 20, 444, 1959.
45. Pshenitsyn, N. K. and Ivonina, O. M., *Zavod. Lab.*, 24, 1185, 1962.
46. Pollock, E. N. and Zopatti, L. P., *Anal. Chim. Acta,* 32, 418, 1965.
47. Kolthoff, I. M. and Langer, A., *J. Am. Chem. Soc.*, 62, 211, 1940.
48. Peshkova, V. M. and Gallay, Z. A., *Zh. Anal. Khim.*, 7, 152, 1952.
49. Burger, K., Syrek, G., and Farsang, G., *Acta Chim. Scand. Sci. Hung.*, 49, 113, 1966; *Anal. Abstr.*, 14, 7492, 1967.
50. Pflaum, R. T. and Cook, L. E., *J. Chromatogr.*, 50, 120, 1970.
51. Kassner, J. L., Ting, S. F., and Grove, E. L., *Talanta,* 7, 269, 1961.
52. Barling, M. M. and Banks, C. V., *Anal. Chem.*, 36, 2359, 1964.
53. Votov, R. C. and Banks, C. V., *Anal Chem.*, 21, 1320, 1949.
54. Marov, I. N., Gureva, E. S. and Peshkova, V. M., *Zh. Neorg. Khim.*, 15, 3039, 1970; *Chem. Abstr.*, 74, 16235u, 1971.
55. Banks, C. V. and Kyrd, E. K., *Anal. Chim. Acta,* 10, 129, 1954.
56. Warshawsky, A., *Talanta,* 21, 624, 1974.

57. Cacho Palomar, J. F., *Rev. Acad. Cienc. Exactas, Fis-Quim. Nat. Zaragoza,* 31, 91, 1976; *Anal. Abstr.,* 33, 1A13, 1977.
58. Frierson, W. J., Patterson, N., Harrill, H., and Marable, N., *Anal. Chem.,* 33, 1096, 1961.
59. Frierson, W. J. and Marable, N., *Anal. Chem.,* 34, 210, 1962.
60. Soules, D., Holland, W. J., Stupavsky, S., and Notenboom, H. R., *Mikrochim. Acta,* 787, 1970; *Mikrochim. Acta,* 565, 1971; *Mikrochim. Acta,* 247, 1972; *Mikrochim. Acta,* 187, 1973.
61. Belcher, R., Ghanaim, S. A., and Townshend, A., *Talanta,* 21, 191, 1974.
62. Gorbach, G. and Pohl, F., *Mikrochem. Mikrochim. Acta,* 38, 258, 1951.
63. Simonsen, S. H., Christopher, P., and Burnett, H. M., *Anal. Chem.,* 26, 681, 1954; *Anal. Chem.,* 27, 1336, 1955.
64. Yamamoto, Z., Ueda, K., and Ueda, T., *Nippon Kagaku Zasshi,* 89, 288, 1968.
65. Reddy, R. B. S. and Rao, S. B., *Curr. Sci.,* 48, 298, 1979; *Chem. Abstr.,* 90, 214687m, 1979.
66. Melaven, A. D. and Whetsel, K. B., *Anal. Chem.,* 20, 1209, 1948.
67. Einaga, H. and Ishii, H., *Benseki Kagaku,* 18, 439, 1969.
68. Savostina, V. M., Shpigun, O. A., Parmenova, V. A., and Peshkova, V. M., *Zh. Anal. Khim.,* 32, 556, 1977.
69. Kozlicka, M. and Wojtowicz, M., *Fresenius Z. Anal. Chem.,* 257, 191, 1971.
70. Desai, K. K., Naik, N. D., and Naik, H. B., *J. Indian Chem. Soc.,* 54, 414, 1977.
71. Saxena, S. B., Agarwal, R. N., and Gupta, A., *J. Indian Chem. Soc.,* 54, 644, 1977.
72. Sindliwani, S. K. and Singh, R. P., *Anal. Chim. Acta,* 55, 409, 1971; *Talanta,* 20, 248, 1973.
73. Banks, C. V. and Barnum, D. V., *J. Am. Chem. Soc.,* 80, 3579, 1958.
74. Suzuki, M., *Muki Oyo Hishoku Bunseki,* Vol. 4, Kyoritsu, Tokyo, 1975, 40.
75. Feigl, F., Anger, V., and Oesper, R., *Spot Test in Inorganic Analysis,* Elsevier, Amsterdam, 1972.
76. Kuse, S., Motomizu, S., and Toei, K., *Nippon Kagaku Kaishi,* 1611, 1973; *Anal. Chim. Acta,* 70, 65, 1974.

DIAMINOBENZIDINE AND RELATED REAGENTS

(1)

(2)

(3) X = H
(4) X = Cl
(5) X = NO$_2$

Synonyms

Aromatic o-diamines covered in this section are listed in Table 1, together with their synonyms.

Source and Methods of Synthesis

All are commercially available. They are prepared by the reduction of corresponding nitro compounds, i.e., (1) from 3,3'-dinitrobenzidine,[1] (2) from 2,3-dinitronaphthalene,[2] (3) from o-nitroaniline,[3] and (4) from 4-chloro-2-nitroaniline[4] with tin and hydrochloric acid. (5) is obtained by the reduction of 2,4-dinitroaniline with ammonium sulfide.[5]

Analytical Uses

Highly selective reagents for Se.[M1,R1,R2] (1) and (3) are used as photometric reagents in visible and UV region, respectively. (2) is used for fluorimetric finish, and (4) and (5) are for gas chromatographic finish.

Properties of Reagents

3,3'-Diaminobenzidine (1) is usually supplied as tetrahydrochloride which is a colorless crystalline powder, mp 328 to 330° (decomposition). It darkens under light;[6,7] hence it should be kept in a dark cool place under nitrogen. It is easily soluble in water, but insoluble in nonpolar organic solvents.

2,3-Diaminonaphthalene (2) is a colorless crystalline powder in pure state, but commercial samples are often a yellow or grey-brown powder due to air oxidation, mp 190 to 191°,[8] and is almost insoluble in cold water or alcohol, but fairly soluble above 50°. It should be stored in a dark cool place.

o-Phenylenediamine (3) is a colorless crystalline powder which easily darkens in the air, mp 101 to 103°, pb 257° (or 143° at 68 Torr) and is easily soluble in water and alcohol. It should be stored in a dark cool place; pKa_1(H$_2$L^{2+}) = 0.86 and pKa_2(HL$^+$) = 4.75 (μ = 1.0, 20°).[R1]

5-Chlorophenylenediamine (4) is a colorless crystalline powder which is more resistant to air oxidation than (1) to (3), mp 72 to 74°. It is easily soluble in water, and the aqueous solution is also stable; pKa_1(H$_2$L^{2+}) = −0.11 and pKa_2(HL$^+$) = 4.16 (μ = 0.1, 20°).[R1]

5-Nitrophenylenediamine (5) is a pale yellow crystalline powder which is fairly stable against air oxidation, mp 198 to 200°. It is easily soluble in water; pKa(HL$^+$) = 3.07 (μ = 1.0, 20°).[R1]

Table 1
AROMATIC o-DIAMINES

Reagent	Diamine	Synonyms	Molecular formula and mol wt
(1)	3,3'-Diaminobenzidine, tetrahydrochloride	3,3',4,4'-Biphenyltetramine, tetrahydrochloride, DAB	$C_{12}H_{14}N_4 \cdot HCl \cdot 2H_2O$, 396.14
(2)	2,3-Diaminonaphthalene	DAN	$C_{10}H_{10}N_2$, 158.20
(3)	1,2-Phenylenediamine	o-Phenylenediamine	$C_6H_8N_2$, 108.14
(4)	5-Chloro-1,2-phenylenediamine		$C_6H_7N_2Cl$, 142.59
(5)	5-Nitro-1,2-phenylenediamine		$C_6H_7N_3O_2$, 153.14

Complexation Reaction and Properties of Complexes
Reaction with Metal Ions

Aromatic o-diamines behave as *N,N*-donating bidentate ligand to form colored complexes with some metal ions, especially with the platinum metals. For example, (3) forms a blue ML_2-type chelate with Pt(II) (λ_{max}, 703 nm, $\varepsilon = 9.8 \times 10^4$) in hot neutral solution,[9,10] and (1) forms a ML-type chelate with V in pH 2 to 3 (λ_{max}, 470 nm; $\varepsilon = 3310$).[11] However, they are of no particular practical importance as chromogenic reagents for metal ions.

Reaction with Selenium(IV)

Hoste first found that (1) reacts with Se(IV) in acidic solution to form a yellow complex called "piaselenol" which can be extracted at pH 6 to 7 into organic solvent, such as benzene, toluene, chloroform, butanol, and ethylacetate.[1,12] The reaction may be written as below.[R1,13]

The resulting piaselenol is yellow and can be determined photometrically in the visible region, as illustrated in Figure 1.[6] Later, various aromatic o-diamines were found to form piaselenol, and the reagents such as (2) to (5) are also recommended as selective reagents for Se(IV). In contrast to the piaselenol of (1), the absorption maximum of that of (3) lies in an UV region, necessitating the performance of photometry in an UV region. In the case of (2), the fluorescence maximum of the piaselenol shifts to the longer wavelength than that of the free reagent as illustrated in Figure 2, enabling the highly sensitive fluorimetric determination of Se(IV).[8] The absorption and fluorescence spectral characteristics of the piaselenols are summarized in Tables 2 and 3, respectively.

The piaselenol from (3) to (5) is fairly volatile, so that it can be analyzed by gas chromatography. Ultratrace amount of the piaselenol of (4) and (5) can be determined with the use of the electron capture detector.[19]

Purification and Purity of Reagents

(1) — Slightly deteriorated material can be purified by dissolving in water, followed by precipitation with concentrated hydrochloric acid. The precipitates are dried over solid sodium hydroxide. It is not so easy to purify the darkened material.

(2) — The reagent solution of low fluorescent blank should be prepared daily. Recrystallized material from water (0.05 g) was dissolved in 50 mℓ of 0.1 *N* hydrochloric acid with heating (50°). After cooling, the solution was extracted twice

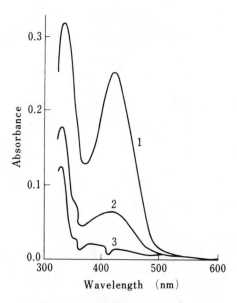

FIGURE 1. Absorption spectra of 3,3'-diaminobenzidine and piaselenol in toluene: (1) 25 μg Se in 10 mℓ of toluene; (2) 5 μg Se in 10 mℓ of toluene; (3) 3,3'-Diaminobenzidine (1) in toluene; reagent concentration, 0.5% in water. (From Cheng, K. L., *Anal. Chem.*, 28, 1738, 1956. With permission.)

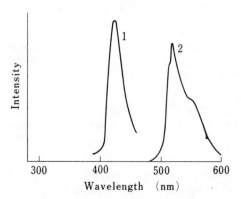

FIGURE 2. Fluorescence spectra of 2,3-diaminonaphthalene and its piaselenol. (1) 2,3-Diaminonaphthalene (λ_{ex}, 336 nm); (2) Piaselenol (λ_{ex}, 377 nm). (From Cuker, P., Walzcyk, J., and Lott, P., *Anal. Chim. Acta,* 30, 473, 1964. With permission.)

with 10-mℓ portions of decalin to remove fluorescent impurities and spun in a centrifuge.[14,15]

(3), (4) and (5) — The reagents can be purified by recrystallization from appropriate solvents.[9,16] The purity of the reagents, except (1) can be evaluated by observing their melting points.

Table 2
SPECTRAL CHARACTERISTICS OF PIASELENOLS[R1,17]

Reagent	Condition	Piaselenol (in toluene) λ_{max} (nm)	$\varepsilon(\times 10^4)$	D(pH 2)	Range of determination (ppm Se)	Interferences
(1)	Reaction at pH 2 ∿ 3, (50°, 50 min), extraction at pH 6∿7	335	0.78[a]		5∿26	Ce(IV), Cu(II), Fe(III), V(V), SO_4^{2-}, CrO_4^{2-}, and MnO_4^-
		420	1.99[b]			
(2)	pH 1.7 ∿ 2.2, 120 min	380 (378)	1.18 (4.1)[c]		∿4	Cu(II), Te(IV), NO_2^- NO_3^- and oxidant
(3)	pH 1 ∿ 2.5, 120 min	335	1.78	154	∿2.5	Bi(III), Fe(III), Mo(IV), Sb(III), and Sn(IV)
(4)	pH <2.3, 90 min	341 (340)	1.84 (4.90)[c]	2580	∿2.5	Fe(III), Mo(VI), Sn(IV), and V(V), but Fe(III) and Mo(VI) can be masked with EDTA
(5)	pH <2, 150 min	350	1.55	367		

[a] Value for single extraction.
[b] Value for triple extraction.
[c] Values in parentheses are those observed on the isolated piaselenols.[8]

Table 3
FLUORESCENCE CHARACTERISTICS OF PIASELENOLS

Reagent	Piaselenol λ_{ex} (nm)	λ_{em} (nm)	Solvent	Range of determination (ppm Se)	Ref.
(1)	420	550 ∿ 600	Toluene	0.02 ∿ 0.4	18
(2)	377	520	Cyclohexane	0.001 ∿ 0.1	8
(4)	400	448	Cyclohexane	0.01 ∿ 0.1	8

Analytical Uses

The only practical application of these reagents is for the determination of Se(IV). Since selenium of the other oxidation states does not react with these reagents, Se in the samples has to be reduced or oxidized to Se(IV). One of the convenient ways to oxidize Se^{2-} and S° is the use of redox buffer solution (10^{-3} *M* bromine plus 2×10^{-3} *M* bromide) after wet digestion of the sample.[19] The reduction of Se(VI) to (IV) is effected with Ti(III).

Diaminobenzidine (1) is recommended for the photometry in the visible region. The optimum condition for the reaction with Se(IV) is pH 2 to 3 at 50°, but extraction of piaselenol has to be carried out at pH 6 to 7 to effect the deprotonation of the free amino group. Multiple extractions of piaselenol with toluene are necessary to attain quantitative separation (∿60% for single extraction).[6] Sulfate ion interferes the determination by precipitating the reagent as insoluble sulfate, but can be masked by adding excess ammonium chloride.[20] Common cations can be masked with EDTA, Fe(III) can be masked with fluoride or phosphate, and Cu(II) can be masked with oxalate.[1,6,12]

o-Phenylenediamine (3) is also used as chromogenic reagent for Se(IV), but in an UV region.[21] However, excess reagent is not extracted into toluene, since the extraction of piaselenol is carried out at pH 1 to 2. Furthermore, the reagent is not interfered with sulfate, and the sensitivity is twice as high than that with (1).

Ultratrace determination of Se can be attained either by fluorimetry with (2)[31] or by gas chromatography with (3) to (5). In the fluorimetry, it is important to use the purified diaminonaphthalene (2) solution with low fluorescent blank at observing wavelength (520 nm). In the gas chromatography, the sample solution (pH 1) is treated with a large excess of reagent (4) or (5) for 1 hr. The piaselenol is extracted with 1 ml of toluene, and a 5-μl aliquot is subjected to gas chromatography at the following condition.[19]

> Glass column, 1-m length, 4-mm bore, filled with 60 to 80 mesh Chromosorb® W containing 15% SE-30; temperature of the column and the electron capture detector, 200°; flow rate of carrier gas, 20 ml/min helium (in the case of (4)[22] and 44 ml/min nitrogen (in the case of (5)).[19]

Common cations, if not large excess, do not interfere, as the complexes are not volatile. As small as 0.15 μg of Se(IV) per milliliter toluene can be determined. This method has been applied for the determination of ultratrace amount of Se in sulfuric acid,[19] tellurium metal,[23] sea water,[24] copper salts,[25] plant materials,[26] and milk.[27]

Other Reagents with Related Structure

Aromatic o-diamines which have been investigated as the reagent for Se(IV) are listed below.

> 4-Methoxy-o-phenylenediamine[28]
> 4,5-Dichloro-o-phenylenediamine[29]
> 5-Dimethylamino-o-phenylenediamine[30]
> 5-Methylthio-o-phenylenediamine[30]
> N-(2-Hydroxypropyl)-o-phenylenediamine[32]

MONOGRAPH

M1. Pribil, R., *Analytical Applications of EDTA and Related Compounds*, Pergamon Press, New York, 1972.

REVIEWS

R1. Kawashima, T., Determination of Selenium, *Sci. Repor. Kagoshima University*, (Japan), No. 17 1968; No. 21 1972.

R2. Murashova, V. I. and Sushkova, S. G., Photometric and extraction-photometric methods for determining selenium and tellurium, *Zh. Anal. Khim.*, 24, 729, 1969; *J. Anal. Chem. USSR*, 24, 572, 1969.

R3. Shimoishi, Y. and Toei, K., Determination of a trace selenium by gas chromotography, *A & R (Japan)*, 17, 433, 1979.

REFERENCES

1. Hoste, J., *Anal. Chim. Acta*, 2, 402, 1948.
2. Goldstein, H. and Streuli, M., *Helv. Chim. Acta*, 20, 522, 1937.
3. Ter-Minasyan, L. E., *Zh. Fiz. Khim.*, 27, 719, 1953; *Chem. Abstr.*, 48, 55f, 1953.
4. Ulman, F. and Mauthner, F., *Ber.*, 36, 4027, 1904.
5. Gottlieb, J., *Ann.*, 85, 27, 1953.
6. Cheng, K. L., *Anal. Chem.*, 28, 1738, 1956.
7. Iwasaki, I., Kishioka, A., and Yoshida, Y., *Bunseki Kagaku*, 10, 479, 1961.
8. Cukar, P., Walzcyk J., and Lott, P. F., *Anal. Chim. Acta*, 30, 473, 1964.
9. Golla, E. D. and Ayres, G. H., *Talanta*, 20, 199, 1973.
10. Sen Gupta, J. G., *Anal. Chim. Acta*, 23, 462, 1960.
11. Cheng, K. L., *Talanta*, 8, 658, 1961.
12. Hoste, J. and Gillis, J., *Anal. Chim. Acta*, 12, 158, 1955.
13. Barcza, L., *Mikrochim. Acta*, 967, 1964.
14. Parkar, C. A. and Harvey, L. G., *Analyst*, 87, 558, 1962.
15. Tamari, Y., Hiraki, K., and Nishikawa, Y., *Bunseki Kagaku*, 28, 164, 1979.
16. Toei, K. and Ito, K., *Talanta*, 12, 773, 1965.
17. Tanaka, M. and Kawashima, T., *Talanta*, 12, 211, 1965.
18. Watkinson, J. H., *Anal. Chem.*, 32, 981, 1960.
19. Shimoishi, Y. and Toei, K., *Talanta*, 17, 165, 1970.
20. Danzuka, T. and Ueno, K., *Anal. Chem.*, 30, 1370, 1958.
21. Ariyoshi, H., Kiniwa, M., and Toei, K., *Talanta*, 5, 112, 1960.
22. Akiba, M., Shimoishi, Y., and Toei, K., *Analyst*, 100, 648, 1975.
23. Shimoishi, Y., *Bull. Chem. Soc. Jpn.*, 44, 3370, 1971.
24. Shimoishi Y., *Anal. Chim. Acta*, 64, 465, 1973.
25. Shimoishi, Y., *Bull. Chem. Soc. Jpn.*, 48, 2797, 1975.
26. Shimoishi, Y., *Bull. Chem. Soc. Jpn.*, 47, 997, 1974.
27. Shimoishi, Y., *Analyst*, 101, 298, 1976.
28. Kawashima, T. and Ueno, A., *Anal. Chim. Acta*, 58, 219, 1972.
29. Neve, J., Hanocq, M., and Molle, L., *Talanta*, 26, 15, 1979.
30. Demeyere, D. and Hoste, J., *Anal. Chim. Acta*, 27, 288, 1962.
31. Analytical Methods Committee, *Analyst*, 104, 778, 1979.
32. Kasterka, B., *Chem. Anal. (Warsaw)*, 24, 329, 1979; *Anal. Abstr.*, 37, 6B105, 1979.

PORPHYRIN REAGENTS

Synonyms

Reagents covered in this section are listed in Table 1, together with their synonyms. Porphyrin is the term for a group of compounds having porphin ring.

Source and Method of Synthesis

Commercially available. (1) is obtained by refluxing the equimolar mixture of pyrole and benzaldehyde in propionic acid.[1] (2) and (3) are the sulfonation products of (1).[2] Preparation of (4)[1,3] and (5)[4] are similar to that of (1), using corresponding pyridylaldehye in place of benzaldehyde, followed by quaternarization with methyl iodide or tosylmethylate.

Analytical Uses

As highly sensitive photometric reagents for Cd, Cu(II), Fe (III), Hg (II), Mg, Pb, Pd(II), Zn, and some other metals.[R1]

Properties of Reagents

TPP (1) — violet needles, mp 300°, sublimes at 232 to 262° (2×10^{-2} Torr),[19] is stable in the air, is not soluble in water and alcohol, but slightly soluble in acetic acid, and is easily soluble in benzene, chloroform, DMF, and THF. It can be solubilized in water (5×10^{-5} M) with an aid of surfactant such as sodium lauryl sulfonate ($\sim 5\%$). A strong absorption band (Soret band, $\varepsilon \simeq 10^5$) is observed at 438 nm (pH <3.7) or 418 nm (pH >4.5).

TPPS$_4$ (2) and TPPS$_3$ (3) — When (1) is sulfonated by heating in concentrated sulfuric acid, products of various degree of sulfonation are obtained. Pasternak et al.[4] and Yotsuyanagi et al.[2] obtained TPPS$_3$ (3) as a free acid, while Ishii and Kon isolated TPPS$_4$ (2) as an ammonium salt.[5] Both samples seem to behave similarly as an analytical reagent. They are blue crystals, easily soluble in water. The aqueous solution is red-violet at pH 6 (Soret band, λ_{max}, 413 nm; $\varepsilon = 5.1 \times 10^5$) and green at pH 4 ($\lambda_{max}$, 434 nm; $\varepsilon = 5.0 \times 10^5$).[2] Absorption spectra of (3) in visible range are illustrated in Figure 1. The proton dissociation scheme of the reagent can be shown as following, where the ionization of sulfonic groups are disregarded.

$$H_4L^{2+} \underset{pKa_{1,2}}{\overset{pKa_1}{\rightleftharpoons}} H_3L^+ \overset{pKa_2}{\rightleftharpoons} H_2L \overset{pKa_3}{\rightleftharpoons} HL^- \overset{pKa_4}{\rightleftharpoons} L^{2-}$$

$pKa \sim 4.8$ for (2) ($\mu = 0.1$, 25°)[20], and pKa_1 4.86 and pKa_2 4.95 for (3) (25°).[2]

T(-MPy)P (4) and T(4-MPy)P (5) — These are another type of water-soluble porphyrin reagents. Both are green crystals, slowly dissolving in water. (4) is claimed to be more soluble in water than (5). Aqueous solution of (4) is reddish-purple at pH >2 (Soret band, λ_{max}, 417 nm; $\varepsilon = 2.8 \times 10^5$) and green at pH <1 ($\lambda_{max}$ 434 nm; $\varepsilon = 3.8 \times 10^5$ in 3 N H$_2$SO$_4$).[3] Aqueous solution of (5) behaves similarly, at pH 4 to 7 (Soret band, λ_{max}, 422 nm; $\varepsilon = 1.49 \times 10^5$) and in 1 M HCl (λ_{max}, 446 nm; $\varepsilon = 1.93 \times 10^5$).[6] Absorption spectra of (4) and (5) are illustrated in Figures 2 and 3, respectively; pKa_2 2.3 for (4) ($\mu = 0.1$ NaNO$_3$, 25°),[7] $pK_{1,2}$ 2.2, pKa_3 12.9, and $pKa_4 \sim 6$ for (5).[8]

Complexation Reactions and Properties of Complexes

Porphyrin reagents form 1:1 stable chelate with various metal ions. The stability order for divalent metal ions was found to be Pt > Pd > Ni > Co > Cu > Fe > Zn >

Table 1
TETRAPHENYLPORPHIN AND RELATED COMPOUNDS

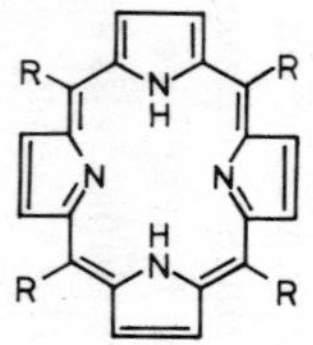

Reagent	Name	R	Synonyms	Formula for isolated reagent	Molecular formula, mol wt
(1)	$\alpha,\beta,\gamma,\delta$-Tetraphenyl-porphin	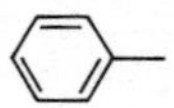	TPP	H_2L	$C_{44}H_{30}N_4$, 614.75
(2)	$\alpha,\beta,\gamma,\delta$-Tetraphenyl-porphin tetrasulfonic acid	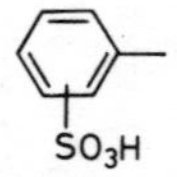	TPPS$_4$	$(H_2L)(NH_4)_4 \cdot 12H_2O$	$C_{44}H_{42}N_8S_4O_{12} \cdot 12H_2O$, 1221.30
(3)	$\alpha,\beta,\gamma,\delta$-Tetraphenyl-porphin trisulfonic acid	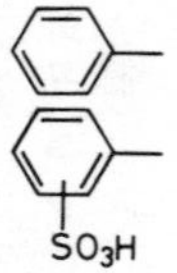	TPPS$_3$	$(H_2L)H_3 \cdot 4H_2O$	$C_{44}H_{30}N_4S_3O_9 \cdot 4H_2O$, 926.98
(4)	$\alpha,\beta,\gamma,\delta$-Tetra(3-N-methylpyridyl) por-phin	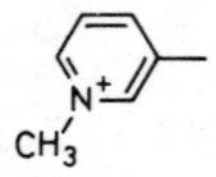	T(3-MPy)P	$(H_2L)(CH_3C_6H_4\text{-}SO_3)_4$	$C_{72}H_{66}N_8O_{12}S_4$, 1363.70
(5)	$\alpha,\beta,\gamma,\delta$-Tetra(4-N-methylpyridyl) por-phin	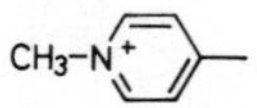	T(4-MPy)P	$(H_2L)(ClO_4)_4$	$C_{44}H_{38}N_8O_{16}Cl_4$, 1076.64

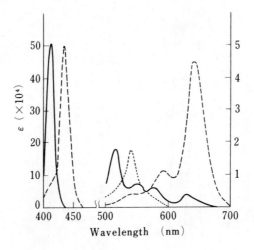

FIGURE 1. Absorption spectra of TPPS₃ (3) in aqueous solution. (1) H₂L (pH 6.5); (2) H₄L²⁺ (pH 2.5); (3) CuL (pH 2.5). (From Itoh, J., Yotsuyanagi, T., and Aomura, K., *Anal. Chim. Acta*, 74, 53, 1975. With permission.)

FIGURE 2. Absorption spectra of T(3-MPy)P (4) in aqueous solution; reagent concentration, 2.7 × 10⁻⁶ *M*. (1) pH >2.5; (2) pH 1.1; (3) 3 *N* H₂SO₄. (From Ishii, H. and Koh, H., *Talanta*, 24, 417, 1977. With permission.)

Mn > Mg > Cd > Sn> Hg> Pb > Ba, regardless of the type of substituents on porphin ring.^{M1} The structure of Cu(II) chelate of TPPS₃ is shown below.

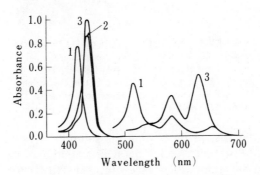

The rate of complex formation is generally rather slow, and especially so when metal ion is complexed at trace concentration level. However, the rate can be increased substantially by the use of auxiliary complexing agent, such as pyridine, 2,2′-bipyridine,

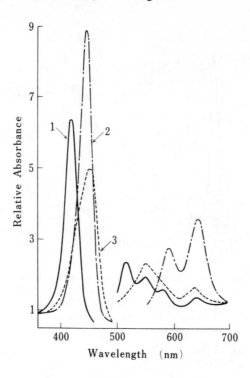

FIGURE 3.   Absorption spectra of T(4-MPy)P (5) in aqueous solution; reagent concentration, 2.3 × 10^{-6} *M*. (1) H_2L; (2) H_4L^{2+} (pH 0.0); (3) HL^- (pH 14.0). (Reprinted with permission from Hambright, P. and Fleischer, E. B., *Inorg. Chem.*, 9, 1757, 1970. Copyright 1970 American Chemical Society.)

cysteine, or hydroxylamine, or by heating the solution at 100°. The metal chelates, once formed, do not dissociate readily by acidification and show strong absorption band (Soret band) in the region where excess reagent (H_4L^{2+}) does not absorb, as illustrated in Figures 1 to 3. This becomes the basis for the ultrahigh sensitive photometric determination (parts per billion level) of trace elements. Excess reagent ((4) and (5)) may also be photodecomposed by irradiation under a fluorescent lamp in the presence of ascorbic acid for 30 min.[9]

Analytical Applications

Metal ions that can be determined with porphyrin reagents ar limited as summarized in Table 2. Most of interfering ions may be masked by proper choice of masking agents, but Cu(II), Hg(II), and Zn interfere seriously. 4-Carboxyphenyl derivative has been also investigated as a reagent for Cd[10] and Cu(II).[11]

Table 2

PHOTOMETRIC DETERMINATION OF METALS WITH PORPHYRIN REAGENTS

Metal ion	Reagent	Condition	λ_{max} (nm)	ε ($\times 10^5$)	Range of determination (ppb)	Interference	Ref.
Zn	(1)	Glacial acetic acid, (room temp., 60 ~ 70 min)	551	0.14	~200	Cu at higher conc	12
Cu(II)	(1)	Solubilized by Na-laurylsulfate, complexation at pH 4.7 + NH_2OH(100°, 3 min), then pH 0.6 ~ 1.2 with H_2SO_4	414	4.7	~140	Ag, Hg, Zn >10 µg, Pd >5 µg	13
Pd(II)	(2)	Complexation at pH 3.6(100°, 15 min), then pH 2.5 with chloroacetic acid	410	2.2	~360	Cu(II), Zn interfere. Other metal can be masked with tartaric acid	5
Cu(II)	(3)	Complexation at pH 4.0(100°, 15 min), then pH 2.5 with chloroacetic acid	434	4.8	6 ~ 60	Zn	2
Cd	(3)	pH 12.5(NaOH), 2,2'-bipyridine (5 min)	432	4.45	~100	After separation of Cd as CdI_4^{2-}	14
Fe(II)	(3)	pH 3.9 ~ 4.2(100°, 15 min)	395	1.4	20 ~ 180	Cu(II), Co(II), Mn(II), Pd(II), Sn(II) and Zn(II)	18
Pb	(3)	pH 10.2(borate), KCN (70°, 5 min)	464	2.75	50 ~ 500	Cr(III), In(III), Mn(II) and Sn(II)	15
Cd	(3)	Complexation at pH 3, L-ascorbic acid(100°, 8 min), then pH 2.5 with chloroacetic acid	411.6	2.15	~250	Cu(II) and Hg(II)	16
Cu(II)	(4)	Complexation at pH 6 (100°, 7 min), then acidify with H_2SO_4 (2 ~ 3 N)[a]	434	3.5	1 ~ 13	Fe(III) and Pd(II) (can be masked with NH_2OH + KI)	3
Cu(II)	(5)	Complexation at pH >4.5 (room temperature), 0.02% N $H_2OH \cdot HCl$	446	2.46	~190	Pd(II) interfere, but can be masked with KI	17

[a] Complexation is instantaneous at room temperature in the presence of hydroxylamine or ascorbic acid.[9,16]

MONOGRAPH

M1. **Falk, J. E.**, *Porphyrins and Metalloporphyrins*, Elsevier, Amsterdam, Netherland, 1964.

REVIEW

R1. **Yotsuyanagi, T. and Itoh, J.**, Highly Sensitive Colorimetric Reagents — Porphyrins, *Kagaku No Ryoiki (Tokyo)*, 31, 146, 1977.

REFERENCES

1. **Adler, A. D., Longo, F. R., Finarelli, J. D., Goldmacher, J., Assour, J., and Korsakoff, K.**, *J. Org. Chem.*, 32, 476, 1967.
2. **Itoh, J., Yotsuyanagi, T., and Aomura, K.**, *Anal. Chim. Acta*, 74, 53, 1975.
3. **Ishii, H. and Koh, H.**, *Talanta*, 24, 417, 1977.
4. **Posternack, R F., Huber, P. R., Boyd, P., Engasser, G., Francesconi, L., Gibbs E., Fasella, P., Venturo, G. C., and L. de C. Hindes**, *J. Am. Chem. Soc.*, 94, 4511, 1972.
5. **Ishii, H. and Koh, H.**, *Nippon Kagaku Zasshi*, 390, 1978.
6. **Hambright, P. and Fleischer, E. B.**, *Inorg. Chem.*, 9, 1757, 1970.
7. **Ishii, H.**, private communication.
8. **Hambright, P. and Chok, P. B.**, *J. Am. Chem. Soc.*, 96, 3123, 1974.
9. **Ishii, H. and Koh, H.**, *Bunseki Kagaku*, 28, 473, 1979.
10. **Ishii, H, Koh, H., and Kawamura, K.**, Abstract of paper, The 37th Annual Meeting of Chemical Society of Japan, 1978.
11. **Ishii, H., Koh, H., and Okuda, Y.**, *Nippon Kagaku Kaishi*, 686, 1978.
12. **Banks, C. V. and Risque, R. E.**, *Anal. Chem.*, 29, 522, 1957.
13. **Ishii, H.**, *Anal. Chim. Acta*, 101, 423, 1978.
14. **Igarashi, S., Itoh, J., Yotsuyanagi, T., and Aomura, K.**, *Nippon Kagaku Kaishi*, 212, 1978.
15. **Itoh, J., Yamahira, M., Yotsuyanagi, T., and Aomura, K.**, *Bunseki Kagaku*, 25, 781, 1976.
16. **Igarashi, S., Yotsuyanagi, T., and Aomura, K.**, *Bunseki Kagaku*, 27, 66, 1978.
17. **Ishii, H. and Koh, H.**, *Bunseki Kagaku*, 26, 473, 1979.
18. **Nomura, T. and Orita, M.**, *Bunseki Kagaku*, 28, 377, 1979.
19. **Honjo, T., Imura, H., Shima, S., and Kiba, T.**, *Anal. Chem.*, 50, 1547, 1978.
20. **Fleischer, E. B., Palmer, J. M., Srivastava, T. S., and Chatterjee, A.**, *J. Am. Chem. Soc.*, 93, 3162, 1971.

CHELATING REAGENTS WITH SULFUR FUNCTIONS

DITHIZONE AND RELATED REAGENTS

$$HS-C \begin{matrix} N=N \\ | \quad | \\ N-N \end{matrix} H$$

C₁₃H₁₂N₄S
mol wt = 256.32

H_2L

Synonyms

1,5-Diphenylthiocarbazone, N,N-Diphenyl-C-mercaptoformazane, Phenylazothio-formic acid 2-phenylhydrazide

Source and Method of Synthesis

Commercially available. It can be prepared by the reaction of carbon disulfide with phenylhydrazine, followed by the careful heating to oxidize the reaction mixture.[1,2]

Analytical Uses

Dithizone, first synthesized by E. Fischer in 1878[1] and named as "Dithizon" by H. Fischer in 1925,[3] is widely used as a highly selective and sensitive extraction photometric reagent for heavy metal ions, such as Cd, Cu, Hg, Pb, and Zn.[4]

Properties of Reagent

Violet-black crystalline powder with metallic luster, mp 165 to 169° (decomposition); sublimes at 40 to 123° (0.02 Torr);[5] practically insoluble in water below pH 7 (5 to 7 .2 × 10⁻⁵ g/ℓ),[M1] but dissolves readily in alkali (pH >7, >20 g/ℓ)[M3] to give an yellow dithizonate ion (HL⁻) (λ_{max}, 470 nm; ε = 2.2 × 10⁴); and is soluble in various organic solvents. Solubilities in selected solvents are listed in Table 1.[M1,M3]

Although dithizone is expected to behave as a dibasic acid (dissociation of thiol proton and imino proton), there is no evidence of dithizone to form L^{2-} in aqueous solution. Among the pKa values reported by various workers, the best one recommended by Irving is 4.47 ± 0.25 which was estimated from the solubility data.[M1]

The distribution of ratio of dithizone between organic and aqueous phases can be expressed and simplified as below if the aqueous phase is basic, in which $[H_2L]_{aq} \ll [HL^-]_{aq}$ and $[H^+] \ll K_a$:

$$D = \frac{[H_2L]_{org}}{[H_2L]_{aq} + [HL^-]_{aq}} = K_D \left(1 + \frac{[H^+]}{K_a} \right) \cong \frac{K_D}{K_a} [H^+] \quad \text{(when } K_D \ll K_a\text{)}$$

The log D-pH plots for some solvents are shown in Figure 1.

Absorption spectra of dithizone (H_2L) in carbon tetrachloride, dithizonate ion (HL⁻) in water, and mercury dithizonate [Hg(HL)₂] in carbon tetrachloride are shown in Figure 2. The spectrum of dithizone in organic solvents shows two well-defined bands which are related to the thione (1) and thiol (2) tautomers, although the assignment of each peak to the respective species is not yet decisive.

(1)

$$S=C \begin{matrix} N=N \\ | \\ N-NH \\ H \end{matrix}$$

$\rightleftharpoons$

$$S-C \begin{matrix} N=N \\ | \quad | \\ N-N \end{matrix} H$$

(2)

Table 1
VISIBLE ABSORPTION SPECTRA OF DITHIZONE IN ORGANIC SOLVENTS[M1,M3]

Solvent	Solubility (g/l, 20°)	λ_{max} (nm) 1st	2nd	ε ($\times 10^3$) 1st	2nd	R (peak ratio)	Color
Ethanol	0.3	596	440	27.0	16.6	1.64	Blue-green
Carbon tetrachloride	0.512	620	450	34.6	20.3	1.70	Green
Dioxane	0.349	617	446	32.9	19.1	1.72	—
Water + dioxane (50:50 v/v)	—	602	446	32.4	16.4	1.98	—
Benzene	1.24	622	453	34.3	19.0	1.80	Deep green
Chlorobenzene	1.43	622	452	34.8	18.3	1.90	Green
Nitrobenzene	—	627	454	31.9	16.7	1.91	—
Chloroform	16.9	605	440	41.4	15.9	2.59	Blue-green

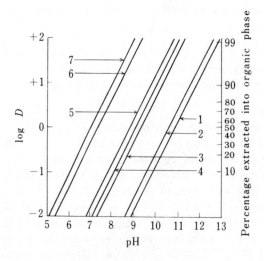

FIGURE 1.   Distribution ratio of dithizone between aqueous and some organic phases. Solvent: (1) dichloroethane; (2) chloroform; (3) chlorobenzene; (4) benzene; (5) carbon tetrachloride; (6) cyclohexane; (7) *n*-hexane. (From Irving, H. M. N. H., *Dithizone*, The Chemical Society, London, 1977. With permission.)

The relative intensities of two peaks vary markedly with the change of solvents, as summarized in Table 1.

Complexation Reaction and Properties of Complexes

Dithizone is S,N-donating ligand, reacting preferentially with soft metal ions. If the reagent is in excess and the aqueous phase is acidic, the cation in the aqueous phase reacts with dithizone dissolved in an organic solvent to give a highly colored primary complex which distributes into the organic phase.

$$M^{n+} + nH_2L_{org} = M(HL)_{n\ org} + nH^+$$

Metal ions which react with dithizone are illustrated in Figure 3.

60
40
20
0
ε (×10³)
3
1
2
400
500
600
700
Wavelength (nm)

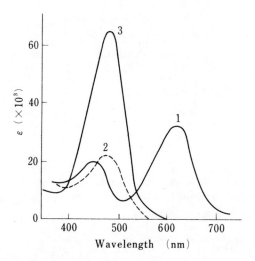

FIGURE 2. Absorption spectra of dithizone (H_2L) and mercury(II) dithizonate in carbon tetrachloride and of dithizonate ion (HL^-) in water. (1) Dithizone (H_2L) in CCl_4; (2) dithizonate (HL^-) in water; (3) mercury(II) dithizonate [$Hg(HL)_2$] in CCl_4. (From Irving, H. N. N. H., *Dithizone,* The Chemical Society, London, 1977. With permission.)

Ia IIa IIIa IVa Va VIa VIIa VIII Ib IIb IIIb IVb Vb VIb VIIb O
H H He
Li Be B C N O F Ne
Na Mg Al Si P S Cl Ar
K Ca Sc Ti V Cr Mn Fe Co Ni Cu Zn Ga Ge As Se Br Kr
Rb Sr Y Zr Nb Mo Tc Ru Rh Pd Ag Cd In Sn Sb Te I Xe
Cs Ba La Hf Ta W Re Os Ir Pt Au Hg Tl Pb Bi Po At Rn
Fr Ra Ac Th Pa U

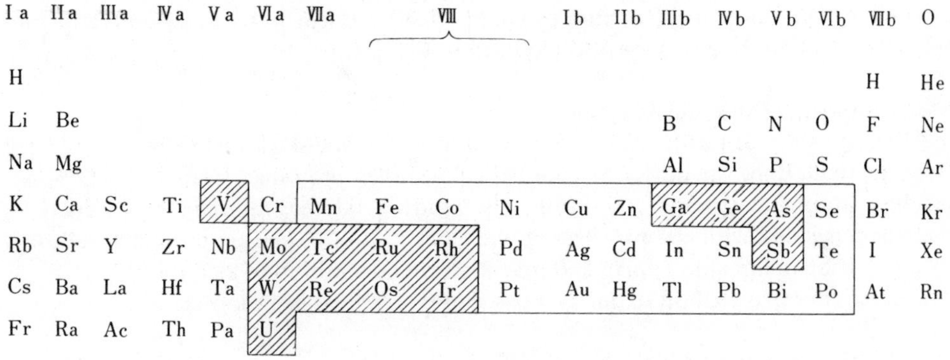

FIGURE 3. Metals extracted by dithizone and precipitated as sulfides from aqueous solution. Metals within the shaded area can be separated as sulfides from aqueous solution and do not normally form extractable dithizonates. Metals within the solid line can be extracted by dithizone from aqueous solution. Some metals (e.g., Cu, Fe, and Hg) form different dithizone complexes according to their oxidation state. The situation with Se is still under discussion. The rare earths give complexes of low stability in 50% aqueous ethanol, and several other metals may form dithizonates in the absence of water and at high reagent concentration. (From Irving, H. M. N. H., *Dithizone*, The Chemical Society. London, 1977. With permission.)

These complexes are readily soluble in organic solvents such as chloroform or carbon tetrachloride. The structure of primary dithizonate for the divalent tetra-coordinate metal ion can be shown as (3).

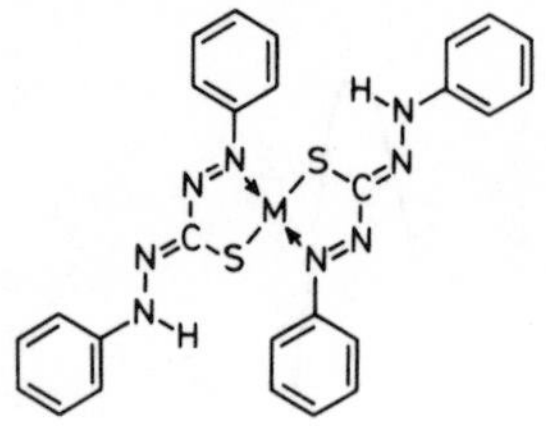

(3)

If the metal is in excess or if the pH is high, metal ions, such as Cu(II), Ag, and Hg(II), give a secondary complex, whose structure can be given as (4) and (5) for divalent and univalent cation, respectively.

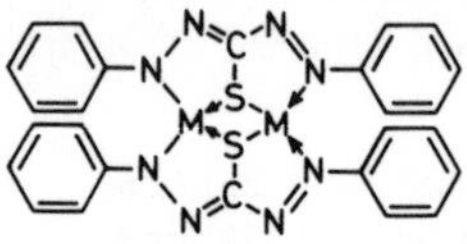

(4) 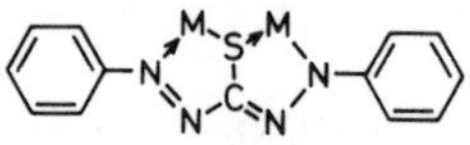   (5)

The spectral characteristics of metal dithizonates are summarized in Table 2, and the absorption spectra of primary and secondary copper(II) dithizonates are illustrated in Figure 4. Overall stability constants of primary complexes, βn, were also determined spectrophotometrically on a very dilute aqueous solution and those values are listed in Table 2. The values of extraction constant on chloroform-water or carbon tetrachloride-water systems (log K_{ex}) are also included in Table 2.

Purification and Purity of Reagent

Although solid dithizone can be kept well in a cool and dark place that is protected from air oxidation, the purity of commercial samples can range from 30 to 90%, depending upon the storage condition. The crude solid reagent may be purified by Soxhlet extraction with ether to remove the more soluble oxidation products, followed by dissolution in hot chloroform and precipitation by addition of alcohol.[2,32]

A more effective method to purify a few grams sample is as follows.

> Filter the concentrated solution of crude sample in carbon tetrachloride through a sintered glass funnel, then shake the filtrate with 0.8 N aqueous ammonia to extract dithizonate ion. Extract an aqueous phase with several portions of carbon terachloride to remove undesired materials. Finally acidify the aqueous solution with sulfuric acid to precipitate pure dithizone. Dry the sample in vacuo.[33-35]

Miligram amounts of dithizone may be purified by chromotographing a benzene solution on a column consisting a mixture of silica gel and celite 545 or by paper chromatography.[7,36,37]

Purity of dithizone can be determined by one of the following methods:

1. Photometric method[38] — The absorbance of carbon tetrachloride solution of known dithizone concentration is observed at 620 nm against a solvent blank. Purity is calculated by using the molar absorptivity of pure dithizone ($\varepsilon = 3.4 \times 10^4$ at 620 nm).

Table 2
CHARACTERISTIC DATA FOR PRIMARY DITHIZONE CHELATES^{M1,M3}

Metal ion	Solvent	Solubility (g/l, 20°)	Absorption spectra				log K_{ex}	log βn in water[7] at 25 ± 1°	Ref.
			Ratio	λ_{max} (nm)	ε (× 10³)	Color[a]			
Ag	CCl₄	1.14	MHL	463	27.2, 29.1	Golden yellow	6.5, 7.18, 7.81, 7.90, 8.94	6.98	8—13
Ag	CHCl₃	1.81×10^2	MHL	420	—	—	5.8	—	8,9
Au(III)	CCl₄	4.87×10^{-3}	M(HL)₂Cl	—	—	—	—	—	53
Au(III)	CHCl₃	4.87×10^{-2}	M(HL)₂Cl	~450	~24	Golden yellow	—	—	
Bi	CCl₄	9.75×10^{-3}	M(HL)₃	490	80.0, 84.6	Red orange	9.54 (KCN), 9.75, 9.98 10.76	32.11	10,14,15
Bi	CHCl₃	9.75×10^{-3}	M(HL)₃	—	—	—	8.7	—	14
Cd	CCl₄	6.23×10^{-3}	M(HL)₂	520	88.0	Rose red	1.58, 2.14	15.10	10,16
Cd	CHCl₃	6.23×10^{-3}	M(HL)₂	520	85.6	Rose red	0.53	—	10
Co(II)	CCl₄	1.14×10^{-1}	M(HL)₂	542	59.2	Red violet	0.1, 1.5~1.59, 1.53	13.97	10,12,13
Co(II)	CHCl₃	1.14	M(HL)₂	—	—	—	—	—	
Cu(II)	CCl₄	2.87×10^{-1}	M(HL)₂	550	45.2	Red violet	9.56, 10.53	19.18	10,12,17
Cu(II)	CHCl₃	1.15	M(HL)₂	—	—	—	6.50	—	10
Cr(III)	Acetone	—	M(HL)	525	29.5	—	—	—	54
Fe(II)	CCl₄	5.66×10^{-3}	M(HL)₂	560	50.0	—	-3.4	8.99	18
Fe(II)	CHCl₃	5.66×10^{-3}	M(HL)₂	560	—	—	-5.0	—	18
Ga	CHCl₃	—	M(HL)₃	—	—	—	-1.3	—	19
Hg(II)	CCl₄	4.98×10^{-2}	M(HL)₂	485	72.2	Orange yellow	26.85, 26.7—26.9	40.3	20,21
Hg(II)	CHCl₃	2.13×10^{-1}	M(HL)₂	—	—	—	—	—	
Hg(II)	CCl₄	9.83×10^{-3}	M₂L₂[b]	515	24	Violet	—	—	
In	CCl₄	7.05×10^{-1}	M(HL)₃	510	87.0	Red	4.84	—	22
In	CHCl₃	8.81×10^{-1}	M(HL)₃	—	—	—	0.6	—	23
Mn(II)	—	—	M(HL)₂	—	—	Brown violet	—	9.55	
Ni	CCl₄	5.69×10^{-1}	M(HL)₂	665	19.2	Brown violet	-1.19, 0.7, 0.63	14.17	10,12,24
Ni	CHCl₃	1.14	M(HL)₂	—	—	—	-2.92	—	10

Table 2 (continued)
CHARACTERISTIC DATA FOR PRIMARY DITHIZONE CHELATES[M1,M3]

Metal ion	Solvent	Solubility (g/ℓ, 20°)	Absorption spectra: Ratio	Absorption spectra: λ_{max} (nm)	Absorption spectra: ε (× 10^3)	Color[a]	log K_{ex}	log βn in water[7] at 25 ± 1°	Ref.
Pb	CCl_4	1.14×10^{-2}	$M(HL)_2$	520	67.0∼72.0	Carmine red	0.38, 0.44, −2.86 (citrate), −3.53 (citrate + CN^-)	14.16	10, 13
Pb	$CHCl_3$	2.87×10^{-1}	$M(HL)_2$	518	63.6	—	−0.89	—	10
Pd(II)	CCl_4	3.09×10^{-1}	$M(HL)_2$	640	33.0	Brown green	42.5	21.78	26
Pd(II)	CCl_4	—	$M(HL)_x$	500	—	Violet	—	—	
Pt(II)	CCl_4 (no extraction with $CHCl_3$)	—	$M(HL)_2$	710	30.0	Brown yellow	—	—	
Sn(II)	CCl_4	—	$M(HL)_2$	520	54	Red	−2	11.99	12
Tl(I)	CCl_4	—	M(HL)	—	—	—	−3.3, −3.5	—	27, 28
Tl(I)	$CHCl_3$	—	M(HL)	505	336	Red	—	—	
Zn	CCl_4	5.76×10^{-1}	$M(HL)_2$	535	92.6, 96.0	Purple red	1.7, 1.8∼2.0, 2.3	13.96	10, 29, 30
Zn	$CHCl_3$	28.8	$M(HL)_2$	530	88.0	Purple red	−1.52, −0.52, 0.64, 0.90	—	16,29—31

[a] Color observed on a 25 μM solution of 1-cm thickness.
[b] Secondary complex.

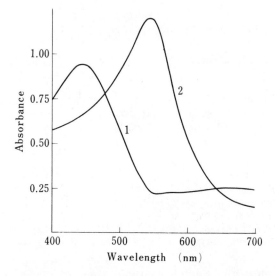

FIGURE 4. Absorption spectra of copper(II) dithizonate in chloroform. (1) Secondary chelate (CuL); (2) primary chelate [Cu(HL)₂]. (From Fischer, H. and Weyl, W., *Wiss. Veroff. Siemens-Werken*, 14, (2) 41, 1935. With permission.)

2. Volumetric method[39] — Standard 50×10^{-6} M silver nitrate solution (25.0 mℓ) is titrated with a chloroform solution of known dithizone concentration (0.125 mg in 100 mℓ), discarding the organic phase after each extraction. Extraction titration is continued until the extract is no longer pure golden-yellow, but has a mixd yellow-green color. Purity is calculated from the volume of dithizone solution consumed in the titration.

3. Peak ratio method[38,39] — The ratio of the intensities of two prominent bands of dithizone ($\lambda_1 = 602$ to 617 nm, $\lambda_2 = 440$ to 454 nm) can be used as a measure of the purity because impurities in dithizone tend to absorb below 550 nm. Although the positions of two bands stay almost unchanged with the change of solvent, the intensity ratio changes strikingly. The values in chloroform and carbon tetrachloride are as below:

$$A_{605}/A_{445} \ (CHCl_3) = 2.59 \ (AnalaR \ specification: > 2.5)$$
$$A_{620}/A_{450} \ (CCl_4) = 1.70 \ (ACS \ specification: > 1.55)$$

Analytical Applications

Use as an Extraction Reagent

The extent of metal ions to be extracted with dithizone into organic phase can be estimated from $K'_{ex}[H_2L]_{org}$, and pH of aqueous phase. The separation of two or more metal ions is attained by selecting the proper pH value of the aqueous phase and the dithizone concentration in organic phase. The selectivity for metal ions can be much more improved by the combined use of auxiliary complexing agents (or masking agents). Some of typical examples are summarized in Table 3. The detailed procedures for the separations of each metal ions by dithizone extraction are described by Sandell[M2] and Iwantscheff.[M3]

Use as Photometric Reagent[M2]

The extraction photometry with dithizone can be carried out in three ways: monocolor photometry, mixed color photometry, and dual-wavelength photometry.

Table 3
SELECTIVITY IN DITHIZONE EXTRACTION[40,41]

Conditions	Metal ions extracted	Ref.
Alkaline region		
Alkaline solutions containing CN⁻	Bi, Pb Sn(II), Tl(I)	
Strongly alkaline solutions containing citrate or tartrate	Ag, Cd, Co, Cu, Ni, Tl	
Slightly alkaline solutions containing bis (2-hydroxyethyl) dithiocarbamate	Zn	42—44
Acidic region		
Slightly acidic solutions containing CN⁻	Ag, Cu, Hg(II), Pd(II)	
Dilute acidic solutions contaning Br⁻ or I⁻	Au, Cu, Pd	
Dilute acidic solutions containing CN⁻ or SCN⁻	Cu, Hg	
Slightly acidic solution (pH 5) containing $S_2O_3^{2-}$	Cd, Pd, Sn(II), Zn	
Slightly acidic solutions (pH 4∼5) containing $S_2O_3^{2-}$	Sn(II), Zn	
Dilute acidic solutions containing EDTA	Ag, Hg	45—47

Monocolor photometry — Sample solution, after adjustment of pH and addition of auxiliary complexing agents if necessary, is shaken with successive portions of a carbon tetrachloride or chloroform solution of dithizone until all of the metal has been extracted. In the final stage of extraction, the color of dithizone solution remains green. The combined extracts are then shaken with a dilute aqueous ammonia to remove excess dithizone, and the resulting solution is subjected to the photometry. Several sources of error are reported in this method. If the alkalinity of aqueous ammonia is too high, some of the metal dithizonate may be decomposed, resulting in a negative error. If the alkalinity is not high enough, the removal of exces dithizone is not complete, resulting in a positive error. Dithizone can be removed more easily from a carbon tetrachloride solution than from a chloroform solution. Beside these, there is a possibility of transforming of a primary complex into a secondary complex on washing with aqueous alkali. Thus, it is practically impossible to remove excess dithizone quantitatively without forming some of secondary complex, and the monocolor method is not recommended for the accurate determination.

Mixed-color photometry — In this method, the organic solution which contains the metal dithizonate and the excess dithizone is subjected to the photometry. Since the absorption peak of many metal dithizonates occurs roughly in a region where a minimum in the absorption curve of free dithizone is observed, the photometry is usually conducted at λ_{max} for the metal dithizonate to be determined, and the absorption due to the excess dithizone is compensated by using a reagent blank in the reference cell. Principal error in this method may be due to the possible loss of dithizone by oxidation during the analytical procedure, but can be reduced to minimal through the use of the reagent blank which is obtained by precisely duplicating the analytical procedure including extraction step.

Dual-wavelength photometry[48] — In the dual-wavelength spectrophotometer, two light beams of different wavelengths are time shared through a single cell, and the difference ΔA between the absorbances at wavelength λ_1 and λ_2 is measured. When λ_1 and λ_2 are chosen at 510 nm and 663.5 nm, respectively, on a mixture of mercury dithizonate and free dithizone, as illustrated in Figure 5,[52] ΔA may be proportional to the concentration of mercury dithizonate. Thus, in this case, the absorption due to free dithizone is automatically compensated, avoiding the use of reagent blank.

The detailed monocolor and mixed-color procedures for the spectrophotometric determination of traces of metal ions using dithizone are described in the monographs by Sandell[M2] or Iwantscheff.[M3]

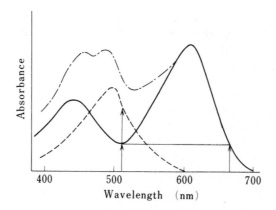

FIGURE 5. Absorption spectra of dithizone and mercury(II) dithizonate [Hg(HL)₂] in aqueous Triton® X-100 solution: ——— , dithizone; — — —, mercury(II) dithizonate [Hg(HL)₂]; — · — · —, their mixture. (From Ueno, K., Shiraishi, K., Togo, T., Yano, T., Yoshida, I., and Kobayashi, H., *Anal. Chim. Acta*, 105, 289, 1979. With permission.)

Notes on the preparation of reagent solution —

Dithizone is soluble in aqueous alkali, but the solution is so unstable that it can not be used for practical purposes. Solutions in carbon tetrachloride or chloroform are recommended, but the purity of the solvents has a great influence on the stability of dithizone solutions and on the absorption spectra of dithizone. Concentrations of dithizone are about 0.001% for the photometry and about 0.01% for the separation by solvent extraction.

Carbon tetrachloride solution is reported to be stable when the solution is overlaid with 10% of its volume of 0.1 N sulfurous acid and stored in a dark cool place.[49] In the case of chloroform solution, the use of well-purified solvent and storage in a dark cool place are more effective than the use of reductants.[49,50] Recently, a stabilized aqueous solution of metal dithizonate, $Zn(HL)_2$, for example, is recommended for the nonextraction dual-wavelength photometry of trace of metal ions, such as Cu(II), Hg(II), or Ag.[51] Metal dithizonate is solubilized with an aid of nonionic surfactant such as Triton X®-100, and dithizone is generated from zinc complex, *in situ*, by acidification of the solution to pH 1, where the freed Zn does not interfere the determination.[52] About 5 mg of zinc dithizonate is treated with a small portion of 20% Triton® X-100 solution in an agate mortar and made to 100 mℓ with the same solvent. The resulting solution is filtered through a membrane filter (pore size 0.45 μm) to obtain a transparent solution. It is kept in a dark cool place.

Dithizone is so sensitive to many common metals that extreme care has to be paid to minimize blanks by using metal free reagent, water, and clean glass-ware in a clean laboratory.

Other Reagents of Related Structure
Substituted Dithizones

The following substituted dithizones have been prepared and their physicochemical properties were investigated for use as an analytical reagent.

2,2′-(or *o,o′*-) dichlorodithizone[55]
4,4′- (or *p,p′*-)dichlorodithisone[56]
2,2′-dibromodithizone[57]
4,4′-dibromodithizone[57]
2,2′-diiododithizone[58]
4,4′-diiododithizone[58]
4,4′-difluorodithizone[59]
2,2′-dimethyldithizone[60]
4,4′-dimethyldithizone[61]
4,4′-disulfodithizone[62]
di (*o*- or *p*-diphenyl)thiocarbazone[63]

Di-α-naphthyl- (6) and Di-β-naphthylthiocarbazone (7)

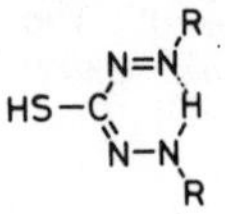

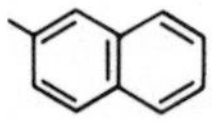

Naphthyl derivatives have higher distribution coefficients than dithizone for reagent alone and metal chelates. Wavelength of absorption maxima for reagent and metal chelates also shift to longer wavelengths than those of dithizone. Absorption spectra of (6) and (7) are shown in Figure 6. The spectral characteristics of the reagents and some metal chelates are summarized in Table 4. While (6) seems to have few advantages over dithizone, (7) may have a definite advantage when metal ions are extracted in the basic range, since the optimum pH ranges for the extraction of metals shift further to basic side with (7) than in the case of dithizone.[66,68] For example, Zn, and Bi can be extracted quantitatively at pH 8 to 10.5 and 12, respectively. Commercial samples of (7) are often impure and may be purified by column chromatography of a chloroform solution on an alumina.[69-71]

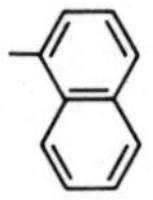

Substituted Thiosemicarbazones

A large numbers of thiosemicarbazones substituted with aromatic or heteroaromatic groups have been synthesized and evaluated as analytical reagents. Although not many fruitful results were obtained with these reagents, details are treated in a review.[R1]

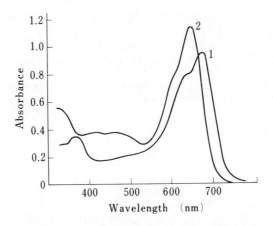

FIGURE 6. Absorption spectra of di(α- or β-naph-thyl)thiocarbazone. (1) Di(α-naphthyl)thiocarbazone, 3.14×10^{-5} M; (2) Di(β-naphthyl)thiocarbazone, 3.93×10^{-5} M. (Reprinted with permission from Math, K. S., Fernando, Q., and Freiser, H., *Anal. Chem.*, 36, 1762, 1964.)

Table 4
SPECTRAL CHARACTERISTICS OF DI-(NAPHTHYL)THIOCARBAZONES AND THEIR METAL CHELATE IN CCl$_4$[65]

Chelates	(6) λ_{max} (nm)	ε ($\times 10^3$)	K_{ex}	(7) λ_{max} (nm)	ε ($\times 10^3$)	K_{ex}	Ref.
Reagent	685	49.3	—	484	—	—	
	—	—	—	658	40	—	
				(650)	(67)		
Ag(HL)	—	—	—	—	—	—	
				(500)	(48)		
Bi(HL)$_3$	535	—	—	—	—	6×10^6	67
				(520)	(114)		
Cd(HL)$_2$	—	—	—	—	—	—	63
				(545)	(107)	(40)	
Cu(HL)$_2$	560	66.3	2×10^8	570	—	—	
				(570)	(78)		
Hg(HL)$_2$	525	51.5	1.4×10^{22}	512	—	—	
				(510)	(86)		
Pb(HL)$_2$	555	—	—	552	—	—	
				(545)	(80)		
Zn(HL)$_2$	560	—	—	568	—	3×10^4	67
				(560)	(121)		

Note: Values in parentheses are observed in CHCl$_3$ solution.[66]

MONOGRAPHS

M1. Irving, H. M. N. H., *Dithizone,* The Chemical Society, London, 1977.

M2. Sandell, E. B., *Colorimetric Determination of Traces of Metals,* 3rd ed., Interscience New York, 1959.

M3. Iwantscheff, G., *Das Dithizon und Seine Anwendung in der Mikro und Spurenanalyse,* 2nd ed., Verlag Chemie, Weinheim, West Germany, 1972.

REVIEW

R1. Singh, R. R., Gary, B. S., and Singh, R. P., Analytical application of thiosemicarbazones and semicarbazones, *Talanta,* 25, 619, 1978.

REFERENCES

1. Fischer, E., *Ann.,* 190, 118, 1878; Fischer, E, and Besthorn, E., *Ann.,* 212, 316, 1881.
2. Billman, J. H. and Cleland, E. S., *J. Am. Chem. Soc.,* 65, 1300, 1943.
3. Fischer, H., *Angew. Chem.,* 42, 1025, 1929.
4. Fisher, H., *Wiss. Veroff. Siemens-Werken,* 4, 158, 1925.
5. Honjo, T., Imura, H., Shima, S., and Kiba, T., *Anal. Chem.,* 50, 1545, 1978.
6. Fischer, H. and Weyl, W., *Wiss Veroff. Siemens-Werken,* 15, 41, 1935.
7. Budesinsky, B. W. and Sagat, H., *Talanta,* 20, 228, 1973.
8. Schweitzer, K. G. and Dyer, F. F., *Anal. Chim. Acta,* 22, 172, 1960.
9. Dyer, F. F. and Schweitzer, K. G., *Anal. Chim. Acta,* 23, 1, 1960.
10. Koroleff, F., Merentutkimuslaitoksen Julkaisu Havsforknings, Institutes Skrift, Helsinki, 145, 5, 1950.
11. Tremillon, B., *Bull. Soc. Chim. Fr. Mem.,* 1156, 1160, 1954.
12. Pilipenko, A. T., *Zh. Anal. Khim.,* 8, 286, 1953.
13. Stary, J., *The Solvent Extraction of Metal Chelates,* Pergamon Press, Oxford, 1964.
14. Busev, A. I. and Bazhanova, L. A., *Russ. J. Inorg. Chem.,* 6, 1128, 1961.
15. Bidleman, T. F., *Anal. Chim. Acta,* 56, 221, 1971.
16. Babko, A. K. and Pilipenko, A. T., *Zh. Anal. Khim.,* 2, 33, 1947.
17. Geiger, R. W. and Sandell, E. B., *Anal. Chim. Acta,* 8, 197, 1953.
18. Marzenko, Z. and Mojski, M., *Chim. Anal. (Paris),* 53, 529, 1971.
19. Pierce, T. B. and Peck, P. F., *Anal. Chim. Acta,* 27, 557, 1962.
20. Breant, M., *Bull. Soc. Chim. Fr. Mem.,* 948, 1956.
21. Kato, T., Takei, S., and Okagami, A., *Bunseki Kagaku,* 5, 689, 1956.
22. May, I. and Hoffman, J. I., *J. Wash. Acad. Sci.,* 38, 329, 1948.
23. Schweitzer, G. K. and Coe, G. R., *Anal. Chim. Acta,* 24, 311, 1961.
24. Freiser, B. S. and Freiser, H., *Talanta,* 17, 540, 1970.
25. Vouk, V. B. and Weber, O. A., *Analyst,* 85, 46, 1960.
26. Minczewski, J., Krasiejko, M., and Marczenko, Z., *Chem. Anal. (Warsaw),* 15, 43, 1970.
27. Pilipenko, A. T., *Zh. Anal. Khim.,* 5, 14, 1950.
28. Schweitzer, G. K. and Davidson, J. E., *Anal. Chim. Acta,* 35, 467, 1966.
29. Schweitzer, G. K. and Honaker, C. B., *Anal. Chim. Acta,* 19, 224, 1958.
30. Irving, H. N. M. H., Bell, C., and Williams, R. J. P., *J. Chem. Soc.,* 356, 1952.
31. Kolthoff, I. M. and Sandell, E. B., *J. Am. Chem. Soc.,* 63, 1906, 1941.
32. Grummitt, O. and Stickle, R., *Ind. Eng. Chem. Anal. Ed.,* 14, 953, 1942.
33. Kägl, J. H. R. and Vallee, B. L., *Anal. Chem.,* 30, 1951, 1958.
34. Cooper, S. S. and Sullivan, M. L., *Anal. Chem.,* 23, 613, 1951.
35. Weber, O. and Vouk, V. B., *Analyst,* 85, 40, 1960.
36. King, K. G. C. and Pruden, G., *Analyst,* 96, 146, 1971.

37. Irving, H. M. N. H., Kiwan, A. M., Rupainwar, D. C., and Sahota, S. S., *Anal. Chim. Acta,* 56, 205, 1971.
38. *Reagent Chemicals, American Chemical Society Specifications,* 4th Ed., 1968.
39. AnalaR Standards for Laboratory Chemicals, 7th Ed., AnalaR Standards Ltd., London, 1977.
40. Fischer, H., *Agnew. Chem.* 46, 442, 517, 1933; *Angew. Chem.,* 47, 685, 1934; *Angew. Chem.,* 50, 919, 1937.
41. Fischer, H. and Leopoldi, G., *Fresenius, Z. Anal. Chem.,* 107, 241, 1936.
42. Serfass, E. J. and Levine, W. S., *Chemist — Analyst,* 36, 55, 1947.
43. Kato, T. and Takei, S., *Bunseki Kagaku,* 2, 208, 1953.
44. Margerum, D. W. and Santacana, F., *Anal. Chim. Acta,* 23, 1, 1960.
45. Fischer, H., *Chemist Analyst,* 50, 62, 1961.
46. Suzuki, N., *Bunseki Kagaku,* 8, 283, 349, 1959.
47. Erdey, L., Rady, G., and Fleps, V., *Acta Chim. Acad. Sci. Hung.,* 5, 133, 1954.
48. Shibata, S., *Angew. Chem. Int. Ed. Engl.,* 15, 673, 1976.
49. Biddle, D. A., *Ind. Eng. Chem. Anal. Ed.,* 8, 99, 1936.
50. Clifford, P. A., *J. Assoc. Offic. Agric. Chem.,* 21, 695, 1938.
51. Watanabe, H. and Miura, J., *Bunseki Kagaku,* 26, 196, 1957.
52. Ueno, K., Shiraishi, K. Togo, T., Yano, T., Yoshida, I., and Kobayashi, H., *Anal. Chim Acta,* 105, 289, 1979.
53. Marczenko, Z. and Krasiejko, M., *Chem. Anal. (Warsaw),* 17, 1201, 1972; *Anal. Abstr.,* 24, 3347, 1973; *Chem. Abstr.,* 79, 13119b, 1973.
54. Hoshi, S., Yotsuyanagi, T., and Aomura, K., *Nippon Kagaku Zasshi,* 619, 1979.
55. Kiwan, A. M. and Kassim, A. Y., *Talanta,* 22, 931, 1975.
56. Busev, A. I. and Bazhanova, L. A., *Zh. Neorg. Khim.* 6, 2805, 1961; *Chem. Abstr.,* 56, 14905g, 1962.
57. Takei, S., *Bunseki Kagaku,* 9, 402, 1960.
58. Takei, S., *Bunseki Kagaku,* 10, 715, 1961.
59. Salihy, A. R. Al. and Freiser, H., *Talanta,* 17, 182, 1970.
60. Takei, S., *Bunseki Kagaku,* 6, 630, 1957; *Bunseki Kagaku,* 10, 708, 1961.
61. Takei, S. and Shibuya, K., *Bunseki Kagaku,* 5, 695, 1956.
62. Arendaryuk, E. N. and Pilipenko, A. T., *Ukr. Khim. Zh.,* 44, 1096, 1978; *Anal. Abstr.,* 36, 6B29, 1979.
63. Takei, S., *Bunseki Kagaku,* 9, 409, 1960.
64. Math, K. S., Fernando, Q., and Freiser, H., *Anal. Chem.,* 36, 1762, 1964.
65. Takei, S., *Bunseki Kagaku,* 9, 288, 294, 1960.
66. DuBois, R. J. and Knight, S. B., *Anal Chem.,* 36, 1316, 1964.
67. Grzhegorshevskii, A. S., *Zh. Anal. Khim.,* 11, 689, 1956; *J. Anal. Chem. USSR,* 11, 737, 1956.
68. Cholak, J., Hubbard, D. M., and Burkey, R. E., *Ind. Eng. Chem. Anal. Ed.,* 15, 754, 1943.
69. Hubbard, D. M. and Scott, E. W., *J. Am. Chem. Soc.,* 65, 2390, 1943.
70. Cooper, S. S. and Kofron, V. K., *Anal. Chem.,* 21, 1135, 1949.
71. DuBois, R. J. and Knight, S. B., *Anal Chem.,* 36, 1313, 1964.

THIOXINE

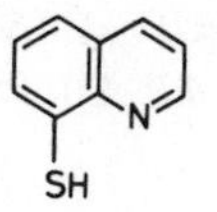

$C_9H_7NS \cdot 2H_2O$
mol wt = 197.25

HL

Synonyms
8-Mercaptoquinoline, 8-Quinolinethiol

Source and Method of Synthesis
Commercially available as dihydrate or hydrochloride. It is prepared by the reduction of quinolyl-8-sulfonylchloride with $SnCl_2$[1] or the reaction of diazotized 8-amino-quinoline with thiourea[2]. As thioxine itself is rather unstable to air oxidation, diquinolyl-8,8'-disulfide, an oxidation product of thioxine, is conveniently used as its precursor. It is stable and commercially available. Thioxine is easily prepared by reducing the disulfide with H_3PO_2[3].

Analytical Uses
Thioxine is a sulfur analogue of 8-quinolinol (Oxine) and behaves as a S,N-donating ligand to form metal chelates with a preference to soft metal ions, such as Ag, As, Au, Bi, Cd, Co, Cu, Fe, Ga, Hg, In, Ir, Mn, Mo, Ni, Os, Pb, Pd, Pt, Re, Rh, Ru, Sb, Se, Sn, Te, Tl, V, W, and Zn.[R1,4] Thioxine has been used as photometric, fluorimetric, and solvent extraction reagents in the analysis of traces of these elements.

Properties of Reagent
Although thioxine itself is a dark blue oil, the dihydrate is dark red needles, mp 58 to 59°, and sublimes at 99 to 177° (2×10^{-2} Torr).[55] It is easily oxidized in air to give diquinolyl-8,8'-disulfide. The disulfide can also be prepared by oxidizing thioxine with H_2O_2. It is a white powder, mp 202 to 204°[6] and is very stable.

Its hydrochloride ($HL \cdot HCl$) and sodium salt ($NaL \cdot 2H_2O$) are yellow powder and should be kept under nitrogen in a dark place, although they are more stable than a free reagent.

Thioxine is slightly soluble in water (0.67 g/ℓ at 20°, pH 5.2),[5] but easily soluble in various organic solvents, such as ethanol (12.5 g/100 mℓ), chloroform, acetone, pyridine, benzene, and toluene. Its aqueous solution is orange-yellow in neutral range, but turns to light yellow in acidic or alkaline range. Its pyridine, quinoline, and ether solutions are blue, while benzene and toluene solutions are brown.

Its hydrochloride or sodium salt[8] is easily soluble in water, aqueous hydrochloric acid, and polar organic solvents, but insoluble in nonpolar solvents.

Quinolyl-8,8'-disulfide is insoluble in water or aqueous alkali, but easily soluble in aqueous mineral acid.[6]

Acid dissociation of thioxine takes place as follows:

$$H_2L^+ \xrightleftharpoons{K_{a1}} H^+ + [HL \quad \text{and} \quad H^+L^-]$$

$$[HL \text{ and } H^+L^-] \xrightleftharpoons{K_{a2}} H^+ + L^-$$

Table 1
ACID DISSOCIATION CONSTANTS OF THIOXINE AND ITS ANALOGUES IN WATER AND 50% AQUEOUS DIOXANE

Reagent	In water[6,7]			In 50% aqueous dioxane[7]	
	K_t	pKa_1	pKa_2	pKa_1	pKa_2
8-Quinolinol	0.04	5.13	9.74	3.97	11.54
Thioxine (8-quinolinethiol)	3.8	2.0	8.36	1.74	9.20
8-Quinolineselenol	1740	−0.08	8.18	0.12	8.50

where HL and H^+L^- denote the neutral (1) and zwitterionic forms (2) of the reagent.

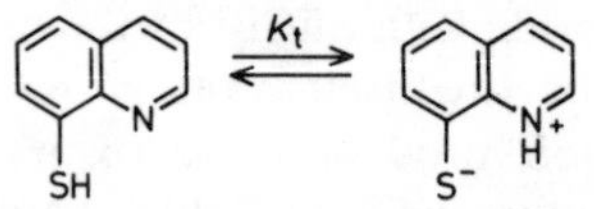

(1)   (2)

Both forms exist as a tautomeric mixture, and the tautomeric constant can be expressed as follows.

$$K_t = \frac{[\text{Zwitterion}]}{[\text{Neutral}]}$$

Observed values on the acid dissociation and the tautomeric constants in water and aqueous dioxane are summarized in Table 1, along with those of 8-quinolinol and 8-quinoline-selenol for comparison.

Absorption spectra of thioxine (H_2L^+, HL, and L^-) in water are illustrated in Figure 1, and the spectral characteristics of each species are summarized in Table 2. The apparent molar absorption coefficient is known to increase when thioxine is oxidized in air.[3]

Thioxine behaves as a solvatochromic dye, changing its spectra with the polarity of solvents because the equilibrium composition of "neutral" and "zwitterion" forms changes with the solvent polarity. Examples are shown in Figures 2 and 3.

Thioxine can be quantitatively extracted from aqueous phase into chloroform in a pH range of 2 to 8.4.[5] The distribution ratio of HL at pH 6.2 ($\mu = 0.15$, 20°) in a system of organic solvent/water are 324 ($CHCl_3$), 250 (BrC_6H_5), 160 (C_6H_6), 81 (CCl_4), and 10.4 (iso-octane).[5]

Complexation Reaction and Properties of Complexes

Thioxine reacts preferentially with soft metal ions, forming insoluble precipitates or colored solutions. The composition of chelates depends on the nature of acid, acidity of the solution, and the presence of anions, as summarized in Table 3.

Stability constants of thioxine chelates in 50% aqueous dioxane are summarized in Table 4. The formation constants of thioxine chelates in the system of water-chloroform:

$$(M^{n+})_{aq} + n(L^-)_{aq} \xrightleftharpoons{\quad K'_{ML_n} \quad} (ML_n)_{org}$$

are also included in the table.

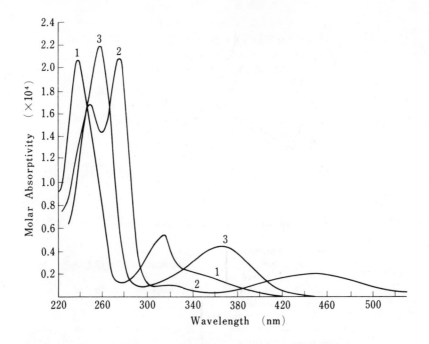

FIGURE 1. Absorption spectra of thioxine in water. (1) pH 0 (H₂L⁺); (2) pH 5.2 (HL); (3) pH 13 (L⁻). (From Bankovsky, Yu. A., Chera, L. M., and Ievinsh, A. F., *Zh. Anal. Khim.*, 18, 668, 1963. With permission.)

Table 2
SPECTRAL CHARACTERISTICS
OF THIOXINE

Species	pH	λ_{max} (nm)	ε (× 10³)
H₂L⁺	0 ~ 2	243	16.57
		316	4.28
HL	5.2	252	14.4
		278	17.97
		316	0.82
		446	1.6
L⁻	12 ~ 14	260	19.29
		368	3.25

Reprinted with permission from Anderson, P. D. and Hercules, D. M., *Anal. Chem.*, 38, 1702, 1966. Copyright 1966 American Chemical Society.

The reactions of thioxine with metal ions in th presence of various masking agents have been investigated in detail.[23] The most effective masking agents against thioxine are concentrated hydrochloric acid (for Ag, Bi, Fe, Hg, Mo, Sb, and Sn), thiourea (for Ag, Au, Cu, Hg, Os, Pt, and Ru), sodium fluoride (for Fe(III) and Sn(IV)), and alkaline potassium cyanide (for Ag, Au, Co, Fe(II), Ir, Ni, Os, Pd, Pt, and Ru). Potassium thiocyanate is a good masking agent for Fe(III) and a moderate amount of Cd and Zn.

Figure 4 shows the elements that can be extracted with thioxine into chloroform.[R1]

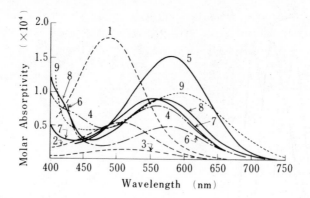

FIGURE 2.   Absorption spectra of thioxine in various organic solvents

(1) — — — —, in methanol ⎫   dye
(2) - - - - -, in ethanol   ⎬   concentration 1.7×10^{-2} *M*
(3) — — — —, in octanol   ⎭   dye
(4) — · — · — ·, in DMF   concentration 3.4×10^{-2} *M*
(5) —————, in pyridine   ⎫
(6) ———————, in dichloroethane ⎱   dye
(7) —————, in chloroform   ⎰ concentration 2.5×10^{-1} *M*
(8) —————, in acetone   ⎭
(9) · · · · · ·, in benzene   conc 3.0 *M*
(From Bankovsky, Yu. A., Cherra, L. M., and Ievinsh, A. F., *Zh. Anal. Khim.*, 18, 668, 1963. With permission.)

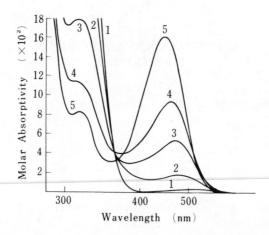

FIGURE 3.   Absorption spectra of thioxine in water-ethanol mixture. (1) 100% ethanol; (2) 80% ethanol — 20% water; (3) 50% ethanol — 50% water; (4) 30% ethanol — 70% water; (5) 100% water. (Reprinted with permission from Anderson, P. D. and Hercules, D. M., *Anal. Chem.*, 38, 1702, 1966. Copyright 1966 American Chemical Society.)

The pH dependencies of percent extraction for some representative metal ions in a water-chloroform system are illustrated in Figure 5. The values of log D_{ML_2} (CHCl$_3$/H$_2$O) at optimum pH are reported to be 4.11 for FeL$_2$, 6.15 for HgL$_2$, 2.78 for PbL$_2$, and 5.92 for ZnL$_2$.[24,25] The distribution ratio may be improved by the combined used of auxiliary complexing agent, such as pyridine or 1,10-phenanthroline.[26] The solubility data of some thioxine chelates in water and chloroform are summarized in Table 5.

Table 3
OPTIMUM CONDITIONS FOR THE PRECIPITATION AND EXTRACTION OF METAL IONS WITH THIOXINE[R1,9]

Metal ion[a]	Precipitation			Extraction		
	Optimum condition	Composition	Color (sublimation temp)[b]	Optimum condition	Solvent	Ref.
As(III)	3 NHCl	$(H_2L)_3MCl_3 \cdot 3H_2O$	Yellow	<3.5 NH_2SO_4	CCl_4	10
As(V)	1 NHCl	NL_5	Yellow	Soluble in conc HCl	—	
Au(III)	HCl	$(H_2L)_3MCl_6 \cdot H_2O$	Yellow	Strong acid	$CHCl_3$	10
Bi	Tartaric acid	ML_3	Yellow	pH 7 ∼ 11.5	$CHCl_3$	
Cd	pH 1 ∼ 11 (NH₃)	ML_2	Orange-yellow	pH >4	$CHCl_3$	11, 12
Cu(II)	4 NHCl∼ 3 NNaOH	$ML_2 \cdot 1/2H_2O$	Brown (173 ∼ 232)	>2.5 NHCl	$CHCl_3$	11, 31
Hg(I)	Acid∼neutral	$ML \cdot HL \cdot 2H_2O$	Red-brown	Strong acid	$CHCl_3$	10
Hg(II)	3 NHCl	ML_2	Yellow (77—180)	—	—	
Ir(III)	—	HL_3	Rose	pH 7.6 ∼ 9	$CHCl_3$	40
Mo(VI)	pH 2 ∼ 3	$Mo_2L_2 \cdot H_2O$	Black-green (186∼257)	Strong acid	$CHCl_3$	
Os(III)	—	HL_3	Blue-violet	pH — 7.5	$CHCl_3$ or DMF	40
Pb	pH >2.6	ML_2	Yellow	pH >4.5	$CHCl_3$ or toluene	11
Pb	pH >2.3(HCl)	MLC1	Dull yellow	—	—	
Pd(II)	1 NHCl	$ML_2 \cdot H_2O$	Orange-red (218 ∼ 249)	4 NHCl∼ pH 11	$CHCl_3$ or DMF	11, 14, 16, 40
Pt(II)	pH 10	ML_2	Brown	pH >1(KI)	$CHCl_3$ or toluene	11, 17, 40
Sb(III)	Tartarate ML_3	$ML_3 \cdot H_2O$	Yellow	pH 5 ∼ 9	$CHCl_3$	
Tl(I)	Weak acid ∼ alkali	ML	—	pH 9 ∼ 14	Toluene	19
V(IV)	Acid	MOL_2	Green	—	$CHCl_3$	R1
W(VI)	5 NHCl∼pH 5	MO_2L_2	Dark grey	5 NHCl∼pH 4	$CHCl_3$	43
Zn	pH 4.5	ML_2	Yellow-orange (144 — 190)	pH >1		11

[a] Other metal ions precipitatel with thioxine include Ag,[10,11] Co(II) (III),[11,55] Cu(I), Fe(II) (III),[55] Ga,[19] In,[19] Mn(II),[11] Ni,[14,15] Rh, Ru,[40] Sb(V),[10,18] Se(IV),[44] Sn(IV)(II), Te(IV),[44] and V(V).

[b] Sublimation temperature at 2×10^{-2} Torr.[55]

Most of the metal-thioxine chelates show single absorption band in the visible range which is useful for the extraction photometry of traces of metal ions. The spectral characteristics of metal thioxinates are summarized in Table 6. Figure 6 illustrates a typical absorption curve of the thioxine chelate.[19] Cadmium and zinc thioxine chelates in chloroform show fluorescence at 535 nm.[3,34]

Purification and Purity of Reagent

As thioxine (free base) is easily oxidized in air (less easily with the hydrochloride and sodium salt), the commercial samples are often partially oxidized to be contaminated with diquinolyl-8,8′-disulfide which is insoluble in water. Although the supernatant liquid of the aqueous solution is often satisfactory for the general purposes, it can be purified by recrystallization from appropriate solvents. However, the best way to obtain pure thioxine is to prepare it from diquinolyl-8,8′-disulfide.[6]

Dissolve 5.5 g disulfide in a mixture of 20 m*l* of concentrated hydrochloric acid and 6 m*l* of 50% hypophosphorous acid and heat the solution under

Table 4
STABILITY CONSTANTS OF THIOXINE CHELATES

Metal ion	log K_{ML}[20,21] (50 v/v % water-dioxane, 27°, $\mu = 0.1$)	log K'_{ML}[22] (Water-CHCl₃ 20 ± 0.5°)	μ and acidity	Composition
Bi	—	45.6	2 NHClO₄	BiL₃
Cd	10.79	—	—	—
Co(II)	—	28.3	0.1	CoL₂(HL)
Cu(II)	12 ~ 14	—	—	—
Fe(III)	—	39.6	0.1	FeL₃
Ga	—	26.3	0.1	GaL₂⁺
Hg(II)	—	47.5	3 NHCl	HgL₂
In	—	41.3	0.1	InL₃
Mn(II)	6.74	28.15	0.1	MnL₂(2HL)
Mo(VI)	—	30.25	1.5 NHCl	MoO₂L₂
Ni	10.95	25.25	0.1	NiL₂
Pb	11.52 ~ 11.85	21.3	0.1	PbL₂
Sb(IV)	—	31.6	5 NHClO₄	SbL₃⁺
Tl(I)	—	6.5	0.1	TlL
V(IV)	—	23.6	0.1	VOL₂
Zn	10.89 ~ 11.05	25.65	0.1	ZnL₂

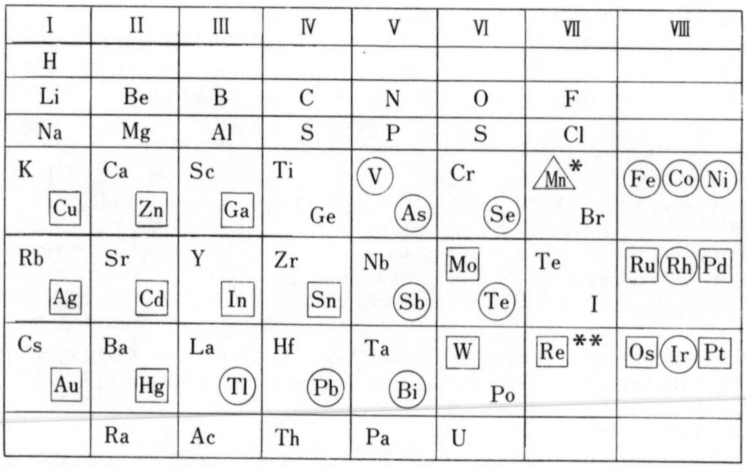

I	II	III	IV	V	VI	VII	VIII
H							
Li	Be	B	C	N	O	F	
Na	Mg	Al	S	P	S	Cl	
K	Ca	Sc	Ti	(V)	Cr	Mn*	Fe Co Ni
Cu	Zn	Ga	Ge	(As)	(Se)	Br	
Rb	Sr	Y	Zr	Nb	Mo	Te	Ru Rh Pd
Ag	Cd	In	Sn	(Sb)	(Te)	I	
Cs	Ba	La	Hf	Ta	W	Re**	Os Ir Pt
Au	Hg	(Tl)	(Pb)	(Bi)	Po		
	Ra	Ac	Th	Pa	U		

FIGURE 4.   Elements that can be extracted with thioxine. □, Extracted from strong acid solutions; O, extracted from weakly acid and neutral solutions; △*, extracted from alkaline solution;[11] **, taken from Ref. 28. (From Kuznetsov, V. I., Bankovsky, Yu. A., and Ievinsh, A. F., *Zh. Anal. Khim.*, 13, 267, 1958. With permission.)

reflux in nitrogen for about 2 hr. Cool and filter the solution and add dropwise excess of saturated sodium hydroxide solution while passing a stream of nitrogen over the solution. Filter yellow precipitates of sodium salt, recrystallize a few times from alcohol-ether, and dry in vacuo.

The sodium salt prepared in this way contains one to two molecules of water of crystallization and is stable for several months under nitrogen.

The purity of thioxine (free base, hydrochloride, or sodium salt) can be assayed by the iodometry of SH group according to the standard procedure or more conveniently by TLC method with CCl₄-iso-propyl alcohol (50:3) as a developing solvent.[41] The purity of diquinolyl-8,8'-disulfide can be checked by observing the melting point.

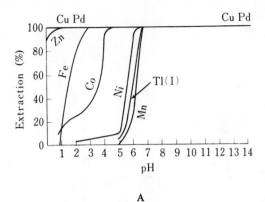

A

B

FIGURE 5. pH Dependence of percent extraction of thioxine chelates in water-chloroform system. (From Kuznetsov, V. I., Bankovsky, Yu. A., and Ievinsh, A. F., *Zh. Anal. Khim.*, 13, 267, 1958. With permission.)

Table 5
SOLUBILITY PRODUCT AND SOLUBILITY OF METAL THIOXINATES[29]

Thioxine chelate	log K_{SP} (μ = 0.25 (NaCl), (20°)	Solubility in CHCl₃ (20°)	Ref.
AgL	—	2.6×10^{-4}	56
BiL₃	−46.31	5.8×10^{-5}	29
CoL₂	−29.60	—	29
FeL₂	—	3.1×10^{-3}	24
		1.7×10^{-2}	29
HgL₂	—	1.4×10^{-1}	24
MnL₂	−15.94	1.1×10^{-3}	29
NiL₂	—	1.2×10^{-4}	29
PbL₂	−26.02	—	29
VOL₂	−25.10	—	29
ZnL₂	−29.44	7.1×10^{-2}	24

Analytical Applications
Use as Extraction Photometric Reagent

Thioxine is used as an extraction photometric reagent for various soft metal ions. The selectivity can be improved by the selection of suitable reaction condition (nature of acids, acidity, auxiliary complexing agent, and masking agent). Some examples are

Table 6
APPLICATION OF THIOXINE AS EXTRACTION PHOTOMETRIC REAGENT

Metal ion[a]	Condition	Metal chelate Ratio	λ_{max} (nm)	ε ($\times 10^3$)	Extraction solvent	Range of determination (ppm)	Interference	Ref.
As(III)	<3.5 NH_2SO_4, KI	MIL_2	380	1.9	CCl_4	~3	ML_3 in large excess of reagent. Al, Hg, Ti give positive error; Ag, Pb give negative	27
Bi	pH 3~13	ML_3	394	17.9	$CHCl_3$	~15	Cd, Co, Cu, Fe, Ni	30
Cd	Pyridine or 1,10-phen.	ML_2	403	8.0	$CHCl_3$	~9	Co, Cu, Fe, and Ni can be masked with KCN	12
Co(III)	pH 4~10, 1,10-phen.	ML_3	465	1.16	$CHCl_3$	~13	Cu, Fe and Ni can be stripped with 4NHCl	32, 45
G(III)	pH 3.5~7	ML_3	450	1.24	$CHCl_3$	0.1~6	Ag, Au, Cd, Co, Cu, Fe, Hg, Mn, Ni, Tl, and V do not interfere	53
Cu(II)	>2.5 NHCl	ML_2	431	7.53	$CHCl_3$	0.6~6	Co, Ni, Sn(II), Sb(III)	31
Ga	pH 6.5~10	ML_3	397	8.4	Toluene	—	—	19
In	pH 5~13	ML_3	407	11.1	Toluene	0.01~1	Bi, Ga, Hg, Mn, Pb, Sb, Sn, Tl, Zn	19
Ir(III)	pH 7.6~9	ML_3	485	9.95	$CHCl_3$	—	—	40
Mo(VI)	2.5 NHCl	MO_2L_2	420	4.5	Toluene	2~15	W(VI) can be masked with H_3PO_4, ascorbic acid	50
Os(III)	pH 4~7.5	ML_3	557	5.8	$CHCl_3$	—	—	40
Pb	pH 5.5~9.5	ML_2	412	11.0	$CHCl_3$	~20	Bi, Ga, Mo, V	25
Pd(II)	<6 NHCl, thiourea	ML_2	505	6.8	$CHCl_3$	~3.3	Cu, Ir, Mo(IV), Rh, Ru, Sn(IV)	16
Pt(II)	pH 2.5~5.0	ML_2	567	7.6	$CHCl_3$	—	—	40
Re	5~11 NHCl	—	438	8.47	$CHCl_3$	1~40	Mo(5000-fold), W(500-fold) do not interfere	28
Sb(III)	pH 7~10	—	382	16.0	$CHCl_3$	~6	Cd, Co, Cu, Fe, Ni	30

[a]   Other metal ions determined with thioxine are Be, Fe(III),[32] Ga, In,[19] Mn(II),[33] Ni,[15] Rh, Ru,[40] Tl(I),[19] V(IV),[19] W(VI),[51] and Zn.[54]

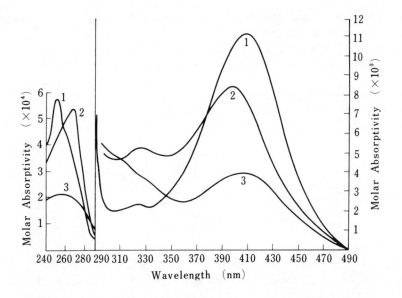

FIGURE 6. Absorption spectra of thioxine chelates in chloroform. (1) GaL₃; (2) InL₃; (3) TlL. (From Bankovsky, Yu. A., Tsirule, Yu. A., and Ievinsh, A. F., *Zh. Anal. Khim.*, 16, 562, 1961. With permission.)

Table 7
APPLICATION OF THIOXINE AS A FLUORIMETRIC REAGENT

Metal ion	Condition	λ_{ex} (nm)	λ_{fl} (nm)	Solvent	Reagent of determination (ppm)	Ref.
Cd	pH 4.5, ascorbic acid	365	515	CHCl₃	0.05 ∼ 1.5	3,12,26
In	pH 2 ∼ 4	365	515	CHCl₃ or MIBK	0.05 ∼ 2.5, Au, Cd, Co, Cu, Ga, Ni, Pd, Sb, V and Zn interfere; elements that can be masked are Ag (thiourea), Fe (ascorbic acid), and Hg(KI)	35
Zn	pH 2.5	365	535	Water-DMF-ethanol (1:1:1)	∼0.01	3, 34

summarized in Table 6. Suitable masking agents were discussed previously. A 0.2% aqueous solution of thioxine hydrochloride (0.1 g in 50 mℓ of 6 N HCl) is recommended, as it is stable for 1 month in a dark, cool place. A methanol solution of thioxine is easily oxidized to disulfide even in a dark cool place and has to be prepared daily.

Use as Fluorimetric Reagent

Some of metal thioxinates fluoresce in organic solvents, and such metals can be determined by fluorimetry. Examples are summarized in Table 7. As the fluorescence intensity is strongly dependent upon the nature of solvent used, the proper selection of solvent is important.[12]

Other Uses

Thioxine has been used as a gravimetric reagent or a titrant in amperometry.[42] Filterpaper impregnated with thioxine was recommended as a stationary phase for precipitation chromatography of metal ions.[36]

Diquinolyl-8,8′-disulfide can be used as a chromogenic reagent for metal if a reducing agent such as hydroxylamine or ascorbic acid is added simultaneously.[37]

Other Reagents with Related Structure

Selenoxine (3), a selenium analogue of thioxine, has been synthesized, and physicochemical properties and its analytical applications were investigated.[R3,38]

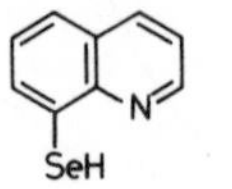

$C_{10}H_7NSe \cdot H_2O$
mol wt = 226.14

(3)

Althiox (4) is a neutral ligand derived from thioxine and is highly specific for Rh (400 nm, $\varepsilon = 7 \times 10^3$), through a bond formation with olefinic group. The coordination structure of Rh chelate (5) can be shown as below.[39,46,47,49] A sulfonated derivative is also used for the same purpose.[48,50]

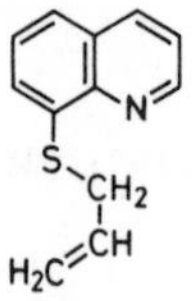

$C_{12}H_{11}NS$
mol wt = 201.24

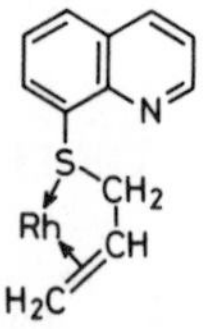

(4)   (5)

REVIEWS

R1. Kuznetsov, V. I., Bankovsky, Yu. A., and Ievinsh, A. F., The analytical application of 8-mercaptoquinoline(thiooxine) and its derivatives, *Zh. Anal. Khim.*, 13, 267, 1958; *J. Anal. Chem. USSR*, 13, 299, 1958.

R2. Bankovsky, Yu. A., Chera, L. M., and Ievinsh, A. F., Study of mercaptoquinoline (thiooxine) and its derivatives, *Zh. Anal. Khim.*, 18, 668, 1963; *J. Anal. Chem. USSR*, 18, 577, 1963.

R3. Honjo, T. and Kiba, T., Solvent extraction reagent, *Bunseki Kagaku*, 21, 676, 1972.

REFERENCES

1. Edinger, A., *Ber.*, 41, 937, 1908.
2. Kealey, D. and Freiser, H., *Talanta*, 13, 1381, 1966.
3. Anderson, P. D. and Hercules, D. M., *Anal. Chem.*, 38, 1702, 1966.
4. Bankovskis, J., Zaruma, D. E., and Mrasovska, M. E., *Izv. Akad. Nauk. Latv. SSR, Ser. Kim.*, 430, 1972; *Anal. Abstr.*, 25, 2946, 1973.
5. Bankovsky, Yu. A., Chera, L. M., and Ievinsh, A. F., *Zh. Anal. Khim.*, 18, 555, 1963.
6. Nakamura, N. and Sekido E., *Talanta*, 17, 515, 1970.

7. Sekido, E., Fernando, Q., and Freiser, H., *Anal. Chem.*, 36, 1768, 1964.
8. Mido, Y. and Sekido, E., *Bull. Chem. Soc. Jpn.*, 44, 2127, 1971.
9. Sekido, E., Fujiwara, I., and Masuda, Y., *Talanta*, 19, 479, 1972.
10. Akimov, V. K., Busev, A. I., Zaitsev, B. E., and Bragina, S. I., *Zh. Obshch. Khim.*, 40, 1331, 1970; *Chem. Abstr.*, 74, 58854a, 1971.
11. Mido, Y. and Sekido, E., *Bull. Chem. Soc. Jpn.*, 44, 2130, 1971.
12. Watanabe, K. and Kawagaki, K., *Bunseki Kagauku*, 23, 1356, 1974.
13. Golubtsova, B. B., *Zh. Anal. Khim.*, 14, 493, 1959.
14. Daiziel, J. A. W., *Analyst*, 89, 411, 1964.
15. Watanabe, K. and Kawagaki, K., *Bunseki Kagaku*, 23, 510, 1974.
16. Kammori, O., Taguchi, I., Takahashi, K., and Koike, T., *Bunseki Kagaku*, 14, 702, 1965.
17. Rudenko, N. P. and Rordyukevich, V. O., *Radiochem. Radioanal. Lett.*, 13, 263, 1973.
18. Yoshimura, C., Noguchi, H., and Hara, H., *Bunseki Karaku*, 13, 1249, 1964.
19. Bankovsky, Yu. A., Tsirule, Ya. A., and Ievinsh, A. F., *Zh. Anal. Khim.*, 16, 562, 1961; *J. Anal. Chem. USSR*, 16, 556, 1961.
20. Corsini, A., Fernando, Q., and Freiser, H., *Anal. Chem.*, 35, 1424, 1963.
21. Sekido, E., Fernando, Q., and Freiser, H., *Anal. Chem.*, 37, 1556, 1965.
22. Bankovsky, Yu. A., Chera, L. M., and Ievinsh, A F., *Zh. Anal. Khim.*, 23, 1284, 1968.
23. Bankovsky, Yu. A., Ievinsh, A. F., and Liepina, Z. E., *Zh. Anal. Khim.*, 15, 4, 1960; *J. Anal. Chem. USSR*, 15, 1, 1960.
24. Veveris, O. E., Bankovksy, Yu. A., and Pelekis, L. L., *J. Radioanal. Chem.*, 9, 47, 1971.
25. Watanabe, K. and Kawagaki, K., *Bunseki Kagaku*, 25, 246, 1976.
26. Kawagaki, K., Watanabe, K., and Yoshida, T., *Bunseki Kiki (Tokyo)*, 14, 530, 1976.
27. Stara, V. and Stary, J., *Talanta*, 17, 341, 1970.
28. Bankovsky, Yu. A., Ievinsh, A. F., and Luksha, E. A., *Zh. Anal. Khim.*, 14, 714, 1959.
29. Kharkover, M. Z., Bankovsky, V. F., Vdovina, V. M., and Gurova, L. P., *Zh. Anal. Khim.*, 25, 30, 1970.
30. Watanabe, K., Horikawa, S., and Kawaguchi, K., *Bunseki Kagaku*, 26, 570, 1977.
31. Bankovsky, Yu. A. and Ievinsh, A. F., *Zh. Anal. Khim.*, 13, 643, 1958.
32. Ganopolsky, V. I., Sharapova, V. S., Kharkover, M. Z., and Sheleg, M. U., *Zh. Anal. Khim.*, 25, 982, 1970.
33. Bankovsky, Yu. A., Ievinsh, A. F., and Luksha, E. A., *Zh. Anal. Khim.*, 14, 222, 1959.
34. Shibata, M. and Kakiyama, H., *Bunseki Kagaku*, 26, 640, 1977.
35. Fujiwara, A. and Kawagaki, K., *Bull. Chem. Soc. Jpn.*, 50, 1460, 1977.
36. Nagai, H., Deguchi, T., and Narahara, S., *Bunseki Kagaku*, 24, 184, 1974.
37. Bankovsky, Yu. A., Ievinsh, A. F., Luksha, E. O., and Bochkans, P. Ya., *Zh. Anal. Khim.*, 16, 150, 1961.
38. Sekido, E. and Fujiwara, I., *Talanta*, 19, 647, 1972.
39. Dedkov, Yu. M., Lovskaya, L. V., and Slotintseva, M. G., *Zh. Anal. Khim.*, 27, 512, 1972.
40. Bankovsky, Yu. A., Mezharaups, G. P., and Ievnish, A. F., *Zh. Anal. Khim.*, 17, 721, 1962.
41. Zbigniew, G., Baranowska, I., Baranowski, R., and Karminski, W., *Chim. Anal. (Paris)*, 14, 1185, 1969; *Anal. Abstr.*, 19, 4037, 1970.
42. Suprunovich, V. I., Usatenko Yu. I., and Velichko, V. V., *Zavod. Lab.*, 36, 652, 1970; *Anal. Abstr.*, 20, 3044, 1971.
43. Awad, K., Rudenko, N. P., Kuznetsov, V. I., and Gudym, L. S., *Talanta*, 18, 279, 1971.
44. Songina, O. A., Zakharov, V. A., Bessarabova, I. M., Rakhimzhanov, P., and Safonov, I. I., *Zh. Anal. Khim.*, 27, 1121, 1972.
45. Watanabe, K. and Kawagaki, K., *Bunseki Kagaku*, 27, 467, 1978.
46. Dedkov, Yu. M., Eliseeva, O. P., Savvin, S. B., and Slotintseva, M. G., *Zh. Anal. Khim.*, 27, 726, 1972.
47. Dedkov, Yu. M., Ermakov, A. N., Kotov, A. V., Lozovskaya, L. V., and Slotintseva, M. G., *Zh. Anal. Khim.*, 27, 1312, 1972.
48. Dedkov, Yu. M., Ermakov, A. N., and Slotintseva, M. G., *Dokl. Akad. Nauk SSR*, 209, 858, 1973; *Anal. Abstr.*, 26, 152, 1974; *Chem. Abstr.*, 79, 13101q, 1973.
49. Dedkov, Yu. M., Ermakov, A. N., Kotov, A. V., and Slotintseva, M. G., *Zh. Anal. Khim.*, 32, 870, 1977.
50. Dedkov, Yu. M., and Slotintseva, M. G., *Zh. Anal. Khim.*, 28, 2367, 1973.
51. Nazarenko, V. A. and Poluektova, E. N., *Zh. Anal. Khim.*, 26, 1331, 1971.
52. Keil, R., *Mikrochim. Acta*, 919, 1973.
53. Usatenko, Yu. I., Suprunovich, V. I., and Velichko, V. V., *Zh. Anal. Khim.*, 29, 807, 1974.

54. Kholmogorov, S. N., Palnikova, T. I., Gribora, L. I., and Perevoshchikova; Yu. D., *Tr. Vses. Nauchno-Issled. Proecktno Inst. Alyumin. Magn. Elektrod. Prom.*, 166, 1971; *Anal. Abstr.*, 24, 684, 1973.
55. Honjo, T., Imura, H., Shima, S., and Kiba, T., *Anal. Chem.*, 50, 1547, 1978.
56. Veveris, O., Bankovskis, J., Pelekis, L., and Pelue, A., *Latv. PSR Zinat. Akad. Vestis. Kim. Ser.*, 511, 1968; *Chem. Abstr.*, 70, 7034t, 1969.

SODIUM DIETHYLDITHIOCARBAMATE AND RELATED REAGENTS

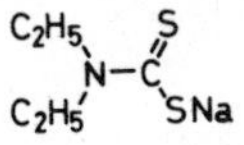

$$C_5H_{10}NS_2Na \cdot 3H_2O$$
$$\text{mol wt} = 225.30$$

NaL (HL)

Synonyms

Cupral, Carbamidat DDTC sodium salt

Source and Method of Synthesis

Commercially available. It is prepared by the reaction of diethylamine and carbon disulfide in aqueous sodium hydroxide.[1]

Analytical Uses

As a precipitating and solvent extraction reagent for soft metal ions. Also used as a photometric reagent for Bi, Cu, Ni, and other metals.

Properties of Reagent

The commercially available reagent is, usually, a monosodium, trihydrate salt, colorless crystals. Anhydrous salt melts at 94 to 96° and is freely soluble in water (35 g/ 100 mℓ, 20°)[2] to give alkaline reaction. Although the solid reagent is stable, the acidic aqueous solutions decompose rapidly. With increasing pH value, the solutions become more stable. A 1% aqueous solution can be kept for a few weeks, and a 0.1% solution can be kept for a week (in an amber bottle). The half-lives of aqueous solutions at pH <2 and pH 5 are 7 sec and 87 min, respectively.[R1,3] It is almost insoluble in nonpolar solvents (0.006 g/100 mℓ in CCl$_4$),[2] but soluble in alcohols.

For the solvent extraction use, diethylammonium salt is preferable because it is soluble in water, as well as in carbon tetrachloride, chloroform, and other organic solvents.[4] Furthermore, in the extraction of metals in acid solution, the bulk of the dithiocarbamate remains in the organic phase, so that the acidic decomposition of the reagent can be avoided.

Diethyldithiocarbamic acid (HL) is a moderately weak acid, pKa = 3.95 (19°);[5] K$_D$ (CCl$_4$/H$_2$O) = 2.4 to 3.4 × 10^2; K$_D$ (CHCl$_3$/H$_2$O) = 2.30 to 2.35 × 10^3.[M3] Absorption spectrum of the sodium salt in an aqueous solution is illustrated in Figure 1.

Complexation Reactions and Properties of Complexes

DDTC behaves as a bidentate univalent anionic ligand, having two donor sulfur atoms, and forms colored precipitates with more than 30 elements at above pH 4. The reactions become more selective in acidic solution, with a preference to soft metal ions. Most of them are uncharged, coordination saturated chelates and can be extracted into organic solvents, such as chloroform or carbon tetrachloride. As an example, the structure of copper(II) chelate is shown below.

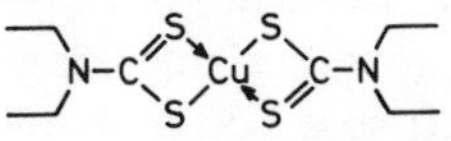

Some of the chelates are colored (Bi, Cu(II), Ni, etc.), so that such elements can be determined photometrically.

Precipitation reactions and extractabilities of metal ions are summarized in Table 1.

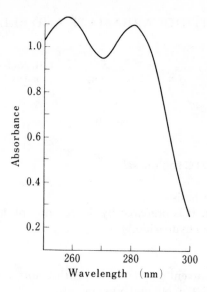

FIGURE 1. Absorption spectrum of Na-DDTC in aqueous solution; pH 6.0; concentration 10 μg/mℓ. (From Kojima, T., Yano, T., and Ueno, K., *Technol. Rep. Kyushu Univ. (Jpn.)*, 45, 209, 1972. With permission.

Selectivity for metal extraction can be improved by the selection of pH range and proper choice of masking agent in the aqueous phase.

As to the stability constants of DDTC chelates, the indirect methods based on the competition reaction with a different metal ion or a different ligand have often been employed to avoid the possible decompositions of ligand and metal chelate in acidic solution. Reported values are summarized in Table 2.

As described above, some of the metal chelates can be extracted into chloroform or carbon tetrachloride. The values for K_{ex} are listed in Table 1. When the metal chelate is highly colored, it can easily be determined photometrically after solvent extraction because the reagent itself does not absorb in the visible range. The spectral characteristics of some of the metal DDTC chelates are summarized in Table 3.

Purification and Purity of Reagent

Sodium diethyldithiocarbamate can be purified by the recrystallization from water to get 99.5% pure material.

A pure sample should give a colorless clear solution when dissolved in water or ethanol (0.1%). Determination of nitrogen by the Kieldahl method could be used for the assay.[19]

Analytical Applications

Use as a Solvent Extraction Reagent

The selectivity of metal extraction can be improved by the proper selection of pH value of aqueous phase and by the use of masking agent. Table 4 summarizes the typical conditions of extraction. It is a common practice to add Na-DDTC to the aqueous phase to precipitate the metal DDTC chelates, followed by the extraction of chelates with organic solvent. The rate of extraction is fairly fast in most cases. The presence of EDTA decreases the extraction rate. Due to the acid instability of DDTC, the extraction from fairly acidic solution is not recommended, although the higher

Table 1

PRECIPITATION AND EXTRACTABILITIES OF METAL ION WITH DDTC

Precipitate			pH range of extraction							Ref.
Metal ion[a]	Ratio	Solubility (condition)	$CHCl_3$[7] (log K_{ex})	CCl_4[7] (log K_{ex})	CCl_4 + KCN[7]	CCl_4 + EDTA[7]	CCl_4 + EDTA + citrate[R1]	CCl_4 + tartrate[R1]	Ethyl-acetate[43]	
Ag(I)	ML	log K_{sp} −20.66 (μ = 0.3)	2.6~5	4~11	4~8	4~11	~9	5~13	3	10
		log K_{sp} −19.6 (0.1 M KNO$_3$, 25°)	(12.6)	(11.90)						9,11,22
As(III)			3~6	4~5.8		4~5.8				
Au(III)	ML$_3$		2~14	4~11	4~8	4~11	~9	5~13	3	13
Bi	ML$_3$	S 2.1×10^{-10} M	2~9	4~11	8~11	4~11		5~13		9,18
			(13.26)	(16.79)						
Cd	ML$_2$	log K_{sp} −22.0 (0.1 M KNO$_3$, 25°)	0~9.5	4~11	8~11	4~6		5~13	3	11
		S 1.8×10^{-7} M	(5.77)	(5.41)						
Co(II)	ML$_2$	log K_{sp} −20.06 (μ = 0.3)	2~14	4~11	8~11	4~6		5~13		8,9,13
		S 1.3×10^{-7} M								10
Cr(III)[b]			(0.78)	(2.33)						9,13,18
			2~6	4~11						
Cu(II)	ML$_2$	log K_{sp} −30.21 (0.2 M NaCl, 25°)	2~14	4~11	4~8	4~11	~9	5~13		12
		S 3.2×10^{-10} M	(11.9)	(13.70)						
Fe(II)	ML$_3$		2~9	(1.28)	8~9	4~6		5~9	9	9,13,18
Fe(III)	ML$_3$		2~9	4~9.8		4~11				
Ga	ML$_3$	log K_{sp} −24.5 (μ = 0.3)		4~5.5	4~8.5	4~11	~9		3	10
Hg(II)	ML$_2$	log K_{sp} −43.79 (0.1 M KNO$_3$, 20°)	2~14	4~11	4~8.5	4~11	~9	5~13	3	15
In	ML$_3$	log K_{sp} −27.06 (μ = 0.3)	4~10	4~10	8~11	4		5~9	3	10
		log K_{sp} −25.0 (0.1 M KNO$_3$, 20°)		(10.34)						M3,9

Table 1 (continued)
PRECIPITATION AND EXTRACTABILITIES OF METAL ION WITH DDTC

Metal ion[a]	Precipitate Ratio	Precipitate Solubility (condition)	CHCl$_3$[7] (log K$_{sx}$)	CCl$_4$[7] (log K$_{sx}$)	CCl$_4$ + KCN[7]	CCl$_4$ + EDTA[7]	CCl$_4$ + EDTA + citrate[7]	CCl$_4$ + tartrate[R1]	Ethyl−acetate[43]	Ref.
Ir(IV)				6∿11	8∿11					13
Mn(II)	ML$_2$	S 1.5×10^{-4} M	2∿9	6∿9	8∿9	4∿11		5∿9	6.5	9
Mo(IV)			2∿9	(−4.42)						
Nb(V)				5∿11		4∿5.5				
Ni	ML$_2$	$\log K_{sp}$ −19.26 ($\mu =$ 0.3); $\log K_{sp}$ −23.1 (0.1 M KNO$_3$, 25°); S 2.3×10^{-7} M	2∿14, (11.18)	4∿11, (11.58)	8∿11	5.5∿11		5∿13		10, 11
Os(IV)				4∿5	8∿11	5∿11				9,15,18
Pb	ML$_2$	$\log K_{sp}$ −21.7 (0.1 M KNO$_3$, 25°); S 3.4×10^{-8} M	3∿9.5	4∿11	8∿11	5∿7		5∿13	acidic	11
Pd(II)	ML$_2$		(7.94)	(7.77), 4∿11	8∿11	4∿11		5∿13		8,9,13
Pt(IV)			2∿9	(>32), 4∿11	8∿11	4∿11				9
Rh(III)			3∿5	4∿11	8∿11					
Ru(III)			2∿14	6∿11	8∿11					
Sb(III)			2∿14	4∿8.5	9∿9.5	4∿9.5	∿9	5∿9		
Se(IV)			3∿5	4∿6.2		4∿6.2		5∿6		
Sn(II)			3∿5	3∿5				5∿6		
Sn(IV)			≃2	4∿5.8		4∿5				
Te(IV)				4∿8.8	8∿8.8	4∿8.8	∿9	5∿9		
Te(VI)				3∿5						
Ti(IV)[c]			2∿14	4∿11	8∿11	4∿6.5	∿9	5∿13	3	38
Tl(I)	ML	$\log K_{sp}$ −10.1 (0.1 M KNO$_3$, 25°)		(−0.53)						M3, 9

Ti(III)	2~14	4~11	8~11	4~11			
U(VI)	6.5~8.5	6.5~8.5	4~8	4~8		3	
V(IV)	2~6	3~5		5~6			
V(V)	2~6	4~5.9					
W(IV)	1~2	5~11				1~1.5	
Zn	2~9	4~11	8~11	5~11	5~13	3	11
ML₂							
log K_{sp} −16.9 (0.1 M KNO₃, 25°)	(2.39)	(2.96)					8,9,13
S 2.5×10^{-6} M							

a Al(III), Be, Ge, Re(VII), Sc, Th, Y, Zr and rare earths are not extracted with $CHCl_3$.

b Cr(VI) can be extracted quantitatively with MIBK at pH 1~5.5, while Cr(III) is extracted at pH >4.[37]

c According to Bode, Ti cannot be extracted with $CHCl_3$, while Rooney reports that 90~100% extraction at pH ≃2 and 50% at pH 5~5.5.[38]

Table 2
STABILITY CONSTANTS OF METAL-DDTC CHELATES

Metal ion	$\log K_{ML}$	$\beta_2 =$ $\log K_{ML} \cdot \log K_{ML2}$	Condition			Ref.
Cu(II)	14.9	28.8	75% ethanol,	20°		14
Hg(II)	22.2	38.1	$\mu = 0.1$, KNO_3,	20°		15
		$(\beta_3\ 39.1)$				
Ni		12.9	$\mu = 0.01$	22°		16
Pb		18.3	$\mu = 0.01$	22°		16
Tl(I)	4.3	5.3	$\mu = 0.1$, KNO_3,	25°		R1,17
Zn		11.4	$\mu = 0.01$	22°		16

Table 3
SPECTRAL CHARACTERISTICS OF COLORED METAL-DDTC CHELATES

Metal ion[a]	Ratio	Color[20]	In CCl₄		In CHCl₃	
			λ_{max} (nm)	$\varepsilon\ (\times 10^3)^{20}$	λ_{max} (nm)	$\varepsilon\ (\times 10^3)^{21}$
Bi	ML_3	Yellow	366	8.62	370	10.5
Co(III)	ML_3	Green	323	23.3		
			367	15.7		
			650	0.55	650	0.52
Cu(II)	ML_2	Brown	436	13.0	440	12.1
Fe(III)	ML_3	Brown	335 ∿ 350	12.7		
			515	2.49	515	2.79
			600	2.05		
Ni	ML_2	Yellow-green	326	34.2		
			393	6.11	395	5.87
Pd(II)	ML_2		305	54.8		
			345 ∿ 350	7.13		
Pt(IV)			355	5.87		
Sb(III)	ML_2		350[b]	3.37		
Te(IV)			428	3.16		
UO_2^{2+}	MO_2L_2	Red-brown	380 ∿ 400[c]		390[c]	3.87

[a] Other metal ions that form colored chelates are Au(III), Cd, Cr(VI), Hg(II), Mo(VI),[41] Mn(II), Os(IV), Pb, Rh, Ru(III), Sn(IV), Ti(IV),[38] Tl(III), and V(V).

[b] Shoulder.

[c] Plateau.

selectivity for metal ions may be expected. In such case, the exchange reaction with metal DDTC chelate, which will be described below, or the use of diethylammonium diethyldithiocarbamate is preferable.

The difference in the values of conditional extraction constants for various metal ions can be utilized for the selective extraction and determination of metal ions. When an aqueous copper solution is shaken with an organic solvent solution of Pb-DDTC chelate, the following reaction will proceed almost quantitatively:

$$Cu^{2+} + (PbL_2)_{org} \longrightarrow (CuL_2)_{org} + Pb^{2+}$$

because $\log K_{ex}$ for PbL_2 and CuL_2 are 7.8 and 13.7 (in CCl_4), respectively. Thus, metal ions can be extracted selectively from the fairly acidic solution where the free DDTC

Table 4

EXTRACTION OF METAL-DDTC CHELATE WITH CCl₄

1. Elements extracted at pH >11
   Ag, Bi, Cd, Co(II), Cu(II), Hg(II), Ni, Pb, Pd(II), Tl(I) (III), and Zn
   Ag, Bi, Cu(II), Hg(II), Pd(II) and Tl(III) can not be masked by EDTA; Bi, Cd, Pb and Tl(I), (III) can not be masked by KCN
2. Elements extracted at pH 9
   Fe(III), In(III), Mn(III), Sb(III), and Te(IV)
   Sb (III) and Te(IV) can not be masked by EDTA; Fe, In, Mn(III) Sb(III), and Te(IV) can not be masked by KCN
3. Elements extracted at pH 6
   As(III), Se(IV), Sn(IV) and V(V)
4. Elements incompletely extracted
   Au, Ba, Ir, Nb(V), Os, Rh, Ru, Pt and U(VI)
5. Other elements not listed above can not be extracted at pH 4 to 11

From Bode, H., *Fresenius Z. Anal. Chem.*, 143, 182, 1954. With permission.

decomposes rapidly. Furthermore, the colorless metal-DDTC chelate can be determined photometrically after the exchange reaction to form the highly colored chelate. Figures 2 and 3 illustrate the possibility of selective extraction of various metal ions from the aqueous solutions of different acidities.[22] The extractions are complete within 2 min in most cases, but SeL_4, CoL_3, and FeL_3 react only slowly.

Use as a Photometric Reagent

DDTC is used as a photometric reagent for Cu(II) and, less often, for Bi, Ni, and some other elements. The spectral characteristics of metal-DDTC chelates are shown in Table 3. The photometry in the visible range is quite easy because the free DDTC does not absorb in the visible range (Figure 1). However, the absorption of excess reagent may interfere in the UV photometry, although the extraction of HL becomes negligible above pH 9. For example, Cu(II) (0.002 to 0.03 mg) can be determined at 440 nm, after extracting as CuL_2 with *n*-butyl acetate at pH 9 in the presence of EDTA. Cations such as Co, Cr, Fe, Mn, and Ni do not interfere.

Other Uses

It is used as a detecting reagent for Cu (detection limit, 0.2 μg and dilution limit, $1:5 \times 10^7$),[23] but the reaction is not selective. The reagent is also used as a precipitating agent for the gravimetry.[M4] The metal-DDTC chelates can also be separated and determined by gas chromatography[24] or HPLC.[25] A polymer with DDTC functional group[39] or silica-gel beads loaded with DDTC[40] has been investigated as an ionselective adsorbent. The metal ions are preconcentrated on the surface of glass on which DDTC is covalently bonded and examined by the X-ray photoelectron spectroscopy for high selectivity and sensitivity.[45]

Other Reagents with Related Structure
Diethylammonium Diethyldithiocarbamate[M3]

C₂H₅＼   S
　　　N-C＜
C₂H₅／   S·[NH₂(C₂H₅)₂]

$C_9H_{22}N_2S_2$
mol wt = 222.41

HL

Colorless crystals, mp 82 to 83°, soluble in water, chloroform, carbon tetrachloride,

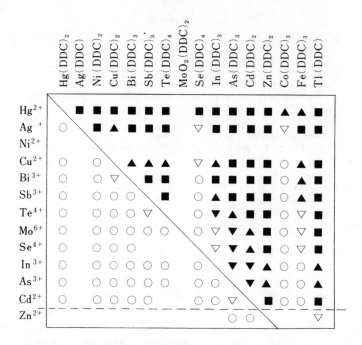

FIGURE 2. Extraction of metals by exchange reaction from 0.1 *N* H₂SO₄ solution; (■), extraction complete within 2 min. (△), extraction complete within 15 min; (▼), extraction complete within 60 min; (▽), extraction is only partial within 60 min. (○), there is no extraction within 15 min. (Reprinted with permission from Wyttenbach, A. and Bajo, S., *Anal. Chem.*, 47, 1813, 1975. Copyright 1975 American Chemical Society.)

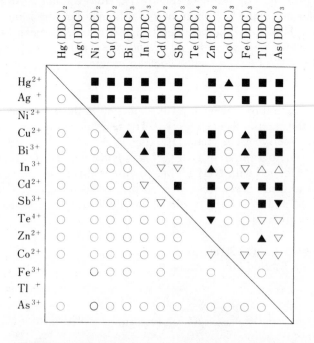

FIGURE 3. Extraction of metals by exchange reaction from the aqueous solution of pH 5 (citrate). (Reprinted with permission from Wyttenbach, A. and Bajo, S., *Anal. Chem.*, 47, 1813, 1975. Copyright 1975 American Chemical Society.)

Table 5
METAL ION EXTRACTION BEHAVIOURS WITH DIETHYLAMMONIUM DIETHYLDITHIOCARBAMATE FROM MINERAL ACID SOLUTION (0.04% CCl₄ SOLUTION)

Solution	Metals quantitatively extracted[a]
10 NHCl	Pd(II), Pt(II)
6 NHCl	As(III), Cu(II), Hg(II), Pd(II), Pt(II), Sb(III), Tl(III)
2 NHCl	Ag, As(III), Bi(III), Cu, Hg(II), Pd(II), Pt(II), Sb(III), Se(IV), Tl(III)
5 NH₂SO₄ or 5 NHCl	Ag, As(III), Bi(III), Cd, Cr(III), Cu(II), Hg(II), In, Mo(VI), Pb, Pd(II), Pt(II), Sb(III), Se(IV), Te(IV), Tl(III)
NH₂SO₄	Ag, Bi(III), Cu(II), Hg(II), In, Mo(VI), Pd(II), Pt(II), Sb(III), Se(IV), Sn(II), Tl(III)

Other metals investigated

Ga (pH 5 ⌇ 6), V(V) (pH 4 ⌇ 5); Mn(II) (pH 6 ⌇ 9); Fe, Co, Nl (pH 3 ⌇ 10); VO₂²⁺ (pH 6.5 ⌇ 7.5)

Note: Aqueous phase (25 ml) was shaken with 25 ml of 0.04% CHCl₃ solution of the reagent for 5 min. In the case of Mo(VI), a mixed solvent (CCl₄ + amyl alcohol, 4:1) was used.

From Bode, H. and Neuman, F., *Fresenius Z. Anal. Chem.*, 172, 1, 1960, With permission.

and other organic solvents. The advantage of this reagent over sodium-DDTC is its solubility in organic solvents because metal ions from mineral acid solution can be extracted with the organic solvent solution of this reagent. During the extraction process, the bulk of the reagent as well as metal chelates are present in the organic phase without acid decomposition.[26] The extraction behaviors from strongly acid solutions are summarized in Table 5. The extractability from the neutral or basic solution is almost similar to that of the sodium salt.

Ammonium Pyrrolidinedithiocarbamate[M3,R3]

$$
\begin{array}{c}
CH_2{-}CH_2 \quad \; S \\
\quad | \qquad\quad N{-}C\!\!\nparallel \\
CH_2{-}CH_2 \quad S{\cdot}NH_4
\end{array}
\qquad
\begin{array}{c}
C_5H_{12}N_2S_2 \\
\text{mol wt} = 164.28
\end{array}
$$

HL (NH₄L)

APDC, ammonium pyrrolidine-1-carbodithioate; commercially available; colorless crystals, mp 142 to 144°; easily soluble in water (18.9 g/100 ml, 20°)[2,26] and alcohol, but only slightly soluble in chloroform (0.38 g/100 ml) and carbon tetrachloride (0.12 g/100 ml, 20°).[2] The free reagent (HL) is a weak acid, pKa = 3.29 (μ = 0.01, KCl, 25°),[16] K_D (CHCl₃/H₂O) = 1.1 × 10³. The reagent is much more stable in acidic solution than DDTC. The half-lives at pH 1.0, 3.0, and 7.3 are 63 min, 175 min, and 170 days, respectively.[6,44] Therefore, APDC is recommended as a precipitating and solvent extraction reagent of heavy metal ions from acidic solutions.

In general, the metal extraction behavior with APDC is similar to that with diethylammonium dithiocarbamate in acid range and to that with sodium salt in neutral and alkaline ranges. Table 6 summarizes the extractability of metal ions with APDC from acidic solutions. Extraction constants for some metal ions are also included in the same table. The solubility of metal-APDC chelates is usually lower than that of metal-DDTC. Nonhalogenated solvents, such as MIBK or alkyl acetate, are recommended as an extraction solvent for the preconcentration of trace of heavy metals when atomic absorption spectrophotometry is employed for the metal determination.[24,42]

Table 6
METAL ION EXTRACTION BEHAVIOR WITH AMMONIUM PYROLIDINE DITHIOCARBAMATE

Metal ion	Metal extracted (%)[a][27]				log K_{ex} (CHCl$_3$/H$_2$O)[b][28]
	6 N HCl	1 N HCl	0.1 N HCl	pH 2	
As(III)	93	100	100	99	—
Bi(III)	90	100	100	98	15.5
Cd(II)	—	20	36	20	1.0
Co(II)	16	41	100	100	−0.8
Cr(III)	0	0	0	0	—
Cu(II)	99	100	100	100	11.4
Fe(III)	45	66	73	100	—
Mn(II)	Trace	0	0	0	—
Ni(II)	90	100	100	100	—
Pb(II)	—	96	100	100	—
Sb(III)	99	99	100	100	—
Sn(II)	96	96	97	96	—
V(V)	22	70	97	100	—
Zn(II)	22	—	80	95	0.4

[a]   Aqueous 1% APDC solution (5 ml) was added to the solution (26 ml) containing about 1 mg metal, and the chelate was extracted with 20 ml CHCl$_3$ after 2 min.

[b]   24°, $\mu = 0.1$.

Silver Diethyldithiocarbamate

$$\begin{array}{ccc} C_2H_5 & S{\to}Ag{-}S & C_2H_5 \\ \diagdown N{-}C & & C{-}N \diagup \\ C_2H_5 \diagup & S{-}Ag{\bullet}S & \diagdown C_2H_5 \end{array}$$

(C$_5$H$_{10}$NS$_2$Ag)$_2$
mol wt = 512.26

It is a pale yellow crystalline powder, almost insoluble in water, but easily soluble in pyridine or chloroform. As the reagent is sensitive to moisture and light, it should be stored in a desiccator in a dark cool place. Deteriorated sample may be purified by recrystallization from pyridine. When arsine gas bubbles through the pyridine solution of Ag-DDTC, the following reactions occur:

$$AsH_3 + 6AgL + 3B = AsAg_3 \cdot 3AgL + 3L^- + 3HB^+$$
$$AsAg_3 \cdot 3AgL + 3B = 6Ag + AsL_3 + 3L^- + 3HB^+$$

where B represents pyridine or other nitrogen base. The formation of colloidal silver results in the pink coloration of the reaction mixture.[29,32] This reaction can be utilized for the determination of trace of As in various samples. A chloroform solution of Ag-DDTC, containing a nitrogen base such as triethylamine (3×10^{-2} M),[34] has been recommended as a preferable reaction medium because the undesirable odor of pyridine can be avoided. Other nitrogen bases were also proposed to replace pyridine.[29,33,35]

Photometric Determination of As[29]

Reagents — 0.5% Pyridine solution of Ag-DDTC. Pyridine solution can be replaced by 0.5% chloroform solution of Ag-DDTC containing 3×10^{-2} M triethylamine; 15% KI solution; 40% SnCl$_2$ solution; concentrated H$_2$SO$_4$.

Apparatus — The reaction flask such as shown in Figure 4.[30]

Procedure—Place an aqueous sample solution containing 4 to 15 μg of As

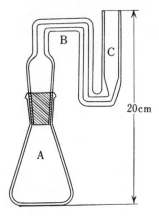

FIGURE 4.   Reaction flask for As determination.

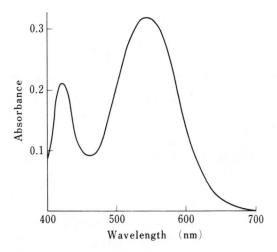

FIGURE 5.   Absorption spectrum of reaction product from Ag-DDTC and arsine. As 10 μg (+10 μg Sb). (From Budesinsky, B., *Microchem. J.*, 24, 80, 1979. With permission.)

in a reaction flask and dilute it to 25 mℓ, add 5 mℓ of concentrated H_2SO_4, 2 mℓ of 15% KI solution, and 0.5 mℓ of $SnCl_2$ solution. Let the mixture stand for 15 to 30 min, with intermittent shaking. Put a plug of glass wool impregnated with lead acetate solution into the neck space of the scrubber tube (to trap S(II) and Se(II)) and transfer 3 mℓ of Ag-DDTC solution into the scrubber. Add 3 g of granular metallic zinc (As free) into the reaction flask and seal the flask immediately with the scrubber. The evolution of arsine will be complete after 60 to 90 min, and the pale yellow solution in the scrubber turns to pink. Transfer the colored solution into a 10-mm cell and observe the absorbance at 540 to 560 nm against a reagent blank. Absorption curve of the colored solution is illustrated in Figure 5.

Interferences[31] — S(II) must be oxidized by nitric acid. Ge, HNO_3, and $HClO_4$ must be removed by repeated evaporation with HCl. Interferences by Cu, Fe, Ni, Sb, Se, Sn, Te, PO_4^{3-}, and many other elements are eliminated by the extraction of $AsCl_3$ from concentrated HCl solution with *p*-xylene or benzene. Metal ions such as Ag, Bi, Cu, Hg, Pb, Pd, Pt, and Sb tend to deposit on the surface of zinc, preventing the dissolution of zinc, but can be masked by KI.

Xanthates

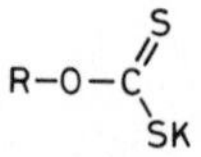

$$R-O-C\overset{\displaystyle S}{\underset{\displaystyle SK}{\big|\big|}}$$

Potassium ethylxanthate ($R=C_2H_5$) and potassium benzylxanthate ($R=C_6H_5CH_2-$) are of practical value as the solvent extraction reagents for metal separations. Their applications in analytical chemistry and mineral floatation are treated in the literature.[M5,36]

MONOGRAPHS

M1. **Morrison, G. H. and Freiser, H.,** *Solvent Extraction in Analytical Chemistry,* John Wiley & Sons, New York, 1957.
M2. **Stary, J.,** *The Solvent Extraction of Metal Chelates,* Pergamon Press, New York, 1964.
M3. **Sandell, E. B. and Onishi, H.,** *Photometric Determination of Traces of Metals,* 4th ed., John Wiley & Sons, New York, 1978.
M4. **Welcher, F. J.,** *Organic Analytical Reagents,* Vol. 3, Von Nostrand, Princeton, N.Y., 1947.
M5. **Rao, S. R.,** Xanthates and Related Compounds, Marcel Dekker, New York, 1971.

REVIEWS

R1. **Hulanicki, A.,** Complexation reactions of dithiocarbamates, *Talanta,* 14, 1371, 1967.
R2. **Halls, D. J.,** Properties of dithiocarbamate, *Microchim. Acta,* 62, 1969.
R3. **Shindo, E. and Morifuji, M.,** Pyrrolidine-dithiocarbamate as an analytical organic reagent, *Kagaku No Ryoiki, (Tokyo),* 21, 206, 1966.

REFERENCES

1. **Chifford, A. M. and Lichty, J. G.,** *J. Am. Chem. Soc.,* 54, 1163, 1932.
2. **Malissa, H. and Gomiscek, S.,** *Anal. Chim. Acta,* 27, 402, 1962.
3. **Aspila, K. I., Sastri, V. S., and Chakrabarti, C. L.,** *Talanta,* 16, 1099, 1969.
4. **Bode, H. and Neumann, F.,** *Fresenius Z. Anal. Chem.,* 172, 1, 1960.
5. **Zahradnick, R. and Zuman, P.,** *Collect. Czech. Chem. Commun.,* 24, 1132, 1959.
6. **Kojima, T., Yamo, T., and Ueno, K.,** *Technol. Rep. Kyushu Univ.,* 45, 209, 1972.
7. **Koch, O. G. and Koch-Dedic, G. A.,** *Handbuch der Spurenanalyse,* Springer-Verlag, Berlin, 1964, 216.
8. **Bajo, S. and Wyttenbach, A.,** *Anal. Chem.,* 51, 376, 1979.
9. **Stary, J. and Kratzer, K.,** *Anal. Chim. Acta,* 40, 93, 1968.

10. Pevtsov, G. A., Manova, T. G., and Raginskaya, L. K., *Zh. Anal. Khim.*, 23, 159, 1967; *J. Anal. Chem. USSR*, 22, 134, 1967.
11. Hulanicki, A., *Acta Chim. Acad. Sci. Hung.*, 27, 41, 1961; *Chem. Abstr.*, 55, 23179e, 1961.
12. Kemula, W. and Hulanicki, A., *Bull. Acad. Pol. Sci. Ser. Sci. Chim.*, 9, 477, 1961; *Chem. Abstr.*, 60, 8699h, 1964.
13. Usatenko, Y. I., Barkalov, V. S., and Tulyupa, F. M., *J. Anal. Chem. USSR*, 25, 1257, 1970.
14. Janssen, M. J., *Rec. Trav. Chim.*, 75, 1411, 1956; *Chem. Abstr.*, 51, 5514b, 1957.
15. Kemura, W., Hulanicki, A., and Nawrot, W., *Rocz. Chem.*, 36, 1717, 1962; *Chem. Abstr.*, 59, 9598g, 1963; *Rocz. Chem.*, 38, 1065, 1964; *Chem. Abstr.*, 62, 13926b, 1965.
16. Scharfe, R. R., Sastri, V. S., and Chakrabarti, C. L., *Anal. Chem.*, 45, 413, 1973.
17. Hulanicki, A., Dissertation, University Warsaw, 1967.
18. Grekova, I. M., *Zh. Anal. Khim.*, 27, 1157, 1972; *J. Anal. Chem. USSR*, 27, 1034, 1972.
19. Czechoslovak Fine Chemicals Standards, Vol. 3, Chemapol, Prague, Czechoslovakia, 1955, 423.
20. Bode, H., *Fresenius Z. Anal. Chem.*, 143, 182, 1954; *Fresenius Z. Anal. Chem.*, 144, 165, 1956.
21. Lacoste, R. J., Earing, M. H., and Wiberley, S. E., *Anal. Chem.*, 23, 871, 1951.
22. Wyttenbach, A. and Bajo, S., *Anal. Chem.*, 47, 1813, 1975.
23. Clark, B. L. and Hermance, H. W., *Ind. Eng. Chem. Anal. Ed.*, 9, 292, 1937.
24. Tavlaridis, A. and Neeb, R., *Fresenius Z. Anal. Chem.*, 293, 211, 290, 1978.
25. O'Laughlin, J. W. and O'Brien, T. P., *Anal. Lett.*, A11, 829, 1978.
26. Gomiscek, S., Lengar, Z., Cerentic, J., and Hudnik, V., *Anal. Chim. Acta*, 73, 97, 1974.
27. Malissa, H. and Gomiscek, S., *Fresenius Z. Anal. Chem.*, 169, 401, 1959.
28. Likussar, W. and Boltz, D. F., *Anal. Chem.*, 43, 1273, 1971.
29. Bode, H. and Hachmann, K., *Fresenius Z. Anal. Chem.*, 229, 261, 1967; *Fresenius Z. Anal. Chem.*, 241, 18, 1968.
30. By courtesy of Dojindo Labs. Ltd., Kumamoto, Japan.
31. Budesinsky, B. W., *Microchem. J.*, 24, 80, 1979.
32. Vasak, V. and Sedivec, V., *Chem. Listy*, 46, 341, 1952; *Chem. Abstr.*, 47, 67e, 1953.
33. Yamamoto, Y., Kumamaru, T., Hayashi, Y., Kanke, M., and Matsui, A., *Bunseki Kagaku*, 21, 379, 1972.
34. Yamamoto, D. and Ueda, T., *Bunseki Kagaku*, 21, 938, 1972.
35. Fujinuma, H. and Itsuki, K., *Bunseki Kagaku*, 28, 627, 1979.
36. Donaldson, E. M., *Talanta*, 23, 417, 1976.
37. Hiiro, K., Owa, T., Takaoka, M., Tanaka, T., and Kawahara, A., *Bunseki Kagaku*, 25, 122, 1976.
38. Rooney, R. C., *Anal. Chim. Acta*, 19, 428, 1968.
39. Dingman, J. F., Jr., Gloss, K. M., Milano, E. A., and Siggia, S., *Anal. Chem.*, 46, 774, 1974.
40. Leyden, D. E. and Luttrell, G. H., *Anal. Chem.*, 47, 1612, 1975.
41. Sarma, V. B. and Suryanarayana, M., *Fresenius Z. Anal. Chem.*, 240, 6, 1968.
42. Kamada, T., Shiraishi, T., and Yamamoto, Y., *Talanta*, 25, 15, 1978.
43. Chernikhov, Yu. A. and Dobkina, B. M., *Zavod. Lab.*, 15, 1143, 1949; *Chem. Abstr.*, 44, 1358g, 1950.
44. Joris, S. J., Aspila, K. I., and Chakrabarti, C. L., *Anal. Chem.*, 41, 1441, 1969.

TOLUENE-3,4-DITHIOL AND RELATED REAGENTS

$C_7H_8S_2$
mol wt = 156.26

H_2L

Synonyms
4-Methyl-1,2-dimercaptobenzene, 3,4-Dimercaptotoluene, Dithiol

Source and Method of Synthesis
Commercially available. It is synthesized by the reduction of toluene-3,4-disulfonylchloride with Sn and HCl.[1]

Analytical Uses
Photometric reagent for Ag, Mo, Re, Sn, Tc, and W; detecting reagent for Sn.

Properties of Reagent
Colorless crystals, mp 31 to 34°, bp 185 to 187° (84 Torr); n_D^{20} 1.6378. It has a characteristic odor of thiol and is very easily oxidized in air to give a yellow oil or white polymer $\{(C_7H_6S_2)_n\}$. It should be kept in a refrigerator under nitrogen. The melting ranges of typical commercial samples are 28 to 31°. It is very slightly soluble in aqueous acid, but is easily soluble in common organic solvents and in dilute aqueous alkali hydroxide. As the aqueous solution is unstable, the working solution should be prepared freshly.[2] Dithiol is a diprotic acid; $pKa_1 = 5.34$ and $pKa_2 = 11.0$ (in water, 25°);[3] $\log K_D(CHCl_3/H_2O) = 4.13$ ($\mu = 0.1$, 25°).[3]

Complexation Reaction and Properties of Complexes
Dithiol reacts with many kinds of soft metal ions, forming colored soluble chelates or precipitates, some of which can be extracted into immiscible solvents, such as chloroform, 1,2-dichloroethane, iso-amyl acetate, or DMF. The color and composition of the chelates at various pH are summarized in Table 1.

A number of the higher valent metal ions, such as Ge(IV), Mo (VI), Pb(IV), Si(IV), Sn(IV), and W(VI), form uncharged coordination saturated chelates with the bidentate dithiolate anion (L^{2-})[7] and can be extracted into immiscible solvents. However, metals of lower valence often form soluble chelates or polymeric chelates which are soluble neither in water nor in organic solvents. The polymeric chelates may be solubilized with an aid of surfactant (Teepol®,[8] Santomerse®,[9,10] gelatine, or Triton® X-100) for the photometric determination of metals.

Zinc forms two types of chelates, ZnL_2^{2-} (1), a coordination-saturated, charged chelate which is water soluble, and ZnL (2), a coordination-unsaturated, uncharged insoluble chelate which is not extractable. However, a charged chelate can be extracted with a suitable cation such as quaternary ammonium ion into the organic phase as an ion-pair.

(1)   (2)

Table 1
COLOR REACTIONS OF DITHIOL WITH METAL IONS[M1]

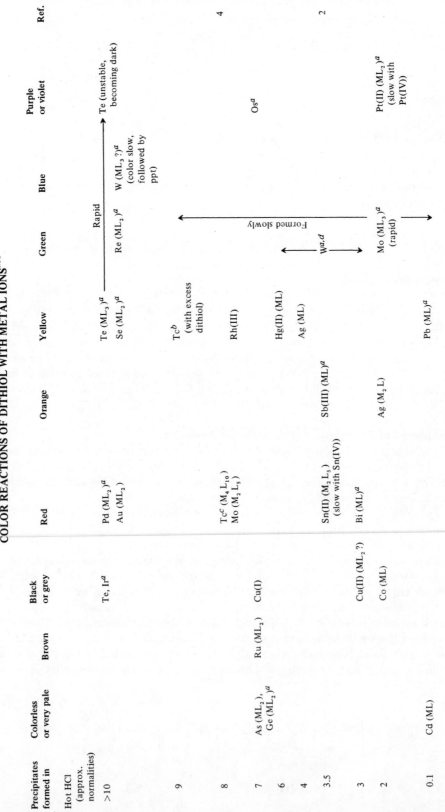

Precipitates formed in Hot HCl (approx. normalities)	Colorless or very pale	Brown	Black or grey	Red	Orange	Yellow	Green	Blue	Purple or violet	Ref.
>10			Te, Ir^a	Pd (ML$_2$)^a Au (ML$_2$)		Te (ML$_3$)^a Se (ML$_2$)^a	Re (ML$_2$)^a	W (ML$_3$?)^a (color slow, followed by ppt)	Te (unstable, becoming dark)	
9						Tc^b (with excess dithiol)				4
8				Tc^c (M$_4$L$_{10}$) Mo (M$_2$L$_5$)						
7	As (ML$_2$), Ge (ML$_2$)^a	Ru (ML$_2$)	Cu(I)			Rh(III)				
6						Hg(II) (ML) Ag (ML)			Os^a	
4										
3.5				Sn(II) (M$_2$L$_5$) (slow with Sn(IV))	Sb(III) (ML)^a					
3			Cu(II) (ML$_2$?)	Bi (ML)^a						
2			Co (ML)		Ag (M$_2$L)		Mo (ML$_3$)^a (rapid)		Pt(II) (ML$_2$)^a (slow with Pt(IV))	2
0.1	Cd (ML)					Pb (ML)^a				

Annotations: "Rapid" (over the >10 yellow–purple reactions); "Formed slowly" and W^{a,d} (green–blue region); Mo (ML$_3$)^a (rapid) formed across green–blue.

Excess dithiol

Ni (ML)^e $\longrightarrow$ Ni (ML$_2$)

Co(III) (ML$_2$) (unstable) $\longleftrightarrow$ (stable)

Acetate buffer — Zn (ML), Ga (ML), In (ML), Sn(II) (ML)

Fe(II) (ML) $\longrightarrow$ (ML$_3$?) (with excess dithiol), Tl (ML)^a, Sn(IV) (ML$_2$)^a

NaOH or NH$_4$OH — Cu(I) and (II)

Pyridine and aqueous pyridine — Fe(III) (ML$_3$?), Sb(III) (ML$_3$), Pd (ML)^f

Mn, yellow-green, Au (ML)^f

Cu(I) (II),^{g} Ru(III), V(V)

a Soluble in organic solvents, such as 1,2-dichloroethane or isoamyl acetate.
b Maximum acidity for Te is not reported.
c With thioglycolic acid.
d Can be extracted with CHCl$_3$.
e Quinoxaline-2,3-dithiol is the preferred reagent for Ni.
f Slightly soluble in water.
g Fluoresces.

From Riddett, N. J., *Organic Reagents for Metals*, Johnson, W. C., Ed., Vol. 2, Hopkin & Williams, Essex, England, 1964, with slight modifications.

Also, an uncharged coordination unsaturated chelate can be converted to an extractable species by replacing the coordinated water of ZnL chelate with an uncharged base such as 1,10-(phenanthroline).

The absorption spectrum of the red-colored Sn(II)-dithiolate dispersed in water is illustrated in Figure 1. Although Sn-dithiolate is soluble in organic solvents, the resulting solution is yellow and is not useful for the photometry.

de Silva observed the absorption spectra of Mo(VI)-dithiolate in various organic solvents.[11] After extracting Mo(VI) with tri-*n*-butylphosphate from the hydrochloric acid solution, it was converted to the dithiolate in various organic solvents. The highest molar absorptivitiy ($\varepsilon = 1.84 \times 10^4$ at λ_{max}, 705 nm) was observed in a mixed solvent of glacial acetic acid-TBP-phosphoric acid (20:5:2 by volume).

Purification and Purity of Reagent

Dithiol can be purified by the vacuum distillation under nitrogen. As the reagent is easily oxidized, it is recommended to keep it in a sealed ampule in nitrogen at a freezing temperature. Alternatively, dithiol is converted to the more stable derivatives, from which dithiol is regenerated *in situ* as needed. Of the following stabilized dithiol derivatives, the last one is most widely accepted in the practical applications.[2,12,13]

(3)   (4)   (5)

Compounds (3) to (5) are white powders which are fairly stable against oxidation, and dithiol can be regenerated by the simple procedure, e.g., by treatment with 1 N NaOH (for 3, 4 and 5) or with mineral acid (for 5).

The purity of free dithiol can be determined by the iodometry of SH groups. The purity of zinc-dithiol can be checked by the EDTA titration of zinc after the digestion of sample with HNO_3-$HClO_4$.

Analytical Applications

Use as a Photometric Reagent

The conditions for the photometric determination of some metal ions using dithiol are summarized in Table 2.

Determination of Sn[10,21] — Heat the sample solution containing 3 to 50 µg of Sn with 0.5 mℓ of H_2SO_4 (1 + 2) in a beaker on a hot plate until fuming. Transfer the content into a volumetric flask (10 mℓ) using 7 mℓ of water. To this add one drop of thioglycolic acid, and after mixing, add 1 mℓ of 0.5% gum arabic and 0.5 mℓ of 0.2% dithiol solution (in 1% NaOH solution containing 1% thioglycolic acid). Heat the mixture on a water bath at 50 ± 5° for 1 min and dilute it to the volume after cooling. Measure the absorbance at 530 nm against a reagent blank. Since As, Bi, Ge, Hg, Mo, Sb, Se, and Te interfere, this method is recommended after separating Sn by distillation as $SnCl_4$.

Determination of Mo(VI)[10] — Adjust the acidity of digested sample solution containing up to 100 µg of Mo to 6 to 8 N in hydrochloric acid and make to the total volume of 25 mℓ. Shake the solution with 5 mℓ of TBP

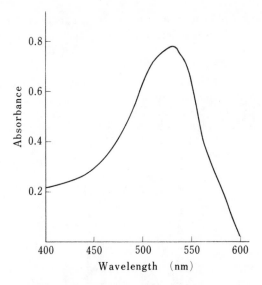

FIGURE 1. Absorption spectrum of Sn-dithiol chelate, Sn 16 ppm. (Reprinted with permission from Farnsworth, M. and Pekola, J., *Anal. Chem.*, 26, 735, 1954. Copyright 1954 American Chemical Society.)

(saturated with hydrochloric acid) for 5 min. After phase separation, transfer a 5-mℓ aliquot of the TBP extract into a 50-mℓ volumetric flask. Add 25 mℓ of glacial acetic acid and 2 mℓ of phosphoric acid. Then, add five drops of dithiol solution (1 g dithiol in 100 mℓ of 10% NaOH containing 1 g of thioglycolic acid) and swirl the mixture. Make the volume up to the mark with glacial acetic acid and allow the solution to stand for 3 hr. Measure the absorbance at 705 nm against water. Up to 1000 μg of Bi, Cd, Co, Cu, Fe, Ni, Pb, Sn, Zn, and W do not interfere.

When zinc-dithiol is used, the following procedure is recommended to prepare the working solution.

Mix 0.2 g zinc-dithiol with 2 mℓ ehtanol and add 100 mℓ of 1% NaOH solution. Then, add 1 mℓ of thioglycolic acid and filter the mixture. Keep the filtrate in a polyethylene bottle and store in a refrigerator.[14]

Use as a Detecting Reagent

As summarized in Table 1, dithiol reacts with wide range of metal ions at various reaction conditions. For the metal ions such as Mo, Sn, and W, the precipitation reactions are so sensitive that 1 to 5 ppm can be detected on a spot plate. In 1 to 2 N HCl, only Pt gives violet precipitate which can be extracted into organic solvents.[22]

Other Reagents with Related Structure

Quinoxaline-2,3-dithiol

$C_8H_6N_2S_2$
mol wt = 194.27

H$_2$L

Table 2
APPLICATION OF DITHIOL AS A PHOTOMETRIC REAGENT FOR METALS

Metal Ion	Condition	Metal chelate λ_{max} (nm)	Metal chelate $\varepsilon(\times 10^4)$	Range of determination (ppm)	Interference	Ref.
Mo (VI)	Extraction with TPB from $6 \sim 8$ N HCl, then glacial acetic acid + H_3PO_4 + dithiol	705	1.84	~ 4	—	11, 16
Mo (VI)	2.5 NH_2SO_4 or 5 N HCl, extraction with CCl_4, (use Zn-dithiol)	$670 \sim 680$	2.8	$0.1 \sim 15$	—	2, 5, 17
Sn (IV)	$0.4 \sim 0.8$ NH_2SO_4 or $0.4 \sim 0.6$ N HCl, thioglycolic acid, Santomerse® 8	530	0.53	$0.25 \sim 16$	Bi, Mo, F⁻, PO_4^{3-}	9, 10
W (VI)	5.4 $NH_2SO_4 + 3.8$ N HCl, extraction with CCl_4	680	2.49	—	W in Nb metal	2
W (VI)	$0.12 \sim 0.36$ N HCl, extraction with butyl acetate	630	—	$1 \sim 10$	W in sea water, sediments, rocks	19, 20

Note: Other metal ions determined with dithiol are Ag,[15] Re,[6] and Tc.[18]

Yellow-brown to red-brown crystals, mp 345° (with decomposition);[23] easily oxidized at higher temperature or when exposed to air;[24] same precautions are needed for the storage of the reagent as dithiol. It is almost insoluble in water (7.2×10^{-5} M at 20°) and is sparingly soluble in common organic solvents[23] and in ammonia, but is slightly soluble in DMF.[25] Sodium salt is easily soluble in water (>27.4 g/100 mℓ).[26] The solution can be stored in a dark and cool place for 1 week; pKa_1 = 6.94 and pKa_2 = 9.91 (μ = 0.01, 20°);[27] K_{sp} = {H$^+$} {HL$^-$} = 1.2×10^{-11} (20°).[M1]

The chelating behavior of the reagent is quite similar to that of dithiol, and it has been used as a photometric reagent for Bi (4 to 5 N H$_2$SO$_4$, 490 nm, ε = 3.4×10^4, 4 ppm),[28] Co(II) (pH 10, 8% ethanol, 472 nm, ε = 3.6×10^4, 0.15 to 1.5 ppm),[29] Cu(II) (pH 2, 625 nm, ε = 2.3×10^3, 0.1 to 1 ppm),[30] Ni (NH$_3$ alkali, 520 nm, ε = 2.7×10^4, 0.03 to 3 ppm),[31] Os(VI) (HCl in DMF-water, 560 nm, ε = 1.73×10^4, 2 to 8 ppm),[25] Pd (1 N HCl in DMF-water, 548 nm, ε = 3.3×10^4, 0.4 to 3 ppm),[32] and Pt(II) (HCl in DMF-water, 624 nm, ε = 2.75×10^4, 1.4 to 5 ppm).[24] Anionic chelates of ML$_2^{2-}$-type with Co, Ni, Pd, and Pt can be extracted with diphenylguanidium, As(C$_6$H$_5$)$_4^+$, P(C$_6$H$_5$)$_4^{+[33,34]}$ or tetrabutylammonium ion[35] into chloroform for the subsequent photometry.

Amino- and nitro-derivatives are investigated as a photometric reagent of Co, Ni,[36,37] and Pd.[38] Also, naphthalene-1,8-dithiol was investigated as a photometric reagent for Pd(II) (2 N HCl, gum arabic, 450 nm, ε = 1.25×10^4, 0.2 to 7 ppm)[39] and as a detection reagent for Co and Ni.[40] Other dithiols, such as o-xylene-4,5-dithiol (for Mo and W)[41] and cis-1,2-dicyanoethylene-1,2-dithiol (for Mo) have been reported.[42]

MONOGRAPHS

M1. Riddett, N. J., Dithiol, in *Organic Reagent for Metals*, Vol. 1, Johnson, W. C., Ed., Hopkin & Williams, Essex, England, 1955; Clark, R. E. D., Stable Dithiol Derivatives, in *Organic Reagent for Metals*, Vol. 2, Johnson, W. C., Ed., Hopkin & Williams, Essex, England, 1964, 85.

M2. Sandell, E. B. and Onishi, H., *Photometric Determination of Traces of Metals*, 4th ed., John Wiley & Sons, New York, 1978.

REFERENCES

1. Mills, C. C. and Clark, R. E. D., *J. Chem. Soc.*, 178, 1936.
2. Hobart, E. W. and Hurley, E. P., *Anal. Chim. Acta*, 27, 144, 1962.
3. Hamilton, H. G. and Freiser, H., *Anal. Chem.*, 41, 1310, 1969.
4. Koyama, M., Emoto, K., Kawashima, M., and Fujinaga, J., *Chem. Anal.*, 17, 679, 1972; *Anal. Abstr.*, 24, 2807, 1973.
5. Tanaka, K. and Takaki, N., *Bunseki Kagaku*, 19, 790, 1970.
6. Kawashima, M., Kayama, M., and Fujinaga, T., *J. Inorg. Nucl. Chem.*, 38, 801, 1976.
7. Fink, F. H., Turner, J. A., and Payne, D. A., Jr., *J. Am. Chem. Soc.*, 88, 1571, 1966.
8. Williams, F. R. and Whitehead, J., *J. Appl. Chem.*, 2, 213, 1952.
9. Farnsworth, M. and Pekola, J., *Anal. Chem.*, 26, 735, 1954.
10. Onishi, H. and Sandell, E. B., *Anal. Chim. Acta*, 14, 153, 1956.
11. de Silva, M. E. M. S., *Analyst*, 100, 517, 1975.
12. Clark, R. E. D., *Analyst*, 82, 182, 760, 1957, *Analyst*, 83, 103, 1958.
13. Clark, R. E. D. and Tamale-Ssali, C. E., *Analyst*, 84, 16, 1959.
14. Yamane, T., Iida, K., Mukooyama, T., and Fukazawa, T., *Bunseki Kagaku*, 19, 808, 1970.
15. Dux, J. P. and Feairheller, W. R., *Anal. Chem.*, 33, 445, 1961.
16. Gilbert, T. W., Jr., and Sandell, E. B., *J. Am. Chem. Soc.*, 82, 1087, 1960.
17. Stanton, R. E. and Hardwick, A. J., *Analyst*, 92, 387, 1967.
18. Miller, F. J. and Thomason, R. F., *Anal. Chem.*, 33, 404, 1961.
19. Chan, K. M. and Riley, J. P., *Anal. Chim. Acta*, 39, 103, 1967.

20. Machlan, L. A. and Hague, J. L., *J. Res. Natl. Bur. Stand.,* 59, 415, 1957; *Chem. Abstr.,* 52, 8836h, 1958.
21. Kawabuchi, K., Kaya, M., and Ouchi, Y., *Bunseki Kagaku,* 15, 543, 1966.
22. Clark, R. E. D., *Analyst,* 61, 242, 1936.
23. Morrison, D. C. and Furst, A., *J. Org. Chem.,* 21, 470, 1956.
24. Ayres, G. H. and McCrory, R. W., *Anal. Chem.,* 36, 133, 1964.
25. Janota, H. F. and Choy, S. B., *Anal. Chem.,* 46, 670, 1974.
26. Beezer, A. E. and Slawinski, A. K., *Talanta,* 18, 837, 1971.
27. Chernomorchenko, L. L., Akhmetshin, A. G., and Chuiko, V. T., *Zh. Anal. Khim.,* 25, 231, 1970.
28. Chernomorchenko, L. I. and Butenko, G. A., *Zavod. Lab.,* 39, 1448, 1973; *Anal. Abstr.,* 27, 698, 1974.
29. Dalziel, J. A. W. and Slawinski, A. K., *Talanta,* 15, 367, 1968.
30. Burke, R. W. and Deardorff, E. R., *Talanta,* 17, 255, 1970.
31. Skoog, D. A., Lai, M. G., and Furst, A., *Anal. Chem.,* 30, 365, 1958.
32. Ayres, G. H. and Janota, H. F., *Anal. Chem.,* 31, 1985, 1959.
33. Ryabushko, O. P., Pilipenko, A. T., and Krivokhizhina, L., *Ukr. Khim. Zh.,* 39, 1274, 1973; *Anal. Abstr.,* 27, 700, 1974; *Org. Reagenty Anal. Khim., Tezisy Dokl, Vses. Konf., 4th.,* 2, 6, 1976; *Chem. Abstr.,* 177046s, 1977.
34. Pilipenko, A. T., Ryabushko, O. P., and Krivokhizhina, L. A., *Ukr. Khim. Zh.,* 42, 1077, 1976; *Chem. Abstr.,* 86, 100127r, 1977.
35. Ryabushko, O. P., Pilipenko, A. T., and Krivokhizhina, L. A., *Ukr. Khim. Zh.,* 41, 1084, 1975; *Chem. Abstr.,* 84, 83688p, 1976.
36. Ohira, K., Kidani, Y., and Koike, H., *Bunseki Kagaku,* 23, 658, 1974.
37. Bhaskare, C. K. and Jagadale, U. D., *Fresenius Z. Anal. Chem.,* 278, 127, 1976, *Anal. Chim., Acta,* 93, 335, 1977.
38. Bhaskare, C. K. and Jagadale, U. D., *Anal. Lett.,* 10, 225, 1977.
39. Sugimoto, M. and Furuhashi, A., *Bunseki Kagaku,* 23, 1078, 1974.
40. Price, W. B. and Smiles, S., *J. Chem. Soc.,* 2372, 1928.
41. Hoyer, E., Mueller, H., and Wagler, H., *Wiss. Z. Karl-Marx-Univ. Lpz.,* 21, 47, 1972; *Anal. Abstr.,* 23, 2442, 1972.
42. Chakrabarti, A. K. and Bag, S. P., *Talanta,* 19, 1187, 1972.

BISMUTHIOL II

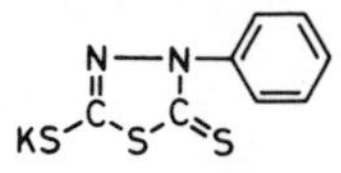

$C_8H_5N_2S_3K$
mol wt = 264.32

HL (KL)

Synonyms

3-Phenyl-5-mercapto-1,3,4-thiadiazol-2-thione, potassium salt, Bismuthone, Bismuthol

Source and Method of Synthesis

Commercially available. It is prepared by the reaction of phenylhydrazine with carbon disulfide in potassium hydroxide solution.[1]

Analytical Uses

Precipitating and gravimetric reagent for soft metal ions. Photometric reagent for Bi, Se, Te, and some other elements.

Properties of Reagent

Bismuthiol II is usually supplied as a potassium salt, mp 243 to 248° (with decomposition at 250 to 255°),[M1,R1] and is colorless or pale yellow needles. The reagent is fairly stable, but is slowly oxidized to give a disulfide. It is easily soluble in water and alcohol, but insoluble in common organic solvents. Free thiol (HL) can be obtained as colorless needles (mp 90 to 91°)[R1] by acidifying the aqueous solution of potassium salt. The free thiol is more readily oxidized than potassium salt and is insoluble in water or ligroin (low boiling point), but is easily soluble in organic solvents, p$Ka \simeq 3$ (20°).[8] The aqueous solution (L⁻) (λ_{max}, 335 nm; $\varepsilon = 1.1 \times 10^4$) and chloroform solution (HL) (λ_{max}, 330 nm) are colorless.[R1]

Complexation Reactions and Properties of Complexes

Bismuthiol II reacts with wide variety of soft metal ions in aqueous solution to form yellow to red-brown precipitates.[3] The reactivity with metal ions varies with the pH of solution and the nature of masking agents, as summarized in Table 1. In these reactions, Bismuthiol II is considered to behave as a unidentate univalent anion, forming the complexes such as shown below.[11]

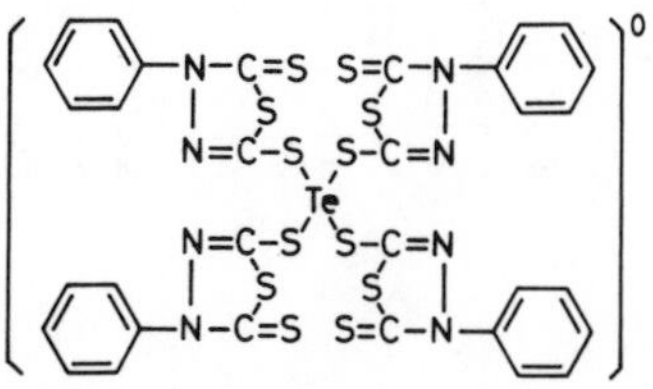

Such uncharged complexes can be extracted into chloroform. Since the absorption sepectra of the complexes are very similar to that of free reagent as illustrated in Figure 1, the excess reagent has to be stripped by washing with water at pH 7.5 prior to the photometric determination of metals.

Purification and Purity of Reagent

Bismuthiol II (potassium salt) is usually contaminated with the disulfide after stor-

Table 1

REACTION OF BISMUTHIOL II WITH METAL IONS[R1]

Metal ion	Ratio	0.2 N HCl	pH 4.5, citrate	pH 9, citrate	pH 9, citrate + EDTA	pH 9, citrate + KCN	Ref.
Ag	ML	Yellow ppt (X)[a]	Yellow ppt (X)	Yellow ppt (X)	Yellow ppt (X)	n.p.	
Al		n.p.[b]	n.p.	n.p.	n.p	n.p	
As(III)	ML$_3$	Yellow ppt (Y)[c]	n.p.	n.p.	n.p.	n.p.	
As(V)		White ppt (X)	n.p.	n.p.	n.p.	n.p.	
Au(IV)		Yellow ppt (X)	Yellow ppt (X)	Yellow ppt (X)	Yellow ppt (X)	n.p.	
Bi	ML$_3$ or ML$_2$Cl	Orange-red ppt (X)	Orange-red ppt (X)	n.p.	n.p.	n.p.	M1
Cd	ML$_2$	White (X)	White (X)	White (X)	n.p.	n.p.	
Co(II)	n.p.	n.p.	n.p.	n.p.	n.p.	n.p.	
Cu(II)	ML$_2$	Deepyellow (X)	Deepyellow (X)	Deepyellow (X)	n.p.	n.p.	4
Fe(III)		Yellow ppt (X)	n.p.	n.p.	n.p.	n.p.	
Hg(II)	ML$_2$	Yellow ppt (X)	Yellow ppt (X)	Yellow ppt (X)	Yellow ppt (X)	n.p.	5
Mn(II)		n.p.	n.p.	n.p.	n.p.	n.p.	
Mo(VI)		n.p.	n.p.	n.p.	n.p.	n.p.	
Ni		—	—	—	—	—	
Os	ML$_8$						15
Pb	ML$_2$	Yellow ppt (X)	Yellow ppt (X)	Yellow ppt (X)	n.p.	Yellow ppt (X)	
Pd(II)	ML$_2$	Red-brown ppt (X)	Red-brown ppt (X)	Red-brown ppt (X)	Red-brown ppt (X)	n.p.	6
Pt(IV)	ML$_4$	Yellow ppt (X)	Yellow ppt (X)	Yellow ppt (X)	Yellow ppt (X)	White ppt (X)	7
Re(VII)	ML$_2$ or ML$_4$	Yellow ppt[d] (X)	—	—	—	—	
Ru(III)		Red-brown ppt (X)	Red-brown ppt (X)	n.p. ppt (X)	n.p.	n.p.	
Sb(III)	ML$_3$	Yellow ppt (X)	n.p.	n.p.	n.p.	n.p.	8
Se(IV)	ML$_4$	Yellow ppt (Y)	n.p.	n.p.	n.p.	n.p.	
Se(VI)		n.p.	n.p.	n.p.	n.p.	n.p.	
Sn(II)		Yellow ppt (X)	n.p.	n.p.	n.p.	n.p.	
Te(IV)	ML$_4$	Yellow ppt (Y)	Yellow ppt (Y)	n.p.	n.p.	n.p.	
Te(VI)		n.p.	n.p.	n.p.	n.p.	n.p.	
Tl(I)	ML	Yellow ppt (X)	Yellow ppt (X)	Yellow ppt (X)	n.p.	Yellow ppt (X)	
U(VI)		n.p.	n.p.	n.p.	n.p.	n.p.	

V(V)	Deep-yellow ppt (X)	n.p.	n.p.	n.p.	n.p.
W(VI)	n.p.	n.p.	n.p.	n.p.	n.p.
Zn	White ppt (X)	n.p.	n.p.	n.p.	n.p.

[a] (X): precipitate is extractable into chloroform.
[b] n.p.: no precipitation occurs.
[c] (Y): chloroform extract is yellow.
[d] When reduced with Sn(II) or Ti(III).

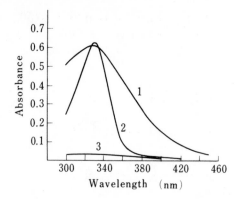

FIGURE 1.   Absorption spectra of Bismuthiol
II and Te chelate in chloroform. (1) Te complex
in chloroform; (2) Bismuthiol II in chloroform;
(3) Blank (after back extraction of Bismuthiol II
into water). (From Yoshida, H. and Hikime, S.,
Bunko Kenkyu, 14, 131, 1965. With permission.)

age for long period, but can be purified by the recrystallization from hot ethanol. The
reagent is assayed by the iodometry of SH group according to the standard procedure.

Analytical Applications

Use as a Precipitating Reagent

As shown in Table 1, the precipitation reactions of Bismuthiol II can be made fairly
selective to the soft metal ions. The precipitations have, in most cases, well-defined
compositions and are convenient for the gravimetric finish after drying at $\sim 100°$.[M1,R1,2]

The precipitation reactions are, in some instances, so sensitive that they are useful
for the spot tests for metal ions (Bi, red precipitation in HNO_3; dilution limit, 1:6 ×
10^6).[9,16]

Use as a Photometric Reagent

The most important application of Bismuthiol II is the extraction photometric determination of Te(IV). The Bismuthiol complex (TeL_4) is formed in acidic condition, then
is extracted into chloroform or benzene for the subsequent photometry. As the absorption spectrum of the excess reagent overlaps considerably with that of Te complex, the
following procedures are proposed for the determination of Te.

1.   Complex formation at pH 4.15, extraction of both complex and excess reagent
     at the same pH, then measure the absorbance of TeL_4 at 416 nm where the ab-
     sorbance of HL is negligible[10]
2.   Complex formation at pH 2.0 to 2.3 (citrate) for 30 min, extraction of TeL_4 at
     pH 6.5 where the extraction of HL is negligible, then measure the absorbance at
     335 nm[11]
3.   Complex formation and extraction of both TeL_4 and HL at 3 N HCl or pH 3.5
     (citrate), back extract L^- with a buffer solution (pH 7 to 8), then measure the
     absorbance at 330 nm

Sandell's sensitivities of the first, second, and third procedures are 0.026 $\mu g/cm^2$ (ε =
4.9 × 10^3),[10] 0.005 (ε = 2.8 × 10^4),[11] and 0.0036 (ε = 3.5 × 10^4),[12] respectively, but
As(III), Fe(III), Sb(III), and Se(IV) interfere seriously.

Elements, such as As(V) (3 to 4 N HCl, 335 nm, ε = 1.62 × 10^4 in $CHCl_3$),[13] Os(VIII) (pH 3, 770 to 780 nm, $\varepsilon = 1.17 \times 10^4$),[14] Pd(II) (0.75 N HClO$_4$, 450 nm, $\varepsilon = 4.6 \times 10^3$ in TBP),[15] Re(VIII) (6 to 7 NHCl, 360 nm, $\varepsilon = 2.2 \times 10^4$ in iso-pentyl alcohol-acetone-water (1:1:3)[7] Se(pH 7.5, 330 nm, $\varepsilon = 3.16 \times 10^4$),[8] are also determined with Bismuthiol II.

Other Reagents with Related Structure

Other Bismuthiol Reagents

The following mercaptothiazole derivatives are reported:

Bismuthiol I; 3,5-Dimercapto-1,3,4-thiadiazole (1);[15-17]

3-(2'-Naphthyl)-5-mercapto-1,3,4-thiadiazol-2-thione (2).

The latter reagent has been used for the determination of Bi (325 nm, $\varepsilon = 2.97 \times 10^4$ in CHCl$_3$),[18] Pd (345 nm, $\varepsilon = 2.1 \times 10^4$ in cyclohexane),[19] and Te (326 nm, $\varepsilon = 3.0 \times 10^4$ in CHCl$_3$).[20]

3-(2'-Pyridyl)-5-mercapto-1,3,4-thiadiazol-2-thione (3) has been used for the determination of Bi (pH 2 to 3, 340 nm, $\varepsilon = 2.9 \times 10^4$ in CHCl$_3$),[21] Pd, Se, and Te.[22]

(1)

(2): R =

(3): R =

2-Mercaptobenzothiazole (4), 2-Mercaptobenzimidazole (5), and 2-Mercaptobenzoxazole (6)

(4): X = S
(5): X = NH
(6): X = O

2-Mercaptobenzothiazole (Mertax, Thiotax, Captax, ® MBT), C$_7$H$_5$NS$_2$, mol wt = 167.24, is a colorless or pale yellow crystalline powder, mp 179 to 182°.[23] It is insoluble in water, but is soluble in aqueous alkali and common organic solvents, including ethanol (2 g/100 mℓ; λ_{max}, 327 nm),[24] benzene (1 g/100 mℓ; 25°), and acetone (10 g/100 mℓ); p$Ka_1 = 7.22$ and p$Ka_2 = 11.4$ ($\mu = 0.1$, 20°).[25] = 7.22 and p$Ka_2 = 11.4$ ($\mu = 0.1$, 20°).[25]

It reacts with wide variety of soft metal ions to form yellow to yellow-orange insoluble complexes and has been used as a gravimetric reagent for Ag, Au, Bi, Cd (K$_{sp}$ = 8.9×10^{-9}), Ir, Ni (K$_{sp} = 4.9 \times 10^{-15}$), Pb (K$_{sp} = 6.1 \times 10^{-14}$), Pt, Rh, and Ru[26] and as an extraction photometric reagent for Au, Bi, Ir, Os, Pd, Pt, Rh, and Ru after extraction with chloroform.[27,28]

Some other substituted mercaptobenzothiazoles have been investigated for use as an extraction photometric reagent.[29,30]

2-Mercaptobenzimidazole (5), C$_7$H$_6$N$_2$S, mol wt = 150.20, glistening colorless platelets when recrystallized from 95% ethanol, mp 303 to 304°,[36] p$Ka = 9.97$ ($\mu = 0.1$, 20°),[25] is slightly soluble in water and soluble in alcohol. The complexing behavior of this reagent is almost similar to that of 2-mercaptobenzothiazole, and it has been used as an extraction photometric reagent for Bi (350 nm, $\varepsilon = 3.4 \times 10^4$ in ethanol-$CHCl_3$),[31] Hg(II) (pH 4.5, extracted as $HgL_2 \cdot$(Bromocresol Purple)$_2$, 410 nm, $\varepsilon = 3.8 \times 10^4$ in $CHCl_3$-iso-PrOH (20:1))[32] and Se (325 nm, $\varepsilon = 1 \times 10^4$ in BuOH-$CHCl_3$).[33]

1-Benzyl-2-mercaptobenzimidazole has been used for the determination of Te(IV) (440 nm, $\varepsilon = 2.95 \times 10^4$, 1 to 5 ppm).[34]

2-Mercaptobenzoxazole (6), C_7H_5ONS, mol wt = 151.18, colorless crystals, mp 193 to 195°,[36] is insoluble in water, but is soluble in aqueous alkali and common organic solvents; p$Ka_1 = 6.58$ and p$Ka_2 = 11.46$. This reagent has been recommended as an extraction photometric reagent for Os(III) (pH 5, 630 nm, $\varepsilon = 4 \times 10^3$) and Ru(III) (pH 5, 460 nm, $\varepsilon = 2 \times 10^3$).[35]

MONOGRAPHS

M1. **Welcher, F. J.**, *Organic Analytical Reagents*, Vol. 3, Von Nostrand, Princeton, N.Y., 1947.
M2. **Pribil, R.**, *Analytical Applications of EDTA and Related Compounds*, Pergamon Press, Oxford, England, 1972.

REVIEW

R1. **Yoshida, H. and Hikime, S.**, Bismuthiol II as an analytical reagent, *Bunko Kenkyu*, 14, 131, 1965.

REFERENCES

1. **Rush, M.**, *Ber.*, 27, 2510, 1894.
2. **Dubky, J. V. and Tartilek, J.**, *Fresenius Z. Anal. Chem.*, 96, 412, 1934, *Chem. Obzov.*, 9, 142, 171, 173, 205, 1934.
3. **Majumdar, A. K., Singh, B. R., and Chakrabartty, M. M.**, *Fresenius Z. Anal. Chem.*, 154, 262, 413; *Fresenius Z. Anal. Chem.*, 155, 1, 7, 81, 85; *Fresenius Z. Anal. Chem.*, 156, 103, 265, 1957.
4. **Majumdar, A. K. and Singh, B. R.**, *Fresenius Z. Anal. Chem.*, 161, 81, 1958.
5. **Sedivec, V.**, *Chem. Listy*, 45, 177, 1951; *Collect. Czech. Chem. Commun.*, 16, 398, 1951.
6. **Majumdar, A. K. and Chakrabartty, M. M.**, *Anal. Chim. Acta*, 19, 482, 1958.
7. **Lazarev, A. I. and Lazareva, V. I.**, *Zh. Anal. Khim.*, 32, 751, 1977; *J. Anal. Chem. USSR*, 32, 597, 1977.
8. **Yoshida, H., Taga, M., and Hikime, S.**, *Bunseki Kagaku*, 14, 1109, 1965.
9. **Majumdar, A. K.**, *J. Ind. Chem. Soc.*, 21, 347, 1944.
10. **Jankovsky, J. and Ksir, O.**, *Talanta*, 5, 238, 1960.
11. **Cheng, K. L.**, *Talanta*, 8, 301, 1961.
12. **Yoshida, H., Taga, M., and Hikime, S.**, *Talanta*, 13, 185, 1966.
13. **Minami, T., Yamamoto, Y., and Ueda, S.**, *Bunseki Kagaku*, 27, 419, 1978.
14. **Majumdar, A. K. and Bhowal, S. K.**, *Anal. Chim. Acta*, 62, 223, 1972.
15. **Tomioka, H. and Terajima, K.**, *Bunseki Kagaku*, 22, 264, 1973.
16. **Dubsky, J. V., Okac, A., Okac, B., and Trtilek, J.**, *Fresenius Z. Anal. Chem.*, 98, 184, 1934.

17. Zaidi, S. A. A. and Islam, V., *Indian J. Chem.*, 15A, 473, 1977; *Chem. Abstr.*, 87, 157866f, 1977.

18. Busev, A. I., Simonova, L. N., and Gaponyuk, E. I., *Zh. Anal. Khim.*, 23, 59, 1968.

19. Busev, A. I., Simonova, L. M., Shishkov, A. N., and Toleva, A. D., *Zh. Anal. Khim.*, 29, 1134, 1974.

20. Busev, A. I., Simonova, L. N., and Zayukova, N. D., *Zh. Anal. Khim.*, 22, 1850, 1967; *Tr. Kom. Anal. Khim. Akad. Nauk SSSR*, 17, 218, 1969; *Chem. Abstr.*, 72, 128316m, 1970.

21. Busev, A. I., Simonova, L. N., Sedova, A. A., and Samkaeva, L. T., *Zh. Anal. Khim.*, 29, 1986, 1974.

22. Busev, A. I. and Simonova, L. N., *Geokhim. Anal. Metody Izuch. Veshchestv. Sostava Osad. Porod Rud*, 2, 86, 1974; *Chem. Abstr.*, 87, 77833c, 1977.

23. Weyker, R. G. and Ebel, R. H., U.S. Patent, 2,730,528; *Chem. Abstr.*, 51, 1289c, 1957.

24. Terada, K., Inoue, A., Inamura, J., and Kiba, T., *Bull. Chem. Soc. Jpn.*, 50, 1060, 1977.

25. Hopkala, H. and Przyborowski, L., *Ann. Univ. Mariae Curie-Skalodowska Sect. D*, 26, 57, 1971; *Chem. Abstr.*, 79, 26814z, 1973.

26. Galkina, E. I., and Ragdasarov, K. N., and Manko, V. I., *Fiz. Khim. Methody Anal. Kontrolya Proizvod. Mezhvuz. Sb.*, 2, 21, 1976; *Chem. Abstr.*, 87, 161124e, 1977.

27. Satake, M. and Suzuki, T., *Fukui Daigaku Kogakubu Kenkyu Hokoku*, 27, 1, 1979; *Chem. Abstr.*, 91, 116732y, 1979.

28. Diamantatos, A., *Anal. Chim. Acta*, 66, 147, 1973; *Anal. Chim. Acta*, 67, 317, 1973.

29. Vasilenko, L. F., Nemodruk, A. A., and Solozhenkin, P. M., *Izv. Akad. Nauk Tadzh. SSR Otd. Fiz. Mat. Geol. Khim. Nauk*, 117, 1975; *Chem. Abstr.*, 85, 83722r, 1976.

30. Korotkaya, E. D., Klibus, A. Kh., Pochinok, V. Ya., and Gaiduk, S. K., *Ukr. Khim. Zh.*, 42, 755, 1976; *Anal. Abstr.*, 32, 2B35, 1976.

31. Uvarova, K. A., Usatenko, Yu. I., and Chagir, T. S., *Zh. Anal. Khim.*, 28, 693, 1973.

32. Kovalenko, A. A., Uvarova, K. A., Usatenko, Yu. I., and Zubtsova, T. I., *Zh. Anal. Khim.*, 32, 270, 1977.

33. Busev, A. I., *Talanta*, 11, 485, 1964.

34. Melchekova, Z.E., Mukrushina, G. A., and Bendnyagina, N. P., *Zavod. Lab.*, 42, 257, 1976; *Anal. Abstr.*, 31, 4B162, 1976.

35. Lonakina, L. N., Ignateva, T. I., and Busev, A. I., *Zh. Anal. Khim.*, 33, 527, 1978.

36. Van Allan, J. A. and Deacon, B. D., *Org. Synth.*, IV, 569, 1963.

THIOTHENOYLTRIFLUOROACETONE

$C_8H_5OS_2F_3$
mol wt = 228.24

HL

Synonyms

1,1,1-Trifluoro-4-mercapto-4-(2-thienyl)-but-3-en-2-one, STTA, Monothio-TTA

Source and Method of Synthesis

Commercially available. It is prepared by passing H_2S gas in an ethanolic hydrochloric acid solution of 2-thenoyltrifluoroacetone (TTA).[1]

Analytical Uses

Used as an extraction reagent and a chromogenic reagent for soft metal ions.

Properties of Reagent

Red needles, mp 61 to 62°[2] (or 72 to 74°)[3]; is easily soluble in benzene and carbon tetrachloride and is slightly soluble in dioxane, DMF, and ethanol. Of the three possible tautomeric isomers, it exists almost entirely as a thioenol form; pKa (SH) = 3.96,[4] log K_D (CCl_4/H_2O) = 3.49, ($CHCl_3/H_2O$) = 4.29, and (C_6H_6/H_2O) = 4.30.[5] The absorption spectrum of STTA is illustrated in Figure 1 (λ_{max}, 370 nm; ε = 1.60 × 10⁴ in xylene).[6]

Similar to other thiol reagents, it is easily oxidized in the air; hence STTA has to be stored in an ampoule under nitrogen and in a dark cool place.

Complexation Reactions and Properties of Complexes

STTA reacts with soft metal ions to form reddish chelates which are fairly stable in the air and are easily soluble in nonpolar solvents like benzene, carbon tetrachloride, or petroleum ether, but not so soluble in dioxane, DMF, or ethanol. The physical properties of some of STTA chelates are summarized in Table 1. The values of λ_{max} of STTA metal chelates vary from metal to metal and from solvent to solvent.

STTA has been used as an extraction reagent and a photometric reagent for soft metal ions. Absorption spectra of STTA and Hg(II)-STTA chelate are illustrated in Figure 1. The extraction equilibria on metal-STTA chelates were investigated on various solvents in comparison with the TTA extraction,[5,7] and some equilibrium data are summarized in Table 2. For Hg(II) chelate, log β_2 = 33.0 and log K_{ex} ($CHCl_3/H_2O$) = 19.6.[8]

Some metal-STTA chelates are known to sublime without decomposition, and the temperature of sublimation at 1.3 Torr are Ni (185°), Cu(II) (195°), Pb (150°), Zn (150°), Co(II) (140°), Cd (165°), and Pd (109°).[1] They may be separated by gas chromatography.

Purification and Purity of Reagent

STTA is easily oxidized in a usual storage condition and has to be purified before use. STTA can be purified either by recrystallization from benzene or by dissolution in petroleum ether, extraction into 1 M NaOH solution, acidification of aqueous phase with 1 to 6 M HCl solution, back-extraction into petroleum ether, and final evaporation of the solvent to get red crystals.

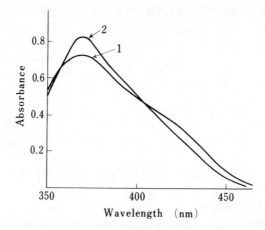

FIGURE 1.   Absorption spectra of STTA and its Hg chelate in xylene. (1) Reagent 5×10^{-5} M, $\varepsilon = 1.6 \times 10^4$; (2) Hg-STTA (47 µg Hg), $\varepsilon = 3.0 \times 10^4$. (From Hashitani, H. and Katsuyama, K., *Bunseki Kagaku*, 19, 355, 1970. With permission.)

Table 1
PROPERTIES OF METAL-STTA CHELATES[1]

Composition	Color of solid	λ_{max} (nm)[a]	$\varepsilon (\times 10^3)$[a]	λ_{max}(nm)[b]
HL	Red	355	8.45	—
NiL$_2$	Brown-red	358	7.75	373, 450
CuL$_2$	Olive-green	338	5.60	400
ZnL$_2$	Yellow-green	374	4.22	395
CoL$_2$	Brown-black	362	7.52	375
PbL$_2$	Orange	345	3.70	373
CdL$_2$	Yellow	364	4.35	383
PdL$_2$	Brown-red	369	6.95	415
HgL$_2$	—	—	—	370

[a]   Solvent, absolute ethanol.
[b]   Solvent, xylene.

Table 2
DISTRIBUTION RATIOS OF STTA CHELATES AT FIXED LIGAND ION CONCENTRATION[5]

	log D[a]		
Solvent/water	CuL$_2$ at [L$^-$] = 10^{-10} M[b]	ZnL$_2$ at [L$^-$] = 10^{-6} M	CoL$_3$ at [L$^-$] = 10^{-6} M
Benzene	0.78	0.24	1.65
Carbon-tetrachloride	0.84	−0.17	2.19
Chloroform	0.76	0.51	−0.30
Cyclohexane	−0.80	−1.95	0.84

[a] $D = [ML_2]_{org}/[M^{2+}]_{aq}$.
[b] $[L^-] = [HL]_{org\ init.}/[\{(K_D(HL) + 1)[H^+]/K_a\} + 1]$, where $[HL]_{org\ init.}$ is an initial reagent concentration in the organic phase, and K_D (HL) is a distribution coefficient of STTA(HL).

The purity may be checked by running TLC on a silica-gel plate (0.5 mm) with a mixture of chloroform-n-hexane (3:7) as an elution solvent. Rf = 0.8.[9]

Determination of S is another criteria of purity.

Analytical Applications

Use as a Photometric Reagent

Some representative examples on the use of STTA as a photometric reagent for metals are summarized in Table 3. The advantage of STTA as a photometric reagent is its selectivity for soft metal ions, however, the molar absorptivities of metal chelates are not so high ($\sim 10^3$). Moreover, as the absorption peak of the metal chelates overlaps with that of free reagent (see Figure 1), the excess reagent has to be back-extracted before photometry, or the observing wavelength has to be chosen so that the reagent blank is low.

Use as an Extraction Reagent

With the use of STTA, various combinations of mixtures of metals can be separated by the solvent extraction under the proper choice of pH in the aqueous phase.

For example, Co(II) and Ni can be extracted quantitatively at pH 5.5 to 6.0, then Ni is back-extracted into the aqueous phase at pH 1.5, leaving Co in the organic phase.[21] For the mixture of Fe(II), Fe(III), and Cu(II), only Cu is extracted into cyclohexane solution of STTA at pH 2 (acetate), then Fe(II), (III) is extracted at pH 5 to 6 in the presence of pyridine and hydroxylamine as FeL_2.[3]

The use of auxiliary ligand is often effective for the separation of metals, whose $pH_{1/2}$ values are fairly close. Separation of Zn and Cd by STTA extraction is difficult, as the $pH_{1/2}$ values (5.0×10^{-3} M STTA in dichloroethane) for each element are 4.5 and 4.9, respectively. However, in the presence of 2,2'-bipyridine, only Cd is likely to form CdL_2(bipy), and at the same time, it is expected as a masking agent for Zn, enabling the preferential extraction of Cd at pH 3.6 (with 0.1 M NaCl) or pH 3.0 (with $NaNO_3$).[28]

Other Uses

Metal ions can be separated through their STTA chelates by various means, such as fractional sublimation,[1] TLC,[3] extraction chromatography,[29] or liquid chromatography on polystyrene beads.[30] It is also suggested as an extraction reagent for heavy metal ions in the determination by atomic absorption photometry.[31]

Other Reagents with Related Structure

Monothio Derivatives of β-Diketone

Many monothio derivatives of β-diketone have been prepared and evaluated as a selective extraction reagent and as a extraction photometric reagent for soft metal ions.[R1,R2]

Monothio-acetylacetone,[39] monothio-benzoylacetone,[32] monothio-dibenzoylmethane,[33] monothio-benzoyltrifluoroacetone,[34] monothio-dipivaloylmethane, monothio-heptafluorobutanoylpivaloylmethane,[35] and 3-thionaphthoyltrifluoroacetone[36] are such examples.

Monoseleno derivatives of TTA,[38] selenoylacetone, and selenoyltrifluoroacetone[37] have also been investigated.

Salicylideneamino-2-thiophenol (SATP): $C_{13}H_{11}NOS$, mol wt = 229.30

Table 3
APPLICATIONS OF STTA AS A PHOTOMETRIC REAGENT

Metal ion[a]	Condition	Metal chelate			Solvent	Range of determination (ppm)	Remarks	Ref.
		Ratio	λ^b (nm)	$\varepsilon (\times 10^3)$				
Ag	pH 0.5 ~ 1.0	ML	460	3.5	CCl₄– BuOH	1 ~ 1.2	Bi, Hg, In, Pd, Zn seriously interfere	10
Cd	pH 11.5, 1,10-phen(X)	ML₂X	370	34.3	Xylene	1 ~ 35	Co(II), Cu(II), Hg(II), Mn(II), Zn can be masked with CN⁻	11
Cu(II)	pH 2 ~ 5	ML₂	490	6.0	CCl₄	1 ~ 12	Cd, Fe(III), Ni, Rh(III), Ru(III), citrate interfere	12
Hg(II)	5 NH₂SO₄, strip excess HL with borate (pH 11.5)	ML₂	370	16	Xylene	0.2 ~ 8	Ag, Pb, Pd, Sn(II), CN⁻ interfere	6, 13
Pb	pH 6.5	ML₂	480	4.1	CCl₄	1 ~ 10	Pt(IV) interferes	14
Pd(II)	pH 5.6 ~ 6.0	ML₂	410	3.7	Xylene	0 ~ 3	Cu(II), CN⁻, S²⁻ interfere	15
UO₂²⁺	pH 5.5 ~ 6	ML₂	490	—	CCl₄–BuAc (7:3)	4.5 ~ 54(U)	Co, Cu, Ni, Pb, Pd interfere	16
V(IV)	pH 5.0	—	450	3.5	CCl₄–BuOH (1:1)	~11	Be, Co, Nb, Ni, Pt, Ru, Ti(IV) interfere	17
Zn	pH 7.0 ~ 7.5	ML₂	450	6.7	CCl₄	0.4 ~ 30.3	Be, Cr(II), Fe(III), In interfere	18

[a] Other elements determined by STTA-photometry include: Au,[19] Bi,[20] Co(II),[21] Fe,[3] Ga,[40] In,[22] Mn(II),[23] Ni,[24] Np,[41] Tl(III),[25] CN⁻[26] (with Hg-STTA) and S²⁻[27] (with Cu-STTA).

[b] Observing wavelength which does not always coincide with λ_{max}.

Colorless needles, mp 133 to 136°; is insoluble in water, is slightly soluble in dilute acids to give a colorless solution, and is soluble in alkali hydroxide or carbonate to give an yellow solution. It is also soluble in common organic solvents and is easily hydrolyzed in an acidic solution. At pH 2, SATP reacts with various heavy metal ions to form an insoluble yellow or colorless chelates, but only Sn(IV) and (II) form an extractable yellow complex (λ_{max}, 415 nm; $\varepsilon = 1.61 \times 10^4$ in benzene). Thus, the reaction is fairly selective for Sn, and SATP has been recommended as an extraction photometric reagent for Sn.[42] In the presence of an auxiliary ligand like pyridine or 1,10-phenanthroline or in the extraction with donating solvent like chloroform, however, some other metal ions can also be extracted for the photometry; Cu(pH 4.0 to 12.3, pyridine, extraction with benzene, 415 nm, $\varepsilon = 1.1 \times 10^4$),[43] Ni(pH 9.8, 420 nm, $\varepsilon = 1.26 \times 10^4$),[44] and In (pH 4.6 to 6.0, 1,10-phenanthroline, extraction with CHCl$_3$, 425 nm, $\varepsilon = 1.1 \times 10^4$).[45]

REVIEWS

R1. Cheston, S. H. H., Livingston, S. E., Lockyer, T. N., Pickles, V. A., and Shannon, J. S., Thio derivatives of β-diketones and their metal chelates. I. Some monothio β-diketones and their nickel(II) chelates, *Aust. J. Chem.*, 18, 673, 1965.

R2. Honjo, T. and Kiba, T., Solvent extraction reagent, *Bunseki Kagaku*, 21, 676, 1972.

R3. Watanabe, K. and Kawaguchi, S., Thio-β-diketone metal complex, *Kagaku (Kyoto)*, 27, 991, 1972.

REFERENCES

1. Berg, E. W., *Anal. Chim. Acta*, 36, 372, 1966.
2. Purity Specification of Organic Reagents, Dojindo Laboratories Ltd., Kumamoto, Japan.
3. Honjo, T., Fujioka, Y., Itoh, H., and Kiba, T., *Anal. Chem.*, 49, 2241, 1977.
4. Uhlemann, E. and Muller, H., *Z. Chim.*, 8, 185, 1968, *Chem. Abstr.*, 69, 30669s, 1968.
5. Honjo, T., Honda, R., and Kiba, T., *Anal. Chem.*, 49, 2246, 1977.
6. Hashitani, H. and Katsuyama, K., *Bunseki Kagaku*, 19, 355, 1970.
7. Shinde, V. M. and Khopkar, S. M., *Chem. Ind.*, 1785, 1967; *Chem. Abstr.*, 67, 113351s, 1967.
8. Yokoyama, A., Nakanishi, N., and Tanaka, H., *Chem. Parm. Bull.*, 20, 1856, 1972; *Chem. Abstr.*, 77, 144512f, 1972.
9. Muller, H. and Rother, R., *Anal. Chim. Acta*, 66, 49, 1973.
10. Solanke, K. R. and Khopkar, S. M., *Mikrochim. Acta*, 41, 1976.
11. Deguchi, M. and Kiyokawa, N., *Eisei Kagaku*, 22, 308, 1976; *Chem. Abstr.*, 87, 62049c, 1977; *Anal. Abstr.*, 32, 5B60, 1977.
12. Shinda, V. M. and Khopkar, M., *Anal. Chem.*, 41, 342, 1969.
13. Shinde, V. M. and Khopkar, M., *Indian J. Chem.*, 11, 485, 1973; *Anal. Abstr.*, 26, 1440, 1974; *Chem. Abstr.*, 79, 73212m, 1973.
14. Skki, S. and Khopkar, S. M., *Bull. Chem. Soc. Jpn.*, 45, 167, 1972.
15. Deguchi, M. and Hamamura, T., *Bunseki Kagaku*, 28, 575, 1979.
16. Solanke, K. R. and Khopkar, S. M., *Chem. Anal. (Warsaw)*, 17, 1175, 1972; *Anal. Abstr.*, 24, 3406, 1973.
17. Solanke, K. R. and Khopkar, S. M., *Talanta*, 21, 245, 1974.
18. Solanke, K. R. and Khopkar, S. M., *Bull. Chem. Soc. Jpn.*, 46, 3082, 1973.
19. Akki, S. B. and Khopkar, S. M., *Indian J. Chem.*, 10, 125, 1972; *Anal. Abstr.*, 23, 3740, 1972.
20. Solanke, K. R. and Khopkar, S. M., *Anal. Lett.*, 6, 31, 1973.
21. Honjo, T., Unenoto, T., and Kiba, T., *Bunseki Kagaku*, 23, 203, 1974.
22. Solanke, K. R. and Khopkar, S. M., *Anal. Chim. Acta*, 66, 307, 1973.

23. Solanke, K. R. and Khopkar, S. M., *Fresenius Z. Anal. Chem.*, 275, 286, 1975.
24. Mulye, R. R. and Khopkar, S. M., *Sep. Sci.*, 7, 605, 1972; *Anal. Abstr.*, 25, 206, 1974.
25. Patil, P. S. and Shinde, V. M., *Talanta*, 24, 696, 1977.
26. Yashiki, M. and Deguchi, M., *Bunseki Kagaku*, 20, 1192, 1971.
27. Deguchi, M. and Kitamura, A., *Bunseki Kagaku*, 29, 527, 1978.
28. Akaiwa, H., Kawamoto, H., and Tsutsumi, Y., *Bunseki Kagaku*, 27, 447, 1978.
29. Honjo, T. and Kiba, T., *Bull. Chem. Soc. Jpn.*, 45, 185, 1972.
30. Suzuki, N., Suzuki, J., and Saitoh, K., *J. Chromatogr.*, 177, 166, 1979.
31. Deguchi, K., Okumura, I., Deguchi, M., and Kinoshita, T., *Bunseki Kagaku*, 26, 507, 1977; *Bunseki Kagaku*, 27, 125, 1978.
32. Murti, M. V. R. and Khopkar, S. M., *Chem. Ind.*, 33, 1978; *Anal. Abstr.*, 35, 1B34, 1978.
33. Uhlemann, E. and Muler, H., *Anal. Chim. Acta*, 41, 311, 1968.
34. Rao, G. N. and Chouhan, V. S., *Indian J. Chem. Sect. A*, 16, 177, 1978; *Anal. Abstr.*, 35, 5B79, 1978.
35. Burton, R. C. and Sweet, T. R., *Anal. Chim. Acta*, 64, 273, 1973.
36. Johnson, J. R., Holland, W. J., and Gerard, H., *Microchim. Acta*, 608, 1972.
37. Yurev, Yu. K. and Menzentsova, N. N., *Zh. Obshch. Khim.*, 31, 1449, 1961; *Chem. Abstr.*, 55, 23491g, 1961.
38. Honjo, T., *Chem. Lett.*, 481, 1974.
39. Uden, P. C., Nonnemakar, K. A., and Geiger, W. E., *Inorg. Nucl. Chem. Lett.*, 14, 161, 1978.
40. Dhond, P. C. and Khopkar, S. M., *Indian J. Chem. Sect. A*, 15, 934, 1977; *Anal. Abstr.*, 35, 2B62, 1978.
41. Ramanujam, A., Ramakrishna, V. V., and Patil, S. K., *Sep. Sci. Technol.*, 14, 13, 1979; *Anal. Abstr.*, 37, 4B83, 1979.
42. Gregory, G. R. E. C. and Jeffery, P. G., *Analyst*, 92, 293, 1967.
43. Ishii, H. and Einaga, H., *Nippon Kagaku Zasshi*, 91, 734, 1970.
44. Sarkar, A. M. and Malek, A., *Nucl. Sci. Appl. Ser. B*, 10, 91, 1977; *Chem. Abstro.*, 90, 97011y, 1979.
45. Fujiwara, A., Watanabe, K., and Kawagaki, K., *Bunseki Kagaku*, 27, 645, 1978.

THIO-MICHLER'S KETONE

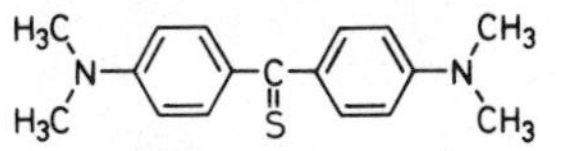

$$H_3C \text{—} N \text{—} C_6H_4 \text{—} \overset{\parallel}{\underset{S}{C}} \text{—} C_6H_4 \text{—} N(CH_3)_2$$

$$C_{17}H_{20}N_2S$$

mol wt = 284.42

Synonyms

4,4′-Bis(dimethylamino)thiobenzophenone, TMK

Source and Method of Synthesis

Commercially available. It is prepared by the reaction of Auramin with H_2S.[1]

Analytical Uses

As a highly sensitive chromogenic reagent for Hg, Pd, and the noble metals.

Properties of Reagent

It is a dark red fine crystalline powder, mp 202°,[1] and is insoluble in water, but soluble in alcohol to give an intense yellow solution (λ_{max}, 457 nm; $\varepsilon = 2.92 \times 10^4$ in 30 vol % n-propanol).[2] The solution of the reagent as well as its complexes are light sensitive, but the n-propanol solution can be kept for 40 days in the dark at room temperature.

Complexation Reactions and Properties of Complexes

Thio-Michler's ketone was first suggested as an indicator in the titration of Hg with the standard NaCl solution,[3] but has been recently suggested as a photometric reagent for Hg and other soft metal ions, such as Au and Pd.[R1] These metal ions react with Thio-Michler's ketone at pH 3 in a medium of aqueous alcohol to give red-purple complexes which can be extracted into iso-amylalcohol. The reactions with these metal ions may be written as below, where n is 2, 3, and 4 for Au, Hg(II), and Pd(II), respectively.[4]

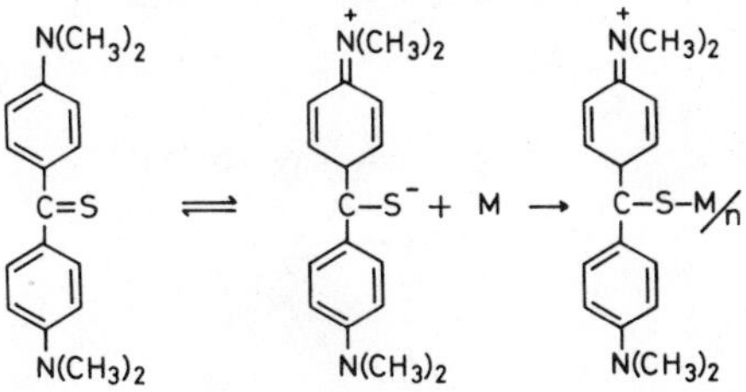

However, these numbers do not always indicate the composition of the complexes because the partial loss of the reagent by oxidation with the metal ion may also occur (Au(III) → Au(I)).[5]

The absorption spectra of the reagent and Hg-complex are illustrated in Figure 1 (λ_{max}, 457 nm; $\varepsilon = 2.92 \times 10^4$).[2]

The complexes with Hg(II) and Pd(II) are so stable that they are not affected by EDTA. Hence, EDTA can be used as an effective masking agent in the determination of Hg and Pd with this reagent.[4]

Purification and Purity of Reagent

Thio-Michler's ketone can be purified by the recrystallization from hot ethanol or tritrating with a small amount of chloroform, followed by filtration and washing with cold ethanol.[1]

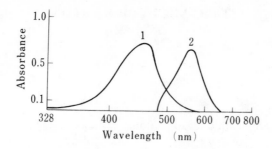

FIGURE 1.   Absorption Spectra of Thio-Michler's ketone and Hg complex in aqueous-*n*-propanol (30 vol %
n-propanol). (1) Reagent, 5×10^{-5} *M*; (2) Hg(II) complex, 5×10^{-6} *M*; pH 5.8. (From Ackermann, G. and
Roder, H., *Talanta*, 24, 99, 1977. With permission.)

The purity of the reagent can be checked by observing the TLC on a mixture of
alumina and silica-gel (1:1) with benzene as an eluent. A pure material should give one
spot.[2]

Analytical Applications

The main application of Thio-Michler's ketone is for the photometric determination
of Hg and other soft metal ions. The photometry was carried out directly either on
the aqueous alcoholic solutions or after the extraction with higher alcohol. The typical
examples are summarized in Table 1. The sensitivity for Hg is reported to be the highest
among the mercury reagents, Sandell's sensitivities being 0.0013 to 0.0023 $\mu g/cm^2$
($\varepsilon = 15.1$ to 8.8×10^4) with Thio-Michler's ketone, whereas they are 0.003 to 0.004
$\mu g/cm^2$ ($\varepsilon = 6.7$ to 5.0×10^4) with dithizone.[2]

Thio-Michler's ketone is also used as a detection reagent of metal ions, such as Pd
(red-purple; dilution limit, $1:10^7$), Hg(II) (blue-green; dilution limit, $1:5 \times 10^6$), Ag
(red-purple; dilution limit, $1:5 \times 10^6$), Cu(I) (orange-pink; dilution limit, $1:10^6$), Au
(blue-purple; dilution limit, $1:10^6$), and Pt (red-purple).[3]

Table 1

APPLICATIONS OF THIO-MICHLER'S KETONE AS A PHOTOMETRIC REAGENT

Metal ion	Condition	Complex			Extraction solvent	Remarks	Ref.
		Ratio	λ_{max} (nm)	$\varepsilon(\times10^4)$			
Ag	pH 3.0	—	520	8.85	iso-AmOH	Low selectivity	6
Au(III)	pH 3.0, 50% alcohol, pH 3.0	1:2	545	—	Aqueous	—	5
		1:2	545	15.0	iso-AmOH	Negative error with EDTA; CN^- masks the reaction	5
Hg(II)	pH 3.2, 20~60% alcohol	1:3	560	11.7	Aqueous	EDTA can be used as a masking agent	4
Hg(II)	pH 5.8, 30% n-propanol	1:1	560	15.1	Aqueous	0.007 to 1.5 ppm; highest sensitivity for Hg; Ag, Au(III), Pd(II), and Sn(II) interfere	2
Hg(II)	pH 3.2	1:3	550	8.8	iso-AmOH	—	4
Pd(II)	pH 3.0	1:4	520	21.2	Aqueous	—	4
Pd(II)	pH 3.0	1:4	520	16.0	iso-AmOH	EDTA can be used as a masking agent	4
Pd(II)	pH 2.8		520	13.0	n-BuOH	Ag, Au, Hg, Pt, Rh, Ru, and Tl interfere	7

REVIEW

R1. **Pilipenko, A. T., Ryabushko, O. P., Savranskii, L. I., Kribokhizhina, L. O., and Matsibura, G. S.,** Thio-Michler's ketone as an analytical reagent, *Visn. Kiiv. Univ. Ser. Khim.,* 22, 19GD, 1972 Abstr. 22G9; *Anal. Abstr.,* 25, 1439, 1973.

REFERENCES

1. **Tarbell, D. C. and Wystrach, V. P.,** *J. Am. Chem. Soc.,* 68, 2110, 1964.
2. **Ackermann, G. and Roder, H.,** *Talanta,* 24, 99, 1977.
3. **Gehauf, B. and Goldenson, J.,** *Anal. Chem.,* 22, 498, 1950.
4. **Cheng, K. L. and Goydish, B. L.,** *Microchem. J.,* 10, 158, 1966.
5. **Cheng, K. L. and Lott, P. F.,** *Proceedings of International Symposium on Microchemical Techniques,* John Wiley & Sons, New York, 1962, 317.
6. **Cheng, K. L.,** Spectrophotometry and fluorometry, in *Trace Analysis: Physical Methods,* Morrison, G. M., Ed., John Wiley & Sons, New York, 1965.
7. **Tsukahara, I.,** *Bunseki Kagaku,* 28, 253, 1979.

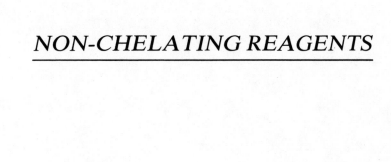

NON-CHELATING REAGENTS

TRI-N-BUTYL PHOSPHATE

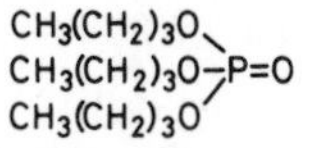

$$CH_3(CH_2)_3O$$
$$CH_3(CH_2)_3O-P=O$$
$$CH_3(CH_2)_3O$$

$C_{12}H_{27}O_4P$
mol wt = 266.32

Synonym
TBP

Source and Method of Synthesis
Commercially available. It is prepared by the reaction of n-butanol with $POCl_3$.

Analytical Uses
It has been widely used as a solvent extraction reagent for various multivalent metal ions.

Properties of Reagent
The organophosphorous extraction reagents can be considered as the derivatives of phosphoric acid.

(1)	(2)	(3)	(4)
Trialkyl phosphate (TBP)	Alkyldialkyl phosphonate	Dialkylalkyl phosphinate	Phosphine oxide (TOPO)

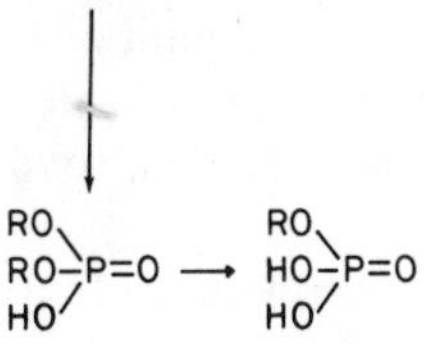

(5)	(6)
Dialkyl phosphate (HDEHP)	Monoalkyl phosphate

Since the first introduction of TBP in 1949 as a solvent extraction reagent for Ce, Th, and U from nitric acid solution,[M1] various kinds of organophosphorous reagents have been investigated for use as an extraction reagent.[R1-R3] The extraction of metal ion with these phosphorous reagents is based on the formation of the coordinatively solvated salt such as $UO_2(NO_3)_2$ $(TBP)_2$ through p = O group.

The structural change from (1) to (4) results in the increase of polarity of p = O group, so that the extractability increases in that order. TOPO (4) is a very effective extraction reagent of this group.

The hydrolysis of trialkylphosphate (1) gives monobasic diester (5) and dibasic monoester (6), of which the former is more important in analytical chemistry. As an example, di(2-ethylhexyl) phosphoric acid (HDEHP) has been widely used as an extraction reagent for cations.

In the following, TBP, HDEHP, and TOPO will be treated separately.

TBP is a colorless viscous liquid, bp 152 to 154° (10 Torr)[3] or 143 to 145° (5 Torr),[9] d_4^{25}, 0.975[3]; n_D^{25}, 1.4215, is slightly soluble in water (4.2 g/100 ml, 16°),[1] and is miscible with most common organic solvents. Polar solvents are not suitable to dilute TBP for the extraction purpose because the strong interaction between the solvent molecule and TBP reduces the extractability with TBP. Hence, nonpolar solvents, such as CCl_4, kerosene, iso-octane, and benzene, are often used as a diluent to control the extractability and the viscosity of the organic phase. $D(CCl_4/H_2O) = 2$ to 3×10^2 (at 1 to 12 M HNO_3 and 0.00732 M TBP).[2]

Extraction Behavior with TBP

Extractions of metal ions with TPB are often carried out from a fairly acidic solution with hydrochloric acid or nitric acid. For example, Fe(III) is extracted as $H[FeCl_4 \cdot (TBP)_2]$ and $FeCl_3(TBP)_3$ from 6 and 2 M HCl, respectively.[M1] Hence, the change in acidity greatly affects the extractability. The increased acidity may also depress the hydrolysis of multivalent metal ions, resulting in the higher extractivity.

In the extraction of UO_2^{2+} from nitric acid solution, the reaction can be written as below.[M1]

$$UO_2{}^{2+} + 2NO_3{}^- + 2TBP_{org} \rightleftharpoons UO_2(NO_3)_2(TBP)_2{}_{org}$$

$$K_{ex} = \frac{[UO_2(NO_3)_2(TBP)_2]_{org}}{[UO_2{}^{2+}][NO_3{}^-]^2[TBP]_{org}^2}$$

The distribution ratio of U(VI) can be given as

$$D = \frac{[UO_2(NO_3)_2(TBP)_2]_{org}}{[UO_2{}^{2+}]} = K_{ex}[NO_3{}^-]^2[TBP]_{org}^2$$

Thus, the extractability can be improved by using the extractant of higher TBP concentration and by extracting from the aqueous phase of higher nitrate ion concentration. The nitrates, such as $NaNO_3$ or $Al(NO_3)_3$, are often used to improve the extractability. These salts also act as a salting out agent.

The distribution ratios of elements in the extraction with undiluted TBP from HCl or HNO_3 solutions of various acidities are summarized in Table 1. When extracting from the nitric acid solution, TBP forms an acid adduct ($TBP \cdot HNO_3$) with nitric acid, resulting in the decrease of extractability due to the lowering of TBP concentration available for the extraction reaction.[6] The prolonged contact of TBP with acid solutions also causes the partial hydrolysis of TBP, causing the change of extraction properties. A solution of 70% HNO_3 is reported to attack TBP slowly at room temperature.

TBP is also useful as a synergestic reagent for the extraction of the coordination unsaturated uncharged metal chelates, as exemplified by the following reaction:[6]

$$Ln^{3+} + 3HL + 2TBP_{org} \xrightarrow{K_{ex}} Ln(L)_3(TBP)_2{}_{org} + 3H^+$$

where Ln^{3+} and HL represent lanthanide ion and β-diketone, respectively. The values of K_{ex} for some β-diketones are summarized in Table 2.

TBP can also be used as a stationary phase in the reversed-phase partition chromatography of metal ions.[8]

Purification of Reagent

The main contaminants in the commercial samples are organic pyrophosphates, mono- and dibutylphosphates, and butanol which can be removed by treating first with 8 N HCl, then with hot 0.4% NaOH or 5% Na_2CO_3 solution, and finally repeatedly with water. The moist sample was dried at 30° under vacuum.[1,7,10]

Table 1

DISTRIBUTION RATIO OF METAL ION ON THE EXTRACTION WITH
UNIDILUTED TBP FROM ACID SOLUTION[4,5]

	log D									
	HCl (N)					HNO₃ (N)				
Metal ion	1	4	6	9	12	1	3.5	6	11	14
Ag	1.1	0.6	−0.2	−1.0	−1.4	—	—	—	—	—
As(III)	−0.7	0.5	1.3	1.7	1.6	−0.8	−1.0	−1.5	−1.7	−1.7
Au	3.9	4.1	4.3	4.6	4.3	3.9	3.2	2.2	0.5	0.2
Bi	1.3	0.3	0.7	−1.0	−1.4	0.3	−0.2	−0.7	−1.4	−1.2
Cd	0.2	1.4	1.3	0.8	0.1	—	—	—	—	—
Fe(III)	1.5	3.7	3.9	3.9	4.0	−2.3	−2.3	−2.0	−0.4	1.1
Ga	0.1	2.9	3.1	3.2	2.5	—	—	—	—	—
Ge	−1.4	0.4	1.8	1.7	2.0	—	—	—	—	—
Hg	1.8	1.8	1.6	0.9	0.2	1.0	0.0	−0.5	−0.7	−0.6
In	0.6	2.3	2.3	1.7	0.8	−2.7	−3.3	−3.0	−1.7	−0.4
Mo	0.5	2.5	2.5	2.4	2.4	−1.7	−1.8	−2.0	−0.8	
									0.4	
Nb	0.4	1.1	2.6	3.4	3.4	−0.8	0.4	0.7	1.2	1.7
Os	0.6	1.0	0.8	0.6	0.3	1.3	0.8	0.6	0.4	0.4
Pt	0.7	1.3	1.3	0.5	0.0	0.2	−0.8	−1.7	−2.5	−2.3
Pu(VI)	—	—	—	—	—	1.5	1.8	2.0	1.8	—
Re	2.1	2.5	1.8	1.1	0.4	1.2	0.2	−0.5	−1.0	−1.0
Ru	−0.8	−0.5	−0.6	−0.8	−1.2	0.8	0.2	−0.5	−1.9	—
Sb	—	1.6	2.6	3.1	2.6	—	—	—	—	—
Sc	−3.2	−1.3	0.4	2.5	3.0	−0.6	0.0	0.3	1.5	2.4
Se(IV)	−0.7	−0.6	−0.4	1.1	2.7	—	—	—	—	—
Sn	1.0	—	0.4	0.4	0.6	—	—	—	—	—
Ta	2.7	2.2	2.6	3.6	3.7	—	—	—	—	—
Te(IV)	−0.3	2.6	2.7	2.7	2.9	—	—	—	—	—
Ti	−2.7	−1.7	−1.2	−0.3	1.4	−2.2	−2.0	−1.5	—	0.6
T1(III)	3.7	4.0	4.1	4.0	3.2	−0.7	−1.0	−1.2	−2.0	−1.8
Th	−3.4	−2.7	−1.7	0.0	1.2	0.7	1.7	1.8	2.5	2.6
U(VI)	−0.5	1.6	1.9	1.8	—	1.3	2.1	2.3	—	—
V	−1.3	0.8	1.5	1.8	1.5	−1.7	−1.9	—	−1.2	−0.4
W	2.4	2.5	2.2	2.1	1.8	1.4	0.5	−0.5	−1.0	−0.8
Y	−3.0	−2.9	−2.2	−1.7	−0.6	−1.0	−0.4	0.0	0.6	1.9
Zn	0.8	1.4	1.3	0.4	−0.1	—	—	—	—	—
Zr	−2.0	−1.0	0.8	3.7	3.5	—	—	—	—	—

Table 2

SYNERGISTIC EXTRACTION OF LANTHANIDES
WITH β-DIKETONE AND TBP IN CYCLOHEXANE[6]

	Extraction reagents[a]					
	HFA - TBP		TAA - TBP		FHD - TBP	
Metal ion	$\log \beta_2$	$\log K_{ex}$	$\log \beta_2$	$\log K_{ex}$	$\log \beta_2$	$\log K_{ex}$
Eu	10.84	5.05	—	−2.22	10.00	10.06
Nd	10.50	4.35	—	−2.77	9.96	9.95
Tm	10.76	4.63	—	−2.62	10.20	10.47

[a] Key for β-diketones: HFA, Hexafluoroacetylacetone; TAA, Trifluoroac-
etylacetone; FHD, Decafluoroheptanedione.

[b]
$$\beta_2 = \frac{[ML_3 \cdot (TBP)_2]_{org}}{[ML_3]_{org}[TBP]^2} \qquad K_{ex} = \frac{[ML_2 \cdot (TBP)_2]_{org}[H^+]^3}{[M^{3+}][HL]^3[TBP]^2_{org}}$$

MONOGRAPH

M1. **Sandell, E. B. and Onishi, H.,** *Photometric Determination of Traces of Metals,* 4th ed., John Wiley & Sons, New York, 1978.

REVIEWS

R1. **Coleman, C. F., Blake, C. A., Jr., and Brown, K. B.,** Analytical potential of separations by liquid ion exchange, *Talanta,* 9, 297, 1962.

R2. **Green, H.,** Recent uses of liquid ion exchangers in inorganic analysis, *Talanta,* 11, 1561, 1964.

R3. **Green, H.,** Use of liquid ion-exchangers in inorganic analysis, *Talanta,* 20, 139, 1973.

REFERENCES

1. **Alcock, K., Grimely, S. S., Healy, T. V., Kennedey, J., and McKay, H. A. C.,** *Trans. Faraday Soc.,* 52, 39, 1956.
2. **Umezawa, H. and Hara, R.,** *Anal. Chim. Acta,* 25, 360, 1961.
3. **Nishi, T. and Asano, A.,** *Kogyo Kagaku Zasshi,* 66, 1424, 1428, 1966.
4. **Ishimori, T., Watanabe, K., and Nakamura, E.,** *Bull. Chem. Soc. Jpn.,* 33, 636, 1966.
5. **Ishimori, T. and Nakamura, E.,** JAERI Rept. - 1047, 1963.
6. **Michell, J. W. and Banks, C. V.,** *Talanta,* 19, 1157, 1972.
7. **Irving, H. and Edgington, D. N.,** *J. Inorg. Nucl. Chem.,* 10, 306, 1959.
8. **Bark, L. S., Duncan, G., and Graham, R. J. T.,** *Analyst,* 92, 347, 1967.
9. **De, A. K. and Rahama, S.,** *Talanta,* 11, 601, 1964.
10. **Peppard, D. F.,** *J. Inorg. Nucl. Chem.,* 4, 326, 1957.

DI (2-ETHYLHEXYL) PHOSPHORIC ACID

$$\left[CH_3(CH_2)_3-\overset{\overset{\displaystyle CH_2CH_3}{|}}{\underset{\underset{\displaystyle H}{|}}{C}}-CH_2-O \right]_2 \!\!\overset{\overset{\displaystyle O}{\|}}{P}-OH$$

$C_{16}H_{35}O_4P$
mol wt = 322.42

HL

Synonyms

DEHP, D₂EHPA, HDEHP

Source and Method of Synthesis

Commercially available.

Analytical Use

The reagent behaves as a liquid cation exchanger, for the extraction of cations.[R1,R2] It is also used as a stationary phase in the reversed-phase partition chromatography.[1,2,4]

Properties of Reagent

It is a viscous liquid, $d^{25} = 0.975$, bp $155°$ (1.5×10^{-2} Torr),[16] is very slightly soluble in water, but is easily soluble in common organic solvents. The reagent behaves as a monobasic acid, $pKa = 1.4$. It exists as a dimer in nonpolar organic solvents, log K (dimerization) = 4.7 in n-octane,[3] but as a monomer in alcohols, log K_D (n-octane/water) = 3.3.[3]

Extraction Behavior with HDEHP

A solution of HDEHP in nonpolar solvents, such as heptane, octane, toluene, and carbon tetrachloride, is used for the extraction of Ce(IV), Hf, Mo(VI), Nb, Sc, Th, Ti, U(VI), Zr, and lanthanide and actinide elements from HCl, HNO₃, and HClO₄ solutions.[4,5]

In general, HEDHP exists as a dimer in nonpolar solvents, and a large number of metal ions at low concentration are extracted in accordance with the following reaction:

$$M^{n+} + n(HL)_{2\ org} \rightleftharpoons M(HL_2)_{n\ org} + nH^+$$

thus, the extractability is highly dependent upon the acidity of the aqueous phase.[6] However, the formation of more complex species are reported in some instances. For example, alkaline earth metals are extracted as $M(HL_2)_2 (HL)_2$ into a benzene solution of HDEHP,[7] lanthanide metals are extracted as $ML_3(HL)_3$ from moderate acid solution into a benzene or heptane solution,[8] and Th is extracted as $ThL_2(HL_2)_2$ from HClO₄ or HCl solution of moderate acidity into a toluene solution.[9]

When metal ion is extracted with an alcohol solution of HDEHP, the reaction can be given as follows:

$$M^{n+} + nHL_{org} \rightleftharpoons ML_{n\ org} + nH^+$$

as HDEHP exists predominantly as a monomer.

In some cases, the undissociated HDEHP behaves as a solvating agent like TBP to extract metal ions as $MX_n(HL)_2$. The extractions of Sc, Ti, and Zr from strongly acid solutions (where X is Cl^-, NO_3^-, or ClO_4^-) are such examples. Gallium is also extracted as $HGaCl_4 (HL)_2$ from HCl solution ($\sim 1\ M$) containing 50% CaCl₂.[10]

Table 1
DISTRIBUTION RATIO OF METAL ION ON THE EXTRACTION WITH 50% HDEHP IN TOLUENE[2]

Metal ion	log D HCl (*M*)		
	0.01	0.1	1
Ac	2.3	−0.8	−3.7
Ag	0	−1.0	−2.5
Am	4	1.0	−2.0
Bi	—	2.2	−2.7
Cu	0	−1.8	−3.9
Fe	4	3.2	0.2
La[a]	4	1.0	−2.8
Mo	—	1.2	1.4
Ni	—	−1.8	−3.0
Np	—	>4	3.3
Os	1	0.9	0.6
Pa	4	>4	>4
Pb	1.3	−0.7	−2.9
Pm	4	1.5	−1.5
Sb	—	1.2	0.3
Sc	4	>4	>4
Sn	—	0.5	−0.1
Th	4	>4	>4
Tm	4	>4	1.2
U	—	>4	3.1
Y	—	>4	1.3
Zn	2	0.3	−1.7
Zr	>2	>2	>2

[a] Distribution ratios (log D) with undiluted HDEHP from 2 *N* HNO_3 are reported as follows: La (−4.9), Sm (−2.9), Ho (−1.0), and Lu (0.6).[15]

The distribution ratio of metal ions in the extraction with 50% HDEHP solution in toluene from HCl solutions of various acidities is summarized in Table 1. As the distribution ratio is also dependent on the nature of respective metal ion (ionic size and charge),[11] HEDHP and other analogues of dialkylphosphates have been widely used as a stationary phase in reversed phase partition chromatography for the separation of lanthanides[12] and actinides.[13,14]

Purification and Purity of Reagent

The commercial samples are often contaminated with the monoester, polyphosphates, pyrophosphate, 2-ethylhexanol, and some metalic impurities,[15,16] and it is recommended to purify it before use as an analytical reagent. The crude HEDHP can be purified by stirring with 16 *M* HCl for 16 hr at 60°.[4] When a colorless material is needed for use in extractive scintillators, it can be purified through copper salt as follows.[16]

Shake a 0.5 M HEDHP solution in toluene with a 10% excess NaOH solution dissolved in saturated sodium sulfate solution for 5 min. After phase separation, shake an organic phase with a 15% excess of 0.5 M copper sulfate solution for 5 min to convert NaL to CuL_2. Concentrate the organic phase by a rotary vacuum evaporator to 0.6 of the original volume, then treat the residue with excess acetone to precipitate CuL_2. After washing the precipitate with acetone and drying, treat it with toluene and 4 M H_2SO_4 to get the toluene solution of HL. Wash the organic layer with 4 M H_2SO_4 several times to remove copper, then with water. Finally evaporate toluene and entrained water in vacuum.

The purity of HDEHP can be determined by the potentiometric titration with a standard alkali solution.[17]

Other Reagents with Related Structure
Dialkylphosphate
A large number of dialkylphosphates have been investigated as the extraction reagent for metals and as ion-selective electrode sensor materials. Some representative reagents are listed below.[R1]

1. Di (n-alkylphenyl)phosphate[18]
2. Aminoalkylphosphonic acid[19]
3. α-Hydroxy-α-dibutylphosphonylpropionic acid[20]

Sulfur Analogues of Organophosphorous Reagents
Monothio- and dithio-alkylphosphates have been investigated as a more selective extraction reagent. Some examples are listed below.

R = n-butyl etc.
Dialkylphosphorothioic acid[21]

R = iso-octyl etc.
Trialkyl thiophosphate[24]

R = n-butyl, etc.
Dialkylphosphorodithioic acid[22,23]

R = n-octyl, etc.
Trialkylphosphine sulfide[25]
(sulfur analogue of TOPO)

MONOGRAPH

M1. **Sandell, E. B. and Onishi, H.**, *Photometric Determination of Traces of Metals,* 4th ed., John Wiley & Sons, New York, 1978.

REVIEWS

R1. **Coleman, C. F., Blake, C. A., Jr., and Brown, K. B.**, Analytical potential of separations by liquid ion-exchange, *Talanta,* 9, 297, 1962.

R2. **Green, H.**, Recent uses of liquid ion-exchangers in inorganic analysis, *Talanta,* 11, 1561, 1964.

R3. **Green, H.**, Use of liquid ion-exchangers in inorganic analysis, *Talanta,* 20, 139, 1973.

REFERENCES

1. Blake, C. A., Crouse, D. J., Coleman, C. F., Brown, K. B., and Kelmers, A. D., ORNL-2172, U.S. Atomic Energy Commission, 1957, 113; *Chem. Abstr.*, 51, 8391e, 1957.
2. Kimura, K., *Bull. Chem. Soc. Jpn.*, 33, 1038, 1960.
3. Kolarik, S., *Collect. Czech. Chem. Commun.*, 32, 311, 1967.
4. Peppard, D. F., Mason, G. W., Maier, J. L., Driscoll, W. J., and Metta, D. N., *J. Inorg. Nucl. Chem.*, 4, 334, 344, 1957; *J. Inorg. Nucl. Chem.*, 38, 2077, 1976.
5. Pierce, T. B. and Peck, P. F., *Analyst*, 88, 217, 1963.
6. Baes, C. F., Jr., *J. Inorg. Nucl. Chem.*, 24, 707, 1962.
7. Peppard, D. F., Mason, G. W., McCarthy, S., and Johnson, F. D., *J. Inorg. Nucl. Chem.*, 24, 321, 1962.
8. Hirashima, Y., Yamamoto, Y., Takagi, S., Amano, T., and Shiokawa, J., *Bull. Chem. Soc. Jpn.*, 51, 2890, 1978.
9. Peppard, D. F., Mason, G. W., and McCarty, S., *J. Inorg. Nucl. Chem.*, 13, 138, 1960.
10. Levin, I. S., Balakireva, N. A., and Novoseltseva L. A., *Zh. Anal. Khim.*, 29, 1095, 1974.
11. Dyrssen, D. and Hay, L. D., *Acta Chem. Scand.*, 14, 1091, 1100, 1960.
12. Pierce, T. B. and Peck, P. F., *Nature*, 194, 84, 1962; *Nature*, 195, 597, 1962.
13. Kosyakov, V. N., Yakovlev, N. G., and Kazakova, G. M., *Tesisy Dokl. Konf. Anal. Khim. Radioakt. Elem.*, 9, 1977; *Chem. Abstr.*, 91, 13048a, 1979.
14. Svantesson, I., Hangstrom, I., Persson, G., and Liljenzin, J. O., *Radiochem. Radioanal. Lett.* 37, 215, 1979.
15. Tsubota, H. and Watari, K., *Nippon Kagaku Zasshi*, 87, 1106, 1966.
16. McDowell, W. J., Rerdue, P. T., and Case, G. N., *J. Inorg. Nucl. Chem.*, 38, 2127, 1976.
17. Stewart, D. S. and Crandall, H. W., *J. Am. Chem. Soc.*, 73, 1377, 1951.
18. Moody, G. J., Tomas, J. D. R., Nassory, N. S., Griffiths, G. H., and Craggs, A., *J. Inorg. Nucl. Chem.*, 34, 3042, 1972; *Analyst*, 103, 68, 1978, *Analyst*, 104, 412, 1979.
19. Wozniak, M. and Nowgrocki, G., *Talanta*, 26, 633, 644, 1135, 1979.
20. Toropova, V. F., Miftakhova, A. Kh., Guryenova, I. V., Zimin, M. G., and Rudovik, A. N., *Zh. Obsch. Khim.*, 40, 2172, 2177, 1970; *Chem. Abstr.*, 74, 80425h, 80431g, 1971.
21. Handley, T. H., *Anal. Chem.*, 35, 991, 1963.
22. Handley, T. H. and Dean, J. A., *Anal. Chem.*, 34, 1212, 1962.
23. Zucal, R. H., Dean, J. A., and Handley, T. H., *Anal. Chem.*, 35, 988, 1963.
24. Handley, T. H. and Dean, J. A., *Anal. Chem.*, 32, 1878, 1966.
25. Elliot, D. E. and Banks, C. V., *Anal. Chim. Acta*, 33, 237, 1965.

TRI-n-OCTYLPHOSPHINE OXIDE

$$\left. \begin{array}{c} \text{n-}C_8H_{17} \\ \text{n-}C_8H_{17} \\ \text{n-}C_8H_{17} \end{array} \right\rangle P{=}O \qquad \begin{array}{c} C_{24}H_{51}OP \\ \text{mol wt} = 386.65 \end{array}$$

Synonym
TOPO

Source and Method of Synthesis
Commercially available. It is prepared by the oxidation of trioctyl phosphine with nitric acid[1] or Grignard's reaction of octyl bromide with $POCl_3$.[29]

Analytical Uses
It is frequently used as an uncharged ligand for the extraction of the coordination unsaturated metal complexes.

Properties of Reagent
It is colorless crystals, mp 59.5 to 60°,[2] bp 180 to 205° (1 Torr).[29] It is stable in the air and can be kept for many months at room temperature without deterioration. It is easily soluble in cyclohexane (35.6 g/100 mℓ, 25°)[3] and other organic solvents. It can be purified by recrystallization from cyclohexane.

Reaction with Metal Ions
Metal ions of oxidation states III, IV, V, and VI can be extracted from hydrochloric or nitric acid with a solution of TOPO into inert solvents, such as hexane, cyclohexane, or toluene. The extracted species are considered to be a coordinatively solvated chloride or nitrate, and the extent of extraction is very much dependent upon the nature and acidity of the acid. The distribution ratios of various metals in the TOPO extraction from hydrochloric acid and nitric acid are summarized in Table 1.

At higher mineral acid concentrations, TOPO in organic solvent also extracts the acid to form TOPO·HA or TOPO·2HA, decreasing the extraction coefficient of metal above a certain acid concentration. The order of extractability of acid is $HNO_3 >$ $HClO_4 > HCl$ up to 2 M acid. However, the addition of indifferent metal salt (alkali nitrate in the extraction from nitric acid, for example) improves the extractability of metals by salting out and mass-action effect, as exemplified in Figure 1.

TOPO has also been used as an auxiliary uncharged ligand to improve the extractability of so-called "coordination unsaturated uncharged metal complexes". Such synergistic effect of TOPO can be found very often in the extraction of multivalent metal ions, rare earths, and actinide ions. Some of the examples are summarized in Table 2.

Other Reagents with Related Structure
Tri n-butylphosphine Oxide, TBPO, (n-C_4H_9)$_3$ PO, $C_{12}H_{27}OP$, mol wt = 218.33

It is a colorless hygroscopic solid, mp 62 to 64°,[1] bp 127 to 133° (1 Torr),[29] and is soluble in water (5.6 g/100 mℓ, 25°, but solubility decreases with increasing temperature.)[20] and in common organic solvents. Solubility of water in TBPO is 37.5 g/100 mℓ at 25°.[20] D(toluene/water) = 3.8 (0.5 N HCl), 3.0 (1.04 N HCl), and 1.92 (2.07 N HCl). In general, TBPO behaves similarly to TOPO and has been used in the metal extractions.[21,22] The extraction behavior of various metal ions from hydrochloric acid with a 1% TBPO solution in toluene has been investigated.[2]

Table 1
EXTRACTION OF METALS WITH 5% TOPO IN TOLUENE FROM ACID SOLUTIONS[2]

Metal ion	HCl (M)				HNO₃ (M)			
	1	4	8	12	1	6	11	15
As	—	−0.1	1.3	1.3	—	—	—	—
Au(III)	4.8	4.8	4.9	4.4	2.8	0.3	−0.6	−0.9
Bi	2.1	−0.1	−1.5	—	0.6	−3.1	—	—
Fe(III)	0.3	1.2	1.2	1.2	−2.3	−2.2	−1.2	−0.3
Ga	−1.2	3.3	3.8	3.3	−3.1	−4.5	−4.3	−4.0
Hf	—	—	—	—	1.3	1.0	1.0	0.8
Hg(II)	1.1	1.5	1.1	−0.5	0.1	−0.7	−0.4	−0.3
In	1.4	1.7	1.9	0.4	—	—	—	—
Mo(VI)	1.6	2.8	1.3	0.8	−0.6	−1.7	−0.5	0.8
Nb	−0.2	1.8	2.6	2.5	−0.7	0.0	0.4	0.7
Os	1.4	1.4	1.4	1.6	−0.3	−0.6	−0.7	0.3
Pd	−0.3	0.1	−0.7	−1.7	—	—	—	—
Pt	−0.1	1.2	0.7	−1.2	—	—	—	—
Re	1.0	1.3	0.9	−0.5	—	—	—	—
Sb	−2.2	0.0	0.4	0.3	—	—	—	—
Sc	−1.3	1.3	1.9	−0.2	0.3	−0.4	−0.3	−0.4
Se	−1.6	−1.3	−0.1	1.0	—	—	—	—
Sn(IV)	2.0	2.0	2.3	1.4	−0.7	−0.4	−0.1	0.5
Tc(VII)	1.5	2.1	1.6	1.0	—	—	—	—
Th	−1.7	0.4	1.6	0.2	2.0	1.6	1.2	0.1
Ti	—	−0.6	1.4	2.0	−1.6	−1.2	−0.6	0.7
U(VI)	3.1	4.3	3.7	1.3	3.3	2.4	1.4	−0.3
V(V)	−1.0	0.0	0.8	0.8	—	—	—	—
W	0.2	0.8	1.0	0.8	—	—	—	—
Zn	1.9	2.1	0.6	−1.1	—	—	—	—
Zr	−0.8	2.4	2.0	2.4	1.0	—	1.7	1.7

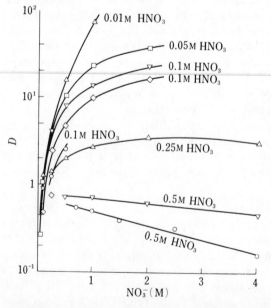

FIGURE 1.   Extraction of metals with TOPO from HNO₃ solution at various NO₃⁻ concentration. (From Watanabe, K. and Ono, R., *J. Nucl. Sci. Technol.*, 1, 75, 1964. With permission.)

Table 2

SOME EXAMPLES ON THE SYNERGISTIC EXTRACTION WITH TOPO

Metal ion	Condition	Extraction solvent	Remarks	Ref.
Actinides	$0.1 \sim 0.5$ M HNO, 0.05 M PMBP	Benzene or cyclohexane	Extraction of Am,Bk,Cf,Cm [Ac(PMBP)$_3$ · (TOPO)$_2$]	5
Rare earths	pH >3, TTA or other β-diketones	Benzene or toluene	Extracted as Ln(β-diketone)$_3$ · (TOPO), $\sim$ 2	6—9
Co(II)	pH $5 \sim 9$, BPA or β-diketones	Hexane or cyclohexane	Extracted as Co(β-diketone)$_2$ · TOPO; Cu,Fe,Mn, Ni,citrate, EDTA interfere	10, 11
Mn(II)	pH 4.75, 0.07 HFA	Cyclohexane	Extracted as Mn(HFA)$_2$ · (TOPO)$_2$	12
Ra(II)	pH $4.6 \sim 8.5$, 0.1 M TTA or PMBP	Hexane or cyclohexane	Extracted as Ra(TTA)$_2$ · (TOPO)$_2$	13, 14
Th	pH $3.0 \sim 3.5$, 0.08 M benzoic acid or salicyclic acid	Chloroform	Fe(III) can be masked with ascorbic acid	15
Ti(IV)[a]	$7 \sim 10$ N HCl, TTA	1,2-Dichloroethane	Extracted as Ti(TTA)(TBPO)	16
(VI)	$1 \sim 5$ M HClO$_4$ or H$_3$PO$_4$, TTA or alkylphosphoric acid	Kerosene or cyclohexane	Extracted as UO$_2$(TTA)$_2$ · (TOPO)	17, 18
Zn	1 M HClO$_4$, 0.1 M AA, TAA, or HFA	CCl$_4$	TOPO forms more stable adducts than does TBP	19

[a] TBPO is used.

Tris(2-ethylhexyl)phosphine   Oxide,   TEHPO, $(CH_3(CH_2)_3CH(C_2H_5)-CH_2)_3PO$, $C_{24}H_{51}OP$, mol wt = 386.65

It is a colorless viscous liquid, bp 205 to 213° (2 to 3 Torr);[29] d = 0.88[28] and is easily soluble in common organic solvents. TEHPO behaves similarly to TOPO and has been used in the metal extraction.[23]

Miscellaneous Reagents

Diphosphine oxides of the following general structure have been investigated as a metal extractant.[24,25]

$$-\overset{|}{\underset{\|}{P}}-(CH_2)_n-\overset{|}{\underset{\|}{P}}- \qquad n = 2 \sim 6$$
$$\quad O \qquad\qquad\quad O$$

These bidentate ligands are expected to be more effective in the metal extraction than with the use of simple phosphine oxide.

Tri-*n*-butylphosphine sulfide[26,27] and triphenylarsine oxide[28] have been investigated as an analytical reagent.

REVIEW

R1. White, J. C. and Ross, W. J., Separation by Solvent Extraction with Tri-*n*-octylphosphine Oxide, NAS-NS 3102, U.S. Atomic Energy Commission, 1961, 56; *Chem. Abstr.,* 55, 21989a, 1961.

REFERENCES

1. Davies, W. C. and Jones, W. J., *J. Chem. Soc.,* 33, 1929.
2. Ishimori, T. and Nakamura, E., *JAERI Rept.,* 1047 (1963); Ishimori, T., Kimura, K., Fujino, T., and Murakami, H., *Nippon Genshiryoku Gakukaishi,* 4, 117, 1962.
3. Zingaro, R. A. and White, J. C., *J. Inorg. Nucl. Chem.,* 12, 315, 1960.
4. Watanabe, K. and Ono, R., *J. Nucl. Sci. Technol.,* 1, 75, 1964.
5. Chmutova, M. K., Kocketkova, N. E., Prybylova, G. A., and Myasoedov, B. F., *Zh. Anal. Khim.,* 25, 710, 1970; *Zh. Anal. Rhim.,* 27, 678, 1972; *Zh. Anal. Khim.,* 28, 2340, 1973.
6. Taketatsu, T. and Banks, C. V., *Anal. Chem.,* 38, 1524, 1966; Taketatsu, T. and Sato, A., *Anal. Chim. Acta,* 108, 429, 1979.
7. Honjyo, T., *Bull. Chem. Soc. Jpn.,* 42, 995, 1969.
8. Shigematsu, T., Matsui, M., and Wake, R., *Anal. Chim. Acta,* 46, 101, 1969.
9. Aly, H. F. and El.-Naggar, A. H., *Microchem. J.,* 18, 405, 1974.
10. Shigematsu, T. and Honjyo, T., *Bunseki Kagaku,* 18, 68, 1979.
11. Le Roux, H. J. and Fouche, K. F., *J. Inorg. Nucl. Chem.,* 34, 747, 1972.
12. Mitchell, J. W. and Ganges, R., *Talanta,* 21, 735, 1974.
13. Ghose, A. K., Sebesta, F., and Stary, J., *J. Radioanal. Chem.,* 24, 345, 1975.
14. Sebesta, F. and Havlik, B., *J. Radioanal. Chem.,* 24, 337, 1975.
15. Povlova, M. and Mareva, S., *Izv. Geol. Inst. Bulg. Akad. Nauk, Ser. Geokhim. Mineral. Petrogr.,* 22, 17, 1973; *Anal. Abstr.,* 29, 4B95, 1975.
16. Roland, G., Pondant, M., and Duyckaerts, G., *Anal. Chim. Acta,* 85, 331, 1976.
17. Awwal, M. A., Haider, S. Z., and Sultana, R., *Nucl. Sci. Appl. Ser. B,* 8, 53, 1975; *Chem. Abstr.,* 87, 44837d, 1977.

18. Akiba, K., *J. Radioanal. Chem.*, 36, 153, 1977.
19. Sekine, T. and Ihara, N., *Bull. Chem. Soc. Jpn.*, 44, 2942, 1971.
20. Higgins, C. E. and Baldwin, W. H., *Anal. Chem.*, 32, 233, 236, 1960.
21. Rhee, C. T., *Doehan Hwakak Hwoejee*, 7, 245, 1963; *Chem. Abstr.*, 61, 7766g, 1964.
22. Irving, H. and Edgington, D. N., *Chem. Ind.*, 77, 1961.
23. Ross, W. J. and White, J. C., *Anal. Chem.*, 33, 424, 1961.
24. Kamin, G. J., O'Laughlin, J. W., and Banks, C. V., *J. Chromatogr.*, 31, 292, 1967.
25. Chmutova, M. K., Nesterova, N. P., Koiro, O. E., and Myasoedov, B. F., *Zh. Anal. Khim.*, 30, 1110, 1975.
26. Hitchcock, R. B., Dean, J. A., and Handley, T. H., *Anal. Chem.*, 35, 254, 1963.
27. Handley, T. H., *Talanta*, 12, 893, 1965.
28. Giround, R., NP-18856, U.S. Atomic Energy Commission, 1970, 95.
29. Blake, C. A., Brown, K. B., and Coleman, C. F., ORNL-1964, U.S. Atomic Energy Commission, 1955, 106.

BENZENEARSONIC ACID AND ITS DERIVATIVES

AsO_3H_2

(1) R = H
(2) R = 2-NO$_2$
(3) R = 2-NH$_2$
(4) R = 4-OH
(5) R = 4-OH, 3-NO$_2$

Synonyms

Reagents covered in this section are listed in Table 1, together with their synonyms.

Source and Methods of Synthesis

All commercially available. (1) is prepared from aniline through diazotization, followed by the reaction with sodium arsenate.[1] (2) is prepared from o-nitroaniline in a similar manner.[2] (3) is prepared by the reduction of (2).[3] (4) is prepared from phenol by arsonation with sirupy arsenic acid at 160°.[4] (5) is prepared from 4-oxybenzenearsonic acid by nitration with mixed acid.[5]

Analytical Uses

As a precipitating reagent for the group separation, gravimetry, and spot test detection of various metal ions. Substitution with various groups at the different positions on the aromatic ring of the mother compound gives rise to the different spectrum of metal ions to be precipitated.

Properties of Reagents

Benzenearsonic acid (1) is colorless prisms, mp 158 to 162° (with decomposition) and is soluble in water (30 g/ℓ, 25°, and 240 g/ℓ, 35°). It is also soluble in 95% ethanol (150 g/ℓ, 26° and 550 g/ℓ, 68°), but not soluble in chloroform; $pKa_1 = 3.47$ and $pKa_{a2} = 8.48$ (22°).[6]

2-Nitrobenzenearsonic acid (2) is a yellow crystalline powder, mp 224 to 225° (with decomposition), and is easily soluble in water; $pKa_1 = 3.37$ and $pKa_2 = 8.54$ (22°).[6]

2-Aminobenzenearsonic acid (3) is colorless or pale yellow needle-like crystals, mp 154 to 156° (with decomposition) and is easily soluble in water and alcohol. It is also soluble in acetic acid and only slightly soluble in ether; $pKa_1 \approx 2$ and $pKa_2 = 8.66$ (22°).[6]

4-Hydroxybenzenearsonic acid (4) is a white odorless powder, mp 175 to 180° (with decomposition), and is freely soluble in water and soluble in alcohol; $pKa_1 = 3.89$, $pKa_2(OH) = 8.37$, and $pKa_3 = 10.0$ (22°).[6]

4-Hydroxy-3-nitrobenzenearsonic acid (5) is white needles or amber yellow prisms which decompose upon heating and is only slightly soluble in cold water, but moderately soluble in hot water. It is very soluble in methanol, ethanol, and glacial acetic acid, but insoluble in ether.

Reactions with Metal Ions

Benzenearsonic acid (1) and its analogues (2) to (5) react with various metal ions in acidic media to form precipitates. The precipitation is often quantitative, and the arsonate group is functioning as a bidentate ligand. Table 2 summarizes the precipitation reactions with metal ions. It is possible to precipitate the respective metal ions by selecting the proper acidity or pH. This is the basis for the separation of metal ions by precipitation or the spot test detection of metal ions. Some of the examples on the separation of metal ions are summarized in Table 3.

Table 1

BENZENEARSONIC ACID AND ITS DERIVATIVES

Reagent	Substitution	Name of compound	Synonyms	Molecular formula	Mol wt
(1)	—	Benzenearsonic acid	Phenylarsonic acid	$C_6H_7O_3As$	202.04
(2)	2-NO_2	2-Nitrobenzenearsonic acid	o-Nitrophenylarsonic acid	$C_6H_6NO_5As$	247.04
(3)	2-NH_2	2-Aminobenzenearsonic acid	o-Aminophenylarsonic acid, o-arsanilic acid	$C_6H_8NO_3As$	217.06
(4)	4-OH	4-Hydroxybenzenearsonic acid	p-Hydroxyphenylarsonic acid	$C_6H_7O_4As$	218.04
(5)	4-OH, 3-NO_2	4-Hydroxy-3-nitrobenzene arsonic acid	Nitrophenolarsonic acid	$C_6H_6NO_6As$	263.04

Table 2
PRECIPITATION OF METAL IONS WITH BENZENEARSONIC ACID AND ITS ANALOGUES IN ACIDIC MEDIA

Metal ion	(1)[7]		(2)[7]		(3)[a8]		(4)[7]		(5)[7]	
$NHCl^b$										
Bi	0.03	(0.003)	0.1	(0.1)	0.1	$(-)^c$	0.4	(0.3)	0.01	(0.004)
Fe(III)	0.1	(0.07)	0.005	(0.003)	0.1	$(-)^c$	0.003	(0.002)	0.2	(0.07)
Nb(V)	10	$(9)^d$	10	(9)	—		9	(8)	10	(8)
Sb(III)	0.6	(0.3)	2.0	(1.6)	—		0.6	(0.3)	1.1	(0.7)
Sb(V)	2.2	(1.9)		—	0.1	$(-)^e$	2.5	(2.2)	2.1	(1.7)
Sn(II)	1.4	(0.5)	1.7	(0.7)	0.1	$(-)^c$	0.7	(0.6)	2.0	(0.02)
Sn(IV)	2.0	(1.4)	1.8	(0.7)	0.1	$(-)^c$	0.5	(0.05)	1.8	(1.1)
Ta(V)	12	$(8)^d$	7	(6)	—		7	(6)	8	(7)
Th	0.3	(0.01)	0.07	(0.07)	0.1	$(-)^c$	0.2	(0.1)	0.5	(0.06)
Ti(IV)	9	(8)	1.8	(1.1)	0.1	$(-)^c$	6	(6)	8	(7)
U(VI)	0.2	(0.03)	0.2	(0.2)	0.1	$(-)^c$	0.4	(0.3)	1.0	(0.7)
Zr	6	$(6)^d$	0.6	(0.1)	0.1	$(-)^c$	4.9	(3.7)	6	(6)
pH^b										
Ag	4.4	(5.1)	5.1	(5.4)	5	$(-)^c$	4.1	(5.5)	3.2	(4.5)
Cd	3.5	(4.5)	7.2	(7.6)	3	$(-)^c$	4.4	(5.4)	1.8	(2.3)
Ce(III)	3.4	(3.5)	2.5	(3.4)	3.5	$(-)^c$	3.3	(3.5)	3.2	(3.7)
Co(II)	6.2	(6.7)	8.2	(9.2)	4	$(-)^f$	6.5	(7.4)	7.0	(7.6)
Cr(III)		—	4.0	(7.1)	3	$(-)^c$	4.5	(5.7)	3.5	(7.5)
Cu(II)		—	5.3	(5.6)	1	$(-)^c$	3.0	(3.4)	2.9	(3.4)
La	3.3	(3.5)	5.9	(6.5)	3.5	$(-)^c$	3.2	(3.3)	3.0	(3.4)
Mn(II)	2.1^g	(5.0)		—	3	$(-)^c$	5.0	(6.2)	1.5	(4.6)
Ni		—		—	5	$(-)^e$	7.0	(7.7)		—
Pb	2.5	(3.3)	2.5	(2.9)	3	$(-)^g$	1.5	(3.1)	1.8	(2.5)
Zn	3.5	(4.5)	4.3	(4.9)	4	$(-)^c$	4.4	(4.7)	2.9	(3.2)

[a]   The following metals are not precipitated with (3): As(III) (V), Ba, Cr(VI), Mo(VI), Se(IV) (VI), Sr, Te(IV) (VI), Tl(I), V(V), and W(VI).[8] (3) forms red-violet soluble chelate with Os(VI) (VIII).[15]

[b]   The numbers for each element indicate the acidity or pH at which the precipitation began, and those in parentheses indicate the point at which the precipitation is completed.

[c]   The precipitation is quantitative.

[d]   The log K_{sp} for $TaOH(L)_2 \cdot H_2O$, -61.9; $NbOH(L)_2 \cdot H_2O$ -67.9; ZrL_2, -35.6.[14]

[e]   The precipitation is not quantitative.

[f]   The precipitation is quantitative at boiling temperature.

[g]   Beyond these acidities, a part of the precipitates dissolved.

[h]   Beyond these acidities, the precipitates dissolved in a more basic region.

Table 3
SOME EXAMPLES OF METAL SEPARATIONS BY
BENZENEARSONIC ACID

Reagent	Metal	Condition	Remarks	Ref.
(1)	Pa	$5\ NH_2SO_4 + 0.1\ M$ $H_2C_2O_4$	From Al, Bi, Fe, Sb, Sn, Th, Ti,[a] U, Zr	9, 16
	Po	$5\ NH_2SO_4 + H_2C_2O_4$ or HF	From Pa[a]	9, 16
	Sn(IV)	5% HCl	Al, Bi, Cd, Co, Cu, Fe, Ni, Pb, Th, Zn, Zr interfere	10
	Ta	pH $0.7 \sim 2.5$, tartarate + EDTA (80°)	From Bi, Sn(IV), Th, W, Lanthanides; Nb, Ti, Zr interfere	11
	Ta	pH $2.0 \sim 3.0$, EDTA	From Nb; Ba, Pb, Sr, Ti, Zr interfere	12
(4)	Ti	$1.5\ NHCl + NH_4SCN$	From Al, Ba, Be, Ca, Cr, Mg, Mn, Mo, Ni, Te, U, V, Zn	10
	Zr(Hf)	$2\ NHNO_3 + H_2O_2$	Only Sn(IV) interferes	13

[a]   Extraction with iso-amyl alcohol.

REFERENCES

1. Palmer, C. S. and Adams, R., *J. Am. Chem. Soc.*, 44, 1356, 1922.
2. Kashima, K., *J. Am. Chem. Soc.*, 47, 2207, 1925.
3. Johnson, J. R. and Adams, R., *J. Am. Chem. Soc.*, 45, 1312, 1923.
4. *Organic Synthesis*, Vol. 1, 490.
5. Conant, J. B., *J. Am. Chem. Soc.*, 41, 435, 1919.
6. Pressman, D. and Brown, D. H., *J. Am. Chem. Soc.*, 65, 540, 1943.
7. Nakata, H., Kusaba, Y., and Kikkawa, S., *Bull. Chem. Soc. Jpn.*, 35, 1611, 1962.
8. Pietsoh, R., *Mikrochim. Acta*, 954, 1955.
9. Myasoedov, B. F., Palshin, E. S., and Palei, P. N., *Colloq. Int. C. N. R. S.*, 154, 293, 1966; *Chem. Abstr.*, 65, 9810b, 1966.
10. Holzbecher, Z., Divis, L., Kral, M., Sucha, L., and Vlacil, F., *Handbook of Organic Reagents in Inorganic Analysis*, Ellis Horwood Chishester, England, 1976.
11. Petrovsky, V., *Collect. Czech. Chem. Commun.*, 30, 1727, 1965.
12. Majumdar, A. K. and Mukherjee, A. K., *Anal. Chim. Acta*, 21, 330, 1959.
13. Kyrs, M., Pistek, P., and Selucky, P., *Collect. Czech. Chem. Commun.*, 32, 747, 1967.
14. Tsykhansky, V. D., Nazarenko, V. A., Shergina, N. I., and Konusova, V. V., *Zh. Anal. Khim.*, 25, 97, 1970.
15. Gangopadhyay, S. and Gangopadhyay, P. K., *Fresenius Z. Anal. Chem.*, 286, 251, 1977.
16. Myasoedov, B. F., Palshin, E. S., and Molochnikova, N. P., *Zh. Anal. Khim.*, 23, 1491, 1968.

NON-COORDINATING REAGENTS

REDOX REAGENTS

INDIGO CARMINE (5,5′-INDIGOTIN DISULFONIC ACID, DISODIUM SALT, $C_{16}H_8O_8S_2Na_2$, mol wt = 466.35)

It is a dark-blue powder with coppery luster. It is light sensitive, is easily soluble in water (approximately 1 g/100 mℓ, 25°) to give a blue or bluish-purple solution, is slightly soluble in alcohol, and is practically insoluble in common organic solvents.

The aqueous solution of Indigo Carmine behaves as a reversible redox system, being reduced to a colorless leuco-acid as shown below:

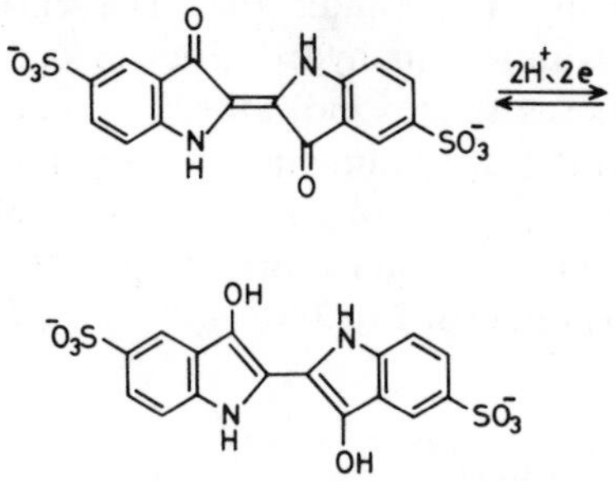

In a solution of pH < 8, the redox potential can be given by

$$E_m = 0.291 + \frac{RT}{F} \ln [H^+]$$

Indigo Carmine is a good indicator in the titration with strong reducing agents, such as Cr(II),[1] Sn(II),[2] and Ti(III),[3] in a fairly acidic solution. Indigo Carmine has also been used as a reducing agent for the titration of oxidizing species, such as MnO_4^-, $Fe(CN)_6^{3-}$, or Fe(III).[4] In this case, Indigo Carmine acts as its own indicator.

Another interesting application is for the photometric determination of dissolved oxygen in water. Reduced Indigo Carmine reacts rapidly with oxygen at pH 11.5 to yield a blue oxidized dye (λ_{max}, 620 nm).[5]

2,6-Dichloroindophenol, Sodium Salt (2,6-Dichloro-N-(p-hydroxyphenyl)-p-benzoquinone imine, Sodium Salt, Tillman's Reagent, $C_{12}H_6Cl_2NO_2Na$, mol wt = 290.08)

It is a dark green powder, containing up to two molecules of water and is freely soluble in water and alcohol. The aqueous solution is deep blue and turns to red by acids, pKa = 5.7.[M1] The aqueous solution is fairly unstable even in a dark place. A 0.1% solution is stable only for 3 weeks; hence it is recommended to use solid mixtures of 1:500 with NaCl.

The reversible redox reaction can be given as below:

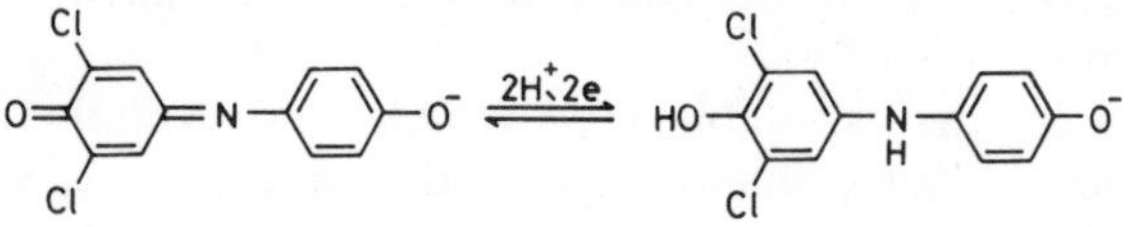

Blue (pH > 6)
Red (pH < 5)   Colorless

The alkaline blue color is far more intense than the acid red color; hence the usable pH range of the reagent is limited to 6.3 to 11.4. The reagent tends to decompose at pH > 11.4, causing the drifting of potentials. $E_o = 0.668$ V (30°).[M1]

2,6-Dichloroindophenol has been used most frequently in the various titrimetries involving ascorbic acid.[6,7] The reagent of oxidized form has also been suggested as an anionic dye in the ion-pair extraction of colorless hydrophobic cations such as cationic surfactants (pH 8 to 9, extraction with nitrobenzene, λ_{max}, 650 nm, 0.5 to 2.2 ppm).[8]

Bindschedler's Green, Leuco Base (4,4′-Bis(dimethylamino)diphenylamine, $C_{16}H_{21}N_3$, mol wt = 255.36)

Bindschedler's Green of dyestuff grade (C.I. 76125) which contains zinc chloride does not function properly as a redox indicator.[9] The leuco base obtained by the reduction of commercial dye with sodium hydrosulfide in ammoniacal sodium hydroxide solution was found to behave as a redox indicator,[10,11] and it is commercially available. The leuco base is a green fluffy crystalline powder which is stable over a long period of time and is slightly soluble in water to give a colorless solution which turns to deep blue upon oxidation. The redox reaction which is shown below is reversible and gives stable potentials over the pH range of 2 to 9.5.

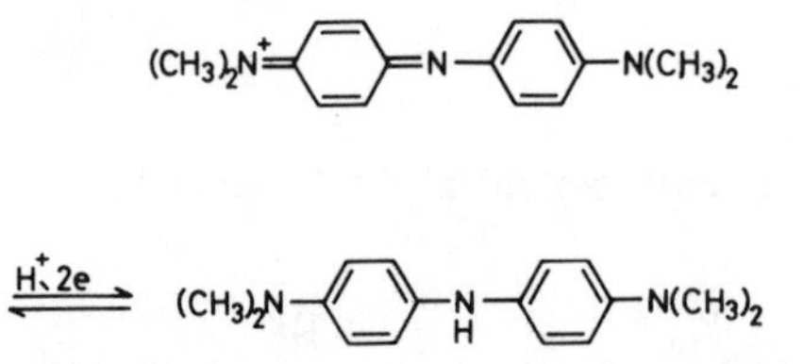

The reagent tends to decompose at higher pH range. $E_o = 0.680$ V (30°).[M1]

Bindschedler's Green can be used over a wider range of pH than most indophenol indicators, but has some drawbacks. It is unstable at extremes of pH and not very soluble in water.

It has been used as an indicator in the titrimetries involving Fe(III). For example, it is recommended as an indicator in the chelatometric titration of Fe(III) with EDTA (pH 2.5 to 3.5, from green to red-orange) and in the back titration of Cr(III) with EDTA and Fe(III).[11] It is also used for the catalytic determination of ultratrace amount of Fe(III). The rate of oxidation of the leuco base with H_2O_2 is accelerated by the presence of trace of Fe(III), being linearly related only with Fe(III) concentration. The resulting Bindscheldler's Green (pH 5.2; λ_{max}, 725 nm; $\varepsilon = 6.9 \times 10^4$) is determined photometrically after 10 min. As low as 0.005 to 0.1 $\mu g/25$ mℓ of Fe can be determined.[12]

Another interesting application is for the use of the oxidized form as a cationic dye in the ion-pair extraction of colorless anions such as $[HgBr_3]^-$.[13]

Neutral Red (3-Amino-7-dimethylamino-2-methylphenazine Hydrochloride, $C_{15}H_{16}N_4 \cdot HCl$, mol wt = 288.78)

It is a dark green powder, is soluble in water (4%) and alcohol (1.8%) to give a red solution (λ_{max} 533 nm, in 50% ethanol), and is also soluble in cellosolve (3.75%) and ethyleneglycol (3.0%), but practically insoluble in aromatic solvents, pKa (NH$^+$) = 6.7.

The redox reaction which is shown below is reversible, and the colorless reduced form is rapidly oxidized by air.[14]

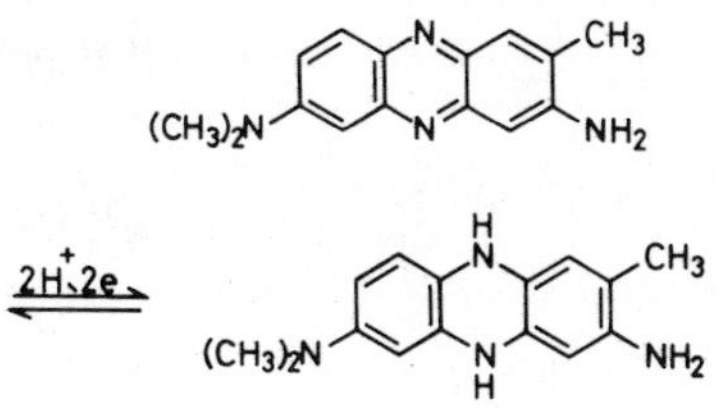

Red

Colorless

$$\xrightarrow{2H^+ 2e^-}$$

When the colorless solution is kept under an air-free condition at pH 5.3, a yellow-green fluorescence is rapidly developed.[14] The formation of the fluorescent material is pH dependent (only slowly at pH 2.7 and hardly at all at 8.2) and causes potentials to be erratic and to drift rapidly. Accordingly, Neutral Red is only usable as an indicator in the pH range where the fluorescent material is not formed. $E_o = 0.240$ V (30°).[M1]

Safranin T (**1**) and Phenosafranin (**2**) behave similarly.

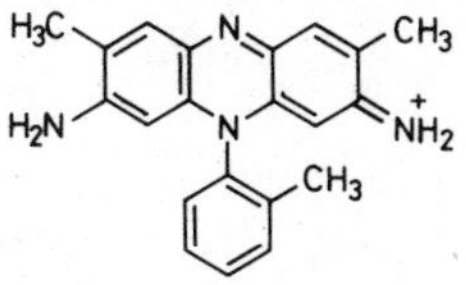

(1)
$E_o = 0.235$ V

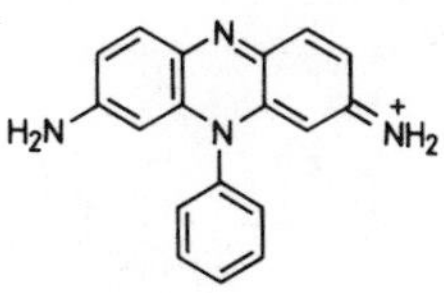

(2)
$E_o = 0.280$ V

The redox potentials of azine derivatives are so low that they are of use only as indicators in the titrations with strong reducing agents, such as Cr(II),[15] Ti(III),[16] and V(II).[17, 18]

Gallocyanine ($C_{15}H_{12}N_2O_5$, mol wt = 300.27)

It is a green crystal, is practically insoluble in cold water, is slightly soluble in hot water, and is soluble in alcohol and glacial acetic acid. It is soluble in alkali carbonate to give a red solution and is soluble in concentrated hydrochloric acid to give a blue solution which becomes red when diluted with water. Gallocyanine is amphoteric, so that it is not very soluble in water within the range of pH 4 to 5.

The redox reaction of Gallocyanine may be written as below:

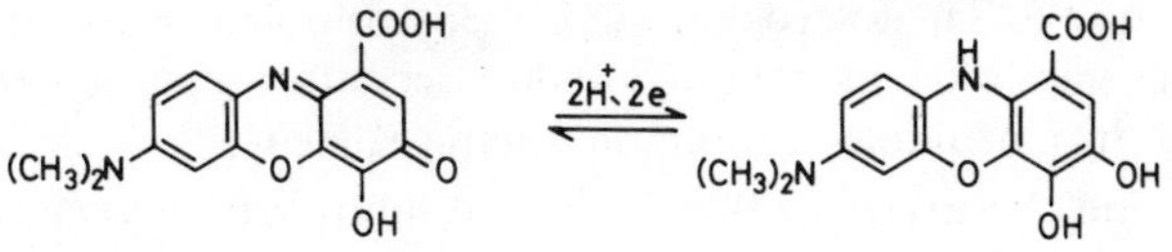

$$\xrightleftharpoons{2H^+ 2e^-}$$

Red-violet, pH < 4 [H_2L]
Blue, pH 5.5 to 8 [HL^-]
Red-violet, pH > 8 [L^{2-}]

Colorless

The reaction is reversible at pH 5.5 to 10. No E_o value is reported, but it shows a potential range close to that of Methylene Blue.[M1]

Methylene Blue (3,7-Bis(dimethylamino)-phenazothionium chloride, C.I. Mordant Blue 52, $C_{16}H_{18}ClN_3S \cdot 3H_2O$, mol wt = 373.90)

It is a dark green odorless crystal with bronze luster or crystalline powder and is easily soluble in water (4 g/100 mℓ), ethanol (1.5 g/100 mℓ), and chloroform, but is insoluble in ether.

The redox reaction of Methylene Blue can be shown as below:

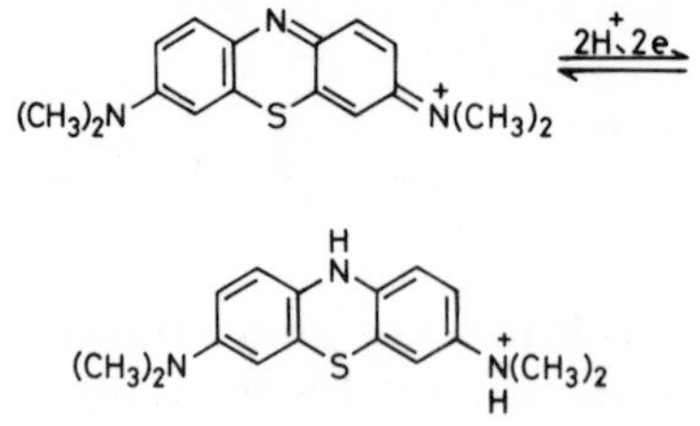

The reaction is reversible over the pH range of 1 to 13. The colorless leuco Methylene Blue (Methylene White) is light sensitive, and the colorless solution, left open to light, but in the absence of air, rapidly redevelops the blue color of Methylene Blue. E_o = 0.532 V (30°).[M1]

Other thiazine derivatives, like Thiazine Blue (3) and Toluidine Blue (4) behave similarly.

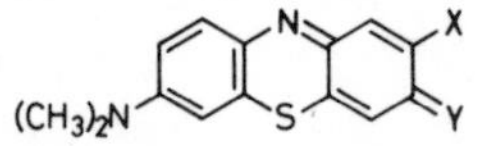

(3) X = H, Y = N$^+$(C$_2$H$_5$)$_2$
(4) X = CH$_3$, Y = N$^+$H$_2$

Methylene Blue has been widely used as a redox indicator in various titrations, but the recent ineresting application of the thiazine dyes is for the use as a cationic dye in the ion-pair extraction of colorless anions. For examples, ReO$_4^-$ is extracted with Methylene Blue into dichloroethane from H$_2$SO$_4$ solution for the subsequent photometry (0.1 to 2.0 N H$_2$SO$_4$; 658 nm; ε = 1.0 × 10^5; 0 to 1.5 ppm).[19] Many other elements, such as Au, B, Bi, Ce, Cu, Ga, Ge, Hg, In, Sb, Se, Sn, Ta, Tl, U, Zn,[R1,20] and Pd,[21] can be determined through their anion derivatives. A trace amount of organic anions like anionic surfactants can also be determined similarly by the chloroform extraction of the Methylene Blue ion-pair.[22]

Variamine Blue B, Hydrochloride (*N*-(*p*-Methoxyphenyl)-phenylenediamine, Hydrochloride, C$_{13}$H$_{14}$ON$_2$·HCl, mol wt = 250.73)

As the free base is unstable, most of the commercial samples are supplied as a hydrochloride. The pure salt should be white crystals, but the commercial samples are pale blue, due to the air oxidation during storage. It is light sensitive and should be kept in an amber bottle. The hydrochloride is easily soluble in water to give a colorless solution which becomes blue on standing, and a precipitate separates from the solution within a few days. It is insoluble in common organic solvents.

The redox reaction of Variamine Blue in an acid solution may be shown as below:

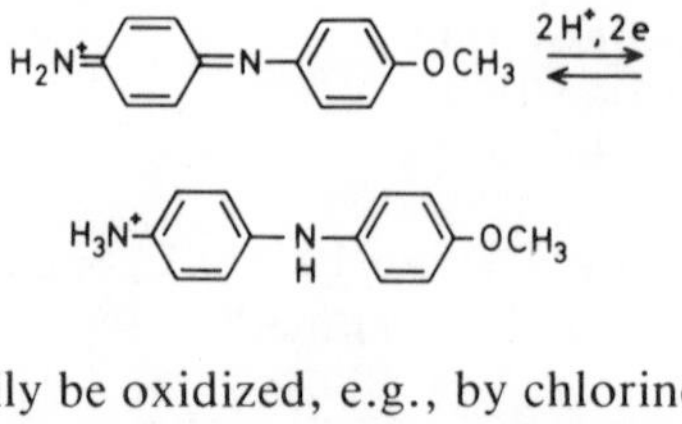

The colorless cation can easily be oxidized, e.g., by chlorine or bromine to give violet-red iminoquinone. The iminoquinone combines reversibly with another molecule of the reduced species to yield an intensely blue colored quinhydrone. The color change is so sharp that it has been accepted as one of the most important redox indicators.[23]

The quinhydrone decomposes at pH < 1.5 to give a product without indicator properties. Also, above pH 6.5, the iminoquinone itself dissociates into a colorless product. Thus, the above redox reaction is only reversible in the pH range of 1.5 to 6.5.

The redox potential of the system can be given by:

$$E_m = E_o + \frac{RT}{2F} \ln \frac{[H^+]^3 + K_{a(red)} [H^+]^2}{[H^+] + K_{a(ox)}}$$

where $E_o = 0.712$ V and $K_{a(red)}$ and $K_{a(ox)}$ are the acid dissociation constants of 4-amino group of reduced and oxidized species, respectively, $pKa_{a(ox)} = 6.6$ and $pKa_{a(red)} = 5.9$[M1]

Variamine Blue B has found wide applications as an indicator for the standard redox titrations in acid range, as well as for a number of precipitation and chelatometric titrations, where the first excess of one of the reagents can oxidize or reduce variamine Blue B at the end point.

The standard redox titrations with this indicator include the titration with Ce(IV),[26] Cr(VI),[24] V(V),[25] and ascorbic acid.[23,24] In the titration of strong oxidizing agents, the indicator should be added near the end point, otherwise the indicator may be destroyed.

In the chelatometric titration of Fe(III), the indicator is oxidized to give a blue solution at the beginning, but is reduced to colorless (or pale yellow due to the presence of Fe-EDTA) by the first excess of EDTA at the end point. The reduction of the indicator is effected by the presence of trace of Fe(II) according to the following equation:

$$E = E_o + \frac{RT}{F} \ln \frac{[Fe(III)]}{[Fe(II)]}$$

in which [Fe(III)] becomes infinitesimal after the end point, as EDTA preferentially complexes Fe(III).[23]

Many other applications as a redox indicator are described in the literature.[M1,24] For indicator use, a dry mixture with NaCl or Na_2SO_4 (1:100) is recommended.

3,3′-Dimethylnaphthidine (4,4′-Diamino-3,3′-dimethyl-1,1′-binaphthyl, $C_{22}H_{20}N_2$, mol wt = 312.41)

Freshly recrystallized sample from ethanol or ligroin is colorless prisms, mp 213°. Commercial samples have melting ranges 207 to 210 or 213 to 215° and gray or grey-brown in color due to the air oxidation during the storage. It is easily soluble in benzene or acetic acid (∼1%), is fairly soluble in ethanol, and is slightly soluble in petroleum ether. It is also slightly soluble in water to give a colorless or pale yellow solution which gradually turns to red-violet upon air oxidation.

The redox reaction may be shown as below:

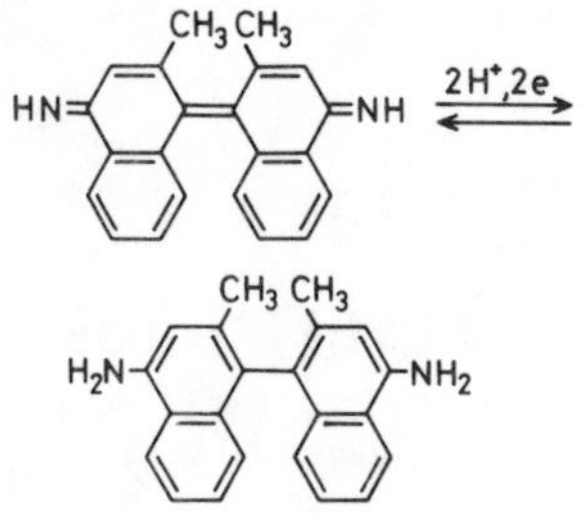

The absorption spectrum of the oxidized form is independent of the nature of the oxidant, but depends on the medium (λ_{max}, 550 nm, in 1 M H_2SO_4 and 560 nm in 4 M H_2SO_4, $\varepsilon = 3.0 \times 10^4$ for both cases).[27,28] As the oxidized form is unstable and decolorizes easily, the direct determination of the formal potential is difficult. The formal potentials obtained by the direct titration with Ce(IV) at 18° are 0.776 V (pH 0), 0.726 V (pH 0.5), and 0.711 V (pH 1.0).[29]

Dimethylnaphthidine has been recommended as a redox indicator in the precipitation titration of Cd and Zn with $Fe(CN)_6^{3-}$.[30] The principle of the function of indicator is as follows. Titrating solution of $Fe(CN)_6^{3-}$ contains a trace amount of $Fe(CN)_6^{4-}$ as impurity. At the initial stage of titration, $Fe(CN)_6^{4-}$ is precipitated with Zn(II), while $Fe(CN)_6^{3-}$ is not. Thus, the oxidation potential of $Fe(CN)_6^{3-}/Fe(CN)_6^{4-}$ couple becomes high enough to oxidize the indicator. Beyond the equivalence point, $Fe(CN)_6^{4-}$ become excess, depressing the potential low enough to reduce the indicator.

The same principle can also be applied to the chelatometric titration of Zn. Each drop of 1% indicator solution (in glacial acetic acid) and 1% $K_3Fe(CN)_6$ aqueous solution are added to the sample solution which is buffered to pH 5 and titrated with EDTA.[31]

Besides the use as a redox indicator, dimethylnaphthidine has been used as a photometric and detecting reagents for oxidizing species, such as V(V),[32] dissolved oxygen in water,[33] or residual chlorine in tap water (0.05 to 1 ppm).[34]

A sulfonated dimethylnaphthidine can also be used for the same purposes.[35,36]

Other reagents with similar redox behavior include N-substituted benzidines and 3,3'-disubstituted benzidines, all of which give good color changes in the potential range of 0.7 to 1.1 V. For example, o-tolidine (3,3'-dimethylbenzidine) has long been used as a photometric reagent for the determination of chlorine (pH 7: 625 nm; $\varepsilon = 3.4 \times 10^4$; 0.6 to 6 ppm).[37]

MONOGRAPH

M1. Bishop, E., Ed., *Indicators,* Pergamon Press, Oxford, England, 1972.

REVIEW

R1. Matsuo, T., Spectrophotometric determination of metals and non-metals with basic dyes, *Bunseki Kagaku,* 21, 671, 1972.

REFERENCES

1. Tandon, J. P. and Mehrotra, R. C., *Fresenius Z. Anal. Chem.,* 158, 20, 1957; *Fresenius Z. Anal. Chem.,* 187, 410, 1962.
2. Henze, G. and Geyer, R., *Fresenius Z. Anal. Chem.,* 200, 434, 1964.
3. Holness, H. and Cornish, G., *Analyst,* 67, 221, 1942.

4. Korenman, I., *Mikrochemie,* 18, 31, 1935.
5. Loomis, W. F., *Anal. Chem.,* 26, 402, 1954; *Anal. Chem.,* 28, 1347, 1956.
6. Birch, T., Harris, L., and Ray, S., *Biochem. J.,* 27, 590, 1933.
7. Svehla, G., Koltai, L., and Erdey, L., *Anal. Chim. Acta,* 29, 442, 1963.
8. Sakai, T., Tsubouchi, M., and Azechi, Y., *Bunseki Kagaku,* 25, 675, 1976.
9. Shine, H. J., Snell, R. L., and Trisler, J. C., *Anal. Chem.,* 30, 383, 1958.
10. Weiland, H., *Ber.,* 48, 1087, 1915.
11. Wehber, P., *Fresenius Z. Anal. Chem.,* 149, 161, 241, 1956; *Fresenius Z. Anal. Chem.,* 150, 186, 1956.
12. Hirayama, K. and Sawaya, T., *Nippon Kagaku Zashi,* 1401, 1976.
13. Tsubouchi, A., *Anal. Chem.,* 42, 1037, 1970.
14. Clark, W. M. and Perkins, M. E., *J. Am. Chem. Soc.,* 54, 1228, 1932.
15. Cooke, W. D., Hazel, F., and McNabb, W. M., *Anal. Chim. Acta,* 3, 656, 1949.
16. Murty, B. V. S. R. and Rao, G. G., *Talanta,* 8, 438, 1961.
17. Ellis, C. M. and Vogel, A. I., *Analyst,* 81, 693, 1956.
18. Mittal, R. K., Tandon, J. P., and Mehrotra, R. C., *Fresenius Z. Anal. Chem.,* 189, 330, 1962.
19. Nagai, H. and Onishi, H., *Bunseki Kagaku,* 21, 1590, 1972.
20. Tarayan, V. M., Ovsepyan, E. N., and Artsruni, V. Zh., *Zavod. Lab.,* 35, 1435, 1969; *Chem. Abstr.,* 72, 96380j, 1970.
21. Kuroda, R., Yoshikuni, N., and Kamimura, Y., *Anal. Chim. Acta,* 60, 71, 1972.
22. Standard Methods for the Examination of Water and Waste Water, American Public Health Association, New York, 1970.
23. Erdey, L. and Bodor, E., *Fresenius Z. Anal. Chem.,* 137, 410, 1953.
24. Erdey, L., *Chemist-Analyst,* 48, 106, 1959.
25. Erdey, L., Vigh, K., and Boder, E., *Acta Chim. Acad. Sci. Hung.,* 7, 293, 1955.
26. Jach, Z., Pacovsky, J., and Svach, M., *Fresenius Z. Anal. Chem.,* 154, 185, 1957.
27. Ladayi, L., Vajda, M., Bendly, M., and Farsang, G., *Acta Chim. (Budapest),* 73, 258, 1972; *Chem. Abstr.,* 77, 125596v, 1972.
28. Belcher, R., Lyle, S. J., and Stephen, W. I., *J. Chem. Soc.,* 4454, 1958.
29. Belcher, R., Nutten, A. J., and Stephen, W. I., *J. Chem. Soc.,* 3857, 1952.
30. Belcher, R., Nutten, A. J., and Stephen, W. I., *J. Chem. Soc.,* 1520, 1951; *J. Chem. Soc.,* 3444, 1952.
31. Brown, E. G. and Hayes, T. J., *Anal. Chim. Acta,* 9, 6, 1953.
32. Bannard, L. G. and Burton, J. D., *Analyst,* 93, 142, 1968.
33. Fadrus, H. and Maly, J., *Analyst,* 96, 591, 1971.
34. Belcher, R., Nutten, A. J., and Stephen, W. I., *Anal. Chem.,* 26, 772, 1954.
35. Frumina, N. S. and Kulapina, E. G., *Iz. Vyssh. Ucheb. Zaved. Khim. Khim. Tekhnol.,* 17, 1422, 1974; *Anal. Abstr.,* 30, 5H31, 1976.
36. Gowda, H. S. and Shakunthala, R., *Analyst,* 103, 1215, 1978; *Anal. Chim. Acta,* 97, 385, 1978.
37. Johnson, J. D. and Overby, R., *Anal. Chem.,* 41, 1744, 1969.

CATIONIC DYES

Analytical Uses of Cationic Dyes

Highly colored cationic species like R^+ or HR^+ form an extractable ion-pair with anionic metal complexes of the halide, thiocyanate, and other inorganic (NO_3^-, CN^-, etc.) or organic anionic ligands (8-hydroxyquinoline, salicyclic acid, benzoic acid, etc.). In the solvent extraction process, the excess colored cation is left in the aqueous phase, making it possible to determine the anionic metal complex by measuring the color intensity of the ion-pair in the organic phase. Based on this principle, numerous numbers of procedures have been reported on the determination of traces of elements.

Highly colored cationic species can be the cationic complexes like tris(bipyridine)-Fe(II) ($[FeL_3]^{2+}$), bis(Cuproine)-Cu(I), ($[CuL]^+$) or the cationic dyes which will be discussed in this section.

In the ion-pair extraction with the cationic dye, the proper selection of anionic ligand, cationic dye, and the extraction solvent is essential to attain the higher selectivity of the extraction.

As to the cationic dye, a singly charged dye with hydrophobic moiety is desirable for the effective extraction. The cationic dyes of higher charge are generally much less effective in forming the ion-pair with anionic metal complexes. The dyes should be stable even in a strongly acid aqueous solution because the extraction is often carried out from a strongly acid solution. The dye should also have a high molar absorptivity to attain the higher photometric sensitivity. Some of the determination can be made fluorimetrically if the dye itself is fluorescent.

In the ideal working condition, the anionic metal complex is almost quantitatively extracted as an ion-pair with a cationic dye, without simultaneous extraction of more than a trace of excess dye.

Cationic Dyes of Analytical Importance
Rhodamine B (Rhodamine C, Tetraethylrhodamine, C.I. Basic Violet 10, C.I. 45170)

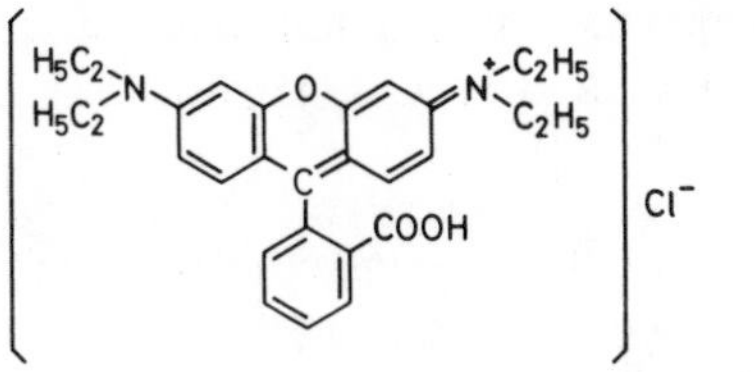

$C_{28}H_{31}N_2O_3Cl$
mol wt = 479.02

It is usually supplied as chloride which is a green to red-violet crystalline powder. It is easily soluble in water (1.2 g/100 mℓ)[R4], ethanol, and cellosolve to give a bluish-red solution with strong yellow fluorescence. It is slightly soluble in chloroform, acetone, and 1 M HCl (0.11 g/100 mℓ).[M1] In benzene and ether solution, Rhodamine B exists as the colorless lacton form as shown below which also shows a weak blue fluorescence.[1]

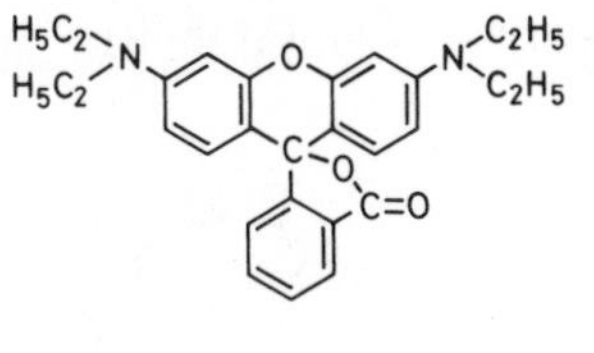

In a polar solvent such as alcohol, acetone or water, the lacton ring opens to form a zwitterion structure ($R^{\pm}$), showing an intense violet color (λ_{max}, 553 nm; $\varepsilon = 1.1 \times$

10^5). The protonation and dimerization equilibria of the dye molecule in the aqueous solution have been investigated in details by Sandell.[M1] The spectral characteristics of the aqueous solution of the chloride ($RH^{\pm}$ Cl^-) are λ_{max}, 556 nm; $\varepsilon = 1.1 \times 10^5$; for RH^+ (pH 1 to 3, violet with yellow fluorescence) and λ_{max}, 494 nm; $\varepsilon = 1.5 \times 10^4$; for RH_2^{2+} (pH -1 to 0, orange).[1] The existence of dimeric form can be neglected at the concentration of 10^{-5} M or less; pKa ($[R^+]$ $[H^+]/[RH^+]$) = 3.2 (0.1 M Cl^-, 25°).[1]

Other Rhodamine Dyes

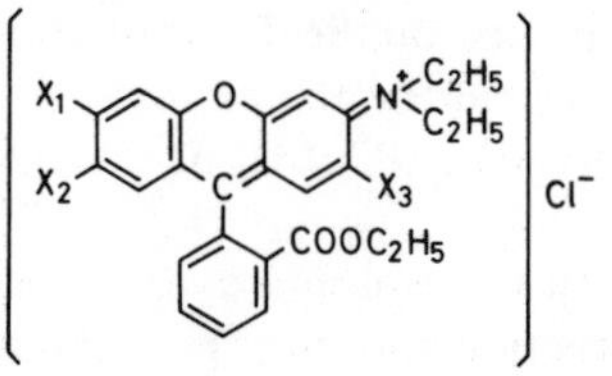

Rhodamine 6G — (Rhodamine 6Zh, C.I. Basic Red 1, C.I. 45160), X_1=NH(C_2H_5), X_2=X_3=CH_3, is a bright bluish-pink powder and is soluble in water (5.4 g/100 ml)[R4] to form a scarlet red solution with a greenish fluorescence.

Rhodamine 3GO — (C.I. Basic Red 4, C.I. 45215), X_1=NH$_2$, X_2=CH_3, X_3=H, is a bright bluish-pink powder.

Rhodamine 4G — (C.I. 45166), X_1=NH(C_2H_5), X_2=X_3=H.

Rhodamine 3C — (Rhodamine 3B, C.I. Basic Violet 11, C.I. 45175), X_1=N(C_2H_5)$_2$, X_2=X_3=H, Ethylester of Rhodamine B, is a bright redish-violet powder and is soluble in water to form a red-violet solution with a brownish-red fluorescence.

These dyes are ethyl ester derivatives of Rhodamine and are expected to be more hydrophobic than Rhodamine B in the ion-pair extraction.

Structurally, these dyes exist as R^+ in the aqueous solution, while Rhodamine B exists as RH^+ only in an acidic solution (pH < 3). A proton adds to the R^+ ion in relatively diluted acid solution to form RH^{2+}. The pKa values of RH^{2+} determined in a sulfuric acid solution are Rhodamine 6G(-1.1), 3GO(-0.40), 4G(-0.21), and 3C(-0.02).[2]

Rhodamine S — (C.I. Basic Red 11, C.I. 45050).

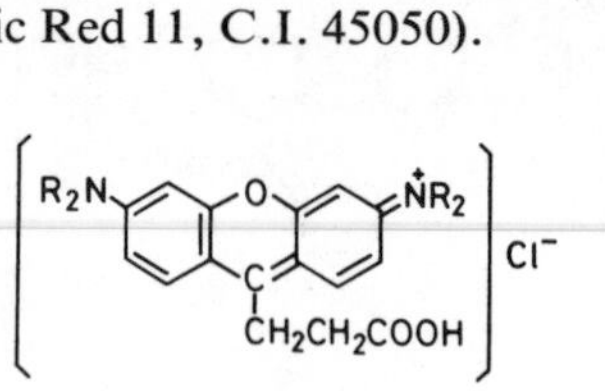

R = CH₃, $C_{20}H_{23}N_2O_3Cl$
mol wt = 374.87
R = C₂H₅, $C_{22}H_{31}N_2O_3Cl$
mol wt = 406.95

R = CH₃ or C₂H₅

It is a pink powder with a green luster, is soluble in water to form a red solution with an yellow fluorescence, and is slightly soluble in ethanol and in concentrated sulfuric acid to form a bluish-yellow solution with a strong green fluorescence. It is claimed to be superior to Rhodamine B and triphenylmethane dyes for the determination of Sb,[3] with regard to the sensitivity and stability.

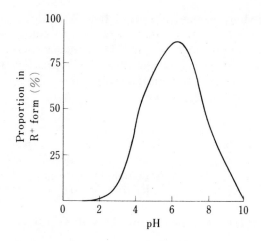

FIGURE 1. Variation of proportion of R⁺ with pH in Brilliant Green solution after 5 hr. (From Fogg, A. G., Burgess, C., and Burns, D. T., *Analyst*, 95, 1012, 1970. With permission.)

Brilliant Green (C.I. Basic Green 1, C.I. 42040)

$C_{27}H_{29}N_2Cl$
mol wt = 392.97

It is a bright green powder and is usually supplied as chloride or sulfate which is soluble in water (9.7 g/100 mℓ)^{R4} to give a green solution (λ_{max}, 625 nm; $\varepsilon \approx 10^5$).[4] The mono-cationic species (R⁺), which is effective in the ion-pair extraction, predominates in a neutral region as illustrated in Figure 1 because it converts to RH²⁺ in acidic range to give an yellow-red solution and to ROH (pale green) in alkaline range. However, the conversion of R⁺ to RH²⁺ and ROH is so slow that almost complete extraction can be expected over a wider pH range (2 ∿ 8) when the extraction is made immediately after adding the alcoholic dye solution.[4,5]

Malachite Green — (C.I. Basic Green 4, C.I. 42000) is a bright blue-green powder. It is a tetramethyl homologue of Brilliant Green and behaves similarly to the parent dye. It is very soluble in water (4.0 g/100 mℓ)^{R4} to give a green solution (λ_{max}, 620 nm; $\varepsilon \approx 10^5$) and soluble in alcohol. The aqueous solution turns to yellow below pH 2.

Crystal Violet (Gentian Violet, C.I. Basic Violet 3, C.I. 42555)

$C_{25}H_{30}N_3Cl$
mol wt = 407.98

It is a bright blue-violet powder, mp 194 to 196°. It is usually supplied as a chloride and is soluble in water (0.1 g/100 mℓ),[R4] ethanol, and chloroform to form a violet solution, but practically insoluble in ether. The aqueous solution is violet (λ_{max}, 541 and 591 nm; $\varepsilon \approx 10^5$), but turns to orange-brown in concentrated sulfuric acid due to the formation of H_2R^{3+} (λ_{max}, $\sim$ 425 nm). Sulfide and cyanide react with Crystal Violet to cause the color weakening.[M1]

Other Triphenylmethane Dyes

Methyl Violet (C.I. Basic Violet 1, C.I. 42535) is a bluish-violet powder. It is a mixture of the hydrochlorides of the more highly methylated pararosanilines, containing principally N-tetra-, penta- and hexamethyl derivatives. It is soluble in water (3.3 g/100 mℓ)[R4] and alcohol to give a violet solution. The color changes from violet to yellow at pH 3.2 to 0.15.

Rosaniline (Fuchsine) — (C.I. Basic Violet 14, C.I. 42510)

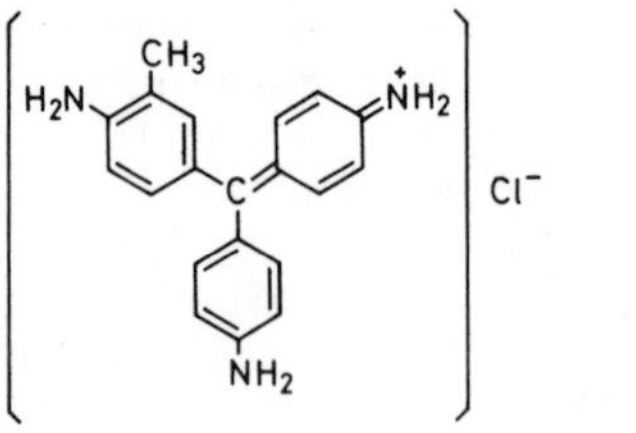

$C_{20}H_{20}N_3Cl$
mol wt = 337.85

The free base is brownish-red crystals, decomposes at 186°, but it is often supplied as a hydrochloride which is metallic green lustrous crystals and decomposes above 200°. It is soluble in water (0.27 g/100 mℓ)[R4] and alcohol to give a carmin red solution (λ_{max}, 543 nm; $\varepsilon = 9.3 \times 10^4$ in ethanol) and in aqueous acid or alkali hydroxide to give an almost colorless solution.

Methyl Green — (C.I. Basic Green 5, C.I. 42590) is prepared by the reaction of Methyl Violet with ethyl bromide. It is green powder which is soluble in water (4.2 g/ 100 mℓ).[R4] The color of the aqueous solution changes from yellow to blue at pH 0.2 to 1.8.

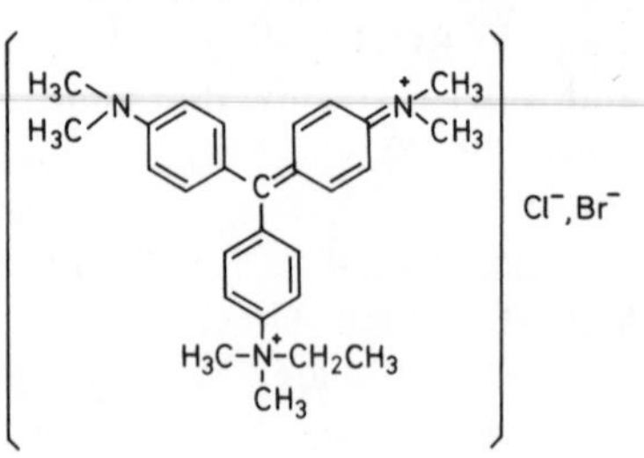

$C_{27}H_{31}N_3ClBr$
mol wt = 512.92

Victoria Blue B — (C.I. Basic Blue 26, C.I. 44045)

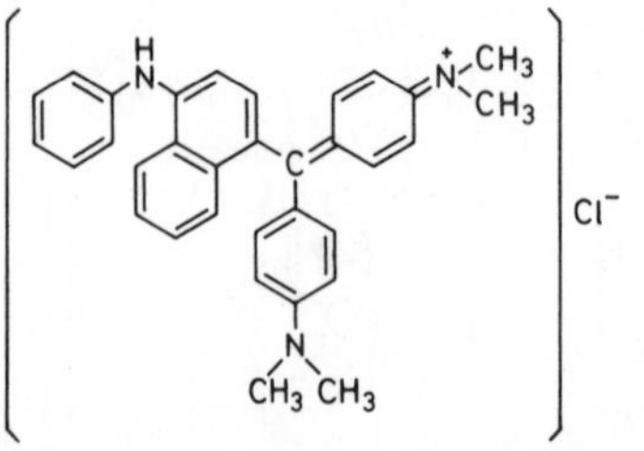

$C_{33}H_{32}N_3Cl$
mol wt = 506.09

It is a bright blue powder and is soluble in water (6.5 g/100 mℓ).[R4] The aqueous solution is blue (R$^+$; λ_{max}, 635 nm; $\varepsilon = 7 \times 10^4$) at pH > 1 and yellow (RH^{2+}, λ_{max}, 475 nm) at pH < 1.[M1]

Victoria Blue 4R — (C.I. Basic Blue 8, C.I. 42563)

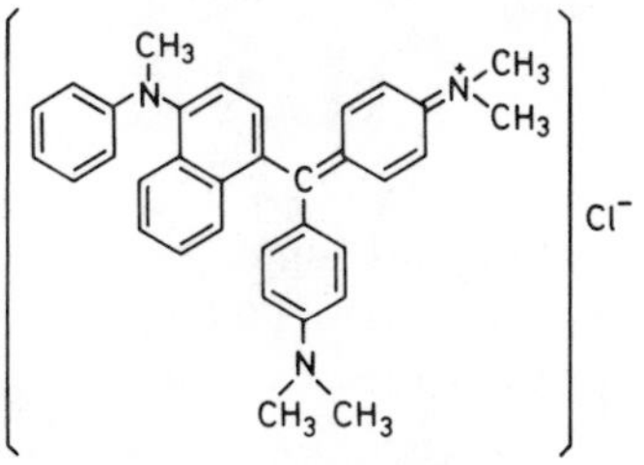

$C_{34}H_{34}N_3Cl$
mol wt = 520.12

It is a reddish blue powder. It is very soluble in hot water and ethanol to give a blue solution (λ_{max}, 624 nm, $\varepsilon = 5 \times 10^4$)[12] and is very soluble in concentrated sulfuric acid to give a yellow-brown solution which, when diluted, turns to green, then yellowish-brown.

Miscellaneous Dyes

Meldola's Blue — (9-dimethylaminobenzo-α-phenazoxinium chloride, C.I. Basic Blue 6, C.I. 51175)

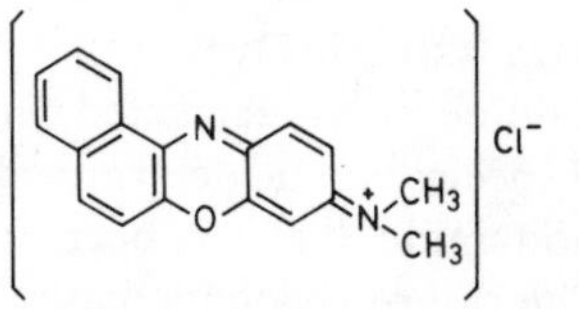

$C_{18}H_{15}N_2OCl$
mol wt = 310.78

It is a black-violet powder. It is insoluble in common organic solvents, but soluble in ethanol and water to give a bluish-violet solution (λ_{max}, 570 nm). Besides the use as a cationic dye in the ion-pair extraction, it is also used as an electron carrier in the enzymatic assay of LDH in serum.[6]

Safranine T — (Safranine O, C.I. Basic Red 2, C.I. 50240)

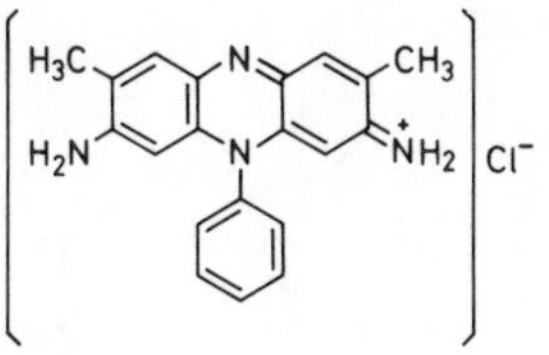

$C_{20}H_{19}N_4Cl$
mol wt = 350.85

It is a bright bluish pink powder and is soluble in ethanol to give a red solution with a yellowish red fluorescence. At low acidities (<1 M H$^+$), the unprotonated form of the dye, R$^+$ (red, λ_{max},520 nm), predominates in the solution. With increasing acidity, the protonated form, RH^{2+} (blue; λ_{max}, 593 nm), predominates,[M1] reaching a maximum in 7 M HCl. With the further increase of acidity, the color of the solution fades out because of the formation of a colorless RH$_2^{3+}$ form;[7] pKa_1(R$^+$ + H$^+$ $\rightleftharpoons$ RH^{2+}) = -0.42 and pKa_2(RH^{2+} + H$^+$ $\rightleftharpoons$ RH$_2^{3+}$) = -3.96.[8]

Methylene Blue —

This dye is described under redox reagents (p.454).

Antipyrine dyes (Chrompyrazols)[M1,R4]

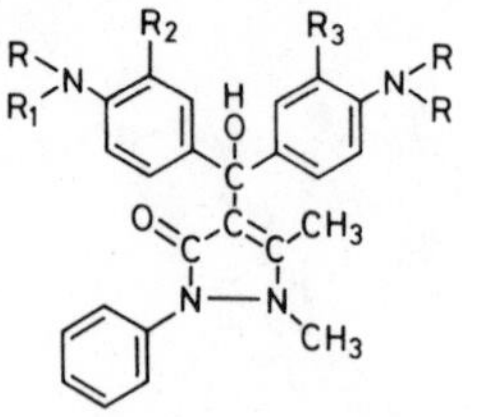

R=CH₃, C₂H₅, Br, etc.
R₁=CH₃, C₂H₅, CH₂C₆H₅, etc.
R₂=H, NO₂, etc.
R₃=H, NO₂, etc.

These can be considered as triphenylmethane analogues in which one of the phenyl groups is replaced by the antipyrine group. The typical dyes may be such as those listed above.[M1] Many elements including boron[9] have been determined with antipyrine dyes.

Absorption Photometric Determination with Cationic Dyes

Some important applications of cationic dyes in the ion-pair extraction photometric determination of elements are summarized in Table 1. A more detailed review on this subject can be found in the literature,[R3,R4] and many other procedures may be developed by the selection of metal ion, anionic ligand, cationic dye, and the extraction solvent.

Fluorimetric Determination with Cationic Dyes

The ion-pairs with Rhodamine dyes extracted in the organic phase fluoresce strongly, so that the extracted species can be determined by the fluorimetry. The strong fluorescence displayed by Rhodamine dyes has been attributed to the rigidity of their planer molecular structure. The fluorescence increases in the order of Rhodamine B < Rhodamine 3B < Rhodamine 4G < Butyl Rhodamine B < Rhodamine 6G, while stability and extractabilty of the ion-pair increase in the order of Rh B < Rh 6G < Rh 3B < BuRh B.[R5] Some representative examples are summarized in Table 2.

Table 1

APPLICATION OF CATIONIC DYES IN THE ION-PAIR EXTRACTION PHOTOMETRY OF ELEMENTS

Cationic dye	Element	Condition	Ion-pair Composition	λ_{max} (nm)	ε (×10⁴)	Extraction solvent	Ref.
Rhodamine B	Au	0.8 N HCl, saturated with NaCl	$(AuCl_4)^-·R^+$	565	9.7	Benzene	10
	Ca	pH 8.9~9.2, TTA(X)	$(CaX_3)^-·R^+$	544	5.6	Benzene	11
	Cd	4 N H₂SO₄, 0.2 N Br⁻	$(CdBr_3·TBP)^-R^+$	575	10.2	Benzene-TBP(4:1)	12
	Ge	0.5 M H₂SO₄, 0.04 M Oxalate	(Molybdogermanic acid)⁻R⁺	560	36.8	Acetone^a	13
	In	4.5 N H₂SO₄, 2 M KBr	$(InBr_4)^-·R^+$	557	5.7	Benzene-iso-propyl ether-acetyl acetone (6:3:1)	14
	Mo	0.8 N HCl, 1.8 N H₂SO₄ 0.8% SCN⁻	$[MoO(SCN)_5]^{2-}·2R^+$	600	10.9	Aqueous^b	15
	Re	pH 0.7~0.8 (H₂SO₄ or H₃PO₄)	$(ReO_4)^-·R^+$	560 ~ 565	11.4	Benzene-iso-butyl acetate (1:1)	16
	Sb	6 N HCl	$(SbCl_6)^-·R^+$	560	9.7	Isopropyl ether	17
	Si	0.5 N HNO₃	$(SiMo_{12}O_{40})^{4-}·4R^+$	555	50.0	Ethanol^a	18
	U	pH 4.5, benzoic acid(X), EDTA	$(UO_2X_3)^-·R^+$	555	10.3	Benzene-ether-hexone	19
	Rare earths	pH 10.5 ~ 10.6, benzoic acid(X)	$(LnX_4)^-·R^+$	548	—	Toluene	20
Ethyl rhodamine B	Zr (Hf)	0.1 N HCl or H₂SO₄, picramine (X)	$(MX_5)^{2-}·2R^+$	560	60.0	Acetone^a	21
Rhodamine S	Sb	2.4 ~ 2.6 N HCl	$(SbCl_6)^-·R^+$	532	8.4	Iso-propylether	3
Rhodamine 6G	Bi	2.85 N H₂SO₄, 0.185 M NaBr	$(BiBr_5)^{2-}·2R^+$	530	15.0	Ethanol^a	22
	Co	pH 6.3, 5,7-di-chloro-8-hydroxyquinoline(X)	$(CoX_3)^-·R^+$	542	7.5	Benzene	23
	Fe(II)	pH 3.9 ~ 5.4, 10⁻³ M 4-Chloro-2-nitrophenol(X), EDTA	—	558	9.0	Toluene	24

Table 1 (continued)

APPLICATION OF CATIONIC DYES IN THE ION-PAIR EXTRACTION PHOTOMETRY OF ELEMENTS

Cationic dye	Element	Condition	Ion-pair Composition	λ_{max} (nm)	ε (×10⁴)	Extraction solvent	Ref.
Rhodamine bG	Ge	pH 5 ~ 6, Alizarine complexone (ALC)	$[Ge(ALC)_3]^{-5} \cdot 5R^+$	520	29.0	CCl₄-CHCl₃ (1:4)	25
	Hg	0.15% KI, 1% gelatin, EDTA, hydrazine	$(HgI_4)^{2-} \cdot 2R^+$	575	7.0	Aqueousᵇ	26
	Mn(II)	pH 7.2 ~ 8.0, 5,7-dichloro-8-hydroxyquinoline(X)	$(MnX_3)^- R^+$	540	7.0	Benzene	27
Brilliant Green	Ag	1 NH_2SO_4, 8×10⁻³ MI⁻	$(AgI_2)^- R^+$	640	4.8	Benzene	28
	Au	0.5 NHCl	$(AuCl_4)^- R^+$	650	10.1	Toluene	12
	B	pH 2.5 ~ 3.5, 2,4-dinitro-1,8-naphthalenediol	—	637	10.3	Toluene	29
	Ge	3,5-dinitro-pyrocatechol(X)	$(GeX_3)^{2-} \cdot 2R^+$	625	14.1	CCl₄	30
	Sb	0.25 ~ 1.5 NHCl	$(SbCl_6)^- R^+$	640	≅10.3	Toluene	31
	Tl	0.15 NHCl	$(TlCl_4)^- R^+$	630	10.6	Iso-propylether	32
	W	pH 1.5 ~ 3.0, dinitro-pyrocatechol(X)	$(WX_3)^{2-} \cdot 2R^+$	640	13.2	CHCl₃	33
Malachite Green	U	pH 6, benzoic acid(X)	$(UO_2X_3)^- R^+$	635	8.3	Cyclohexane	34
	Zn	pH 5 8, 0.6 MSCN⁻	$[Zn(SCN)_4]^{2-} \cdot 2R^+$	626 ~ 632	10.0	Benzene or CCl₄-Cyclohexane (5:1)	35
Crystal Violet	As	2 $NHNO_3$, oxalate	$(Molybdoarsenate)^{3-} R_3^+$	582	32.0	Cyclohexane-toluene (3:2~4:2)	36
	Mo	3.5 ~ 4 NH_2SO_4, 0.2~0.4 MSCN⁻	$[MoO (SCN)_5^{2-}] \cdot 2R^+$	595	23.0	Ethanolᵃ	37
	Sn	0.65 NHCl	$(SnCl_6)^{2-} \cdot 2R^+$	595	8.5	4-Heptanone	38
	Si	2 $NHNO_3$, oxalate	$(Molybdosilicate)^{2-} \cdot 2R^+$	582	14.0	Cyclohexane-toluene (3:2~4:1)	36
	Tl	pH 1.0, citrate	$[TlCl_3 \cdot H_2Cit.]^- R^+$	610	25.0	Benzene	39
	V	pH 4.6 ~ 5.1, PAR	$(VO_2PAR) R^+$	585	11.0	Benzene-MIBK (3:2)	40
	NO₃⁻	pH 5 ~ 7 (KH_2PO_4)	$(NO_3^-) R^+$	595	1.3	Chlorobenzene	41
	ABS	0.1 MNaCl, perchlorate	$(ABS)^- R^+$	615	3.5	Benzene	42

Cationic dye	Element	Condition	Ion-pair	λ (nm)		Extraction solvent	Ref.
Rosaniline	Au	pH 1.0	$(AuCl_4)^-\cdot R^+$	550~555	8.0	Buthylacetate	43
Methyl Violet	Au	0.1 N HCl	$(AuCl_4)^-\cdot R^+$	600	11.5	Trichloroethylene	44
Ethyl Violet	Sb	pH 1.0 (HCl)	$(SbCl_6)^-\cdot R^+$	595	8.3	Iso-amylacetate	45
	Tl	pH 1.0, citrate	$[TlCl_3\cdot H_2Cit.]^-\cdot R^+$	610	290	Benzene	39
Methyl Green	Hg	0.1 MCl^- or $Br^-(X)$	$(HgX_4)^{2-}\cdot 2R^+$	640	13.1	Benzene	46
	Re	pH 2.7~5.2	$(ReO_4)^-\cdot R^+$	640	10.9	Benzene	47
Victria Blue B	In	—	$(InBr_4)^-\cdot 2R^+$	630	12.0	Benzene	M1
Victria Blue 4 R	Te(IV)	9~10 NH_2SO_4, Br^-	$(TeBr_6)^{2-}\cdot 2R^+$	602	8.0	Benzene or nitrobenzene	48
Meldola Blue	Ta	H_2SO_4, NH_4F	$(TaF_6)^-\cdot R^+$	580	2.9	Chlorobenzene	49
Nile Blue	Re	1 NH_2SO_4	$(ReO_4)^-\cdot R^+$	642	7.2	Chlorobenzene	50
Capri Blue	Ga	6 N HCl	$(GaCl_4)^-\cdot R^+$	655	5.6	Benzene	51
	Ta	0.15 NH_2SO_4, 0.22 N HF	$(TaF_6)^-\cdot R^+$	660	10.7	$CHCl_3$	52
Methylene Blue	B	pH 3.2, F^-	$(BF_4)^-\cdot R^+$	645	6.5	Dichloroethane	53
	Cu	pH 5.5~6.0, 0.01 M KCN	$[Cu(CN)_2]^-\cdot R^+$	657	9.8	Dichloroethane	54
	Pd	pH 4.0~5.5, NaN_3	$Pd[NaN_3]R$	653	5.8	$CHCl_3$	55
Janus Green B	Te	1.4 N HBr	$(TeBr_6)^{2-}\cdot 2R^+$	610	1.2	Benzene-acetone (1:2)	56

a   The ion-pair with cationic dye was floated as a solid precipitate, collected, and finally dissolved in the solvent indicated for the subsequent photometry. The unusual high sensitivity may be attributed to the quantitative transfer of the ion-pair complex into the organic solvent and to the higher dye to metal ratio in the ion-pair complex.

b   The photometry was conducted without solvent extraction, since the spectral shift was observed upon ion-pair complex formation in aqueous phase.

Table 2

APPLICATIONS OF CATIONIC DYES IN THE ION-PAIR EXTRACTION FLUORIMETRY OF ELEMENTS[R5]

Cationic dye	Element	Condition	Ion-pair		Extraction solvent	Range of determination (ppm)	Interference	Ref.
			λ_{ex} (nm)	λ_{em} (nm)				
Rhodamine B	As	1.2 N HCl, 2×10⁻⁵ $M KIO_3$	565	590	Benzene	0.06~0.3	Cu,Fe,Ga,Sn	
	Au	3.8 N HBr, 0.5 N KBr	366	580	Benzene-ether	0.05~5	Many cations	
	Hg	4.5 NH_2SO_4, 0.2 N Br^-	560	590	Benzene	0.2~1.5	Many cations	
	In	3~6 NH_2SO_4, 2 N KBr	560	580	Benzene-5% acetone	0.02~0.4	Many cations	

Table 2 (continued)
APPLICATIONS OF CATIONIC DYES IN THE ION-PAIR EXTRACTION FLUORIMETRY OF ELEMENTS[R5]

Cationic dye	Element	Condition	Ion-pair λ_{ex} (nm)	Ion-pair λ_{em} (nm)	Extraction solvent	Range of determination (ppm)	Interference	Ref.
	La	pH 5 ~ 6, 2-phenylquinoline-4-carboxylic acid	555	590	Benzene	—	Other rare earths	
	Mo	1.8 NH_2SO_4, 0.8 N HCl, 0.8% SCN^-	350	578	Aqueous	0.002 ~ 2	None	15
	Re	pH 0.8 (H_2SO_4), EDTA	—	590	Iso-butylacetate-benzene (1:1)	0.002 ~, 0.8	W(750-fold), Mo (50-fold), V (40-fold) do not interfere	57
	Sn	2 N HBr	560	580	Benzene	0.1 ~ 2	Au,Cr,Hg,In,Tl,W	
	Te	5.7% HCl	560	590	Benzene-acetone (2:1)	0.03 ~ 0.06	Many cations	58
	Th	pH 1.8, 5,7-dinitro-8-hydroxyquinoline	560	580	Benzene	0.007 ~ 1.5	—	
	Tl	5 N HCl, 5 N HBr	366	580	Benzene	0.0002 ~	Au,Fe,Hg,Sb,SC,N⁻	
	UO_2^{2+}	pH 5, benzoic acid	366	590	Benzene	0.25 ~ 400	Many cations	
	Zn	pH 3.5~5.0, SCN^-	366	580	Ether	0.2 ~ 5	—	
	PO_4^{3-}	1 N HCl, molybdophosphate	350	575	$CHCl_3$	0.02 ~ 0.3	As,Cr,V	
Butyl Rhodamine B	Ag	6 NH_2SO_4, 0.1 N Br⁻	565	590	Benzene	0.1 ~ 1.0	Many cations	
	Au	0.5 NH_2SO_4, 0.1% KBr, 0.0025% crystal violet	560	595	Benzene	0.001 ~ 0.3	Sb,Tl	
	B	0.1 NH_2SO_4, 0.05 NF⁻	336	590	Benzene	0.01 ~ 0.05	Many cations	
	Ta	1 NH_2SO_4, 0.2% F⁻, Brilliant green	560	595	Benzene	0.02 ~ 0.4	Cr,Nb,Re,Ti,Zr	
	Te	11 NH_2SO_4, 0.07 N Br⁻	366	590	Benzene-butylacetate	0.004 ~ 0.2	As, Bi, In, Sn, Tl	
Rhodamine 6G	Ag	4.2 NH_2SO_4, 0.17 NBr⁻	535	560	Benzene	0.05 ~ 0.5	Many cations	
	In	13 NH_2SO_4, 0.2 NKBr	530	570	Benzene	0.001 ~	Many cations	
	Sb	6 NH_2SO_4, 0.5 NNaCl	532	555	Benzene	0.1 ~ 2	Many cations	

MONOGRAPH

M1. **Sandell, E. B. and Onishi, H.,** *Photometric Determination of Traces of Metals,* 4th ed., John Wiley & Sons, New York, 1978.

REVIEWS

R1. **Blyum, I. A. and Povlova, N. N.,** Extraction-photometric methods of analysis with application of basic dyes, *Zavod. Lab.,* 29, 1407, 1963; *Chem. Abstr.,* 60, 3463g, 1964.

R2. **Fogg, A. G., Burgess, C., and Burns, D. T.,** Use of basic dyes in the determination of anions, particularly as a means of determining antimony, thallium and gallium, *Talanta,* 18, 1175, 1971.

R3. **Matsuo, T.,** Spectrophotometric determination of metals and nonmetals with basic dyes, *Bunseki Kagaku,* 21, 671, 1972.

R4. **Marczenko, Z.,** Nichtchelatierende Farbstoffe als basis empfindlicher specktrophotometrischer Methoden zur Bestimmung von Elementen, *Mikrochim. Acta,* 651, 1977.

R5. **Haddad, P. R.,** The application of ternary complexes to spectrofluorometric analysis, *Talanta,* 24, 1, 1977.

R6. **Motomizu, S., Iwachido, T., and Toei, T.,** Ion association reagents and their application, *Bunseki,* 234, 1980.

REFERENCES

1. **Ramette, R. W. and Sandell, E. B.,** *J. Am. Chem. Soc.,* 78, 4872, 1956.
2. **Zorov, N. B., Golovina, A. P., Alimarin, I. P., and Khvatkova, Z. M.,** *Zh. Anal. Khim.,* 26, 1466, 1971; *J. Anal. Chem. USSR,* 26, 1310, 1971.
3. **Jablonski, W. Z. and Watson, C. A.,** *Analyst,* 95, 131, 1970.
4. **Fogg, A. G., Burgess, C., and Burns, D. T.,** *Analyst,* 95, 1012, 1970.
5. **Fogg, A. G., Willcox, A., and Burns, D. T.,** *Analyst,* 101, 67, 1976.
6. **Burd, J. F. and Usatequi Gomez, M.,** *Clin. Chim. Acta,* 46, 223, 1973.
7. **Burgess, C., Fogg, A. G., and Burns, D. T.,** *Analyst,* 98, 605, 1973.
8. **Pilipenko, A. T., Volkova, A. I., and Shevchenko, T. L.,** *Zh. Anal. Khim.,* 29, 983, 1974.
9. **Busev, A. I., Yakovlev, P. Ya., and Kozina G. V.,** *Zh. Anal. Khim.,* 22, 1227, 1967.
10. **Onishi, H.,** *Microchim. Acta,* 9, 1959.
11. **Poluektov, N. S. and Beltyukova, S. V.,** *Zh. Anal. Khim.,* 25, 2106, 1970.
12. **Kish, P. P. and Balog, I. S.,** *Zh. Anal. Khim.,* 32, 482, 1977.
13. **Ganago, L. I. and Prostak, I. A.,** *Iz. Vyssh. Ucheb. Zaved. Khim. Khim. Teknol.,* 14, 1165, 1971; *Chem. Abstr.,* 76, 20902t, 1972.
14. **Garcic, A. and Sommer, L.,** *Collect. Czech. Chem. Commun.,* 35, 1047, 1970.
15. **Haddad, P. R., Alexander, P. W., and Smythe, L. E.,** *Talanta,* 22, 61, 1975.
16. **Lebedeva, P. and Tarayan, V. M.,** *Zh. Anal. Khim.,* 30, 1403, 1975.
17. **Tanaka, M. and Kawahara, M.,** *Bunseki Kagaku,* 10, 185, 1961.
18. **Golkowska, A. and Pszonicki, L.,** *Talanta,* 20, 749, 1973.
19. **Moeken, H. H. Ph. and van Neste, W. A. H.,** *Anal. Chim. Acta,* 37, 480, 1967.
20. **Mishchenko, V. T. and Tselik, E. I., and Alekseeva, N. S.,** *Ukr. Khim. Zh.,* 43, 1322, 1977; *Anal. Abstr.,* 34, 6B72, 1978.
21. **Dedkov, Yu. M. and Podvigina, T. I.,** *Zh. Anal. Khim.,* 32, 437, 1977.
22. **Chwastowska, J. and Pruszkowski, K.,** *Chem. Anal. (Warsaw),* 21, 525, 1976; *Anal. Abstr.,* 32, 2B120, 1977.
23. **Minczewski, J., Chwastowska, J., and Lachowicz, E.,** *Chem. Anal. (Warsaw),* 21, 373, 1976; *Anal. Abstr.,* 32, 2B177, 1977.
24. **Toei, K., Motomizu, S., and Korenaga, T.,** *Analyst,* 101, 974, 1976.
25. **Flyantikova, G. V. and Korolenko, L. I.,** *Zh. Anal. Khim.,* 30, 1349, 1975.
26. **Ramakrishna, T. V., Aravamudan, G., and Vijayakumar, M.,** *Anal. Chim. Acta,* 84, 369, 1976.

27. Minczewski, J., Chwastowska, J., and Lachowicz, E., *Chem. Anal. (Warsaw)*, 18, 199, 1973; *Chem. Abstr.*, 79, 73239p, 1973.
28. Busev, A. I. and Shestidesyatnaya, N. L., *Zh. Anal. Khim.*, 29, 1138, 1974.
29. Kuwada, K., Motomizu, S., and Toei, K., *Anal. Chem.*, 50, 1788, 1978.
30. Nazarenko, V. A., Levedeva, N. V., and Vinarova, L. I., *Zh. Anal. Khim.*, 27, 128, 1972.
31. Fogg, A. G., Jillings, J., Marriott, D. R., and Burns, D. T., *Analyst*, 94, 768, 1969.
32. Marzenko, Z., Kolowska, H., and Mojski, M., *Talanta*, 21, 93, 1979.
33. Nazarenko, V. A., Poluektova, E. N., and Shitareva, G. G., *Zh. Anal. Khim.*, 28, 1966, 1973.
34. Dubey, S. C. and Nadkarni, M. N., *Talanta*, 24, 266, 1977.
35. Kish, P. P., Zimomrya, I. I., and Zolotov, Yu. A., *Zh. Anal. Khim.*, 28, 252, 1973.
36. Babko, A. K. and Ivashkovich, B. M., *Zh. Anal. Khim.*, 27, 120, 1972.
37. Ganago, L. I. and Ivanova, I. F., *Zh. Anal. Khim.*, 27, 713, 1972.
38. Ducret, L. and Maurel, H., *Anal. Chim. Acta*, 21, 79, 1959.
39. Motsuo, T. and Funada, S., *Bunseki Kagaku*, 12, 521, 1963.
40. Minczewski, J., Chwastowska, J., and Mai, P. T. H., *Analyst*, 100, 708, 1975.
41. Yamamoto, Y., Uchikawa, S., and Akabori, K., *Bull. Chem. Soc. Jpn.*, 37, 1718, 1964.
42. Hedrich, C. E. and Berger, B. A., *Anal. Chem.*, 38, 791, 1966.
43. Tarayan, V. M. and Mikaelyan, D. A., *Arm. Khim. Zh.*, 22, 369, 1969; *Chem. Abstr.*, 71, 77008g, 1969.
44. Ducret, L. and Maurel, H., *Anal. Chim. Acta*, 21, 74, 1959.
45. Matsuo, T., Kuroyanagi, Y., and Meguro, S., *Bunseki Kagaku*, 12, 515, 1963.
46. Tarayan, V. M., Ovsepyan, E. M., and Lebedeva, S. P., *Zh. Anal. Khim.*, 26, 1745, 1971.
47. Tarayan, V. M., Vartanyan, S. V., and Eliazyan, L. A., *Zh. Anal. Khim.*, 24, 1040, 1969.
48. Kish, P. P. and Kremeneve, S. G., *Zh. Anal. Khim.*, 25, 2200, 1970.
49. Pilipenko, A. T. and Tu, N. D., *Ukr. Khim. Zh.*, 34, 1291, 1968; *Chem. Abstr.*, 70, 102777w, 1969.
50. Gagliardi, E. and Fuesselberger, E., *Microchim. Acta*, 385, 1972.
51. Matsuo, T., Funada, S., and Suzuki, M., *Bull. Chem. Soc. Jpn.*, 38, 326, 1965.
52. Elinson, S. V., Nevzorov, A. N., Belogortseva, M. V., Mirzoyan, N. A., and Mordvinova, S. N., *Zh. Anal. Khim.*, 29, 1234, 1974.
53. Ducret, L., *Anal. Chim. Acta*, 17, 213, 1957.
54. Koh, T., Aoki, Y., and Suzuki, Y., *Anal. Chem.*, 50, 881, 1978.
55. Kuroda, R., Yoshikuni, N., and Kamimura, Y., *Anal. Chim. Acta*, 60, 71, 1972.
56. Guseinov, I. K., Bagbanly, I. L., Bagbanly, S. I., and Rustamov, N. Kh., *Doki. Akad. Nauk Azerb. SSR*, 27, 29, 1972; *Anal. Abstr.*, 24, 119, 1973.
57. Grigoryan, L. A., Levedeva, S. P., and Tarayan, V. M., *Arm. Khim. Zh.*, 28, 540, 1975; *Anal. Abstr.*, 30, 4B136, 1976.
58. Nishikawa, Y., Hiraki, K., Naganuma, T., and Niina, S., *Bunseki Kagaku*, 19, 1224, 1970.

LONG-CHAIN ALKYLAMINES

General Consideration

The ammonium cations derived from the secondary and tertiary amines which are substituted with bulky aliphatic groups are important extraction reagents for metals. Usually, metals are extracted from acidic solution as complex anions into a solution of alkylamine diluted with innert solvent or oxygenated solvent like MIBK.

Metal ions, such as Ag, Au, Bi, Cd, Co(II), Fe(III), Ga, In, Ir(IV), Pt, and Zn, are extracted from HCl solution as a $MCl_n{}^{n-m}$-type complex anion. For example, the equilibria involved in the extraction of Fe(III) from the HCl solution into the inert solvent solution of a tertiary amine can be written as below.[1]

$$R_3N \cdot HCl \ + \ FeCl_4{}^- \ \rightleftharpoons \ [R_3NH]^+[FeCl_4]^- \ + \ Cl^-$$

$$2R_3N \cdot HCl \ + \ FeCl_4{}^- \ \rightleftharpoons \ [R_3NH]^+[FeCl_4]^- \cdot R_3N \cdot HCl \ + \ Cl^-$$

At higher HCl concentration, the extraction coefficient may decrease due to the formation of $R_3NHCl \cdot HCl$ or $R_3NHCl_2{}^-$. In such case, alkali chloride is added as a source of Cl^-.

From HNO_3 solution, metal ions, such as Np(IV), Pu(IV, VI), Th, U(VI), and rare earths, can be extracted as a $M(NO_3)_n{}^{n-m}$-type complex.

Extraction from other acids such as H_2SO_4 or a mixture of HCl-HF is also reported in some limited cases.

The extraction coefficient is affected by many factors. The stability of complex metal anions and the charges on the anion complexes are the main reasons for the differences in metal extractabilities. There is a working hypothesis that the extraction of anions is favored by increase in size and decrease in charge. Metal extractability varies also with the amines and with the solvents. In general, primary and secondary amines are less effective extractants than tertiary amines, in which extractability increases with the chain length of the alkyl group and decreases with the branching of the alkyl chain.[2]

The extractability also depends on the nature of solvent. Although not many systematic works have been done to generalize the relationship between the nature of solvents and the extractability, the following result is reported on the extraction of Th from 6 M HNO_3 into a solution of 0.1 M tri-n-octylamine in various solvents, $CHCl_3$, D = 0.03; benzene, 0.40; toluene, 0.47; nitrobenzene, 0.60; CCl_4, 1.0; n-hexane, 3.7; kerosene, 4.6; and cyclohexane, 4.8.[3]

Tri-n-Octylamine: TOA, TNOA

$$\begin{array}{l} CH_3(CH_2)_7 \\ CH_3(CH_2)_7\text{-}N \\ CH_3(CH_2)_7 \end{array} \qquad \begin{array}{l} C_{24}H_{51}N \\ \text{mol wt} = 353.67 \end{array}$$

It is a colorless viscous oil and is insoluble in water, but is easily soluble in common organic solvents. Trace levels (ppm $\sim$ ppb) of TOA dissolved in water can be determined by the ion-pair extraction with dichromate ion from H_2SO_4 solution into chloroform, followed by the photometric determination of Cr(VI) in the organic phase with diphenylcarbazide.[4]

The material supplied as Alamine 336®* is a mixture of tertiary amines with these

---

* Trade name of General Mills, Inc., Minneapolis, Minn.

straight-chain alkyl groups (mainly octyl and decyl) and contains 90 to 95% tertiary amine, average mol wt being 392.

The extraction of metal ions with TOA from various acid solutions are reviewed in several articles [R1-R4] and Table 1 summarizes some examples.

Tri-iso-Octylamine, TIOA

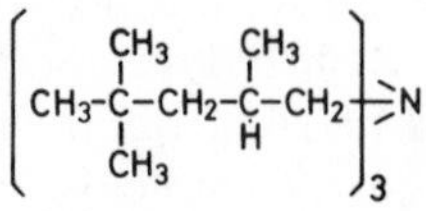

$$C_{24}H_{51}N$$
$$\text{mol wt} = 353.67$$

It is a colorless oil. Commercial materials are a mixture of tertiary amines of dimethylhexyl (3,5-, 4,5-, and 3,4-), methylheptyl, and some other alkane chains.

Similarly to TOA, TIOA is used for the extraction of metal ions from various acid solutions. Extraction behavior of metal ions with 5% TIOA in xylene from 0 to 12 N HCl,[28,29] 0 to 15 N HNO$_3$,[29] and 0.1 N H$_2$SO$_4$ solutions[30] have been investigated in detail. Some of the results on a 5% TIOA in xylene-HCl system are summarized in Table 2. The typical examples of the separation of metals by TIOA extraction are listed in Table 3.

Long-Chain Secondary Amines

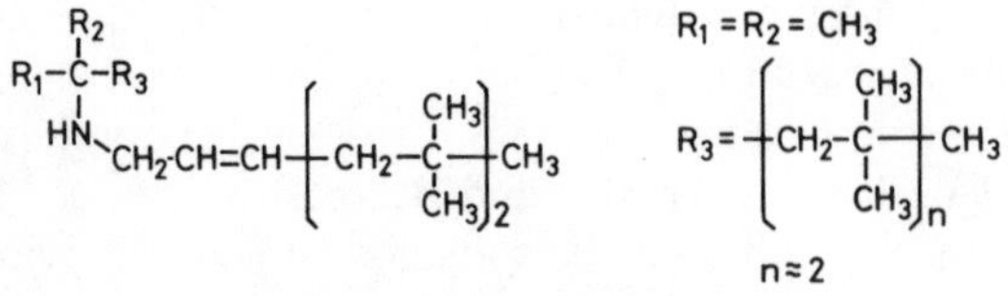

Amberlite® LA-1

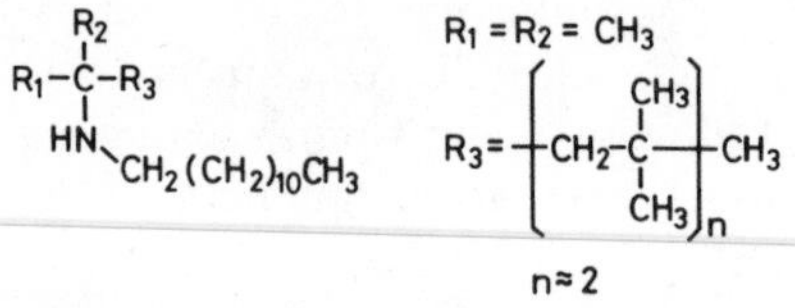

Amberlite® LA-2

Amberlite®* LA-1 (D^{25} = 0.840) and LA-2 (D^{25} = 0.830) are an amber-colored viscous oil. These are commercially available as free amines and are reported to be the mixture of long-chain secondary amines, in which the total carbon numbers of R$_1$ + R$_2$ + R$_3$ are 11 to 14. The average molecular weights are 372 (LA-1) and 374 (LA-2). They are easily soluble in various organic solvents, but are insoluble in water. Solubility in 1 N H$_2$SO$_4$ is 15 to 20 mg/l for LA-1 and ～0 for LA-2.

Extraction behavior of metal ions with 10% Amberlite® LA-1 in xylene from HCl, HNO$_3$, or H$_2$SO$_4$ solution has been investigated at tracer level.[28,29,30,39-42] Some of the results are summarized in Table 4.

These amines can also be used as a stationary phase for the reversed-phase partition chromatography[43] and paper chromatography.[44]

---

*   Amberlite® is a trade name of Rhom & Haars Co., Philadelphia, Pa.

Table 1
SOME EXAMPLES OF METAL EXTRACTION WITH TOA

Metal ion	Condition	Solvent	Separated from	Extractability and remarks	Ref.
Am(III)	0.1 N HNO$_3$, 11.1 M LiCl	Benzene	—	log D = 0.7	5
Am(III)	0.02 N HCl, 11.1 M LiCl	CCl$_4$	—	log D = −0.1	5
Au	5 M H$_2$SO$_4$	CHCl$_3$	Ag,Co,Cu,Fe,Ni,Pt	λ_{max}, 325 ~ 330 nm	6
Cd	0.5~6 N H$_2$SO$_4$, 2 M HaCl	MIBK	Large amount of Al,Cu,Ni,U do not interfere, but, Fe interferes seriously	For the detection of Cd by atom. abs. photometry	7
Bi	0.4 N HCl, 0.01 M KI	Benzene	U	88% extraction; for the determination of Bi by atom. abs. photometry	8
Ce(IV)	0.01~10 M HCl, HNO$_3$, or H$_2$SO$_4$, 0.1 M 4-(5-nonyl)pyridine-N-oxide	Xylene	Th, Lantanides	After oxidation of Ce(III) with K$_2$Cr$_2$O$_7$	9
Co(II)	8 N HCl	Xylene	Ni	93% extraction; no extraction at <2 N	10
Cr(III)	0.1~2.0 N HCl, 4.75 M KSCN	CCl$_4$	—	Extracted as Cr(H$_2$O)$_2$(SCN)$_4$(TOA); >96% extraction (35°)	11
Cu(II)	6.5 N HCl	Xylene	—	92% extraction	10
Eu	0.02 N HCl, 11.1 M LiCl	CCl$_4$	—	log D = −1.8	5
Fe(II)	>4 N HCl	Xylene	—	100% extraction	10
Hf	pH 1.7, 8.9 × 10^{-4} Moxalate(X)	Xylene	—	Extracted as Hf(X)$_4$, D = 199	12
Hg(II)	0.075 N H$^+$, 0.1 M SCN$^-$	CCl$_4$	—	99% extraction	11
Ga	pH 2, tartarate (X)	CHCl$_3$	In	Extracted as Ga(X)$_3$(TOA)$_3$	13
Ir(IV)	6 N HCl	Benzene	Rh	98% extraction, Ir can be stripped with NH$_3$	14
Mg	pH 10.5—12.5, Eriochrome® Black T	CHCl$_3$		100% extraction	15

Table 1 (continued)

SOME EXAMPLES OF METAL EXTRACTION WITH TOA

Metal ion	Condition	Solvent	Separated from	Extractability and remarks	Ref.
Pa(V)	0.5 NHCl, 2 MNaSCN or 0.2~1.5 MHSCN	Xylene (containing 3 ~ 5% amyl alcohol)	Pa(IV)	Pa(IV)	16
Pd(II)	SnCl$_2$, 1.5 NHCl	Benzene	Co,Cr,Cu,Fe,Ir,Mn, Mo,Ni,Os,Pb,Ru,V, W,Zn	Extracted as Pd chlorostanate; λ_{max}, 410 nm; $\varepsilon \simeq 10^4$	17
Pu(VI)	1 ~ 14 NHNO$_3$	Xylene	—	—	18
Rh	SnCl$_2$, 7 NHCl	Benzene	Ir, Pt	λ_{max}, 415 nm	19
Tc(VII)	pH 1~7 (HNO$_3$)	Octyl alcohol	—	D_{max} occurs at pH 3, addition of NaNO$_3$, lowers	20
Te(IV)	2 NHBr	Xylene	Nb, Zr interfere	Extracted as Te(Br)$_6$(TOA)$_2$, λ_{max}, 452 nm; $\varepsilon = 4 \times 10^3$	21
Tl(III)	0.1 NHBr	Benzene	Cd, In, Pb, Zn	Determination of Tl (1 μg); λ_{max} (with I$^-$) 400 nm	22
U	6 NHCl	Xylene	—	60% extraction with U(IV), but 95% with U(VI)	23
U(VI)	6 NHCl, 0.5 MH$_2$SO$_4$ 10$^{-0.3}$ MCH$_3$COO$^-$	Xylene CHCl$_3$	—	U(IV) is not extracted log D > 3	23 24
V(IV)	0.1 NHCl, 9 MLiCl	Benzene	—	Extracted as VO(OH)Cl$_2$ (TOA), D = 0.5 (20°)	25
Zn	1 ~ 6 NHCl	Xylene or benzene	Co,Ni	98% extraction in xylene, log D $\simeq$ 2.5 in benzene	10, 26
Zr	0.2 MH$_2$SO$_4$	Benzene CHCl$_3$	—	D ≥300 (20°) D ≤20(20°)	27

Table 2
EXTRACTION OF METALS WITH 5% TIOA IN XYLENE FROM ACID SOLUTION

Metal ion	log D							
	HCl (N)[28]				HNO$_3$ (N)[29]			
	1	4	8	12	1	6	11	15
Ag	1.8	0.5	−1.0	−2.5	−2.0	−1.8	−1.1	−0.1
Au(III)	4.3	4.7	4.3	3.5	2.6	1.8	0.5	0.2
Bi	2.9	1.8	0.0	−1.9	−0.5	−1.1	−2.4	−1.7
Cd	2.1	2.2	1.6	−0.5	−1.2	−2.0	−2.5	−3.2
Ce(IV)	—	—	—	—	0.1	0.3	−0.7	−1.5
Fe(III)	0.9	1.4	1.9	1.3	—	—	—	—
Ga	−0.3	2.9	2.0	0.7	—	—	—	—
In	0.5	1.7	1.9	0.2	—	—	—	—
Mo	−0.8	1.2	1.5	0.3	—	—	—	—
Os	2.9	2.8	2.8	2.8	0.2	−0.2	−0.6	−0.5
Pa	−2.5	1.0	3.2	3.4	−1.7	−0.7	−1.2	−1.3
Pd	—	—	—	—	0.6	−0.5	−1.3	−0.5
Pt	2.3	2.4	1.7	0.0	−0.1	−0.2	−0.5	−0.4
Re	2.7	2.2	1.3	−0.5	0.9	−0.8	−1.4	−1.3
Tc	2.3	2.2	1.7	1.2	1.5	−0.2	—	—
Te	−0.7	1.6	1.4	1.1	—	−2.3	−2.7	−1.0
Tl(I)	−0.6	−0.6	−0.5	2.8	0.5	0.7	0.4	0.4
U(VI)	1.0	1.5	1.9	1.2	—	—	—	—
W	0.9	0.7	0.5	0.2	0.4	−0.2	−1.0	−1.3
Zn	2.2	2.6	1.7	0.2	—	—	—	—

Long-Chain Primary Amine

Primene®* JM-T

It is an amber-colored viscous oil (D^{25} = 0.845) and is commercially available as a free amine. It is a mixture of long-chain alkylamines, in which the total carbon number of R$_1$ + R$_2$ + R$_3$ is 17 to 23,[R1,R3] and the average mol wt is 311 to 315. It is easily solble in various organic solvents, but is insoluble in water. Solubility in 1 N H$_2$SO$_4$ is 50 mg/ℓ.

The extraction behavior of metal ions with 10% Primine® JM-T in xylene from HCl or H$_2$SO$_4$ solution has been investigated at a tracer level in details. The metal ions that can be effectively extracted from HCl solution are Au(III) (1 to 8 M), Fe (8 to 12 M), Sb(IV) (4 to 12 M), and Pa, Se(IV) (12 M), Cd, Hg(II), Tl, and Zn (2 to 4 M);[45] those from H$_2$SO$_4$ solution are In,Y(0.05 M),[30] Am, Bi, Ce, Eu, Hf,[47] In,[46] La, Lu, Mo, Nb, Np, Pa, Pm, Ru, Sc, Tb, Tc, Th, U, W, and Zr (0.05 to 0.5 M);[30,47] and those from HNO$_3$ solution are Sn[48] and U(IV).[49]

---

* Primene® is a trade name of Rhom & Haas Co., Philadelphia, Pa.

Table 3
SOME EXAMPLES OF METAL EXTRACTION WITH TIOA

Metal ion	Condition	Solvent	Separated from	Extractability and remarks	Ref.
Am	10^{-3} NHCl, 11.8 MLiCl	Xylene or methylene chloride	Lantanides	$96 \sim 97\%$ extraction, but extractability of Ce,Eu,Pr,Th,Y is low ($20 \sim 35\%$)	31
Au	$4 \sim 10$ NHCl or HBr (x = Br or Cl)	CCl$_4$	Ir,Rh	100% extraction, extracted as Au(X)$_4$(TIOA)$_3$, λ_{max} (with Cl$^-$), 325 nm; $\varepsilon = 5.8 \times 10^3$; λ_{max} (with Br$^-$), 300 nm; $\varepsilon = 2.2 \times 10^4$	34
Mo(VI)	$0.01 \sim 0.1$ NHCl, 6 MNaCl	1,2-Dichloroethane or CCl$_4$	—	Extracted as MoO$_2$Cl$_4$(TIOA)$_2$ or Mo$_4$O$_{13}$ (TIOA)$_2$, 100% extraction	32
Np(IV)	4 NHNO$_3$	Xylene	Am,Cm,Pu, fission products	90% extraction (25°)	33
Pd	$4 \sim 6$ NHCl or HBr	CCl$_4$	Ir,Rh	100% extraction, Au,Pt interfere seriously; λ_{max} (with Cl$^-$), 467 nm; $\varepsilon = 1.4 \times 10^3$ λ_{max} (with Br$^-$), 345 nm; $\varepsilon = 9 \times 10^4$	34
	4 NHCl, $6 \sim 8$ M LiCl or 4 NHBr, 10 MKBr	CCl$_4$	Pt	Pt is not extracted in presence of LiCl (10 M), but 30% extraction in the presence of KBr (10 M)	34
Pt	$4 \sim 8$ NHCl or HBr	CCl$_4$	Ir,Rh	100% extraction; no extraction in the presence of LiCl or KBr	34
Pu(VI)	6.4 NHCl, 0.01 MK$_2$Cr$_2$O$_7$	Xylene	Th or fission products	99.8% extraction	35
Ti(IV)	1.87 NH$_2$SO$_4$, 2.75×10^{-2} M H$_2$C$_2$O$_4$(X) + $0.27 \sim 0.54$ M gallic acid(Y)	Toluene	—	Extracted as Ti(Y)$_3$ (X) (TIOA)$_4$	36
U(VI)	$4 \sim 10$ NHCl	Xylene or MIBK	Th or fission products		35
W(VI)	SnCl$_2$, HCl, HSCN	iso-AmOH-CCl$_4$ (1:1)	—	λ_{max}, 400 nm	37
Zn	$1 \sim 3$ NHCl	MIBK	Ba,Co,Ir,Mn, Nb,Sb	$96 \sim 99\%$ extraction; Ag, Cd,Fe,In,Sn interfere	38

Other Amines [R1,R3,R4]

Other amines employed in the metal extractions include trilaurylamine,[50] Alamine 336®[45] tribenzylamine, di-tridecylamine, and di-(2-ethylhexyl)amine.[51]

MONOGRAPH

M1. Sandell, E. B. and Onishi, H., *Photometric Determination of Traces of Metals,* 4th ed., John Wiley & Sons, New York, 1978.

Table 4
EXTRACTION OF METALS WITH 10% AMBERLITE LA-1 IN XYLENE FROM ACID SOLUTIONS

	log D							
Metal ion	HCl (N)[28]				HNO₃(N)[29]			
	1	4	8	12	1	6	11	15
Ag	1.3	0.5	−1.0	—	−2.0	−2.1	−0.8	0.2
Au	2.6	3.0	2.7	2.0	2.1	1.4	0.3	0.6
Cd	1.6	1.6	1.0	−0.5	—	—	—	—
Fe(III)	−0.1	1.5	1.6	1.7	—	—	—	—
Ga	−0.9	2.5	3.9	3.1	—	—	—	—
Hf	−3.6	−3.3	−2.0	0.3	−0.2	−0.7	0.9	0.9
Hg(II)	1.5	1.5	1.4	0.2	0.4	−0.1	−0.4	−0.4
In	−0.7	0.5	0.7	−0.8	—	—	—	—
Nb	−2.4	−2.0	1.2	1.2	−2.5	−1.2	0.6	0.8
Np(VI)	−2.1	−2.9	0.4	1.2	0.7	−0.4	−0.2	−0.3
Os	1.2	1.2	0.7	0.5	0.4	−0.3	0.0	−0.1
Pa	−2.4	−1.1	2.3	3.2	−0.1	0.2	1.2	1.3
Pt	1.8	1.2	0.3	−0.6	0.5	0.2	0.0	0.0
Re	2.5	2.0	0.7	−0.2	1.0	−0.3	−0.7	−0.7
Sn(IV)	0.6	1.5	1.5	0.3	−1.7	−0.7	−0.5	−0.4
Tc	2.3	1.6	1.6	1.2	2.6	0.7	−0.3	−0.2
Te(VI)	−0.8	1.0	1.3	1.0	—	—	—	—
U	—	—	—	—	−1.0	−0.4	−0.3	−0.9
W	0.4	0.9	0.3	0.1	0.6	−0.2	−0.6	−0.5
Zr	−2.3	−2.3	−1.2	−0.5	−0.2	0.4	0.8	0.7

REVIEWS

R1. Coleman, C. F., Blake, C. A., and Brown, K. B., Analytical potential of separations by liquid ion exchange, *Talanta*, 9, 297, 1962.

R2. Green, H., Recent uses of liquid ion exchangers in inorganic analysis, *Talanta*, 11, 1561, 1964.

R3. Green, H., Use of liquid ion-exchangers in inorganic analysis, *Talanta*, 20, 139, 1973.

R4. Sekine, T. and Hasegawa, Y., Solvent extraction of metal ions with various amines, *Bunseki*, 627, 1978.

REFERENCES

1. Good, M. L. and Srivastava, S. C., *J. Inorg. Nucl. Chem.*, 27, 2429, 1965.

2. Sato, T., *Anal. Chim. Acta*, 45, 71, 1969.

3. Sato, T., *Anal. Chim. Acta*, 43, 303, 1968.

4. Florence, T. M. and Farrar, Y. J., *Anal. Chim. Acta*, 63, 255, 1973.

5. Spivakov, B. Ya., Myasoedov, B. F., Shkinev, B. M., Kochetkova, N. E., and Chmatova, M. K., *Proc. Int. Solvent Extr. Conf.*, 3, 2577, 1974; *Chem. Abstr.*, 83, 153291j, 1975.

6. Adam, J. and Pribil, R., *Talanta*, 18, 405, 1971.

7. Kuroda, T., Tsukahara, I., and Shibuya, S., *Bunseki Kagaku*, 20, 1137, 1971.

8. de Moraes, S. and Abrao, A., *Anal. Chem.*, 46, 1812, 1974.

9. Ejaz, M., *J. Radioanal. Chem.*, 36, 409, 1977.

10. Nakagawa, G., *Nippon Kagaku Zasshi*, 82, 1042, 1964.
11. McClellan, B. E., Meredith, M. K., Parmelee, R., and Beck, J. P., *Anal.Chem.*, 46, 306, 1974.
12. Yakabe, K., Kato, H., and Minami, S., *J. Inorg. Nucl. Chem.*, 37, 1973, 1975.
13. Kaplunova, A. M., Ershova, N. I., and Bolshova, T. A., *Zh. Anal. Khim.*, 33, 1940, 1978.
14. Kanert, G. A. and Chow, A., *Anal. Chim. Acta*, 69, 355, 1974.
15. Dziomko, V. M., Ivanov, O. V., Avilina, V. N., Ivashenko, A. K., and Kazakova, T. S., *Proc. Int. Solvent Extr. Conf.*, 2, 1893, 1974; *Chem. Abstr.*, 83, 121574n, 1975.
16. Nekrasova, V. V., Palshin, E. S., and Myasoedov, B. F., *Zh. Anal. Khim.*, 28, 1519, 1973.
17. Khattak, M. A. and Magee, R. J., *Anal. Chim. Acta*, 35, 17, 1966.
18. Keder, W. E., Sheppard, J. C., and Wilson, A. S., *J. Inorg. Nucl. Chem.*, 12, 327, 1960.
19. Holzbecher, Z., Divis, L., Kral, M., Sucha, L., and Ulacil, F., *Handbook of Organic Reagents in Inorganic Analysis*, Ellis Horwood Chishester, England, 1976.
20. Volk, V. I., Rozen, A. M., and Barabash, A. I., *Radiokhimiya*, 18, 243, 1976; *Chem. Abstr.*, 84, 170416s, 1976.
21. Suzuki, T., Sotobayashi, T., and Koyama, S., *Bunseki Kagaku*, 23, 999, 1974.
22. Tsukahara, I., Sakakibara, M., and Yamamoto, T., *Anal. Chim. Acta*, 83, 251, 1976.
23. Hodara, I. and Balouka, I., *Anal. Chem.*, 43, 1213, 1971.
24. Shmidt, V. S. and Rybakov, K. A., *Radkokhimiya*, 16, 580, 1974; *Chem. Abstr.*, 82, 48198r, 1975.
25. Sato, T., Ikoma, S., and Nakamura, T., *J. Inorg. Nucl. Chem.*, 39, 395, 401, 1977; *Nippon Kagaku Kaishi*, 716, 1979.
26. Sato, T. and Kato, T., *J. Inorg. Nucl. Chem.*, 39, 1205, 1977.
27. Sato, T. and Kato, T., *J. Inorg. Nucl. Chem.*, 36, 2585, 1974.
28. Ishimori, T., Akatsu, E., Tsukuechi, K., Kobune, T., Usuba, Y., Kimura, K. K., Onawa, G., and Uchiyama, H., JAERI Rept.-1106 (1966); *Chem. Abstr.*, 65, 6638a, 1966.
29. Ishimori, T. and Nakamura, E., JAERI Rept.-1047 (1963); *Chem. Abstr.*, 64, 15070f, 1966.
30. Ishimori, T., Akatsu, E., Wen-Pi Cheng, Tsukuechi, K., and Osakabe, T., JAERI Rept.-1062 (1964); *Chem. Abstr.*, 62, 13922e, 1965.
31. Moore, F. L., *Anal. Chem.*, 33, 748, 1961.
32. Vieux, A. S., Rutagengwa, N., Noki, V., and Basosila, L., *Inorg. Chem.*, 15, 722, 1976, *J. Inorg. Nucl. Chem.*, 39, 645, 1977; *Rev. Chim. Miner.*, 14, 387, 1977; *Chem. Abstr.*, 88, 79895u, 1978.
33. Schneider, R. A., *Anal. Chem.* 34, 522, 1962.
34. Mirza, M. Y., *Talanta*, 27, 101, 1980.
35. Moore, F. L., *Anal. Chem.*, 30, 908, 1958.
36. Biswas, S. P., Krishnamoorthy, T. S., and Venkateswarlu, Ch., *Indian J. Chem.*, 12, 865, 1974; *Chem. Abstr.*, 82, 48204q, 1975.
37. Vieux, A. S., Bassolila, I., and Rutagengwa, N., *Hydrometallurgy*, 2, 351, 1977; *Anal. Abstr.*, 35, 4B144, 1978.
38. Mirza, M. Y., Ejaz, M., Sani, A. R., Ullah, S., Rashid, M., and Samdani, G., *Anal. Chim. Acta*, 37, 402, 1967.
39. Moore, G. E. and Kraus, K. A., *J. Am. Chem. Soc.*, 75, 1460, 1953.
40. Nakagawa, G., *Nippon Kagaku Zasshi*, 81, 444, 747, 1255, 1533, 1960.
41. Nakagawa, G., *Nippon Kagaku Zasshi*, 9, 821, 1960.
42. Suzuki, T. and Sotobayashi, T., *Nippon Kagaku Zasshi*, 12, 916, 1963.
43. Przeszlakowski, S. and Flieger, A., *Talanta*, 26, 1125, 1979.
44. Przeszlakowski, S. and Flieger, A., *Talanta*, 23, 844, 1976.
45. Vir Singh, Om. and Tandon, S. N., *J. Inorg. Nucl. Chem.*, 36, 439, 1974; *J. Inorg. Nucl. Chem.*, 37, 609, 1975; *Sep. Sci.*, 10, 359, 1975; *Chem. Abstr.*, 83, 184232v, 1975.
46. Watanabe, H. and Akatsuka, K., *Nippon Kagaku Zasshi*, 89, 280, 1968.
47. El-Yamani, I. S., Farah, M. Y., and Abd El-Aleim, F. A., *Talanta*, 25, 523, 1978.
48. Schade, W., *Chem. Technol. (Leipzig)*, 29, 499, 1977; A.A., 34, 4B99, 1978.
49. Schmid, E. R. and Juenger, E., *J. Radioanal. Chem.*, 13, 349, 1973.
50. Patil, S. K., Swarup, R., Ramaniah, M. V., and Srinivason, N., *Radiochim. Acta*, 18, 212, 1972; *J. Inorg. Nucl. Chem.*, 38, 1203, 1976.
51. Ivanov, N. A. and Todorova, N. G., *Dokl. Bolg. Akad. Nauk.*, 30, 261, 265, 1977; *Dokl. Bolg. Akad. Nauk.*, 31, 77, 873, 1978; *Anal. Abstr.*, 33, 3B76, 3B101, 1977; *Dokl. Bolg. Acad. Nauk.*, 35, 5B124, 1978; *Dokl. Bolg. Akad. Nauk*, 36, 5B93, 1979.

QUATERNARY AMMONIUM SALTS

General Considerations

Quaternary ammonium salts having one or more long-chain alkyl groups play important roles as an analytical reagent.

The analytical applications are based on the two functions of the quaternary ammonium ions. One is the use as a cationic reagent for the ion-pair extraction of metals as complex anions. The other is the use as a cationic micelle-forming reagent in the photometric determination of metals.

As in the case of the long-chain alkylamines, various types of the quaternary ammonium salts have been used as a cationic reagent for the extraction of metals. Metal ions are extracted as complex anions of inorganic ligands (Cl^-, NO_3^-, SCN^-, CN^-, etc.) or organic anionic ligands. In contrast to the long-chain alkylamines, the metal extractions can be carried out even in neutral or alkaline range, so far as the complex anion is resistant to hydrolysis, because the quaternary ammonium ion does not need protonation for the anion extraction. The equilibria involved in the extraction with the quaternary ammonium ion can be written as below:

$$(m - n)\, R_4N^+Cl^- + ML_n^{m-n} \rightleftharpoons (R_4N)_{m-n}\,(ML_n)^{m-n} + (m - n)\, Cl^-$$

where ML_n^{m-n} represents the complex anion of metal ion (M^{m+}) with the ligand (L^-).

The extractability is affected by many factors as in the case of long-chain alkylamine extractions. Table 1 shows the effect of solvents and quaternary ammonium ions on the extractability of iron (III)-pyrocatechol-4-sulfonate complex. Pyrocatechol-4-sulfonic acid (H_3L) forms a red soluble anionic chelate with Fe(III) ($Fe(HL)_3^{3-}$; λ_{max}, 480 nm) which can be extracted with the quaternary ammonium ion.[1] The extraction is more favored with the increase in the number of long-chain alkyl groups of ammonium ions and with the increase in the dielectric constant of extraction solvents. The extraction is more effective with the quaternary ammonium ions of longer alkyl chain length. It is also reported that the extractivity decreases in the following order:

$$(C_{16}H_{33})\,(CH_3)_3NBr > (C_4H_9)_4NBr > (C_2H_5)_4NBr$$

in the extraction of Sn-Pyrocatechol Violet complex.[2]

The quaternary ammonium ions having one or two long-chain alkyl groups are fairly soluble in water and behave as a surfactant, while the ammonium ions having three long-chain alkyl groups are almost insoluble in water, but soluble in polar or nonpolar organic solvents and behave as a liquid anion exchanger.

As to the selectivity for the extraction of anions or anion complexes, the extraction is, in general, favored by increase in size and decrease in charge. However, many exceptions for this working hypothesis have been reported. For example, in the extraction with the dichloroethane solution of trioctylmethylammonium chloride, the selectivity orders for the EDTA complexes are $FeY(OH)_2^{3-} > FeY(OH)^{2-} > FeY^-$ and $VO_2Y^{3-} > VO_2HY^{2-}$.[49] Irving also found the following selectivity order for the extraction of cyano complex anions with tetrahexylammonium erdmanate in MIBK:[3]

$$M(CN)_2 > ClO_4^- \gg M(CN)_4^{2-} \gg Fe(CN)_6^{3-}$$

while higher selectivity for the extraction of $Fe(CN)_6^{3-}$ than ClO_4^- is reported in the trioctylmethylammonium chloride - chlorobenzene extraction.[4]

Thus, many more experimental results are needed to get the conclusion on the anion

Table 1
EFFECT OF SOLVENTS AND QUANTERNARY AMMONIUM IONS ON THE EXTRACTABILITY OF FE-PYROCATECHOL-4-SULFONATE COMPLEX

| | | Absorbance at 480 nm | | |
| | | | | |
Solvents	Dielectric constants	Trimethylbenzyl ammonium chloride	Tetradecyldimethyl benzylammonium chloride	Dialkylmonomethyl benzylammonium
CCl_4	2.23	—	—	0.623
$CHCl_3$	4.80	—	0.475	0.628
$C_2H_4Cl_2$	10.36	—	0.655	0.625
C_6H_6	2.28	—	0.242	0.625
$C_6H_4Cl_2$	9.93	—	0.632	0.630
$C_6H_5NO_2$	34.82	—	0.585	—

Note: pH: 9.9 to 10.1; Fe(III), 1.00×10^{-4} M; Pyrocatechol-4-sulfonate, 1.50×10^{-3} M; quaternary ammonium salt, 0.010 M; KCl:0.10 M.

From Kohara, H., Ishibashi, N., and Yoshida, A., *Bunseki Kagaku,* 17, 616 (1968). With permission.

selectivity in ion-exchange extraction with quaternary ammonium salt. However, it is likely that the extractability is strongly dependent on the kind of the quaternary ammonium salt and the solvent used.

As stated previously, the quaternary ammonium salts having one or two long-chain alkyl groups behave as a surfactant, forming a cationic micelle at the concentration above CMC (critical micelle concentration). The complexation reactions on the surface of the cationic micelles are quite different from those in the simple aqueous solution,[R1] forming a chelate of higher ligand to metal ratio than in the aqueous system. This effect usually results in the bathochromic shift and the increase of molar absorptivity of the colored metal chelates. The examples will be shown in the following section.

Quaternary Ammonium Salts of Analytical Interests

The quaternary ammonium salts which are frequently used as analytical reagents are listed below.[R1]

1. Tetradecyldimethylbenzylammonium chloride (Zephiramine)®*)
2. Cetyltrimethylammonium chloride (CTMC) and bromide (CTMB)
3. Hydroxydodecyltrimethylammonium bromide (HDTMB)
4. Dialkylmonomethylbenzylammonium bromide (AMBB) and chloride (AMBC)
5. Dodecyloctylmethylbenzylammonium chloride (DOMBC)
6. Trioctylmethylammonium chloride (Aliquat 336S®)

Tetradecyldimethylbenzylammonium Chloride (Myristyldimethylbenzylammonium Chloride, Benzalkonium chloride)

$$\left[CH_3(CH_2)_{13}-\underset{\underset{CH_3}{|}}{\overset{\overset{CH_3}{|}}{N}}-CH_2-\bigcirc \right] Cl^-$$

$C_{23}H_{42}NCl$
mol wt = 368.04

* Trade name of Dojindo Laboratories, Kumamoto, Japan.

Commercially available. Zephiramine® is an analytical grade of this material. It is a colorless or pale yellow powder of aromatic odor and has a very bitter taste and contains a small percent of water. It is hygroscopic and very soluble in water, alcohol, and acetone, is slightly soluble in benzene, and is almost insoluble in ether. The aqueous solution is slightly alkaline to litmus and foams strongly when shaken. CMC 3.7×10^{-4} M^{R1} or 1.5×10^{-3} M^5. D(org. solvent/water): 0.09 (benzene, toluene), 2.4 (1,2-dichloroethane), 11 (chloroform), 0.10 (carbon tetrachloride), and 0.53 (MIBK).[6]

The quaternary ammonium salt can be assayed by the ion-pair extraction photometry using tetrabromophenolphthalein ethylester in 1,2-dichloroethane (pH 7.5; λ_{max}, 610 nm; $\varepsilon = 7.3 \times 10^4$).[7]

CTMC (Cetyltrimethylammonium Chloride) and CTMB (Bromide)

$$\left[CH_3(CH_2)_{15} - \overset{\overset{\displaystyle CH_3}{|}}{\underset{\underset{\displaystyle CH_3}{|}}{N^+}} - CH_3 \right] X^-$$

CTMC (X =Cl)
$C_{19}H_{42}NCl$
mol wt = 320.00
CTMB (X=Br)
$C_{19}H_{42}NBr$
mol wt = 364.45

It is commercially available as a colorless hygroscopic powder and is very easily soluble in water, alcohol, and acetone. The aqueous solution foams strongly when shaken. It behaves similarly to tetradecyldimethylbenzylammonium chloride as a surfactant and as an ion-pair extraction reagent, CMC 1×10^{-4} M.[R1]

Trioctylmethylammonium Chloride (Tricaprylmethylammonium Chloride)

$$\left[CH_3 - \overset{\overset{\displaystyle (CH_2)_7CH_3}{|}}{\underset{\underset{\displaystyle (CH_2)_7CH_3}{|}}{N^+}} - (CH_2)_7CH_3 \right] Cl^-$$

$C_{25}H_{54}NCl$
mol wt = 404.16

Commercially available as Aliquat® 336S* or Capriquat®** which contain small fractions of C_{10} isomer and has an average mol wt of 442. It is a yellow-brown viscous oil, containing a small percent of water, and is almost insoluble in water, but is easily soluble in common organic solvents including kerosene (100 g/100 mℓ, 0 to 60°); D(C_6H_5Cl/H_2O) $\simeq 10$ ([Cl^-] = 10^{-3} M).[4]

The anion-exchange extraction constants $K_{ex}^{QCl} = ([QCl]_{org} \cdot [X^-]/[QX]_{org} \cdot [Cl^-])$ were reported on a chlorobenzene-water system; log $K_{ex}^{QCl} = 1.34$ for Br^-, 1.81 for NO_3^-, 3.32 for I^-, 3.80 for PAR^-, 4.47 for ClO_4^-, and 10.41 for $Fe(CN)_6^{3-}$.[4]

When metal ions are extracted as anionic complexes, extractability is very much dependent upon the acid concentration. Figure 1 illustrates the acid dependencies of metal extraction with Aliquat 336S®[8].

The commercial reagent can be purified by the following procedure.[9]

Dissolve 50 g of the reagent into 100 mℓ of chloroform. Shake the solution with 200 mℓ of 20% sodium hydroxide solution for 10 min, then with 200 mℓ of hydrochloric acid (1 + 1) for 10 min, and then with 200 mℓ of 20% sodium chloride solution for 10 min. Wash the equilibrated solution with a small amount of water and filter it through a dry filter paper.

---

* Trade name of General Mills Inc., Minneapolis, Minn.
** Trade name of Dojindo Laboratories, Kumamoto, Japan.

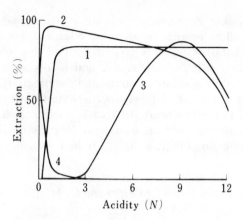

FIGURE 1.   Extraction of various metals with Aliquat 336S® from HCl solution. (1) Fe(III); (2) Zn; (3) Co; (4) Fe(III) from H_2SO_4 solution. Ni can not be extracted at all. (From Okazaki, M., Horiguchi, T., and Watanabe, K., *Nippon Kagaku Zasshi*, 84, 917, 1963. With permission.)

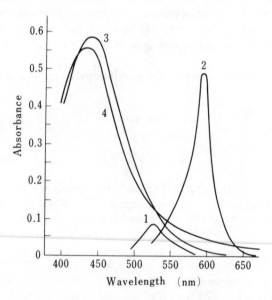

FIGURE 2.   Absorption spectra of Eriochrome® Cyanine R and its Be chelate; pH 6.9. (1) Be chelate without surfactant; (2) Be chelate with surfactant (1.5×10^{-3} *M*); Be, 4.00×10^{-6} *M*; dye, 2.4×10^{-5} *M*; (3) free dye without surfactant; (4) free dye with surfactant (1.5×10^{-3} *M*); dye, 6.0×10^{-5} *M*. (From Kohara, H., Ishibashi, N., and Fukamachi, K., *Bunseki Kagaku*, 17, 1400, 1968. With permission.)

Analytical Applications
Use as an Ion-Pair Extraction Reagent

Highly colored complex anions can be extracted as an ion-pair with the quaternary ammonium ion into the organic solvent for the subsequent photometry. Extractions of various inorganic and organic complex anions have been reported, and some the successful examples are summarized in Table 2.

Table 2
PHOTOMETRIC DETERMINATION OF METALS AFTER ION-PAIR EXTRACTION WITH QUATERNARY AMMONIUM ION

Metal ion	Ligand (L)	Quantemary ammonium ion (X)[a]	Extraction solvent	pH	Ion - pair: Composition	Ion - pair: λ_{max} (nm)	Ion - pair: ε (× 10^4)	Ref.
Al	Calmagite®	Aliquat 336S® (in the presence of EDTA, CN^-)	$CHCl_3$	8.6	$(ML_3)X_n$	570	4.2	10
B	PV	CPC	$CHCl_3$	7.5 ∿ 9	—	630	2.4	11
Be	Chromazurol® S	HDTB	Butanol	6.6 ∿ 7.0	$(ML_2)X_4$	596	5.9	12
Bi	I^-	Zephiramine®	1,2-Dichloroethane	3 ∿ 5.5	$(MI_4)X$	490	1.0	13
Cd	PAN	Aliquat® 336S	$CHCl_3$	>12	$(ML_2)X_n$	555	0.3	14
Co(II)	PAR	Zephiramine® (in the presence of EDTA)	$CHCl_3$	8.3	$(ML_2)X_2$	520	5.9	15
Cr(III)	PAR	Zephiramine® (in the presence of EDTA)	$CHCl_3$	4.8 ∿ 5.2	$(ML_3)X$	540	4.7	16
Cu(II)	Zincon	Zephiramine® (in the presence of EDTA)	$CHCl_3$	7.5 ∿ 9.5	MLX_2	625	2.3	17
Fe(II)	Dinitroso-DMAP	Zephiramine®	$CHCl_3$	6.5 ∿ 9.5	$(ML_4)X$	730	2.9	18
Fe(III)	PAR	Zephiramine® (in the presence of EDTA)	$CHCl_3$	10	$(ML_2)X_2$	522	4.0	15
Hg(II)	MTB + SCN^-	Zephiramine®	$CHCl_3$	6.8 ∿ 7.1	$[ML(SCN)]X$	646	10.2	19
Mg	Eriochrome® Black T	Zephiramine®	1,2-Dichloroethane	11.2 ∿ 12.2	—	690	6.2	20
Mo(VI)	PV	AMBC	$CHCl_3$	0.2 ∿ 0.6	$[ML_2]X_2$	560	6.3	21
Ni	PAR	Zephiramine® (in the presence of EDTA)	$CHCl_3$	9.3	(ML_2X_2)	505	7.7	15
Pd	SCN^-	Zephiramine® (in the presence of EDTA)	$CHCl_3$	0.01 ∿ 0.2 *N* HCl	—	312	3.7	22
Sc	XO	TDEB	Xylene	6.0 ∿ 6.7	$(ML)X_2$	520	2.7	23
Tl(III)	I^-	Zephiramine®	1,2-Dichloroethane	4 ∿ 6	$(ML_4)X$	395	1.2	12
UO_2^{2+}	Arsenazo III	DOMBC	$CHCl_3$	0.8 ∿ 1.2	$(ML)X_n$	655	5.6	24
V(V)	PAR	Zephiramine® (in the presence of EDTA)	$CHCl_3$	6.8	ML_2X_2	560	4.3	15

Table 2 (continued)
PHOTOMETRIC DETERMINATION OF METALS AFTER ION-PAIR EXTRACTION WITH QUATERNARY AMMONIUM ION

Metal ion	Ligand (L)	Quaternary ammonium ion (X)[a]	Extraction solvent	pH	Ion - pair Composition	λ_{max} (nm)	$\varepsilon (\times 10^4)$	Ref.
W(V)	SCN^- ($SnCl_2$)	Zephiramine® (in the presence of EDTA)	$CHCl_3$	4 N HCl	—	405	4.3	25
Zn	TAR	Zephiramine® (in the presence of EDTA)	$CHCl_3$	8.5 ∽ 10.7	$(ML_2)X_2$	550	3.0	26

[a] Key for quaternary ammonium salts: AMBC, dialkylmonomethyl benzylammonium chloride; DOMBC, dodecyloctylmethyl benzylammonium chloride; CPC, cetylpyridinium chloride; HDTB, hexadecyltrimethylammonium bromide; TDEB, tridodecylethylammonium bromide.

Table 3
PHOTOMETRIC DETERMINATION OF METALS IN THE PRESENCE OF CATIONIC SURFACTANT

Metal ion	Ligand	Surfactant[a]	pH	Surfactant Without λ_{max} (nm)	$\varepsilon (\times 10^4)$	With λ_{max} (nm)	$\varepsilon (\times 10^5)$	Ref.
Al	Stilbazo	Zephiramine®	5.0 ∽ 5.2	495	2	540	1.1	28
Al	Erichrome® Cyanine R	CTMC	5.3 ∽ 6.3	535	5.8	587	1.2	29
Be	Chromazurol® S	Zephiramine®	4.7 ∽ 5.3	564	0.2	610	1.0	30
Co	Chromazurol® S	HDTMB + Pyridine	10.6 ∽ 11.5	567	2	654	1.1	31
Cu	Chromazurol® S	Zephiramine® + Pyridine	5.45	600	2	620	1.4	32
Fe(III)	Chromazurol® S	CTMC	3.1 ∽ 3.8	615	3	630	1.5	33

Element	Reagent	Surfactant	pH					Ref.
Ga	Pontachrome® Azul Blue B	CTMC	6.0 ~ 6.6	620	3	680	1.4	34
Ge	Phenylfluorone	CTMC	1.0 ~ 1.5 NHCl	505	8.5	505	1.7	35
Mo(VI)	PR	HDTMB (in the presence of EDTA, F⁻)	5.0 ~ 5.3	500	0.6	587	0.8	36
Ni	Chromazurol® S	HDTMB + Pyridine	10.5 ~ 11.3	550	4	639	1.8	37
Pd	Chromazurol® S	Zephiramine®	5.8 ~ 6.7	610	4	635	1.2	38
Sc	Chromazurol® S	Zephiramine®	5.1 ~ 5.6	510	3	625	1.4	39
Sn(IV)	PV	CTMB	0.5 ~ 4.0	555	6.5	662	0.96	40
Th	Chromazurol® S	CTMC + Pyridine	5.6 ~ 6.1	620	4	631	1.7	41
Ti(IV)	Haematoxylin	CTMB	3.0 ~ 4.0	570	4	655	1.3	42
U(VI)	Eriochrome® Cyanine R	CTMB	5.3 ~ 5.9	560	1.9	605	0.88	43
V(IV)	Chromazurol® S	Zephiramine®	4.3 ~ 5.6	580	3	610	0.68	44
W(VI)	PR	CTMC (in the presence of EDTA)	4.7	560	1.5	576	0.6	45
Zn	Chromazurol® S	Zephiramine®	8.1 ~ 8.4	550 ~ 555	0.7	510	0.44	46
Zr	TAM	Zephiramine®	3.7 ~ 4.5	595	7.5	595	1.1	47
Rare earths	XO	CPB	8 ~ 9	570 ~ 580	3.1 ~ 4.8	600 ~ 610	0.8 ~ 1.1	48

ᵃ Key for surfactant: CTMC, cetyltrimethyl ammonium chloride; CTMB, cetyltrimethyl ammonium bromide; HDTMB, hydroxydodecyltrimethyl ammonium bromide; CPB, cetylpyridinium bromide.

Use as a Cationic Surfactant

As stated previously, the complexation on the interface of cationic micelle results in a substantial gain on the photometric sensitivity, due to the formation of the metal chelates of higher ligand to metal ratio. Figure 2 illustrates the absorption spectra of Eriochrome® Cyanine R and its Be chelate in the absence and in the presence of cationic surfactant (Zephiramine®).[27] While the absorption spectrum of free reagent is not much influenced by the presence of the surfactant, that of Be chelate shows a striking bathochromic shift accompanied by a substantial increase in the absorption intensity. As a result, the higher sensitivity with the less reagent blank can be expected by the use of a cationic surfactant. A fairly large volume of works has been done in this field, and some of the examples are summarized in Table 3.

MONOGRAPH

M1. **Sandell, E. B. and Onishi, H.**, *Photometric Determination of Traces of Metals,* 4th ed., John Wiley & Sons, New York, 1978.

REVIEWS

R1. **Ueno, K.**, Application of cationic surfactants in analytical chemistry, *Bunseki Kagaku,* 20, 736, 1971.
R2. **Nishida, H.**, Ternary complexes; application of cationic surfactants to spectrophotometory, *Bunseki,* 271, 1977.

REFERENCES

1. **Kohara, H., Ishibashi, N., and Yoshida, A.**, *Bunseki Kagaku,* 17, 616, 1968.
2. **Bailey, B. W., Chester, J. E., Dagnall, R. M., and West, T. S.**, *Talanta,* 15, 1359, 1968.
3. **Irving, H. M. N. H. and Damodaran, A. D.**, *Anal. Chim. Acta,* 53, 267, 1971.
4. **Itoh, J., Kobayashi, H., and Ueno, K.**, *Anal. Chim. Acta,* 105, 383, 1979.
5. **Lebetter, J. W., Jr. and Bowen, J. R.**, *Anal. Chem.,* 41, 1345, 1969.
6. **Zenki, M.**, *Bunseki Kagaku,* 25, 552, 1976.
7. **Sakai, T. and Ishida, N.**, *Bunseki Kagaku,* 27, 410, 1978.
8. **Okazaki, M., Horiguchi, T., and Watanabe, K.**, *Nippon Kagaku Zasshi,* 84, 917, 1963.
9. **Adam, J. and Pribil, R.**, *Talanta,* 18, 733, 1971.
10. **Woodward, C. and Freiser, H.**, *Talanta,* 15, 321, 1968.
11. **Serdyuk, L. S., Karaseva, L. B., and Albota, L. A.**, *Zh. Anal. Khim.,* 32, 2361, 1977.
12. **Ishida, R. and Tonosaki, K.**, *Nippon Kagaku Kaishi,* 1077, 1974.
13. **Matsuo, T., Shida, J., and Sasaki, T.**, *Bunseki Kagaku,* 16, 546, 1967.
14. **Akaiwa, H., Kawamoto, H., and Takenouchi, T.**, *Bunseki Kagaku,* 27, 449, 1978.
15. **Yotsuyanagi, T., Yamashita, R., and Aomura, K.**, *Bunseki Kagaku,* 19, 981, 1970.
16. **Yotsuyanagi, T., Takeda, Y., Yamashita, R., and Aomura, K.**, *Anal. Chim. Acta,* 67, 297, 1973.
17. **Sugawara, M., Niiyama, K., and Kambara, T.**, *Bunseki Kagaku,* 22, 1219, 1973.
18. **Toei, K., Motomize, S., and Korenaga, T.**, *Analyst,* 100, 629, 1975.
19. **Nomura, T., Takemura, S., Nakamura, T., and Komatsu, S.**, *Bunseki Kagaku,* 22, 576, 1973.
20. **Fukamachi, K., Kohara, H., and Ishibashi, N.**, *Bunseki Kagaku,* 19, 1529, 1970.

21. Kohara, H., Ishibashi, N., and Abe, K., *Bunseki Kagaku,* 19, 48, 1970.
22. Matsuo, H., Chaki, S., and Akabori, K., *Bunseki Kagaku,* 20, 226, 1971.
23. Shijo, Y., *Nippon Kagaku Kaishi,* 889, 1974.
24. Fukamachi, K., Kohara, H., and Ishibashi, N., *Bunseki Kagaku,* 21, 1165, 1972.
25. Matsuo, H., Chaki, S., and Hara, S., *Bunseki Kagaku,* 17, 752, 1968.
26. Ueda, K., *Anal. Lett.,* A11, 1009, 1978.
27. Kohara, H., Ishibashi, N., and Fukamachi, K., *Bunseki Kagaku,* 17, 1400, 1968.
28. Ozawa, T., *Bunseki Kagaku,* 18, 745, 1969.
29. Shijo, Y. and Takeuchi, T., *Bunseki Kagaku,* 17, 323, 1968.
30. Horiuchi, Y. and Nishida, H., *Bunseki Kagaku,* 18, 180, 1969.
31. Shijo, Y., Takeuchi, T., and Yoshizawa, S., *Bunseki Kagaku,* 18, 204, 1969.
32. Nishida, H. and Nishida,T., *Bunseki Kagaku,* 21, 997, 1972.
33. Shijo, Y. and Takeuchi, T., *Bunseki Kagaku,* 17, 1519, 1968.
34. Uesugi, K. and Shigematsu, T., *Talanta,* 24, 391, 1977.
35. Shijo, Y. and Takeuchi, T., *Bunseki Kagaku,* 16, 51, 1967.
36. Takeuchi, T. and Shijo, Y., *Bunseki Kagaku,* 15, 473, 1966.
37. Shijo, Y. and Takeuchi, T., *Bunseki Kagaku,* 17, 1192, 1968.
38. Horiuchi, Y. and Nishida, H., *Bunseki Kagaku,* 16, 1018, 1967.
39. Horiuchi, Y. and Nishida, H., *Bunseki Kagaku,* 17, 1486, 1968.
40. Dagnell, K. M., West, T. S., and Young, P., *Analyst,* 92, 27, 1967.
41. Shijo, Y. and Takeuchi, T., *Bunseki Kagaku,* 18, 469, 1969.
42. Leong, C. L., *Analyst,* 102, 293, 1977.
43. Otomo, M. and Kodama, K., *Bunseki Kagaku,* 20, 1581, 1971.
44. Horiuchi, Y. and Nishida, H., *Bunseki Kagaku,* 18, 850, 1969.
45. Shijo, Y. and Takeuchi, T., *Bunseki Kagaku,* 22, 1341, 1973.
46. Horiuchi, Y. and Nishida, H., *Bunseki Kagaku,* 17, 756, 1968.
47. Tsurumi, C., *Bunseki Kagaku,* 26, 260, 1977.
48. Otomo, M. and Wakamatsu, Y., *Bunseki Kagaku,* 17, 764, 1968.
49. Irving, H. M. N. H. and Al-Jarrah, R. H., *Anal. Chim. Acta,* 74, 321, 1975.

TETRAPHENYLARSONIUM CHLORIDE AND OTHER ONIUM SALTS

$$C_{24}H_{20}ClAs$$
$$mol\ wt = 418.80$$

L$^+$Cl$^-$

Synonym

TPAC

Source and Method of Synthesis

Commercially available; synthesized by the reaction of phenylmagnesium bromide with triphenylarsonium oxide.[1,2]

Analytical Uses

As a precipitating reagent for anions, especially for oxo-anions and anionic complexes. As these precipitates are extractable into chloroform, it has also been used as an extraction reagent for such anions.

Properties of Reagent

TPAC is obtained as a dihydrate or anhydrous colorless needles, mp 256 to 257°[3] (or 259.5 to 261°).[4] The solid reagent is stable. It is soluble in water (0.99 M as dihydrate, 25°) as well as in chloroform (0.70 M as anhydrous salt).[M1] It polymerizes in chloroform to form dimer ($\beta_2 = [(LCl)_2]/[LCl]^2 = 163$) and tetramer ($\beta_4 = 4.5 \times 10^5$).[4] As a rough measure, only 16% of the reagent is monomeric form in a 0.06 M chloroform solution of TPAC.

In aqueous solution, TPAC exists as the monomer, and the dissociation constant and distribution coefficient for the monomer are reported to be K = $[L^+] \cdot [Cl^-]/[LCl]$ = 0.082 ($\mu = 0.1, 25°$) and K$_D$ (CHCl$_3$/H$_2$O) = 3.7 ($\mu = 0.1, 25°$), respectively.[M1]

Reactions with Anions

Oxo-anions and complex anions of relatively large size form insoluble salts with TPAC (for example, K$_{sp}$ (L·ClO$_4$) = 2.6×10^{-9} and K$_{sp}$ (L·ReO$_4$) = 3.7×10^{-9}, 20°),[5,6] and these salts can be extracted as ion-pairs into chloroform or similar solvents of low polarity. The solubility of some of TPAC salts in organic solvents are summarized in Table 1.[7]

In the medium of moderate acidity, the concentration of arsonium cation is not much affected by the hydrogen ion concentration, so that the extraction ratio is not generally pH dependent. However, in the case of oxo-anions, they may be protonated at higher acidity, resulting in the decrease of extraction ratio. Also, in the case of complex anions, the concentration of anionic ligands greatly influences the extraction ratio. The extraction coefficient of common anions and oxo-anions at different pH ranges are summarized in Table 2.[8] The extraction of chloro complexes have been investigated in detail as a function of HCl concentration. A part of the result is shown in Table 3.[9] The extraction of other complex anions, such as fluoro- and thiocyanato-complexes, was also investigated.

Purification and Purity of Reagent

TPAC can be purified by adding concentrated hydrochloric acid to a saturated aqueous solution of TPAC to precipitate L·Cl·HCl·2H$_2$O, followed by the dissolution

Table 1
SOLUBILITY (S) OF SOME TPAC SALTS IN CHLOROFORM AND 1,2-DICHLOROETHANE

TPAC salt	S (M)	
	Chloroform	1,2-Dichloroethane
L·AuCl$_4$	4.4×10^{-3}	$>2.8 \times 10^{-3}$
L·ClO$_4$	0.8×10^{-3}	1.1×10^{-2}
L$_2$·Cr$_2$O$_7$	—	$>3.5 \times 10^{-?}$
L·I	0.2	—
L·MnO$_4$	0.7×10^{-3}	3.1×10^{-3}
L·ReO$_4$	1.1×10^{-3}	6.1×10^{-3}

(From Alimarin, I. P. and Perezhagin, G. A., *Talanta*, 14, 109, 1969. With permission.)

Table 2
EXTRACTION COEFFICIENTS OF COMMON ANIONS AND OXO-ANIONS WITH TPAC IN A WATER-CHLOROFORM SYSTEM

Anion[a] (X$^-$)	E[b]		Anion (X$^-$)	E	
	pH 0.7 ~ 2.0	pH ~12		pH 1.2 ~ 2.5	pH ~12
Cl$^-$	0.22	0.18	NO$_3^-$	20.3	42.5
Br$^-$	4.56	4.75	NO$_2^-$	—	0.22
I$^-$	>300	>300	SO$_3^{2-}$	—	0.015
SCN$^-$	34.7	32.3	S$_2$O$_3^{2-}$	—	0.017
ClO$_3^-$	>150	>150	PO$_4^{3-}$	~0.01	~0.01
BrO$_3^-$	0.92	0.84	P$_2$O$_7^{4-}$	—	0.003
ClO$_4^-$	>200	>200	CrO$_4^{2-}$	71	0.025
MnO$_4^-$	>300	>300	AsO$_3^{3-}$	0.006	0.05
RhO$_4^-$	>200	>200	TeO$_3^{2-}$	<0.01	0.09
IO$_4^-$	—	0.02			

[a]   The extraction coefficients for the following anions are negligibly small: F$^-$, IO$_3^-$, SO$_4^{2-}$, MoO$_4^{2-}$, WO$_4^{2-}$, VO$_3^-$, BO$_3^{3-}$, AsO$_4^{3-}$, and SeO$_3^{2-}$.

[b]   E = [L$^+$X$^-$]$_{org}$/[X$^-$]$_{aq}$. A 50-mℓ aqueous solution containing 2/3 meq (C$_6$H$_5$)$_4$ AsOH and 1/3 meq anion was shaken with 50 mℓ CHCl$_3$.

(From Bock, R. and Beilstein, G. M., *Fresenius Z. Anal. Chem.*, 192, 44, 1963. With permission.)

of the precipitate into water, neutralizing with sodium carbonate, evaporating to dryness, extracting with chloroform, and finally crystallizing from ethanol by adding ether.[3,10,11]

The aqueous solution of the pure TPAC should be clear and colorless. Impure samples may give somewhat turbid solution which can be clarified by filtration through a filter paper after treating with Celite.[12] The purity of TPAC can be assayed by the extraction photometry with a highly colored anions such as V-PAR chelate (560 nm; ε = 9.3×10^4, 0.1 to 5×10^{-5} MTPAC).[13]

Table 3
EXTRACTION OF METALS
WITH 0.05 M TPAC IN CHCl$_3$
FROM HCl SOLUTIONS[9]

	log D			
Metal ion	HCl (M)			
	2	4	8	12
Ag	2.0	1.3	−0.2	—
As	−1.7	−0.7	1.0	1.2
Au(III)	2.7	2.8	2.8	2.0
Cd	0.3	0.3	−1.2	−1.8
Fe(III)	0.7	1.4	1.6	1.7
Ga	1.0	2.8	2.8	2.8
In	−0.4	0.4	0.8	0.4
Hf	−1.8	−1.9	−1.8	−0.8
Hg(II)	2.3	1.8	0.8	0.0
Mo(VI)	−2.5	−0.5	0.2	0.7
Nb	−2.2	−1.3	0.5	1.1
Np	−1.5	−1.4	−0.6	0.2
Os	1.4	1.3	1.1	1.0
Pa	−2.8	−0.4	1.7	2.5
Re	2.2	2.0	1.4	0.0
Sb(III)	1.6	1.8	0.9	0.8
Se	−1.5	−1.2	0.5	1.6
Sn(IV)	−0.2	0.3	0.7	0.4
Tc	2.8	2.5	1.9	1.5
Tl(III)	2.4	2.4	2.7	2.4
V	−1.1	−0.2	0.6	—
W	0.1	0.2	0.0	−0.4
Zn	−0.4	−0.4	−1.6	−2.0
Zr	−1.9	−2.0	−1.0	−0.2

Analytical Applications
Use as an Extraction Reagent

TPAC is widely accepted as an ion-pair solvent extraction reagent for metals. The conditions for the separation of various elements may be found in Tables 2 and 3. If the anion is highly colored, such elements can be determined by the absorption photometry after extraction. Some examples are summarized in Table 4.

Miscellaneous Uses

TPAC has been used as a precipitating reagent for anions for gravimetry, group separation, or detection. The precipitation is most effective with larger anions of single charge. In general, anions of high extraction coefficient have low water solubility for the quantitative precipitation.

TPAC is also used as a titrant for the precipitation titration of anions with amperometric end point[19] or with anion selective electrode.[12]

Other Onium Salts of Analytical Interest

Tetraphenylphosphonium Chloride, TPPC, C$_{24}$H$_{20}$ClP, mol wt = 374.85

L$^+$Cl$^-$

Table 4
PHOTOMETRIC APPROACHES AFTER EXTRACTION WITH TPAC

Anion	Condition	λ_{max} (nm)	Range of determination (ppm)	Remarks	Ref.
MnO_4^-	pH 6.4, extraction with $CHCl_3$	532	~5	MoO_4^{2-}, Fe(III), Co(II), WO_4^{2-} give negative error, Cr(III), VO_3^- positive	14
ReO_4^-	pH 8 ~ 9, citrate, extraction with $CHCl_3$	255	2.5 ~ 30	Mo does not interfere	15
$MoO_2(SCN)_2^-$ or $MoO(SCN)_2^-$	2 M HCl or 1 M H_2SO_4, KSCN, ascorbic acid, extraction with $CHCl_3$	470	0.5 ~ 9.6	$\epsilon = 1.8 \times 10^4$	16
$W(OH)_2(SCN)_4^-$	KSCN, oxalate, NH_4F, extraction with $CHCl_3$	400 ~ 6	0 ~ 13	Nb can be masked with oxalate, F^-	17 27
$Co(SCN)_4^-$	pH 2.6 ~ 6.8, NH_4SCN, extraction with $CHCl_3$	620	5 ~ 25	Fe(III), Mo(VI), U(VI) are masked with F^-, Cu(II) is masked with I^-	15
$Co(TTA)_2(CH_2ClCO_2)_2^-$	pH 3 ~ 8, TTA, $CH_2ClCOOH$	334	—	$\epsilon = 3.7 \times 10^4$	18

It is a colorless crystalline powder, mp 274 to 278°, and is easily soluble in water. $E_{1/2}$ = -1.790 V (0.1 M KCl). It behaves similarly to TPAC in the precipitation and extraction of anions and has been used for the same purposes. K_{sp} (L·ClO$_4$) = 4.6 × 10^{-9},[6] K_{sp} (L·ReO$_4$) = 2.1 × 10^{-9} (20°),[3] and K_{sp} (L·(C$_6$H$_5$)$_4$B) = 1.1 to 1.4 × 10^{-8}.[20]

The extraction coefficients of TPPC ion-pair in a chloroform-water system are about the same value as those of TPAC ion-pair. The values of E for some anions are Cl$^-$, 0.18; Br$^-$, 3.4; I$^-$, 60; SCN$^-$, > 380; ClO$_3^-$, >100; BrO$_3^-$, 0.5; ClO$_4^-$, >200; MnO$_4^-$, >300; ReO$_4^-$, >600; NO$_3^-$, 5.0; NO$_2^-$, 0.1; and CrO$_4^-$, 25 (pH 2.4), 0.05 (pH 11.5).[21]

The TPPC ion-pair is reported to be more soluble in chloroform than the corresponding TPAC ion-pair.

Tetraphenylstibonium sulfate ((C$_6$H$_5$)$_4$Sb·1/2 SO$_4$) is known to extract F$^-$ into carbon tetrachloride or chloroform.[22]

Triphenylselenonium Chloride[24] and Triethyltelluronium Chloride[23]

Both reagents are a colorless crystalline powder which is easily soluble in water. Both behave similarly to TPAC for the precipitation and extraction of anions.

Beside the uses as extraction and precipitation reagents, the former reagent is used as a detection reagent for Bi which gives red-orange precipitation in the presence of KI (detection limit 0.11 μg, dilution limit 1:9.5 × 10^5).[24]

Several other onium salts, such as 2,4,6-triphenylpyrylium chloride[25] and 1,2,4,6-tetraphenylpyridinium acetate,[26] have been proposed as the reagent for the same purposes.

MONOGRAPH

M1. Sandell, E. B. and Onishi, H., *Photometric Determination of Traces of Metals,* John Wiley & Sons, New York, 1978.

REVIEW

R1. Shinagawa, M. and Matsuo, H., Organic onium compound, its application to analytical chemistry, *Kagaku No Ryoiki,* (Tokyo), 10, 111, 1956.

REFERENCES

1. Willard, H. H., Perkins, L. R., and Blicke, F. F., *J. Am. Chem. Soc.,* 70, 737, 1948.
2. Shriner, R. L. and Wolf, C. N., *Organic Synthesis,* Vol. 4, John Wiley & Sons, New York, 1963, 910.
3. Blicke, F. F. and Monroe, E., *J. Am. Chem. Soc.,* 57, 720, 1935.
4. Fok, J. S., Hugus, Z. Z., and Sandell, E. B., *Anal. Chim. Acta,* 48, 243, 1969.
5. Okubo, T. and Aoki, F., *Nippon Kagaku Zasshi,* 87, 1103, 1966.
6. Okubo, T., Aoki, F., and Teraoka, T., *Nippon Kagaku Zasshi,* 89, 432, 1968.

7. Alimarin, I. P. and Perezhogin, G. A., *Talanta,* 14, 109, 1967.

8. Bock, R. and Beilstein, G. M., *Fresenius Z. Anal. Chem.,* 192, 44, 1962.

9. Ishimori, T. and Nakamura, E., JAERI Report-1047, 1963.

10. Tagliavini, G., *Anal. Chim. Acta,* 40, 33, 1968.

11. Loach, K. W., *Anal. Chim. Acta,* 44, 323, 1969; *Anal. Chim. Acta,* 45, 93, 1969.

12. Baczwk, R. J. and DuBois, R. J., *Anal. Chem.,* 40, 685, 1968.

13. Shiroki, M. and Maric, Lj., *Fresenius Z. Anal. Chem.,* 276, 371, 1975; *Anal. Chim. Acta,* 79, 265, 1975.

14. Richardson, M. L., *Analyst,* 87, 435, 1962.

15. Holzbecher, Z., Divis, L., Kral, M., Sucha, L., and Ulacil, F., *Handbook of Organic Reagents in Inorganic Analysis,* Ellis Harwood, Chishester, England, 1976.

16. Tamhina, B. and Herak, M. J., *Microchim. Acta,* 1, 553, 1976.

17. Affsprung, H. E. and Murphy, J. W., *Anal. Chim. Acta,* 30, 501, 1964.

18. Rahaman, M. S. and Finston, H. L., *Anal. Chem.,* 41, 2123, 1968.

19. Menis, O., Ball, R. G., and Manning, D. L., *Anal. Chem.,* 29, 245, 1957.

20. Nezu, H., *Bunseki Kagaku,* 10, 571, 575, 1961.

21. Ruzicka, J. and Zeman, A., *Talanta,* 12, 997, 1965.

22. Moffett, K. D., Simmler, J. R., and Potratz, H. A., *Anal. Chem.,* 28, 1356, 1956.

23. Shinagawa, M., Matsuo, H., and Sunahara, H., *Bunseki Kagaku,* 3, 204, 1954.

24. Shinagawa, M., Matsuo, H., and Itsushiki, S., *Bunseki Kagaku,* 3, 199, 1954.

25. Kanai, K., Umehara, M., Kitano, H., and Fukui, K., *Nippon Kagaku Zasshi,* 84, 432, 1963.

26. Chadwick, T. C., *Anal. Chem.,* 48, 1201, 1976.

27. Kajiyama, R., Ichihashi, K., and Ichikawa, K., *Bunseki Kagaku,* 18, 1500, 1969.

DIANTIPYRYLMETHANE

$C_{23}H_{26}O_3N_4 \cdot H_2O$
mol wt = 424.50

Synonyms

1,1′-Diantipyrinylmethane, 4,4′-Methylenediantipyrine, DAM, DAPM, MDAP

Source and Method of Synthesis

Commercially available. It is prepared by the condensation of antipyrine with formaladehyde in hydrochloric acid.[1]

Analytical Uses

The protonated species of the reagent is used as a hydrophobic cation in the ion-pair extraction of wide variety of anionic complexes. It is also used as a chromogenic reagent for Ti and some other oxyphilic metals.

Properties of Reagent

It is usually supplied as monohydrate colorless crystals, mp 155 to 157° (with decomposition), and is insoluble in water (0.04%), but easily soluble in chloroform (17.6%, 20°) and less easily soluble in benzene (1.7%) and carbon tetrachloride (0.28%) to give a colorless solution.[2] The reagent is stable in a solid state as well as in solution. It is a weak base and is protonated in acid medium through the carbonyl groups with one or two H^+; pKa (HL^+) = 9.96 (in methylethylketone).[R1]

Behaviors of an Analytical Reagent

Diantipyrylmethane functions as an analytical reagent in two ways, one as a cationic reagent in the protonated forms (HL^+ or H_2L^{2+}) and the other as a coordinating chromogenic reagent for oxyphilic metal ions.

Cationic Reagent

The protonated species of the reagent forms ion-pairs with anionic complexes of Cl^-, Br^-, I^-, SCN^-, NO_3^-, and CN^- with various metals. These ion-pairs are, in general, very slightly soluble in water, but can be extracted into chloroform or dichloroethane. These behaviors have been utilized in the gravimetry of metals and in the separation of metals by the solvent extraction.

The following is a partial list of the ion-pair complexes which have been used in the gravimetry: $(HL)_2CdX_4$,[1] $(HL)_2CoX_4$,[3] $(HL)_2IrX_6$,[4] $(HL)_2OsX_6$,[4] $(HL)_2ReX_6$,[5] $(H_2L)HgX_4$,[6] $(HL)BiX_4$,[7] and $(HL)TlX_4$,[8] where X represents Cl^-, Br^-, I^-, or SCN^-.

The metal ions which form halo- or thiocyanato-complexes, including those listed above, can be separated by the ion-pair extraction with diantipyrylmethane from the metal ions that do not form such anion complexes. For example, Bi, Cd, Cu(II), Ga, In, Sb(III), Sn(IV), Te(IV), and Tl(III) can be extracted from 2.5 to 3 N HCl; Bi, Cd, Pb, Sb(III), and Sn(IV) can be extracted from 3% KI solution acidified with H_2SO_4;[9] Hf, Mo, Nb, Ta, W, and Zr from 0.5 to 4 N HF containing H_2SO_4;[10] and Fe(III), Hf(IV), Hg(II), Pd(II), Sc, Th, Tl(III), and Zr can be extracted from a medium of trichloroacetic acid containing H_2SO_4.[11] Some examples of the separation by the diantipyrylmethane extraction are shown in Table 1.

Table 1

EXTRACTION OF COMPLEX ANIONS WITH DIANTIPYTYLMETHANE

Metal ion				
Extracted	From	Condition	Solvent	Ref.
Ag	—	0.15 N I$^-$ or Cl$^-$, 2 N H$_2$SO$_4$	CHCl$_3$ or dichloroethane	12
CrO$_4$$^{2-}$	Al, Co, Fe, Mn, Ti, alkaline earths	0.3 N HCl, extracted as [(HL)CrO$_3$Cl]	CHCl$_3$	13
Cu(II)	Al, Be, Ce, Co, Cr, Fe, Hf, La, Mg, Mn, Ni, Sc, Ti, V, Zr, alkaline earths	2 ∼ 3 N HCl, ascorbic acid	CHCl$_3$	14
Fe(III)	Al, Ti	NH$_4$SCN, HCl	CHCl$_3$	15
Pb	—	0.4 ∼ 4 N HBr, 0.5∼3 N HCl or H$_2$SO$_4$	CHCl$_3$ or dichloroethane	16
Sc	0.2 ∼ 2g of Ag, Al, Cd, Ce(III), Co(II) Cr(III), Fe(II), Ge, La, Mn(II), Ni, Pb, V, Y, Zn	0.15 M H$_2$SO$_4$, 0.5 M trichloroacetate	CHCl$_3$	17
Sn(IV)	Be, Bi, Cd, Cr(III), Ga, In, Mn(II), Mg, Pb, Zn; also Cu, Fe(III), Hg, V(IV), Zr can be separated by washing the organic phase with a mixture of H$_2$SO$_4$ and oxalic acid	8 ∼ 10-fold excess oxalic acid, 0.5 M H$_2$SO$_4$	CHCl$_3$	18
Te	As(III), Bi, Cu(II), Fe, Ga, In, Ni, Pb, Sb(III), Se, Tl(III), or Cu alloys	HCl(1 + 1)	Tetrachloroethane	19
Ti(III)	Al, Be, Cd, Co, Cr, Mn, Ni, Pb, Zn or Zr	Oxalate, H$_2$SO$_4$	CHCl$_3$	20
V$_6$O$_{17}$$^{4-}$	Various elements	Halides or SCN$^-$	CHCl$_3$ or dichloroethane	21
Zn, Cd	Al, Be, Co, Cr, Fe, La, Mn, Ni, Y, alkaline earths, but Bi, Cu interfere	NH$_4$I, ascorbic acid, 2 ∼ 4 M H$_2$SO$_4$	CHCl$_3$	22
Zr	Hf	I$^-$, 8 ∼ 10 N HCl	Dichloroethane	23
Rare earths	Al, Co, Cr(III), Fe(III), Ga, Mn, Ni, Ti, Zn	KI, ascorbic acid, 0.01 N HCl	CHCl$_3$ or CHCl$_3$-benzene	24, 25

Chromogenic Reagent

Although the exact coordinating structure of the complexes is not known, diantipyrylmethane forms colored soluble complexes with some oxophilic metal ions, such as Fe(III) (orange-red; λ_{max}, 450 nm; $\varepsilon = 5.4 \times 10^3$), Mo(VI) (faint yellow), Ti (yellow), and U(VI) (yellow).

The reaction with Ti(IV) is highly selective and far more sensitive (0.5 to 4 N HCl; [TiL$_3$]$^{4+}$; λ_{max}, 385 to 390 nm; $\varepsilon = 1.5 \times 10^4$; 0.2 to 3 ppm TiO$_2$)[26] than that with H$_2O_2$. This complex can also be extracted into chloroform in the presence of a suitable organic or inorganic anion for the subsequent photometry.[27]

Other Reagents with Related Structure

A large number of diantipyrylalkane derivatives have been prepared and investigated as an analytical reagent.[R1,R2] Some examples are listed below.

1. Diantipyrylmethylmethane (1,1-Diantipyrylethane, R = CH₃): reagent for Sb(III)[28]

2. Diantipyrylpropylmethane (1,1-Diantipyrylbutane, R = C₃H₇): reagent for Ir, Rh, and Ru[29]

3. Diantipyrylbenzomethane (α,α'-Diantipyryltoluene, R = C₆H₅): reagent for Ti(IV) and V(V)[30]

4. Diantipyryl-2-hydroxyphenylmethane (R = o-C₆H₄OH): reagent for Ti(IV)[31]

5. Hexyldiantipyrylmethane (R = n-C₆H₁₃): most soluble in organic solvents, 54% in benzene and 66% in chloroform[2]

6. Disulfodiantipyrylmethane (Bis(1-p-sulfophenyl-2,3-dimethylpyrazol-5-onyl)methane): reagent for Fe and Ti[32]

7. Dithioantipyrylmethane (4,4'-Methylene-bis(1,5-dimethyl-2-phenylpyrazolin-3-thione)): reagent for Au, Bi, Mo,[33] Te,[34] and Tl(III)[35]

REVIEWS

R1. **Zhivopistsev, V. P.**, Use of diantipyrinylmethane in analytical chemistry, *Zavod. Lab.*, 31, 1043, 1965; *Chem. Abstr.*, 63, 12294e, 1965.

R2. **Ishii, H.**, Diantipyrylmethane and its derivatives, *Bunseki Kagaku*, 21, 665, 1972.

REFERENCES

1. **Zhivopistsev, V. P.**, *Zavod. Lab.*, 16, 1186, 1950; *Chem. Abstr.*, 47, 3175a, 1953.

2. **Petrov, B. I.,Zhivopistsev, V. P., Kislitsyn, I. A., and Volkova, M. A.**, *Zh. Anal. Khim.*, 32, 1487, 1977.

3. **Zhivopistsev, V. P.**, *Dokl. Akad. Nauk SSSR*, 73, 1193, 1950; *Chem. Abstr.*, 45, 492i, 1951.

4. **Busev, A. I. and Akimov, V. K.**, *Talanta*, 11, 1657, 1964.

5. **Akimov, V. K., Busev, A. I., and Emelyanova, I. A.**, *Zh. Anal. Khim.*, 25, 1752, 1970.

6. **Busev, A. I. and Khimtibidze, L. S.**, *Zh. Anal. Khim.*, 22, 694, 1967.

7. **Busev, A. I., Saber, S. A., and Akimov, V. K.**, *Zh. Anal. Khim.*, 25, 1124, 1970.

8. **Busev, A. I. and Tiptsova, V. G.**, *Izv. Vyssh. Uchebn. Zaved. Khim. Khim. Tekhnol.*, 3, 69, 1960; *Chem. Abstr.*, 54, 16265e, 1960.

9. **Zhivopistsev, V. P., Petrov, B. I., Selezneva, E. A., and Sibiryakova, N. F.**, *Tr. Kom. Analit. Khim.*, 17, 304, 1969.

10. **Petrov, B. I., and Degtev, M. I.**, *Zh. Anal. Khim.*, 31, 1076, 1976.

11. **Petrov, B. I. and Vilisov, V. N.**, *Zh. Anal. Khim.*, 31, 2298, 1976.

12. **Petrov, B. I., Degtev, M. I., and Zhivopistsev, V. P.**, *Zh. Obshch. Khim.*, 46, 1927, 1976; *Chem. Abstr.*, 85, 198939j, 1976, *Zh. Neorg. Khim.*, 21, 2749, 1976; *Chem. Abstr.*, 86, 9178x, 1976.

13. **Minin, A. A. and Milyutina, L. L.**, *Uch. Zap. Permsk. Gos. Univ.*, 235, 241, 1966; *Anal. Abstr.*, 15, 3906, 1968.

14. Zhivopistsev, V. P. and Aitov, V. Kh., *Uch. Zap Permsk. Gos. Univ.*, 112, 1963; *Anal. Abstr.*, 12, 35, 1965.

15. Zhivopistsev, V. P. and Minin, A. A., *Zavod. Lab.*, 26, 1346, 1960; *Chem. Abstr.*, 55, 16279b, 1961.

16. Petrov, B. I., Degtev, M. I., Zhivopistsev, V. P., and Kondakova, T. A., *Izv. Vyssh. Uchebn. Zaved. Khim. Khim. Tekhnol.*, 20, 681, 1977; *Anal. Abstr.*, 33, 5B109, 1977; *J. Gen. Chem. USSR*, 46, 1867, 1976, *Anal. Abstr.*, 33, 53B110, 1977.

17. Petrov, B. I. and Vilisov, V. N., *Zh. Anal. Khim.*, 31, 2298, 1976.

18. Petrov, B. I., Minina, M. S., Makhnev, Yu. A., and Galinova, K. G., *Zh. Anal. Khim.*, 31, 2142, 1976.

19. Pollock, E. N., *Anal. Chim. Acta*, 40, 285, 1968.

20. Petrov, B. I., Zhivopistsev, V. P., Bobovskaya, E. T., Makhnev, Yu. A., and Ponosov, I. N., *Izv. Vyssh. Uchebn. Zaved. Khim. Khim. Tekhnol.*, 18, 1916, 1975; *Chem. Abstr.*, 85, 56111j, 1976.

21. Zhivopistsev, V. P., Minin, A. A., Milyutina, L. L., Selezneva, E. A., and Aitova, V. Kh., *Tr. Kom. Anal. Khim. Akad. Nauk. SSSR*, 14, 133, 1963; *Chem. Abstr.*, 59, 14556e, 1963.

22. Zhivopistsev, V. P., Aitova, V. Kh., and Selezneva, E. A., *Izv. Vyssh. Uchebn. Zaved. Khim. Khim. Tekhnol.*, 6, 739, 1963; *Anal. Abstr.*, 12, 1644, 1965; *Chem. Abstr.*, 60, 8611g, 1960.

23. Petrov, B. I., *Uch. Zap. Permsk. Gos. Univ.*, 140, 1974; *Anal. Abstr.*, 30, 1B98, 1976.

24. Kalmykova, I. S. and Kukarina, L. V., *Zh. Anal. Khim.*, 33, 909, 1978.

25. Antsiferova, L. M., Kukarina, L. V., Kalmykova, I. S., and Shavrin, A. M., *Zh. Anal. Khim.*, 32, 526, 1977.

26. Minin, A. A. and Erofeeva, S. A., *Uch. Zap. Molotov, Univ.*, 9, 177, 1955; *Chem. Abstr.*, 52, 8826d, 1958; *Uch. Zap. Permsk. Gos. Univ.*, 58, 1958; *Chem. Abstr.*, 54, 15059c, 1960; *Uch. Zap. Permsk. Gos. Univ.*, 97, 1961; *Anal. Abstr.*, 10, 92, 1963.

27. Ishii, H., *Bunseki Kagaku*, 16, 110, 1967.

28. Busev, A. I. and Bogdanova, E. S., *Zh. Anal. Khim.*, 19, 1346, 1964.

29. Shendrikar, A. D. and Berg, E. W., *Anal. Chim. Acta*, 47, 299, 1969.

30. Podchainova, V. N. and Dergachev, V. Ya., *Tr. Ural. Politekh. Inst.*, 148, 106, 1966; *Chem. Abstr.*, 68, 18286m, 1968.

31. Ishii, H. and Higuchi, T., *Bunseki Kagaku*, 23, 1344, 1974.

32. Zhivopistsev, V. P. and Parkacheva, V. N., *Uch. Zap. Permsk. Gos. Univ.*, 136, 141, 1964; *Anal. Abstr.*, 13, 1734, 2953, 1966.

33. Dolgorev, A. V. and Lysak, Ya. G., *Zh. Anal. Khim.*, 29, 1766, 1974.

34. Dolgorev, A. V., Busev, A. I., and Zibarova, Yu. F., *Zavod. Lab.*, 44, 1182, 1978; *Anal. Abstr.*, 36, 6B112, 1979.

35. Skimov, V. K., Zasorina, E. V., Busev, A. I., and Nenning, P., *Z. Chem.*, 17, 186, 1977; *Anal. Abstr.*, 33, 5B58, 1977.

1,3-DIPHENYLGUANIDINE

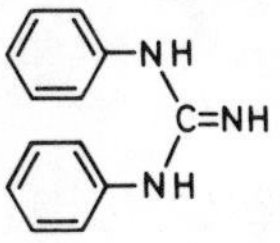

$$C_{13}H_{13}N_3$$
$$mol\ wt = 211.27$$

Synonym
Molaniline

Source and Method of Synthesis
Commercially available. It is prepared by the reaction of diphenylthiourea with ammonia.[1]

Analytical Uses
The protonated form of the reagent (diphenylguanidium ion) is used for the ion-pair extraction of metal ions as complex anions. It is also recommended as a primary material for standardizing acids.[2]

Properties of Reagent
It is colorless crystals, mp 148° (decomposes at 170°), and is only slightly soluble in water, but soluble in alcohol, chloroform, hot benzene, and hot toluene. It is also readily soluble in dilute mineral acids. Aqueous solution is strongly alkaline. The solid reagent can be stored in a glass-stoppered bottle without deterioration, but the aqueous solution does not keep long because the reagent is hydrolyzed slowly to give urea.[M1]

Extraction Behavior with Diphenylguanidine
As diphenylguanidine is a fairly strong base, it exists as a guanidinium ion in acidic as well as in neutral range. Accordingly, the reagent can be used as a cationic reagent for the extraction of wide variety of metal ions in a form of anionic complex. The ion-pair extractions of metals are carried out not only for the separation of metals, but also for the subsequent photometry. In the latter case, a highly colored anionic chelate is extracted with diphenylguanidinium ion, since the reagent itself is colorless. Some of the typical examples of the metal extraction with diphenylguanidine are summarized in Table 1.

MONOGRAPH

M1. Welcher, F. J., *Organic Analytical Reagents,* Vol. 2, Von Nostrand, Princeton, N.J., 1947, 374.

REFERENCES

1. Rathke, B., *Ber.,* 12, 772, 1879.
2. Thoronton, W. A., Jr. and Chirist, C. L., *Ind. Eng. Chem. Anal. Ed.,* 9, 339, 1937.
3. Tananaiko, M. M., Vdovenko, O. P., and Zatsarevnyi, V. M., *Zh. Anal. Khim.,* 29, 1724, 1974.

Table 1

ION-PAIR EXTRACTION OF METALS WITH DIPHENYLGUANIDINE (X)

Metal ion	Condition	Extraction solvent	Extracted species Composition[a]	λ_{max} (nm)	$\varepsilon(\times 10^4)$	Range of determination	Ref.
Al	pH 6.3 ~ 7.9	Butanol	$M(PV)_3X_3$	590	5.6	0.3 ~	3
Be	pH 4.5	Butanol	$M(MTB)X_2$	500	1.3	0.01 ~ 0.7	4
Bi	pH 4.3 ~ 5.0	Aqueous	$M(TAM)X_n$	585	4.6	0 ~ 1.8	5
Co(II)	{Alkaline / Acidic	$CHCl_3$	$M(Quinoxaline-2,3-dithiol)_3X_3$	490 / 520	— / —	— / —	6
Fe(II)	pH 2.5	$CHCl_3$	$M(TAR)X$	505	2.5	—	7
Fe(III)	pH 2.8 ~ 3.2	$CHCl_3$	$M(Indoferron)_2X_2$	635	4.0	0.15 ~ 1	8
Ga	pH 5.7 ~ 6.2	Butanol	$M(Azodye)_2X_2$	495	5.5	0 ~ 1.1	9
Ge(IV)	pH 0.4 ~ 4.5	iso-AmOH	$M(BPR)_2X$	550	3.8	0.15 ~ 1.1	10
La	pH 2.5 ~ 3.5	Butanol	$M(Chlorophosphonazo III)_2X_2$	675	16.0	0.1 ~ 0.8	11
Mo(VI)	pH 2.2	iso-AmOH + $CHCl_3$ (1:1)	$M(Tiron)X_n$	390	—	1 ~ 16	12
Ni	pH 7.7 ~ 7.9	iso-AmOH + $CHCl_3$ (1:1)	$M(Dimethylglyoxime)X_n + I_2$	490	—	0.1 ~ 1	13
Pb	pH 4.8	Butanol	$M(MTB)X$	570	1.63	2 ~ 10	14
Pd(II)	pH 8.8 ~ 9.2	Butanol	$M(Cadion\ IREA)X_2$	500	6	0.02 ~ 0.2	15
Ru(IV)	2 NHCl, $SnCl_2$	$CHCl_3$	—	434	3.5	0.2 ~ 20	16
Sn(IV)	pH $\simeq$ 5	Butanol	$M(PV)_2X$	582	6.7	0.02 ~ 2.5	17
Th	pH 2.5 ~ 3.6	Butanol	$M(XO)_2X_{4-6}$	578	9.3	0.14 ~ 0.83	18
Ti(IV)	pH 4.2 ~ 5.2	Benzyl alcohol	$M(TAM)X_2$	583	3.6	0 ~ 1.2	19
U(VI)	pH 4.6 ~ 4.9	Butanol	$M(Alizarin\ Red\ S)X$	580	0.7	0 ~ 12	20
V(II)	0.42 ~ 2.8 NHCl	$CHCl_3$	$M(SCN)_2X_2$	340	17.0	0.7 ~ 6.7	21
VO^{2+}	pH 2.9 ~ 4.1	iso-AmOH + $CHCl_3$ (1:4)	$M(Tiron)_2X_6$	585	1.0	0 ~ 3.6	22
W(VI)	pH 3.2 ~ 3.7	iso-AmOH + $CHCl_3$ (1:4)	$M(Arylazocatechol)_2\ X_2$[b]	540	5.4	—	23
Zn	pH 6 ~ 7	$CHCl_3$	$M(PAR)X_3$	515	6.7	0.1 ~ 4	24
Zr(Hf)	pH 1 ~ 2	Butanol	$M(XO)_{2\sim4}X_n$	560	3.3	—	25
Rare earths	pH 6.3 ~ 6.7	iso-AmOH	$M(Arsenazo III)_2X_8$	670	5 ~ 7	0.1 ~ 4.7	26

[a] Indicates the combining ratio only, disregarding the state of protonation of the ligand. Key for the ligands: PV, Pyrocatechol Violet; MTB, Methylthymol Blue; TAM, Thiazolylazodimethylaminophenol; TAR, Thiazolylazoresorcinol; BPR, Bromopyrogallol Red; PAR, Pyridylazoresorcinol; XO, Xylenol Orange.

[b] Triphenylguanidine was used as an extractant.

4. Beschetnova, E. T., Anisimova, L. G., and Mataeva, S. S., *Fiz. Khim. Metody Anal. Kontrolya Proizvad.*, 12, 1975; *Chem. Abstr.*, 87, 126626y, 1977.
5. Tsurumi, C. and Furuya, K., *Bunseki Kagaku*, 24, 566, 1975.
6. Pilipenko, A. T., Ryabushko, O. P., and Krivokhizhina, L. A., *Ukr. Khim. Zh.*, 42, 1077, 1976; *Chem. Abstr.*, 86, 100127r, 1977.
7. Biryuk, E. A. and Ravitskaya R. V., *Zh. Anal. Khim.*, 31, 1327, 1976; *J. Anal. Chem. USSR*, 31, 1085, 1976.
8. Wakamatsu, Y. and Otomo, M., *Bunseki Kagaku*, 19, 537, 1970.
9. Wakamatsu, Y., *Bunseki Kagaku*, 28, 385, 1979.
10. Nazarenko, V. A. and Makrinich, N. I., *Zh. Anal. Khim.*, 24, 1694, 1969.
11. Menkov, A. A. and Nepomnyashchaya, N. A., *Zh. Anal. Khim.*, 32, 1409, 1977.
12. Busev, A. I. and Rudzit, G. P., *Zh. Anal. Khim.*, 19, 569, 1964.
13. Zolotov, Yu. A. and Vlasova, G. E., *Zh. Anal. Khim.*, 28, 1540, 1973.
14. Tataev, O. A. and Magaramov, M. M., *Zh. Neorg. Khim.*, 22, 120, 1977; *Chem. Abstr.*, 86, 96824w, 1977.
15. Tataev, O. A. and Gaczhieva, M. D., *Fiz. Kim. Metody Anal. Kontralya Proizvod. Mezhvuz. Sb.*, 2, 156, 1976; *Chem. Abstr.*, 87, 177041m, 1977.
16. Donilova, V. N. and Shilina, G. V., *Ukr. Khim. Zh.*, 44, 421, 1978; *Anal. Abstr.*, 35, 4B178, 1978.
17. Shestidesyatnaya, N. L., Kotelyanskaya, L. I., and Yanik, M. I., *Zh. Anal. Khim.*, 31, 67, 1976.
18. Otomo, M. and Wakamatsu, Y., *Nippon Kagaku Zasshi*, 89, 1087, 1968.
19. Tsurumi, C. and Furuya, K., *Bunseki Kagaku*, 26, 149, 1977.
20. Otomo, M., *Nippon Kagaku Zasshi*, 92, 171, 1971.
21. Verdizade, N. A. and Ragimova, Z. B., *Azerb. Khim. Zh.*, 116, 1978; *Chem. Abstr.*, 90, 114397y, 1979.
22. Wakamatsu, Y. and Otomo, M., *Bull. Chem. Soc. Jpn.*, 47, 761, 1974.
23. Myasoedova, A. S., Ivanov, V. M., and Busev, A. I., *Zh. Anal. Khim.*, 31, 738, 1976.
24. Mamuliya, S. G., Ryatniskii, I. V., and Grigalashvili, K. I., *Ukr. Khim. Zh.*, 44, 410, 1978; *Anal. Abstr.*, 35, 4B46, 1978.
25. Tolmachev, V. N., Goltsberg, I. M., and Konkin, V. D., *Zh. Anal. Khim.*, 22, 950, 1967.
26. Akhmedli, M. A., Granovskaya, P. B., and Melikeva, E. G., *Zh. Anal. Khim.*, 28, 1304, 1973.

SODIUM TETRAPHENYLBORATE

$$C_{24}H_{20}BNa$$
$$mol\ wt\ =\ 342.22$$

Synonyms
Tetraphenylboron, sodium salt, Na-TPB, Kalignost®*, Kalibor®**

Source and Method of Synthesis
Commercially available. It is synthesized by the reaction of phenylmagnesium bromide with BF_3, followed by the hydrolysis in the presence of NaCl.[1]

Analytical Uses
As a precipitating reagent for heavy alkali metal ions (K, Rb, Cs) and ammonium ions, including organic bases.

Properties of Reagent
Pure materials are colorless crystals and easily soluble in water (30 g/100 mℓ) to give a clear colorless solution. It is also soluble in methanol, ethanol, acetone, and ether. However, the deteriorated commercial samples are slightly tan colored and give a turbid solution when dissolved in water. The pure solid reagent is fairly stable and can be kept for many months without decomposition when it is stored in a dark and cool place. The reagent of lower purity tends to deteriorate more quickly, giving the characteristic odors of phenol and biphenyl.

The aqueous solution is stable at pH 7, if protected from direct sun light (0.1% decomposition after 9 weeks at pH 9.9), but very unstable in acidic range because TPB anion is easily attacked by proton to give phenol, biphenyl, and boric acid.

Reaction with Cations
TPB anion forms a white precipitate of very low solubility with univalent metal ions of larger ionic radius (K, Rb, Cs, Tl(I), and Ag) and ammonium ions, including amines and quaternary ammonium ions. On the other hand, salts of Li and Na are easily soluble in water. Solubilities of some of TPB salts are summarized in Tables 1 and 2.

Multivalent metal ions do not give a precipitate, but Cu(II) is reduced to Cu(I) at the sacrifice of TPB and precipitates as Cu(I)-TPB. Other cations, such as Fe(III) or Ce(VI), also decompose TPB ion.

The reactions with Hg(II) proceed as follows:[2,3]

$$B(C_6H_5)_4^- + 4HgCl_2 + 3H_2O \rightarrow 4C_6H_5HgCl + 3H^+ + 4Cl^- + B(OH)_3$$
$$B(C_6H_5)_4^- + 2HgCl_2 + 3H_2O \rightarrow 2(C_6H_5)_2Hg + 3H^+ + 4Cl^- + B(OH)_3$$

The reaction with Hg-EDTA proceeds similarly:[4]

$$B(C_6H_5)_4^- + 4Hg\text{-}EDTA + 3H_2O \rightarrow 4C_6H_5Hg^+ + 3H^+ + 4EDTA + B(OH)_3.$$

These reactions are stoichiometric under a proper condition, so that they can be utilized for the determination of TPB ion.

* Trade name by Heyl Co., West Germany.
** Trade name by Dojindo Laboratories, Kumamoto, Japan.

Table 1
SOLUBILITY (S) OF INORGANIC TPB SALTS[R3]

S
(mol/l)

TPB salt (ML)	Radiochemical method (20°)	Photometric method (25°)	Potentiometric method (20°)	Decomposition temp[a] (°)
Cs⁺	2.9×10^{-5}	2.8×10^{-5}	—	210
K⁺	1.5×10^{-4}	1.8×10^{-4}	1.3×10^{-4}	265
NH₄⁺	—	2.9×10^{-4}	—	130[b]
Rb⁺	4.5×10^{-5}	2.3×10^{-5}	—	240
Tl⁺	3×10^{-5}	5.3×10^{-5}	—	180[b]

[a]   Under a heating rate of 4.5°/min.
[b]   Sublimation temperature.

Table 2
SOLUBILITY (S) AND SOLUBILITY PRODUCTS (K_{sp}) OF
ORGANIC TPB SALTS[R3]

TPB Salt	S g/100 ml (room temp)	S M/l (25°)	K_{sp} (room temp)	mp (°)
Methylamine	—	36.3×10^{-4}	—	209 ~ 211
Ethylamine	—	28.3×10^{-4}	—	163 ~ 169
Propylamine	—	9.0×10^{-4}	—	—
n-Butylamine	—	11.2×10^{-4}	—	133 ~ 134
Dimethylamine	—	16.3×10^{-4}	—	180 ~ 183
Trimethylamine	—	3.9×10^{-4}	—	170 ~ 172
Tetramethylammonium	0.05	0.4×10^{-4}	1.4×10^{-6}	340 ~ 360 (decomposition)
Pyridine	—	2.0×10^{-4}	—	229 ~ 231.5
Atropine	—	ca 10^{-4}	$<2.9 \times 10^{-8}$	141 ~ 144
Choline	0.003	—	—	—
Guanidine	0.14	—	—	173 ~ 177
Histamine	0.01	—	6.8×10^{-12}	131 ~ 135
Histidine	0.24	—	2.5×10^{-5}	183

Organic amines (as ammonium ions) and quaternary ammonium ions also form a 1:1 precipitate with TPB. The relative ease of the precipitation reaction is governed by the basicity and lipophilicity of the base. In the case of amines, the basic ionization constant of 10^{-11} or higher is required for the quantitative precipitation from a 1 to 2 mg/l solution.[5]

Onium compounds other than ammonium ion, such as phosphonium, arsonium, stibonium, bismuthonium, diazonium, and sulfonium ions, can also be precipitated by TPB ion.

Li-TPB is fairly soluble in ether and chloroform, but Na-TPB is less soluble. The solubility increases with decreasing temperature. The alkali-TPB precipitates, like K-TPB, are soluble in polar organic solvents, such as acetone (60 mg/ml in 95% acetone, 42 mg/ml in 100% aceone, 28°),[6] dioxane, DMF, acetonitrile, and pyridine.

Purification and Purity of Reagent

Deteriorated materials give a turbid solution when they are dissolved in water. For most analytical purposes, such solution is still usable after clarifying by treating with a freshly prepared wet aluminum hydroxide gel.

The solid reagent of lower grade can be purified by repeated recrystallization from chloroform or acetone, but with very low yield.

The reagent can be assayed by the gravimetry after precipitating TPB with an excess K. The commercial samples of analytical grade from the reliable sources have a purity of 99.5% up.[7]

Analytical Applications

Use as a Precipitation Reagent for Potassium

Potassium ion can be precipitated quantitatively as K-TPB which is eventually dried at 100° for weighing. The precipitation can be carried out in acidic or alkaline medium, as summarized in Table 3.

In general, the precipitate of good filterability can be obtained in acidic medium, but the quick procedure is necessary as the reagent is unstable in acidic solution. A 3% aqueous solution of Na-TPB is recommended for this purpose. The pH of the solution is adjusted to 8 with NaOH for a long shelf life. Care has to be taken on the use of NaOH because even the analytical grade NaOH often contains a trace of potassium which causes turbidity in the reagent solution. The solution also becomes turbid due to the deterioration of TPB during the storage for long period. Such solution can be clarified by shaking with freshly prepared wet $Al(OH)_3$, followed by filtration.

The volumetric approachs for the potassium determination have been proposed by various workers. The principles are based on the stoichiometric precipitation of K-TPB. In one approach, the precipitates of K-TPB are filtered and dissolved in acetone or DMF, then the TPB ion is determined by various volumetric methods.[4,8-10] In the other approach, a known excess volume of Na-TPB solution is added to the sample solution to precipitate K-TPB salt, then the remaining TPB ion in the supernatant or in the filtrate is determined by the volumetric method.[11-13] The most practical volumetric method for the determination of TPB ion is the titration with a cationic surfactant with Bromphenol Blue or Titan Yellow as an indicator.[14,15] The procedure for the determination of potassium in fertilizers is as follows.[9]

Reagents

1.   1.2% Na-TPB aqueous solution
2.   0.6% Zephiramine® (tetradecyldimethylbenzylammonium chloride) aqueous solution

Both solutions have to be standardized against a standard KCl solution.

Procedure

Place 10- to 15-mℓ sample solution containing not more than 20 mg K_2O in a 100-mℓ volumetric flask and add 3 mℓ of HCl(1:9) and enough water to make 30 mℓ. To this, add 5 mℓ of 37% aqueous formaldehyde, 5 mℓ of 12% NaOH solution (K free), and a known excess of Na-TPB solution (6 mℓ for each 10 mg of K_2O plus 8 mℓ) successively, with constant shaking, then dilute to the volume. Filter the solution through a dry filter paper, and transfer a 50-mℓ aliquot of the clear filtrate into a titration flask and add 8 drops of 0.04% Titan Yellow aqueous solution. Titrate with a standard Zephiramine® solution at 30°. The color change at the end point is from yellow to pale pink.

Table 3
CONDITIONS FOR THE PRECIPITATION OF K-TPB SALT[R3]

Condition	Acidity	Temp (°C)	Filterability	Masking agent	Solubility of TPB Ion	Removal of NH_4^+
Strongly acidic	0.1 N mineral acid	Room temp	Good	None	Fairly unstable	Ignite the residue after evaporation
	0.7 N HCl	0	Good	None		
Weakly acidic	Acetic acid pH 5 ~ 6	Room temp 40 ~ 70	Slightly difficult Good	EDTA and NaF	Stable	
Alkaline	NaOH pH 8 ~ 10	Room temp ~100	Slightly difficult Fairly good	EDTA and NaF	Very stable	Boiling or HCHO

Table 4
DETECTION OF CATIONS WITH TPB[19]

Cation	Limit of detection (μg)	Limit of dilution	Cation	Limit of detection (μg)	Limit of dilution
Ag^+	0.20	1:210,000	Aniline	2.9	1:14,000
Cs^+	0.30	1:100,000[a]	n-Butylamine	2.6	1:16,000
Hg^+	1.1	1:40,000	Di-n-butylamine	0.54	1:78,000
Hg^{2+}	1.4	1:30,000	Methylpyridinium iodde	0.15	1:28,000
K^+	0.13	1:320,000	Pyridine	1.3	1:32,000
NH_4^+	0.11	1:400,000	Quinoline	0.97	1:43,000
Rb^+	0.38	1:110,000	Tetramethylammonium iodide	0.32	1:130,000
Tl^+	1.5	1:28,000	Tetraphenylarsonium chloride	0.32	1:130,000

[a] Values taken from Reference 20.

The precipitates of TPB salt can also be extracted into the organic solvent, such as nitromethane, nitrobenzene, and MIBK. The extractability and log D values (aqueous phase is 0.1 M $NaClO_4$ and 0.01 M Na-TPB) in a system of nitrobenzene/water increase in the order of K (0.89), Rb (1.60), and Cs (2.48).[16]

Other Uses

Na-TPB has been used as a nepherometric reagent for K and other univalent cations in various samples, including biological materials[17] and soils.[18] The conditions for the reproducible results (reagent purity, potassium concentration, temperature, order of addition of reagents, standing time, and interfering elements) have been investigated by many workers.

The precipitation reaction with TPB ion is so sensitive that Na-TPB can be used for the detection of various cations.[19] Some examples are summarized in Table 4. K-TPB is also proposed as an active component in ion selective electrode.[21]

Excellent reviews are available on the analytical applications of Na-TPB.[M1,R1-R3]

Other Reagents with Related Structure

Tetra(p-fluorophenyl)borate, sodium salt, has similar physical properties to Na-TPB, but is known as a selective precipitating reagent for Rb, Cs, Tl(I), and Ag, in the presence of K and NH_4^+.[22]

MONOGRAPH

M1. Flaschka, H. and Barnard, A. J., Jr., Tetraphenylboron (TPB) as an Analytical Reagents, in *Advances in Analytical Chemistry and Instrumentation*, Vol. 1, Reilly, C. N., Ed., Interscience, New York, 1960.

REVIEWS

R1. Barnard, A. J., Jr. and Buechhl, H., Sodium tetraphenylboron, A comprehensive bibliography, *Chemist-Analyst*, 44, 104, 1955; *Chemist-Analyst*, 45, 110, 1956; *Chemist-Analyst*, 46, 16, 1957; and *Chemist-Analyst*, 47, 46, 1958.
R2. Mukoyama, T., The application of sodium tetraphenylboron in analytical chemistry, *Kagaku No Ryoiki, (Tokyo)*, 10, 103, 1965.
R3. Ueno, K., Saito, M., and Tamaoku, K., Sodium tetraphenylborate, its analytical application, *Bunseki Kagaku*, 17, 1548, 1968; *Bunseki Kagaku*, 18, 81, 264, 1969.

REFERENCES

1. Wittig, G. and Raff, R., U.S. Patent 2,853,525; *Chem. Abstr.*, 53, 4211f, 1959.
2. Wittig, G., Keicher, G., and Ruckert, A., *Ann.*, 563, 110, 1949.
3. Motequi, R., Doadrio A., and Serrano, C., *Publ. Inst. Quim. Alonso Barba (Madrid)*, 10, 183, 1956; *Chem. Abstr.*, 51, 11906d, 1957; *An. R. Soc. Esp. Fis. Quim. (Madrid)*, 53, 447, 1957; *Anal. Abstr.*, 5, 1116, 1958; *An. R. Soc. Esp. Fis. Quim.* Ser. 54B, 29, 1958; *Chem. Abstr.*, 52, 11663a, 1958.
4. Flaschka, H. and Sadek, F., *Chemist-Analyst*, 47, 30, 1958.
5. Crane, R. E., Jr., *Anal. Chem.*, 28, 1794, 1956.
6. Scott, A. D., Hunziker, H. H., and Reed, M. G., *Chemist-Analyst*, 48, 11, 1959.
7. Purity Specification of Organic Reagents, Dojindo Laboratories, Kumamoto, Japan.
8. Hahn, F. L., *Fresenius Z. Anal. Chem.*, 145, 97, 1955.
9. *J. Assoc. Off. Agric. Chem.*, 44, 31, 1958.
10. Montequi, R., Doadrio, A., and Serrano, C., *Inform. Quim. Anal. (Madrid)*, 11, 8, 1957; *Chem. Abstr.*, 51, 10312d, 1957.
11. Ievins, A. and Gudriniece, E., *Zh. Anal. Khim.*, 9, 270, 1954.
12. Frank, U. F., *Z. Electrochem.*, 58, 348, 1954; *Z. Electrochem.*, 62, 245, 1958.
13. Suzuki, S., *Bunseki Kagaku*, 10, 837, 1961.
14. Schall, E. D., *Anal. Chem.*, 29, 1044, 1957.
15. Epps, E. M. and Burden, J. C., *Anal. Chem.*, 30, 1882, 1958.
16. Sekine, T. and Dyrssen, D., *Anal. Chim. Acta*, 45, 433, 1969.
17. Sunderman, F. W., Jr. and Sunderman, F. W., *Am. J. Clin. Pathol.*, 29, 95, 1958.
18. Bennett, A. C. and Reed, R. M., *Soil Sci. Soc. Am. Proc.*, 29, 192, 1965; *Chem. Abstr.*, 63, 10604h, 1965.
19. Crane, F. E., Jr., *Anal. Chem.*, 30, 1426, 1958.
20. Amin, A. A. M., *Chemist-Analyst*, 46, 6, 1957.
21. Kataoka, M., Kudoh, M., and Kambara, T., *Denki Kagaku Oyobi Kogyo Butsuri Kagaku, (Tokyo)* 46, 548, 1978; *Chem. Abstr.*, 90, 90916k, 1979.
22. Moore, C. E., Cassaretto, F. P., Posvic, H., and McLafferty, J. J., *Anal. Chim. Acta*, 35, 1, 1966.

ORGANIC REAGENTS FOR ANIONS

CURCUMIN

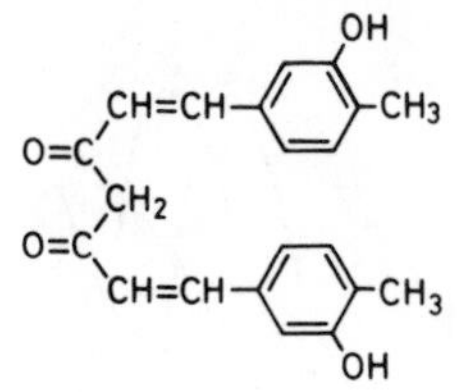

$$C_{21}H_{20}O_6$$
$$\text{mol wt} = 368.39$$

Synonyms

Turmeric yellow, Curcumagelb, Diferuloylmethane, 1,7-Bis-(4-hydroxy-3-methoxy-phenyl)-1,6-heptadiene-3,5-dione

Source

Commercially available. It is obtained from curcuma, the rhizome of *Curcuma Longa L. Zingiberaceae.*

Analytical Uses

Detection of B, Ba, Ca, Hf, Mg, Mo, Ti, U, W, and Zr. Photometric reagent for B. Also used as a spraying reagent for paper chromatography.

Properties of Reagent

It is orange-yellow crystalline powder, mp 183°, is insoluble in water, is very slightly soluble in ether, and is easily soluble in methanol, ethanol, acetone, and glacial acetic acid. It reacts with aqueous alkalis to give a yellow solution.

Although the reagent has β-diketone moiety in its structure, no data is available for the dissociation constant of enolic proton. Figure 1 illustrates the absorption spectra of curcumin in various solution conditions.[1]

Complexation Reaction and Structure of Complexes

Curcumin forms two different colored complexes, rosocyanin (1) and rubrocurcumin (2), with boric acid, depending primarily on whether oxalic acid is absent or present.

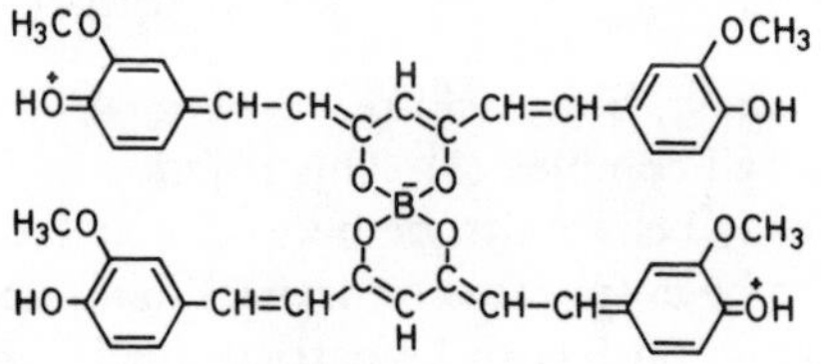

(1)

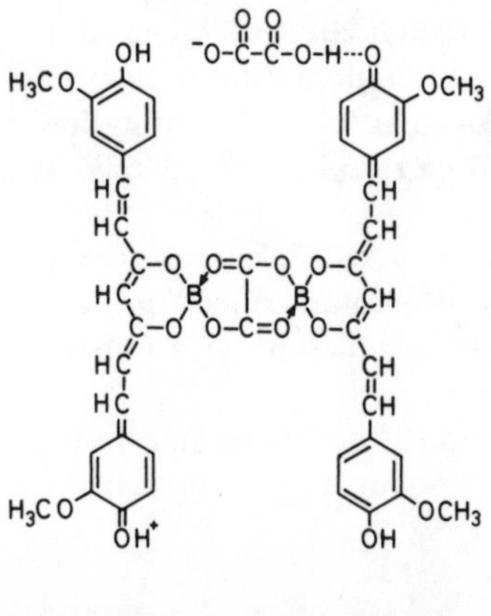

(2)

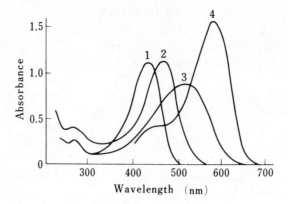

FIGURE 1.   Absorption spectra of Curcumin in various
solutions; reagent, 2×10^{-5} M. (1) In neutral alcohol; (2) in
excess of alkali; (3) in dilute alkali; (4) in phenolic solution
containing acid. (From Spicer, G. S. and Strickland, J. D.
H., *J. Chem. Soc.*, 4644, 1952. With permission.)

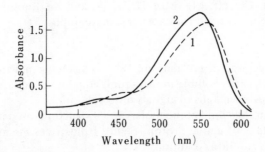

FIGURE 2.   Absorption spectra of Rosocyanin (1) and
Rubrocurcumin (2); B, 2×10^{-5} M; 1-cm cell. (1) Rosocyanin; (2) Rubrocurcumin. (From Spicer, G. S. and Strickland, J. D. H., *J. Chem. Soc.*, 4650, 1952. With permission.)

In the absence of oxalic acid, boric acid reacts with curcumin, when protonated by
a mineral acid, to form a red complex (1). This reaction is rather slow, and although
a certain amount of water is believed to be necessary at the initial stage of reaction,
the reaction mixture has to be evaporated to dryness to attain the complete reaction.[R1]
Alternatively, the color reaction has to be carried out in anhydrous media, such as
sulfuric acid-glacial acetic acid, where the existing water may be destroyed by an addition of propionyl anhydride-oxalyl chloride.[2] The solution of (1) turns to blue-black
when it is made to alkaline. Although curcumin also reacts with Fe(II), Mo, Ti, Ta,
and Zr, their complexes do not turn to black in alkaline conditions. An ethanol solution of (1) is fairly stable and can be kept for 5 days without spectral change when
stored at 0°.[R1,3]

In the presence of oxalic acid, a red 2:2:2 complex (2) is formed. Evaporation of
the reaction mixture to dryness is recommended for maximum color development. The
presence of water again delays the reaction. If a mineral acid is present, the simultaneous formation of (1) is also expected.

Absorption spectra of (1) and (2) are illustrated in Figure 2. Molar absorptivity of
(2) is reported to be 9.3×10^4 at 550 nm.[R1]

Purification and Purity of Reagent

The commercial product is most conveniently purified by the repeated recrystallization from ethanol,[5] until the melting point reaches to 183°. Interestingly, curcumin of synthetic origin is reported to fail the color reaction with boric acid.[R1]

Analytical Applications

Curcumin has been most widely used as a chromogenic reagent for the photometric determination of trace of boron in various materials. Formation of colored complex (1) or (2) is utilized for the photometry. The rosocyanin (1) method is highly sensitive, but the color reaction is very much dependent upon the reaction condition. The rubrocurcumin (2) method is less sensitive than the former, but the procedure is not so involved.

Rosocyanin Method[M1]

The sensitivity of this method is highly dependent upon the presence of water and the amount of excess curcumin remaining in protonated state. Accordingly, it is important to eliminate water and to minimize the absorption due to the excess reagent. In the following procedure which was essentially developed by Uppstrom,[2] water is eliminated by the use of propionic anhydride and excess protonated curcumin is destroyed by acetate ion.

Reagent solutions:

1. Curcumin solution — This solution should be freshly prepared weekly by dissolving 0.125 g of curcumin in 100 mℓ glacial acetic acid and should be stored in a polyethylene bottle.
2. Sulfuric-acetic acid solution — Mix equal volume of concentrated sulfuric acid (98%) and glacial acetic acid.
3. Buffer solution — Mix 90 mℓ of 95% ethanol, 180 g of ammonium acetate, and 135 mℓ of glacial acetic acid. Dilute to 1 ℓ with water.
4. Propionie anhydride
5. Oxalyl chloride

Procedure: Transfer a 1.00 mℓ aqueous sample solution containing 0.2 to 1 μg boron to a polyethylene beaker. Add 2.0 mℓ glacial acetic acid and 5.0 mℓ propionic anhydride and mix. Add dropwise 0.5 mℓ oxalyl chloride and allow to stand for 30 min. Cool to room temperature and to this add 4.0 mℓ sulfuric - acetic acid solution and 40 mℓ curcumin solution, mix thoroughly, and allow to stand for 45 min. Add 20 mℓ buffer solution, mix thoroughly, and cool to room temperature. Measure the absorbance at 545 nm against a reagent blank.

Rubrocurcumin Method[6]

This method is less sensitive than the former, but the color reaction is fast, and it does not need a sulfuric acid medium. The method is suitable for the samples after distillation as methyl borate.

Reagent solutions: Curcumin-oxalic acid solution — Dissolve 0.40 g of curcumin and 50 g of oxalic acid in ethanol (>99%) to make 1 ℓ, and store in polyethylene bottle. The solution should be kept at room temperature for 1 week before use.

Procedure: Place 2.0-mℓ sample of aqueous solution containing 0.1 to 2.0 μg of boron in a platinum dish. To this, add 4.00 mℓ curcumin-oxalic acid solution and mix thoroughly. Evaporate the mixture on a water bath at 55 $\pm$ 3°. Add 25 mℓ ethanol to the dried residue and mix well. After removing insoluble matters by filtration or centrifugation, transfer the clear solution into a 1.00-cm cell and measure absorbance at 550 nm against a reagent blank.

MONOGRAPH

M1. Boltz, D. F. and Howell, J. A., *Colorimetric Determination of Nonmetals,* 2nd ed., John Wiley & Sons, New York, 1978.

REVIEW

R1. Hiiro, K. and Muraki, I., Study on the spectrophotometric determination of boron using organic reagents, *Rept. Gov. Ind. Research Inst. Osaka (Japan),* No. 321 (1964).

REFERENCES

1. Spicer, G. S. and Strickland, J. D. H., *J. Chem. Soc.,* 4644, 1952.
2. Upstrom, L. R., *Anal. Chim. Acta,* 43, 475, 1968.
3. Williams, D. E. and Vlamis, J., *Anal. Chem.,* 33, 1098, 1961.
4. Spicer, G. S. and Strickland, J. D. H., *J. Chem. Soc.,* 4650, 1952.
5. Dible, W. T., Truog, E., and Berger, K. C., *Anal. Chem.,* 26, 418, 1954.
6. Muraki, I. and Hiiro, K., *Nippon Kagaku Kaishi,* 78, 845, 1959.

2-AMINOPERIMIDINE

$C_{11}H_9N_3 \cdot HCl$
mol wt = 219.67

Source and Method of Synthesis

Commercially available as hydrochloride or hydrobromide. It is prepared by the reaction of 1,8-diaminonaphthalene with NH_4SCN.[1]

Analytical Use

As a precipitating and nepherometric reagent for sulfate ion.

Properties of Reagent

The hydrochloride is a greyish-white crystalline powder. It is slightly soluble in water (0.5%, at room temperature), but easily soluble in hot water.[1] Although the reagent is susceptible to oxidation, the solid reagent is fairly stable if it is kept tightly closed in a dark cool place. However, the aqueous solution is stable only for a few days even in a tightly closed container in a dark place.[2] The reagent may be purified by boiling a saturated solution with charcoal, filtering, and leaving the hydrochloride to crystallize.

Reaction with Sulfate Ion

When the aqueous solution of the reagent (saturated at room temperature, 0.5%) is added to the solution containing sulfate ion, the immediate formation of heavy silky white precipitates of the amine sulfate is observed.

The characteristic features of these precipitates are their unusually small particle size (<2 μm), their extremely regular shape in the form of a right-angled cross, and their apparent unwillingness to ripen or aggregate on standing for several hours. These features, together with the very low solubility of 2-aminoperimidinium sulfate, make it ideal for use this reagent as a nepherometric reagent for sulfate ion. As summarized in Table 1, 2-aminoperimidine sulfate is least soluble among the various amine sulfates. Even at 1 ppm of sulfate, the precipitation is readily observed, and it is possible to detect visually as little as 0.05 ppm of sulfate in a test volume of 10 mℓ.

The UV absorption spectrum of the aqueous solution of 2-aminoperimidine hydrochloride is illustrated in Figure 1. A rather broad band at 305 nm ($\varepsilon = 7.23 \times 10^3$) can be used for the photometric determination of 2-aminoperimine cation in the supernatant after precipitating the sulfate ion with a known excess of the reagent. This becomes the basis for the indirect photometry of sulfate (4 to 120 ppm SO_4^{2-}).[2]

Later, Toei proposed the use of colored reagent, 6-(p-acetylphenylazo)-2-aminoperimidine (pH 3.4 to 4.1; λ_{max}, 480 nm; $\varepsilon = 6.1 \times 10^3$) for the similar indirect, but visual photometry of sulfate (0 ∼ 10 ppm).[3]

Analytical Application

The following procedure is recommended for the general purposes.

Procedure for the determination of 0 ∼ 5 ppm of sulfate —

Transfer 1.0 to 5.0 mℓ of the standard sulfate solution (10 ppm) to each of five 10 mℓ volumetric flasks. Dilute to about 5 mℓ with water, add 4 mℓ of the aqueous 0.5% 2-aminoperimidine hydrochloride solution, and dilute the content to the mark. Mix well and leave the suspensions for 5 ∼ 10 min. Transfer to the nepherometer tube and measure the light scattering of each solution. Run on the sample solution in the same way.

Table 1
SOLUBILITIES (S) OF VARIOUS AMINE SULFATES[1]

Amine	S(normal sulfate) (g/ℓ, 25°)
Benzidine	0.098
1,8-Diaminonaphthalene	0.222
4-Amino-4′-chlorobiphenyl	0.155
4,4′-Diaminotoluene	0.059
2-Aminoperimidine	0.020[a]

[a] Measured at 18°.

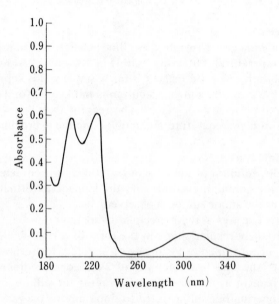

FIGURE 1.   Absorption spectrum of 2-aminoperimidine hydrochloride in aqueous solution; ε = 7230 at 305 nm. (From Jones, P. A. and Stephen, W. I., *Anal. Chim. Acta,* 64, 85, 1973. With permission.)

For 0 to 1 ppm or 0 to 0.5 ppm of sulfate, proceed in exactly the same way but use higher instrumental sensitivities.

The intensity of transmitting light at 600 nm can also be observed instead of measuring the scattering light intensity.[4] The interference by various anions is summarized in Table 2.

Table 2
EFFECT OF INTERFERING ANIONS[1]

Anion	Interference
NO_3^-	Does not interfere at 10- to 100-ppm level
Br^-	Does not interfere up to 10 ppm, but gives 20% positive error at 100 ppm
I^-	Gives 10% positive error at 10- to 100-ppm level
F^-, SiF_4^-	Gives 10% positive error per 1 ppm of F^-, but gives 15% error at 10-ppm level
PO_4^{3-}	Gives 25% positive error at 1 ppm
Cl^-	Does not interfere at 10 ppm, but gives 5 to 15% positive error at 100-ppm level

REFERENCES

1. Stephen, W. I., *Anal. Chim. Acta*, 50, 413, 1970.
2. Jones, P. A. and Stephen, W. I., *Anal. Chim. Acta*, 64, 85, 1973.
3. Toei, K., Miyata, H., and Yamawaki, Y., *Anal. Chim. Acta*, 94, 485, 1977.
4. Tanaka, S., Kusube, T., Iwashita, H., and Hashimoto, Y., *Nippon Kagaku Zasshi*, 1415, 1979.

MONOPYRAZOLONE AND BISPYRAZOLONE

(1)
$C_{10}H_{10}N_2O$
mol wt = 174.20

(2)
$C_{20}H_{18}N_4O_2$
mol wt = 346.39

Synonyms
(1) 3-Methyl-1-phenyl-5-pyrazoline-5-one;
(2) 3,3′-Dimethyl-1,1′-diphenyl-4,4′-bispyrazolin-5,5′-dione

Source and Method of Synthesis
Commercially available. Pyrazolone is now being prepared as a dye intermediate from phenylhydrazine and acetoacetic ester.[1] Bispyrazolone is obtained by refluxing an ethanol solution of monopyrazolone with phenylhydrazine.

Analytical Uses
A mixture of monopyrazolone and bisbpyrazolone is used as a highly sensitive photometric reagent for CN^- and less often for SCN^- and OCN^-.

Properties of Reagents
Monopyrazolone
It is a colorless crystalline powder, mp 128 to 130°. Commercial samples are often pale yellow, but can be used as a reagent for CN^-, is almost insoluble in water, but is fairly soluble in hot alcohol, chloroform, pyridine, and acids. It forms colored complexes with Ag, Co, Cu, and Fe.

Bispyrazolone
It is a colorless or grey-yellow crystalline powder, mp >300°, and is almost insoluble in water and in common organic solvent except pyridine, in which the reagent is fairly soluble.

Reaction with Cyanide Ion
In the determination of cyanide ion by the pyrazolone method, the sample solution is treated with Chloramine T, followed by the reaction with monopyrazolone and bispyrazolone in pyridine to develop a blue color for the photometry. The sequence of the reaction until the color development can be shown in Figure 1. The resulting blue dye can be extracted into n-butanol for the better sensitivity.

The role of bispyrazolone is not certain, but it is indispensable for the maximum color development. A mixing ratio of about 12.5:1 (mono to bis) is recommended.

The unpleasant odor of pyridine may be eliminated by replacing it with DMF containing isonicotinic acid.[3]

Thiocyanate and ammonia interfere seriously, as they are oxidized by Chloramine T to give CNCl and $NHCl_2$, respectively. The latter product is also suggested to react with bispyrazolone to give a red-violet dye (λ_{max}, 545 nm) which can be extracted with trichloroethane after acidification of the aqueous solution (yellow, λ_{max}, 450 nm).[7]

Analytical Application
The recommended procedure for the determination of cyanide is as follows.[4]

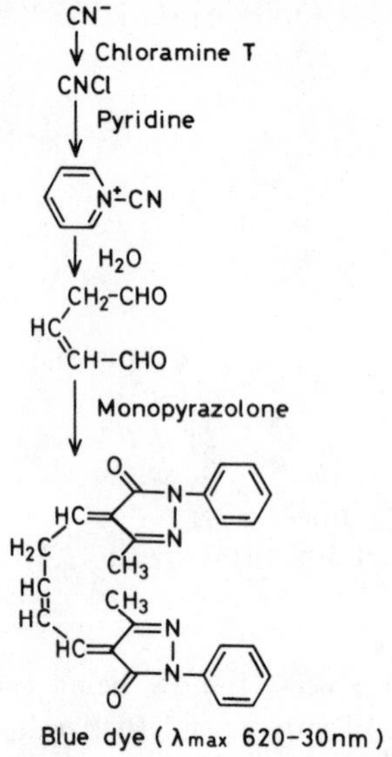

Blue dye (λ_{max} 620-30nm)

FIGURE 1. Sequence of color reactions of pyrazolone reagent with cyanide ion.[2]

Reagent:
1. Pyridine pyrazolone solution. Add monopyrazolone to 125 mℓ of hot water to make a saturated solution. Cool and filter. To the filtrate, add 25 mℓ redistilled pyridine containing 25 mg bispyrazolone. The pyridine solution and the aqueous pyrazolone should be freshly prepared and mixed just before use.*
2. 1% Chloramine T solution. Prepare freshly each day. Phasphate buffer (pH 6.8, 14.3 g Na_2HPO_4 and 13.6 g KH_2PO_4 in 1000 mℓ of water).
3. Standard cyanide solution.

Procedure — Direct photometry: Transfer 1.0 . . . 10.0-mℓ aliquots of the standard cyanide solution to 50-mℓ volumetric flasks. Add 5 mℓ of the buffer and 0.3 mℓ of Chloramine T solution, mix, and allow to stand 1 min. Add 15 mℓ of the pyridine pyrazolone solution, dilute to the volume, mix, and allow to stand stoppered for 30 min. Observe the absorbance at 620 nm against a reagent blank. For the sample containing 1 to 10 μg CN^-, neutralize it to pH 6 to 7 with acetic acid or sodium hydroxide and treat in the same manner as above.

Extraction photometry — Follow the above procedure until the color is fully developed. Transfer the content with rinsings to a 125-mℓ separatory funnel containing exactly 10 mℓ of *n*-butanol and shake for a few min. After phase separation, observe the absorbance of the organic layer at 630 nm against a reagent blank. Thiocyante interferes seriously.

This method can also be applied to the determination of thiocyante (620 nm, 0 to 4 ppm in aqueous solution),[M1] cyanate (450 nm, 0 to 5 ppm in CCl_4),[5] and ammonia

* Premixed solid reagent (mono 12.5:bis 1) is commercially available (Dojindo Laboratories, Kumamoto, Japan). Dissolve 0.27 g in 20 mℓ of pyridine and add 100 mℓ of hot water (75°) to make the pyridine pyrazolone solution. Prepare daily.

(450 nm, 0 to 0.5 ppm in trichloroethylene),[6,7] as these anions behave similarly to cyanide. Nitrate can be determined after reduction to ammonia by alkaline $FeSO_4$.[8] Determination of Vitamin B_{12} (cyanocobalamine) by this method has been reported.[9] Monopyrazolone can also be used as a detection reagent for Ag and Cu.[11]

Other Reagent with Related Structure

Phenazone (2,3-dimethyl-1-phenylpyrazolin-5-one) has been investigated as a reagent for NO_3^-.[10]

MONOGRAPH

M1. Jones, P. D. and Newman, E. J., Bis(3-methyl-1-phenylpyrazol-5-one), in *Organic Reagents for Metals*, Johnson, W. C., Ed., Vol. 2, Hopkin & Williams, Essex, England, 1964, 20.

REFERENCES

1. Knorr, L., *Ann.*, 238, 147, 1877; Höchist Farbenwerke, German Patent 41, 936.
2. Epstein, J., *Anal. Chem.*, 19, 272, 1974.
3. Ishii, K., Iwamoto, T., and Yamanishi, K., *Bunseki Kagaku*, 22, 448, 1973.
4. Boltz, D. F. and Howell, J. A., *Colorimetric Determination of Nonmetals*, John Wiley & Sons, New York, 1978.
5. Krause, J. M. and Mellon, M. G., *Anal. Chem.*, 25, 1188, 1953.
6. Lear, J. B. and Mellon, M. G., *Anal. Chem.*, 29, 293, 1957.
7. Prochazkova, L., *Anal. Chem.*, 36, 865, 1964.
8. Dappenhagen, J. M., *Anal. Chem.*, 30, 282, 1958.
9. Boxer, J. E. and Rickardd, J. C., *Arch. Biochem.*, 30, 372, 382, 392, 1951.
10. Baluja Santos, C., Alvarex Devesa, A., and Cadena Aguiar, J. A., *Quim. Anal.*, 30, 233, 1976; *Anal. Abstr.*, 33, 1B90, 1977.
11. Gehauf, B. and Goldenson, J., *Anal. Chem.*, 27, 420, 1955.

INDEX

INDEX